Phase Structure
of Strongly Interacting Matter

J. Cleymans (Ed.)

Phase Structure of Strongly Interacting Matter

Proceedings of a Summer School on Theoretical Physics, Held at the University of Cape Town, South Africa, January 8–19, 1990

With 190 Figures

Springer-Verlag

Berlin Heidelberg New York London
Paris Tokyo Hong Kong Barcelona

Professor Dr. Jean Cleymans
Department of Physics, University of Cape Town, ZA-7700 Rondebosch, South Africa

Organizing Committee

J. Cleymans (Chairman)
C. A. Engelbrecht
H. G. Miller
R. D. Viollier

ISBN 978-3-642-87823-7 ISBN 978-3-642-87821-3 (eBook)
DOI 10.1007/978-3-642-87821-3

Softcover reprint of the hardcover 1st edition 1990

2155/3140-543210 – Printed on acid-free paper

Preface

The 6th Advanced Course in Theoretical Physics was held at the University of Cape Town, January 8–19, 1990. The topic of the course was "Phase Structure of Strongly Interacting Matter". There were ten invited speakers from overseas, each having up to six hours in which to present his field of research to a relatively small audience of about 50 participants. This allowed for the presentation of a broad, coherent and pedagogical review of the present status of the field. In addition there were several one-hour presentations by local participants.

The main emphasis of the course was on the study of the properties of high-density hot nuclear matter. This field is of particular interest because of the belief that a deconfined quark–gluon plasma could be created in such an environment when the temperature reaches about 200 MeV. In the nuclear regime a so-called "liquid-to-gas" phase transition is expected at a temperature of approximately 10–20 MeV. Both of these topics received ample attention at the school. Owing the nature of the field, there exists much overlapping interest from both the nuclear physics and high-energy particle physics communities. It is hoped that these proceedings will contribute to building a bridge between the two groups.

The course was made possible through generous contributions from:
- Foundation for Research Development (Pretoria)
- University of Cape Town and its Physics Department
- University of Pretoria and its Physics Department
- S.A. Network Computers

The financial support of Professors R.D. Viollier and H.G. Miller is also acknowledged. We furthermore acknowledge contributions from First National Bank and from Standard Bank.

Cape Town
May 1990

Jean Cleymans

Contents

Part I

Lectures on Strongly Interacting Matter and Its Phase Structure

Introduction to Perturbative QCD*

R.L. Thews

Physics Department, University of Arizona, Tucson, AZ85721, USA

1 Introduction

QCD (Quantum Chromodynamics) is a candidate for the theory of strong interactions, and has attained some status as a part of the so-called "Standard Model". I have restricted the scope of these introductory lectures from what is implied in the published title to perturbative aspects of the theory. Although this excludes most of the outstanding problems of the theory yet to be solved, what remains should provide the student with a well-defined and hopefully relevant set of calculations which may be compared with experiment. The level of these lectures is intended for students who have completed a graduate course in quantum mechanics including some relativistic material, and who have some familiarity with the Feynman diagram approach to perturbative calculations. The material has been taken from the following excellent reference books: "Gauge Theories of the Strong, Weak, and Electromagnetic Interactions", by Chris Quigg, Number 56 in the Frontiers in Physics series, Benjamin/Cummings, 1983; "Introduction to Gauge Field Theory", by D. Bailin and A. Love, Graduate Student Series in Physics, IOP Publishing, 1986; "Collider Physics", by Vernon D. Barger and Roger J. N.Phillips, Number 71 in the Frontiers in Physics series, Addison-Wesley, 1987; and "Applications of Perturbative QCD", by Richard D. Field, Number 77 in the Frontiers in Physics series, Addison-Wesley, 1989. Lists of excellent reviews and original references may be found in these books.

The starting point will be a summary of perturbative methods in QED, with emphasis on the standard notation and methods for free particle solutions and Feynman diagrams. The notions of conserved currents and local gauge invariance will be explored for the U(1) case. Then the natural extension to non-abelian gauge groups is shown to lead to self-interacting gauge fields. The specific color SU(3) group of QCD is then used to generate the Feynman rules for future use. The next subject is concerned with the "running coupling constant" of the theory, the very property which allows sensible perturbative calculations in some kinematic regions. Finally, some detailed calculations for application to electron-positron annihilation into hadrons are presented, with examples of regularization schemes for gluon emission processes.

* This work is supported by U.S. Department of Energy Grant DE-FG002-85ER40213

Phase Structure of Strongly Interacting Matter
Editor: J. Cleymans

2 QCD - Definitions and Basic Properties

QCD is a gauge theory, which means that the form of the interaction is specified by invariance of the physical equation under a group of transformations. To understand what this means, it is instructive to look at the classical Maxwell equations of electromagnetism. One can write the equations for the electric and magnetic fields $\vec{E}$ and $\vec{B}$ in terms of a vector and scalar potential $\vec{A}$ and ϕ. Since $\vec{\nabla} \cdot \vec{B} = 0$ one can write $\vec{B} = \vec{\nabla} \times \vec{A}$, and to satisfy $\vec{\nabla} \times \vec{E} = -\frac{\partial \vec{B}}{\partial t}$, one can write $\vec{E} = -\vec{\nabla}\phi - \frac{\partial \vec{A}}{\partial t}$. There remains an ambiguity in specifying these potentials, which comes from the freedom to perform a so-called gauge transformation without changing the values of the physical $\vec{E}$ and $\vec{B}$ fields. We can introduce an arbitrary function $\Lambda(x,t)$ such that under a transformation

$$\vec{A} \rightarrow \vec{A} + \vec{\nabla}\Lambda \tag{1}$$

$$\phi \rightarrow \phi - \frac{\partial \Lambda}{\partial t} \tag{2}$$

no physical quantities are altered. (Note, however, in quantum mechanics phenomena such as the Aharanov-Bohm effect, in which $\vec{A}$ itself has a physical significance).

The Maxwell equation can be written in covariant notation in terms of the field-strength tensor

$$F^{\mu\nu} \equiv \partial^\nu A^\mu - \partial^\mu A^\nu \tag{3}$$

and the corresponding dual tensor

$$\tilde{F}^{\mu\nu} \equiv -\frac{1}{2}\epsilon^{\mu\nu\lambda\rho} F_{\lambda\rho} \tag{4}$$

where $A^\mu \equiv (\phi, \vec{A}), x^\mu \equiv (t, \vec{x})$ and $\partial_\mu \equiv \frac{\partial}{\partial x^\mu}$ transform as 4-vectors, $g^{\mu\nu} \equiv$ diag(1,-1,-1,-1) is the metric tensor, and $\epsilon^{\mu\nu\lambda\rho}$ is the totally antisymmetric tensor in 4 dimensions.

The homogeneous equations becomes $\partial_\mu \tilde{F}^{\mu\nu} = 0$, while the equations with source terms ρ and $\vec{J}$ are written

$$\partial_\mu F^{\mu\nu} = -J^\nu, \tag{5}$$

with

$$J^\nu \equiv (\rho, \vec{J}). \tag{6}$$

In this covariant notation, gauge freedom allows a choice for $\partial_\mu A^\mu$, which then gives an equation for the gauge function

$$\frac{\partial^2 \Lambda}{\partial t^2} - \nabla^2 \Lambda \equiv \Box \Lambda \tag{7}$$

In addition, because of the antisymmetry of $F^{\mu\nu}$, one has $\partial_\nu J^\nu = 0$, which is the continuity equation and indicates conservation of electric charge. The effect of the choice of gauge is readily seen in the field equation when written in the form

$$\partial_\mu F^{\mu\nu} = \partial_\mu(\partial^\nu A^\mu) - \partial_\mu \partial^\mu A^\nu = -J^\nu \tag{8}$$

or

$$\Box A^\nu = J^\nu + \partial^\nu(\partial_\mu A^\mu) \tag{9}$$

It is instructive to derive these equations from a Lagrangian density in a form suitable for use in a quantum field theory. The Lagrangian density is a function of fields and their derivatives $L(\varphi, \partial_\mu\varphi)$ such that the condition of zero variation of the action yields the Euler-Lagrange field equations of the theory.

$$\delta \int_{t_1}^{t_2} dt \int d^3x \mathcal{L} = 0 \Longrightarrow \partial_\mu \frac{\partial \mathcal{L}}{\partial(\partial_\mu \varphi)} = \frac{\partial \mathcal{L}}{\partial \varphi} \tag{10}$$

where one assumes that the fields and their derivatives are sufficienty localized in 3-space that surface integrals vanish at infinity. For example, consider a complex scalar field $\varphi(x,t)$ with Lagrangian

$$\mathcal{L} = |\partial_\mu \varphi|^2 - m^2 |\varphi|^2 \tag{11}$$

This yields the field equations

$$(\Box + m^2)\varphi = (\Box + m^2)\varphi^* = 0 \tag{12}$$

which indicate that momentum space solutions satisfy the usual $E^2 = p^2 + m^2$ energy-momentum relation for free particles.

For spin-$\frac{1}{2}$ particles, one can verify that the Lagrangian $L = \bar{\psi}(i\gamma^\mu\partial_\mu - m)\psi$ yields the Dirac equation $i\gamma^\mu\partial_\mu\psi = m\psi$ for the 4-component spinor $\psi(x)$, with corresponding equation for $\bar{\psi}(x) \equiv \psi^+(x)\gamma^o$. Finally, for the electromagnetic field, the Lagrangian $\mathcal{L} = -\frac{1}{4}F^{\mu\nu}F_{\mu\nu}$ gives the zero-source Maxwell equations $\partial_\mu F^{\mu\nu} = 0$. Note that if one adds a mass term $\frac{1}{2}M^2 A^\mu A_\mu$ to get an equation appropriate for a massive vector field $\partial_\mu F^{\mu\nu} + M^2 A^\mu = 0$ the divergence of the equation implies a gauge choice $\partial_\mu A^\mu = 0$ which is required to eliminate the 4th degree of freedom from the vector field to properly decribe the three spin components. Note therefore that the mass term is not gauge invariant, as can be seen by direct substitution.

Now consider a case where the Lagrangian is invariant under some transformation of fields and/or coordinates. Then we can invoke Noether's Theorem, which indicates that this implies the existence of a conserved current j^μ, $(\partial_\mu j^\mu = 0)$ and a time-independent charge $Q \equiv \int d^3x j^o(t, \vec{x})$. This is simple to show if we restrict ourselves to transformations which affect the fields and their derivatives, but not the coordinates explicitly. Then

$$\delta S = 0 \Longrightarrow 0 = \int d^4x \Big[\frac{\partial \mathcal{L}}{\partial \varphi} - \partial_\mu \frac{\partial \mathcal{L}}{\partial(\partial_\mu\varphi)}\Big]\delta\varphi + \int d^4x\, \partial_\mu \Big[\frac{\partial \mathcal{L}}{\partial(\partial_\mu\varphi)}\delta\varphi\Big] \tag{13}$$

The first integrand is zero by the field equations. In the second, let $\delta\varphi = \chi\delta\lambda$ where λ is a parameter describing the infinitesimal transformation. Then the conserved current is read off directly

$$j^\mu \equiv \frac{\partial \mathcal{L}}{\partial(\partial_\mu\varphi)}\chi \tag{14}$$

As an example, consider the phase invariance in the scalar field case. For

$$\mathcal{L} = |\partial_\mu \varphi|^2 - m^2 |\varphi|^2 \tag{15}$$

let

$$\varphi \rightarrow e^{iq\lambda}\varphi \tag{16}$$

and

$$\varphi^* \rightarrow e^{-iq\lambda}\varphi^* \tag{17}$$

Then

$$\delta\varphi = iq\varphi\delta\lambda \Longrightarrow \chi_1 = iq\varphi \tag{18}$$

$$\delta\varphi^* = -iq\varphi^*\delta\lambda \Longrightarrow \chi_2 = -iq\varphi^* \tag{19}$$

and

$$j^\mu = \partial^\mu\varphi^* \cdot iq\varphi + \partial^\mu\varphi^* \cdot (-iq\varphi*) \tag{20}$$

$$\equiv -iq\varphi^*\overleftrightarrow{\partial^\mu}\varphi, \tag{21}$$

which can be identified with the electric current of the charged particle quanta of the complex scalar field. A similar quantity exists for the Dirac equation: L = $\bar{\psi}(i\gamma^\mu\partial_\mu - m)\psi$.

$$\psi \rightarrow e^{iq\lambda}\psi \Longrightarrow \chi_1 = iq\psi \tag{22}$$

$$\bar{\psi} \rightarrow \bar{\psi}e^{-iq\lambda}\psi \Longrightarrow \chi_2 = -iq\bar{\psi} \tag{23}$$

and since

$$\frac{\partial\mathcal{L}}{\partial(\partial_\mu\bar{\psi})} = 0, \tag{24}$$

one obtains

$$j^\mu = -q\bar{\psi}\gamma^\mu\psi. \tag{25}$$

This type of transformation is called global, since the phase parameter is independent of space. One can also consider what are called local gauge transformations, in which the parameter $\lambda = \lambda(x)$. Obviously the Lagrangian densities we have considered so far are not invariant under such transformations because of the derivative terms $\partial_\mu\varphi$. One can insist on local gauge invariance by introducing what is called a covariant derivative which depends on a gauge field A_μ

$$D_\mu \equiv \partial_\mu + iqA_\mu \tag{26}$$

such that under the transformation $\varphi \rightarrow e^{iq\lambda(x)}\varphi$ one has $D_\mu\varphi \rightarrow e^{iq\lambda(x)}D_\mu\varphi$ which leaves the typical terms in the Lagrangian invariant. This restriction requires

$$(\partial_\mu + iqA'_\mu(x))e^{iq\lambda(x)}\varphi(x) = e^{iq\lambda(x)}(\partial_\mu + iq(A' + \partial_\mu\lambda))\varphi \tag{27}$$

which will work if the gauge field A^μ transforms as $A_\mu \rightarrow A_\mu - \partial_\mu\lambda(x)$. This local gauge invariance then specifies the form of an interaction between the gauge field and the original fields in the free-particle Lagrangian. For the Dirac Lagrangian L $= \bar{\psi}(i\gamma^\mu D_\mu - m)\psi$ one separates the free and interacting parts as L = $L_{free} - J^\mu A_\mu$, with $J^\mu = q\bar{\psi}\gamma^\mu\psi$. To obtain the QED Lagrangian, one only needs to add the free part of the electromagnetic field Lagrangian, since the $J^\mu A_\mu$ interacting part is common to both. Thus

$$\mathcal{L}_{QED} = \bar{\psi}(i\gamma^\mu D_\mu - m)\psi - \frac{1}{4}F^{\mu\nu}F_{\mu\nu} \tag{28}$$

Note again that no photon mass term may be allowed since $A_\mu A^\mu$ cannot be made locally gauge invariant.

The phase invariance we have been looking at is actually invariance under the group of unitary transformations of fields U(1). The global invariance alone ($\lambda =$ constant) implies a conserved current, while local invariance $\lambda = \lambda(x)$ constrains the form of an interaction between the Dirac fermions and the photon. One can extend this invariance principle to more complicated groups than U(1). QCD is such a theory, where the matter fields describe spin-$\frac{1}{2}$ quarks which occur in triplet of color transforming according to the group SU(3), and we insist that the Lagrangian be invariant under such transformations. A general transformation on the Dirac spinor ψ describing the quarks is

$$\psi(x) \rightarrow \psi'(x) = e^{-ig\Sigma_{a=1}^{8} T_a \Lambda_a(x)} \psi(x) \equiv U(x)\psi(x) \tag{29}$$

where $T_a, a = 1, ...8$ are the infintesimal generators of SU(3) (in this case represented by 3 x 3 matrices) which satisfy commutation relations

$$[T_a, T_b] = i f_{abc} T_c \tag{30}$$

where the f_{abc} are called the structure constants of the SU(3) algebra. To construct a Lagrangian which is invariant under this transformation, one introduces 8 vector gauge fields A^μ_a and the corresponding covariant derivatives

$$D^\mu \psi \equiv (\partial^\mu + igT_a A^\mu_a(x))\psi. \tag{31}$$

The invariance requirement

$$D^\mu \psi \rightarrow U(x) D^\mu \psi \tag{32}$$

tells us how the gauge fields must transform. Because the generators T_a do not commute (non-Abelian group) we get some additional terms.

$$(\partial^\mu + igT_a A^{\mu\prime}_a)\psi' = U(x)(\partial^\mu \psi) + (\partial^\mu U)\psi + igT_a A^{\mu\prime}_a (U\psi) = U(\partial^\mu \psi) + igU(T_a A^\mu_a \psi). \tag{33}$$

Regard this as an operator equation acting on an arbitrary ψ, and multiply on the right by U^{-1}, to get

$$T_a A^{\mu\prime}_a = U[T_a A^\mu_a + i/g U^{-1}(\partial^\mu U)]U^{-1} \tag{34}$$

One then considers infintesimal transformations to show that this is satisfied if

$$A^{\mu\prime}_a = A^\mu_a + \partial^\mu \Lambda_a + g f_{abc} \Lambda_b A^\mu_c, \tag{35}$$

where the f_{abc} are the structure constants of the SU(3) algebra. The new non-abelian feature then brings in additional terms via the gauge field kinetic terms which indicate self-interactions of the gauge particles. To generate these kinetic terms, start with the usual field-strength tensor $F^{\mu\nu}_a$ for each gauge field $A^\mu_a(x)$ and form the matrix $F^{\mu\nu} \equiv F^{\mu\nu}_a T_a$. Consider a Lagrangian of the type as for the electromagnetic case.

$$\mathcal{L} = -\frac{1}{4} F^{\mu\nu}_a F_{a\mu\nu} = -\frac{1}{2}\text{Trace}(F^{\mu\nu} F_{\mu\nu}) \tag{36}$$

where the last normalization comes from the fundamental representation of the generators

$$\text{Trace}(T_a T_b) = \frac{1}{2}\delta_{ab} \tag{37}$$

We then need the transformation law $F' = UFU^{-1}$ to make L invariant, but explicit calculation shows that additional terms show up because of the non-abelian nature of the group. One can fix this by adding a commutator term to the field strength tensor

$$F^{\mu\nu} = \partial^\nu A^\mu - \partial^\mu A^\nu + ig[A_\nu, A_\mu] \quad (38)$$

where $A^\mu \equiv \sum_a A^\mu_a T_a$ which leads to

$$F^{\mu\nu}_a = \partial^\nu A^\mu_a - \partial^\mu A^\nu_a - g f_{abc} A^\nu_b A^\mu_c \quad (39)$$

and an SU(3) - invariant total QCD Lagrangian

$$\mathcal{L}_{QCD} = \bar{\psi}(i\gamma^\mu D_\mu - m)\psi - \frac{1}{4} F^{\mu\nu}_a F_{a\mu\nu} \quad (40)$$

The Euler-Lagrange equations then become

$$(i\gamma^\mu \partial_\mu - m)\psi = g T_a \gamma_\mu A^\mu_a \psi \quad (41)$$

and

$$\partial^\nu F^a_{\mu\nu} - g f_{abc} A^\nu_b F^c_{\mu\nu} = g \bar{\psi} \gamma_\mu T_a \psi \quad (42)$$

One is now in a position to generalize the Feynman diagram rules for perturbative graphs in QED to the corresponding manipulations in QCD.

2.1 External particles:

1. For external massless spin-1 bosons (photons) insert the polarization vector $\epsilon_\mu(\lambda)$ for $\lambda = \pm 1$, the physical helicity states. An explicit construction gives $\epsilon_\mu(\lambda = \pm 1) = \mp \frac{1}{\sqrt{2}}(0, 1, \pm i, 0)$ for momentum along the z-axis. They are both transverse, $\epsilon \cdot k = 0$, and normalized to $\epsilon_\mu \epsilon^\mu = -1$. For the massless gluons in QCD, the factors are exactly the same for each of the 8 colors. One then either sums or averages over color in the final or initial states, exactly as for spin.
2. For external spin-$\frac{1}{2}$ fermions (electrons, etc.) with momentum p and spin s, insert the free-particle solutions to the Dirac equation as follows: for a fermion in the initial state a factor $u(p, s)$ on the right; for a fermion in the final state a factor $\bar{u}(p, s)$ on the left; for an anti-fermion in the initial state a factor $\bar{v}(p, s)$ on the left; and for an anti-fermion in the final state a factor v(p,s) on the right. These solutions satisfy

$$\begin{aligned} (\not{p} - m)u(p, s) &= \bar{u}(p, s)(\not{p} - m) = 0; \\ \bar{u} &\equiv u^+\gamma^0 \end{aligned} \quad (43)$$

$$(\not{p} + m)v(p, s) = \bar{v}(p, s)(\not{p} + m) = 0 \quad (44)$$

with normalization

$$\sum_{spin} \bar{u}(p, s)u(p, s) = 2m \quad (45)$$

$$\sum_{spin} \bar{v}(p, s)v(p, s) = -2m \quad . \quad (46)$$

In spin sums for squared amplitudes, one encounters the projection operators

$$\sum_{spin} u(p,s)\bar{u}(p,s) = \not{p} + m \tag{47}$$

$$\sum_{spin} v(p,s)\bar{v}(p,s) = \not{p} - m. \tag{48}$$

In QCD, the spin-$\frac{1}{2}$ fermions are the quarks, and all factors are exactly the same for each of the 3 quark colors. Again one must sum or average over quark color in addition to spins.

2.2 Vertex terms:

In QED there is only one vertex corresponding to the coupling of a photon to a fermion. From the interaction term $q\bar{\psi}\gamma^\mu A_\mu \psi$ one reads off a factor $-ie\gamma^\mu$ for each vertex as in Figure 1. In QCD, there is a corresponding vertex for coupling of gluons to quarks. The interaction term is $g\bar{\psi}\gamma^\mu A_{a\mu}\psi$ from which one gets the factor $-ig\gamma^\mu(T_a)_{ij}$ for the vertex in Figure 2. Note that in this case we must label the quark $i,j = 1,2,3$ and the gluon $a = 1,\cdots,8$ colors. In addition, QCD has the gluon self-interacting terms. Figures 3 and 4 show these vertex functions. For the triple-gluon coupling which comes from the interaction term $g f_{abc}\partial^\nu A_a^\mu A_\nu^b A_\mu^c$, one gets a factor $-g f_{abc} F_{\lambda\mu\nu}(p_1,p_2,p_3)$, where $F_{\lambda\mu\nu}(p,k,q) \equiv (p-k)_\nu g_{\lambda\mu} + (k-q)_\lambda g_{\mu\nu} + (q-p)_\mu g_{\nu\lambda}$ and for the 4-gluon vertex interaction which comes from the square of the $gfAA$ term one inserts $-ig^2 f_{abe} f_{cde}(g_{\lambda\nu}g_{\mu\sigma} - g_{\lambda\sigma}g_{\mu\nu})$ plus two other permutations of pairs of gluons.

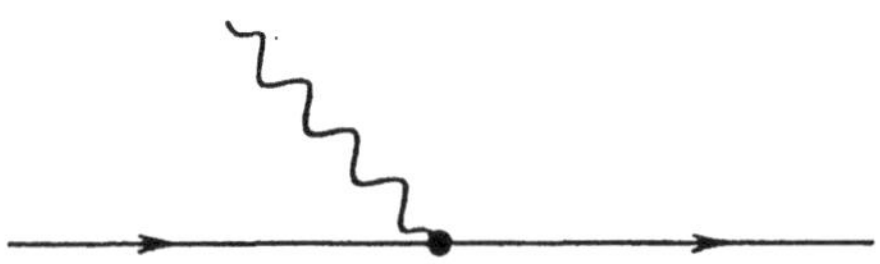

Figure 1: Vertex in QED.

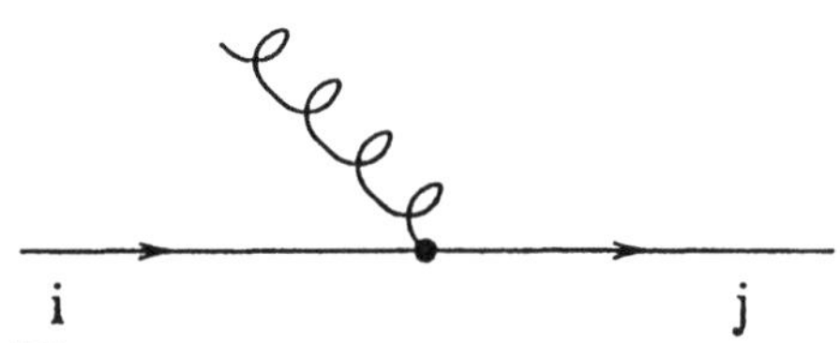

Figure 2: Vertex in QCD.

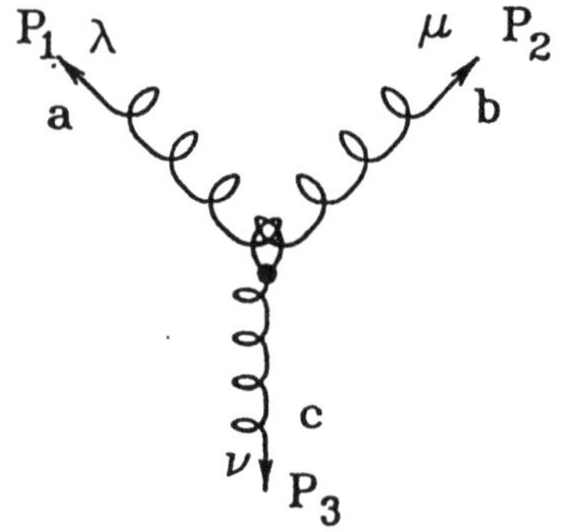

Figure 3: Triple gluon coupling.

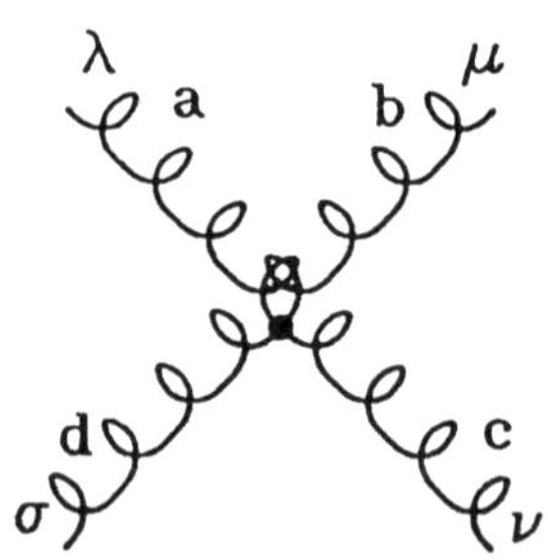

Figure 4: Four gluon coupling.

2.3 Internal lines:

1. For each internal fermion line in QED of four-momentum q, the propagator is

$$\frac{i(\not{q}+m)}{q^2-m^2+i\epsilon} \tag{49}$$

An internal anti-fermion is taken to be the equivalent fermion with $q \to -q$. In QCD, the propagator is the same, but with a multiplicative δ_{ij} where i and j are the color indices at either end of the line.

2. For each internal photon line in QED, one can write the propagator in Feynman gauge as

$$\frac{-ig_{\mu\nu}}{q^2+i\epsilon} \tag{50}$$

where μ and ν are the polarization vector indices at each end of the line. It is possible to use this form which in principle sums over unphysical polarization states of the photon, because gauge invariance insures that these unphysical polarization states will automatically cancel in all physical amplitudes. For QCD, a natural extension would involve just multiplying this propagator by a color delta function. This procedure however, does not give us the desired cancellation of unphysical polarization states in diagrams involving the triple-gluon vertex. One can see how this comes about by looking back at the field equations for the photon. The Feynman gauge is chosen by requiring $\partial_\mu A^\mu = 0$ in the field equation $\Box A^2 + \partial^\nu(\partial_\mu A^\mu) = J^\nu$, thus allowing the inversion of the equation to solve for A^ν in terms of J^ν. One could as easily have chosen a different gauge by inserting a gauge-fixing term in the Lagrangian, $-\frac{1}{2\xi}(\partial_\mu A^\mu)^2$, which leads to the field equations

$$\Box A^\nu - (1-\frac{1}{\xi})\partial^\nu(\partial_\mu A^\mu) = J^\nu \tag{51}$$

and the corresponding propagator

$$\frac{-i[g^{\mu\nu}+(\xi-1)q^\mu q^\nu/q^2]}{q^2+i\epsilon} \tag{52}$$

Note that the gauge-fixing term is itself gauge-invariant under $A_\mu \to A_\mu - \partial_\mu\lambda(x)$ as long as $\Box\lambda(x) = 0$. A similar construction for the QCD Lagrangian runs into trouble, since the non-abelian nature of the gauge group does not allow a simple constraint on the gauge transformation functions $\Lambda_a(x)$ to insure overall gauge invariance. One procedure is to choose a particular gauge to work in, at the expense of writing all expressions in a non-covariant manner. The other alternative is to add to the Lagrangian a color octet of ficticious scalar particles (Faddeev-Popov ghosts) which appear only in closed loops to cancel the unphysical gluon contributions. With the scalar field $\eta_a(x)$ one writes L $_{FP} = \partial_\mu\eta_a(\partial^\mu\eta_a + gf_{abc}\eta_b A^\mu_c)$. This produces a term for the scalar-scalar-gluon vertex, with a factor $gf_{abc}p_\mu$ with p_μ the ghost particle momentum, and also a propagator for the internal ghost line given by

$$\frac{-i\delta_{ab}}{q^2+i\epsilon} \tag{53}$$

2.4 Internal loops and indentical particles:

For each internal loop with momentum k, one performs the usual 4-dimension loop integral $\int d^4k/(2\pi)^4$. A factor of -1 is inserted for fermion loops and a factor of $1/n!$ for boson loops with n identical particles. In QCD the rules are the same, with the additional constraint that ghost loops, although they are bosons, also have the -1 factor as for fermions.

3 Running Coupling "Constants"

In the section we illustrate how interacting field theories can produce effective coupling strengths which vary with the scale of momentum. Calculational tools are introduced to first perform the calculation in QED. Then it is extended to QCD, where the non-abelian nature will manifest itself in producing an effective coupling which becomes very small at large momenta, or small distances. Thus one can find a kinematic region for some physical processes where an argument can be made for a sensible perturbative expansion.

For QED, we start by looking at the higher order terms in the photon propagator, due to virtual fermion loops, as in Figure 5.

$$P_{\mu\nu} = \frac{-ig_{\mu\nu}}{q^2} + -ig_{\mu\mu'} \cdot \prod_{\mu'\nu'} \cdot \frac{-ig_{\nu'\nu}}{q^2} \tag{54}$$

with

$$\prod_{\mu\nu} = -e^2 \int \frac{d^4k}{(2\pi)^4} \frac{\text{Trace}[\gamma_\mu(\not{k} + m)\gamma_\nu(\not{k} - \not{q} + m)]}{(k^2 - m^2 + i\epsilon)((k-q)^2 - m^2 + i\epsilon)} \tag{55}$$

where the trace of γ-matrices comes from matching indices in the closed loop.

Note that the integral formally diverges as $k \to \infty$ (ultraviolet divergence). This is fixed by the usual renormalization procedure, where e is regarded as a "bare" charge, and the physical charge is relaterd to measurable quantities.

To evaluate the loop-integral, we use a parameterization of the propagators

$$\frac{i}{k^2 - m^2 + i\epsilon} = \int_0^\infty dx\; e^{ix(k^2 - m^2 + i\epsilon)} \tag{56}$$

to get

$$\prod_{\mu\nu} = e^2 \int \frac{d^4k}{(2\pi)^4} \int_0^\infty dx \int_0^\infty dy\; e^{ix(k^2 - m^2 + i\epsilon) + iy((k-q)^2 - m^2 + i\epsilon)} * Trace[\;\;] \tag{57}$$

Then one can complete the square in the exponential by defining a new loop momentum $r_\mu \equiv k_\mu - \frac{y}{x+y} q_\mu$ and since the shift is finite $\int \frac{d^4k}{(2\pi)^4} \to \int \frac{d^4r}{(2\pi)^4}$, where we have

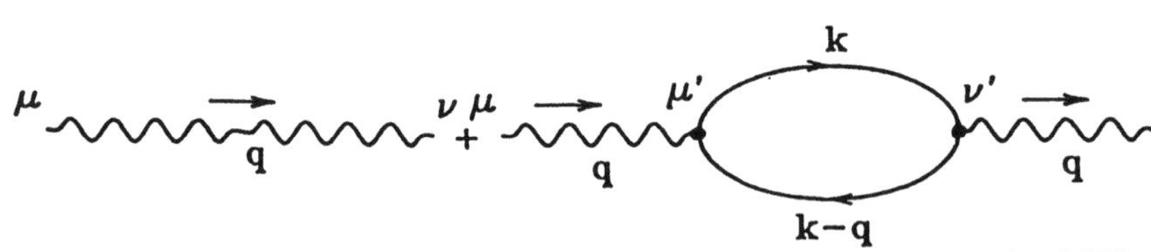

Figure 5: Lowest order corrections to the fermion propagator in QED.

anticipated a procedure which will make the integral finite. The resulting expression is

$$\begin{aligned}\Pi_{\mu\nu} &= 4e^2 \int_0^\infty dx \int_0^\infty dy \;\; e^{-i(x+y)(m^2-i\epsilon)} \;\; e^{\frac{ixy}{x+y}q^2} \\ & \int \frac{d^4r}{(2\pi)^2} e^{i(x+y)r^2} \Big[2r_\mu r_\nu - \frac{x-y}{x+y}(r_\mu q_\nu + q_\mu r_\nu) \\ &- \frac{2xy}{(x+y)^2} q_\mu q_\nu \\ &+ g_{\mu\nu}(m^2 - r^2 + \frac{x-y}{x+y} r\cdot q + \frac{xy}{(x+y)^2} q^2) \Big] \end{aligned} \tag{58}$$

One can perform the r-integrals, using the identities

$$\int \frac{d^4r}{(2\pi)^4} e^{i\lambda r^2} = \frac{1}{16\pi^2 i\lambda^2} \tag{59}$$

and

$$\int \frac{d^4r}{(2\pi)^4} r^2 e^{i\lambda r^2} = \frac{1}{8\pi^2\lambda^3} \tag{60}$$

which comes directly from differentiation with respect to λ. The integrals involving single powers of r_μ momentum vanish identically due to the symmetric integration region, as can be seen by the substitution $r_\mu \to -r_\mu$. The integrals involving $r_\mu r_\nu$ can then be replaced by $\frac{1}{4} g_{\mu\nu} r^2$. The resulting terms in the integrand can then be grouped as follows

$$(g_{\mu\nu}q^2 - q_\mu q_\nu)\frac{2xy}{(x+y)^4} + \frac{g_{\mu\nu}}{(z+y)^2}\left[m^2 - \quad \frac{i}{x+y} - \frac{xyq^2}{(x+y)^2}\right] \tag{61}$$

where the first combination has the correct momentum dependence for the propagator to insure current conservation $q^\mu \Pi_{\mu\nu} = q^\nu \Pi_{\mu\nu} = 0$. Hence the remaining term proportional to $g_{\mu\nu}$ which violates this property must vanish, as can be seen by direct computation. The nonvanishing coefficient of $(g_{\mu\nu}q^2 - q_\mu q_\nu)$ is then

$$\triangle\,(q^2, m^2) = -\;\frac{2i\alpha}{\pi} \int_0^\infty dx \int_0^\infty dy \;\frac{xy}{(x+y)^4} e^{-i(x+y)(m^2-i\epsilon)} e^{\frac{ixy}{x+y}q^2} \tag{62}$$

where we have defined the usual fine structure constant $\alpha \equiv \frac{e^2}{4\pi}$ and the ultraviolet divergence is seen to be present as a log-type singularity in the $x = y = 0$ region. One can use the Pauli-Villars method of regularization in this case. One inserts negative counterterms in the Lagrangian with $m^2 \to \Lambda^2 \gg q^2, m^2$, and the take the limit $\Lambda \to \infty$ for physical quantities. To proceed furthe r, it is convenient to insert a factor of unity in the integrand, written as

$$f(x,y) = \int_0^\infty \delta(\lambda - x - y)d\lambda \cdot f(x,y) \tag{63}$$

and then rescale x and y with a factor of λ to obtain

$$\int_0^\infty \frac{d\lambda}{\lambda}\delta(1-x-y) f(\lambda x, \lambda y) \tag{64}$$

so that the upper limits on x and y are 1. Then the delta-function is used to perform the y-integral, to obtain

$$\Delta\left(q^2,m^2\right) = -\frac{2i\alpha}{\pi}\int_0^1 dx x(1-x)\int_0^\infty \frac{d\lambda}{\lambda} e^{-i\lambda(m^2-i\epsilon)}e^{i\lambda x(1-x)q^2} \tag{65}$$

With the subtraction for regularization, one then gets

$$\bar{\Delta} \equiv \Delta(q^2,m^2) - \Delta(q^2,\Lambda) = \frac{2i\alpha}{\pi}\int_0^1 dx x(1-x)\log(\frac{\Lambda^2}{m^2 - q^2x(1-x)}) \tag{66}$$

for $\Lambda \to \infty$ and $q^2 < 4m^2$.
In scattering reactions the conserved current coupling insures that the $q_\mu q_\nu$ does not contribute, so that the effective one-loop propagator is

$$-\frac{ig_{\mu\nu}}{q^2}\left[1 - \frac{\alpha}{3\pi}\log(\frac{\Lambda^2}{m^2}) + \frac{2\alpha}{\pi}\int_0^1 dx x(1-x)\log(1 - \frac{q^2x(1-x)}{m^2})\right] \tag{67}$$

As $q^2 \to 0$ one can absorb the $1 - \frac{\alpha}{3\pi}\log(\frac{\Lambda^2}{m^2})$ factor into the coupling to renormalize the charge, and then let $\Lambda \to \infty$. In general, the effective coupling for one loop corrections takes the form

$$\alpha_{eff}(q^2) = \alpha(1 + \alpha B(q^2)) \tag{68}$$

where $B(q^2)$ is a divergent quantity. The most divergent graphs are products of these factors, coming from successive insertions of fermion loops in the propagator. One can formally sum the geometric series which results to get

$$\alpha_{eff}(q^2) = \alpha(1 + \alpha B(q^2) + (\alpha B(q^2))^2 + \ldots) = \frac{\alpha}{1 - \alpha B(q^2)}$$

It is instructive to write this as

$$\frac{1}{\alpha_{eff}(q^2)} = \frac{1}{\alpha} - B(q^2) \tag{69}$$

and to then define the renormalization point in QED at $q^2 = 0$, where $\alpha_R \equiv \alpha_{eff}(q^2 = 0)$ is measured to be approximately 1/137 from the Coulomb force law. Then one can write the effective coupling at any other momentum scale as

$$\frac{1}{\alpha_{eff}(q^2)} = \frac{1}{\alpha_R} + B(0) - B(q^2) \tag{70}$$

where no divergent terms appear in the difference

$$B(0) - B(q^2) = -\frac{2}{\pi}\int_0^1 dx x(1-x)\log(1 - \frac{q^2x(1-x)}{m^2}) \tag{71}$$

At large values of $Q^2 \equiv -q^2 > 0$, one obtains the effective coupling

$$\alpha_{eff}(Q^2) = \frac{\alpha_R}{1 - \alpha_R/3\pi\log(Q^2/m^2)} \tag{72}$$

This has the interesting property that as Q^2 increases (small distances) the effective coupling increases. One can interpret this effect in terms of the polarization of the QED vacuum by virtual e^+e^- pairs which lead to screening of the charge at large distances. As one probes smaller and smaller distances, however, the screening effect i s reduced and the effective coupling becomes larger. In fact, no matter how small the coupling at the renormalization point, one can find some large Q^2 for which the coupling becomes large and perturbative calculations no longer make sense. In practice, however, the smallness of m^2 and α_R make these Q^2 values very large. For example, to see an increase of a factor of 2 in the effective coupling requires

$$\begin{aligned} Q^2 &= m^2 e^{\frac{3\pi}{2\alpha_R}} \\ &\approx 10^{273}\ \mathrm{GeV}^2 \quad ! \end{aligned} \tag{73}$$

One can now perform the same calculation for QCD. For the quark loop correction to the gluon propagator, one can take over the QED calculation directly, with the substitution $-ie\gamma^\mu \to -ig\gamma^\mu T_a^{ij}$, such that $\alpha = \frac{e^2}{4\pi}$ is replaced by $\alpha_s \mathrm{Trace}(T_a T_b) = \frac{1}{2}\alpha_s \delta_{ab}$, where $\alpha_s = \frac{g^2}{4\pi}$ plays the role of the QCD coupling strength. The renormalization point is taken at a spacelike point $q^2 = -\mu^2$, and we set the quark masses to zero. The resulting propagator correction is, for $Q^2 = -q^2 \gg \mu^2$,

$$\overline{\prod}_{\mu\nu,ab}^{Quarks} = -i\frac{\alpha_s}{4\pi}\delta_{ab}(q_\mu q_\nu - q^2 g_{\mu\nu}) \cdot \frac{4}{3}\log(\frac{Q^2}{\mu^2}) \cdot \eta_f \tag{74}$$

where the factor of η_f sums over the number of flavors of quarks. There are two entirely gluonic contributions to $\overline{\prod}_{\mu\nu,ab}$. The first comes from the 4-gluon coupling term involving the diagram in Figure 6a, and can be shown to be zero when dimensional regularization is used (Pauli-Villars will not work for closed gluon loops - we introduce the technicalities of dimensional regularization in the next section). The second diagram is nonzero, coming from the gluon loop involving two triple-gluon couplings (Figure 6b). The corresponding calculation gives, in Landau gauge ($\xi = 0$)

$$\begin{aligned} \overline{\prod}_{\mu\nu,ab}^{Gluons} &= i\frac{\alpha_s}{4\pi} f^{acd} f^{bcd} \log(\frac{Q^2}{\mu^2}) \\ &* \ (\frac{11}{6} q_\mu q_\nu - \frac{19}{12} q^2 g_{\mu\nu} + \frac{1}{2}(q_\mu q_\nu - q^2 g_{\mu\nu})) \end{aligned} \tag{75}$$

This contribution by itself does not satisfy the current conservation relations

$$q^\mu \prod_{\mu\nu,ab} = 0 \tag{76}$$

and it is at this point where the need for the ghost particle contribution becomes evident. The relevant diagram is a ghost loop in the gluon propagator (Figure 6c),

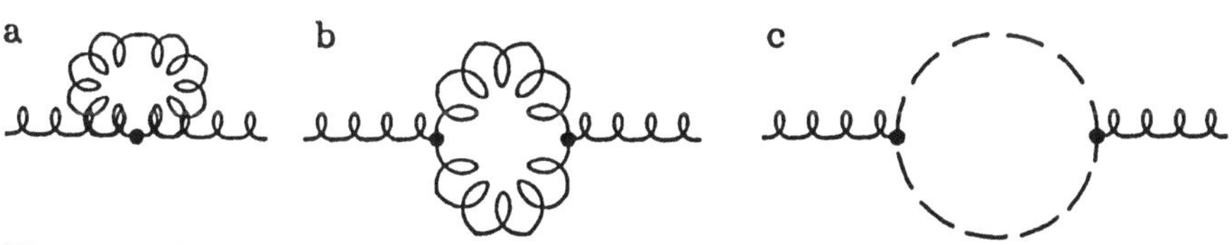

Figure 6: Lowest order corrections to the gluon propagator in QCD.

Figure 7: Correction to the fermion propagator.

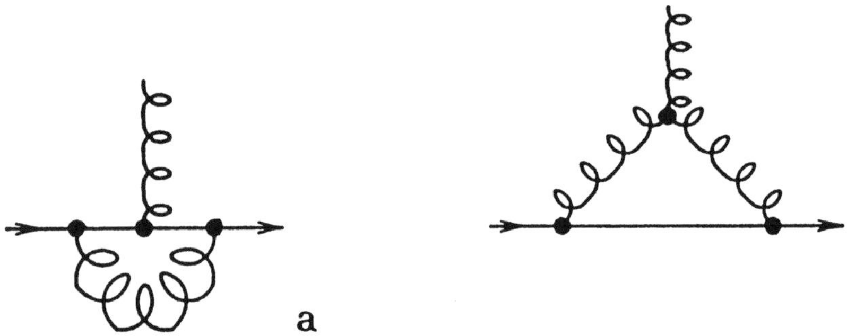

a b

Figure 8: Corrections to the vertex in QCD.

and gives

$$\overline{\prod}^{Ghost}_{\mu\nu,ab} = -i\frac{\alpha_s}{4\pi} f^{acd} f^{bcd} \log(\frac{Q^2}{\mu^2}) \cdot (\frac{1}{6} q_\mu q_\nu + \frac{1}{12} q^2 g_{\mu\nu}) \tag{77}$$

One can perform the sum $f^{acd} f^{bcd} = N\delta_{ab}$ with N = number of colors (3), so that the sum of all three contribution gives a transverse (but gauge-dependent) result for $\xi = 0$:

$$\overline{\prod}^{Total}_{\mu\nu,ab} = i\frac{\alpha_s}{4\pi}\delta_{ab}(q_\mu q_\nu - q^2 g_{\mu\nu}) \log(\frac{Q^2}{\mu^2}) \cdot (\frac{13}{6}N - \frac{2}{3}\eta_f) \tag{78}$$

To get the total contribution to the effective coupling, one must consider all other order α_s loop corrections to a gluon exchange diagram between quarks. The modification to the external quark lines from a single gluon (Figure 7) in this order can be shown to vanish in Landau gauge. The vertex modification comes in two parts, from diagrams in Figure 8.

One obtains (again in Landau gauge) a modification term

$$\triangle \Gamma^\mu_a = -igT_a\gamma^\mu \cdot (-\frac{\alpha_s}{4\pi}) \log(\frac{Q^2}{\mu^2}) \cdot \frac{3}{4}N \tag{79}$$

where the factor of N comes from

$$f^{abc}T^bT^c = \frac{i}{2}NT^a \tag{80}$$

Thus this term comes entirely from the triple-gluon vertex and hence was absent in the QED calculation. To put everything together, one adds one correction for the gluon propagator to two of the vertex correction factors, one on each vertex of the basic gluon exchange diagram, to get the expected

$$\alpha_s(Q^2) = \alpha(1 + \alpha B(Q^2)) \tag{81}$$

with

$$B(Q^2) = \frac{1}{4\pi}\log(\frac{Q^2}{\mu^2})\cdot(\frac{2}{3}\eta_f - \frac{13}{6}N - 2\cdot\frac{3}{4}N). \tag{82}$$

One sees that for $\eta_f < \frac{11}{2}N$ the sign of the log $(\frac{Q^2}{\mu^2})$ term is opposite to that for QED, indicating a net anti-screening. This is obviously due to the contributions involving gluon self–interactions, and can be understood in general terms because the gluons themselves carry color, in contrast to the uncharged photon. The other difference from QED is that the point $Q^2 = 0$ is singular, and one cannot use it as a renormalization point. This property has been included implicitly by the spacelike scale factor $-\mu^2$ in all of the calculations.

The Q^2-dependence is then written (for $N = 3$)

$$\alpha_s(Q^2) = \frac{\alpha_s(\mu^2)}{1 + \frac{\alpha_s(\mu^2)}{12\pi}(33 - 2\eta_f)\log(\frac{Q^2}{\mu^2})} \tag{83}$$

where we have assumed the same geometric series as in QED. It has the property known as Asymptotic Freedom - as $Q^2 \to \infty$, the effective coupling decreases until at some point is may be small enough for perturbative calculations to make sense. Note that $\alpha_s(Q^2)$ is not an independent function of both $\alpha_s(\mu^2)$ and μ^2, but one can define a single parameter Λ_{QCD} (also independent of μ^2) by

$$\Lambda_{QCD} \equiv \mu e^{-\frac{6\pi}{\alpha_s(\mu^2)\cdot(33-2\eta_f)}} \tag{84}$$

and write

$$\alpha_s(Q^2) = \frac{12\pi}{(33 - 2\eta_f)\log(\frac{Q^2}{\Lambda^2_{QCD}})} \tag{85}$$

Thus Λ_{QCD} is the only independent parameter of the theory, and must be determined by experiment. Several experiments indicate that Λ_{QCD} is in the few-hundred MeV range. This produces an $\alpha_s(Q^2 \sim$ few GeV^2)of order unity, consistent with the expectation (hope?) that at typical hadronic length scales the couplings are strong enough to permanently confine color. At the other end of the spectrum, one only needs to go the $Q^2 \sim$ few hundred GeV^2 to get down to $\alpha_s \sim 0.2$ or 0.3, at which point higher order terms may be small and a perturbative solution have some validity.

4 Perturbative QCD Example: $e^+e^- \to$ hadrons

This is the simplest physical process for which QCD perturbative methods may be applied. Since there are no quarks or gluons in the initial state, one can avoid any consideration of the nonperturbative features of the theory by simply summing over all produced hadrons in the final state. Similar applications of perturbative corrections to processes involving electroweak probes always result in additional parameterizations of structure functions or fragmentation functions, in the cases of deep inelastic scattering or production of dileptons in the final state. The purely hadronic collisions must use large momentum transfer triggers in inclusive processes to indicate when

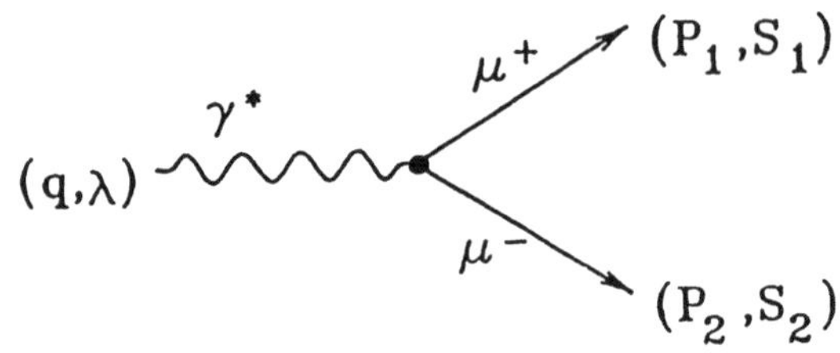

Figure 9: Decay of virtual photon in a muon-pair.

one may be in the perturbative regieme, and again parameterize the nonperturbative effects.

In lowest order in electroweak interactions, the e^+e^- annihilation occurs in two steps. First $e^+e^- \to \gamma^* \to q\bar{q}$, i.e. annihilation through a virtual (timelike) photon and creation of a quark-antiquark pair. Then the $q\bar{q}$ pair hadronizes through the strong QCD interactions at large distance to form the observed hadrons. If we assume exact color confinement, this happens with unit probability, so that we need only to calculate the first process to get the total rate.

For normalization, we first calculate the pure QED rate for virtual photon creation of $\mu^+\mu^-$, according to the diagram in Figure 9. The amplitude is

$$M_o = -ie\bar{u}(p_2, s_2)\gamma_\mu v(p_1, s_1)\epsilon^\mu(\lambda) \tag{86}$$

and the rate is

$$W(\gamma^* \to \mu^+\mu^-) = \int \frac{1}{2Q}|M_o|^2 d\phi_2 \tag{87}$$

where Q is the energy of the virtual photon in its rest frame, $q = (Q, 0, 0, 0)$, $|M_0|^2$ is the square of the amplitude, suitably summed over final spins and averaged over initial spin, and $d\phi_2$ is the element of two-particle phase space for the $\mu^+\mu^-$. In general, for n particles in the final state one has

$$d\phi_n = \prod_{i=1}^{n} \frac{d^3p_i}{(2\pi)^3 2E_i}(2\pi)^4\delta^{(4)}(q - \sum_{i=1}^{n} p_i). \tag{88}$$

One evaluates $|M_0|^2$ in the usual way:

$$|M_0|^2 = -e^2 g^{\mu\nu}\mathrm{Trace}(\not{p}_2\gamma_\mu\not{p}_1\gamma_\nu) \tag{89}$$

where we have neglected the muon masses, and the replacement

$$\sum_\lambda \epsilon^\mu(\lambda)\epsilon^{*\nu}(\lambda) = -g^{\mu\nu} \tag{90}$$

results in choosing the linear combination of transverse and longitudinal polarization of the γ^* which is produced in the e^+e^- annihilation. The final result is

$$|M_0|^2 = 8e^2 p_1 \cdot p_2 = 4e^2Q^2. \tag{91}$$

Since this is a constant, one can integrate the phase space factor separately, to get

$$\int d\phi_2 = \frac{1}{(2\pi)^2}\int \frac{d^3p_1}{2E_1}\int \frac{d^3p_2}{2E_2}\delta^{(4)}(q - p_1 - p_2) = \frac{1}{8\pi} \tag{92}$$

Thus

$$W(\gamma^* \to \mu^+\mu^-) = \alpha Q, \tag{93}$$

with the usual

$$\alpha \equiv \frac{e^2}{4\pi} \tag{94}$$

The same calculation goes through for massless quarks, with the substitution $e \to e_i e$ (where e_i is the quark charge in units of e) and multiplication by a factor of $N = 3$ for the sum over final colors.

$$W(\gamma^* \to q\bar{q}) = 3\alpha e_i^2 Q \tag{95}$$

Thus the famous R ratio is

$$R \equiv \frac{\sigma(e^+e^- \to \text{hadrons})}{\sigma(e^+e^- \to \mu^+\mu^-)} = 3\sum_i e_i^2 \tag{96}$$

the so-called parton model result. The effects of QCD come in only at higher orders, where we consider gluon emission from the quark lines. Before proceeding further, we now calculate the same rate in D spacetime dimensions with $D \neq 4$, in preparation for using the dimensional regularization scheme. Consider one time and D-1 space dimensions. Then the Dirac γ-matrices must satisfy

$$\{\gamma_\mu, \gamma_\nu\} = \gamma_\mu\gamma_\nu + \gamma_\nu\gamma_\mu = 2g_{\mu\nu}I_D \tag{97}$$

where I_D is the unit matrix in spin-space, and $\mu, \nu = 0, 1, \ldots D-1$. One can define $\text{Trace}I_D = 4 + f(D)$ with $f(D = 4) = 0$, and neglect f in all calculations, since we take the limit $D \to 4$ at the end. Then the trace theorems for γ's are the same as for $D = 4$.

$$|M_0|^2 = -3(\text{color})e_i^2 e_D^2 g^{\mu\nu}\text{Trace}(\not{p}_2\gamma_\mu\not{p}_1\gamma_\nu) \tag{98}$$

and

$$g^{\mu\nu}\text{Trace}(\;) = 8p_1 \cdot p_2 - 4p_1 \cdot p_2 g_{\mu\nu}g^{\mu\nu} = -(D-2)4p_1 \cdot p_2 \tag{99}$$

Note that in D dimensions, the QED coupling e_D has a dimension. This is most easily seen by noting that the decay rate into two bodies must still have the dimension of mass, while the phase space factor

$$d\phi_2 = \frac{d^{D-1}p_1}{(2\pi)^{D-1}2E_1}\frac{d^{D-1}p_2}{(2\pi)^{D-1}2E_2}(2\pi)^D\delta^{(D)}(q - p_1 - p_2) \tag{100}$$

acquires extra mass dimensions of M^{D-4}. Hence $e_D \sim M^{-1/2(D-4)}$ and we define the dimensionless $e \equiv e_D M^{1/2(D-4)}$ or $\alpha_D = \frac{e_D^2}{4\pi} = \frac{\alpha}{M^{D-4}}$ in terms of a regularization m ass $\equiv M$. The two body phase space factors are (for massless particles)

$$\frac{d^{D-1}p}{2E} = \frac{E^{D-2}dE}{2E}(\sin\theta_1)^{D-3}(\sin\theta_2)^{D-4}\ldots\sin\theta_{D-3}\prod_{i=1}^{D-2} d\theta_i \tag{101}$$

where θ_i is the angle with respect to axis i in D dimensions. Since the matrix elements we will be dealing with here are independent of angles, we perform the angular integrations first. One may use

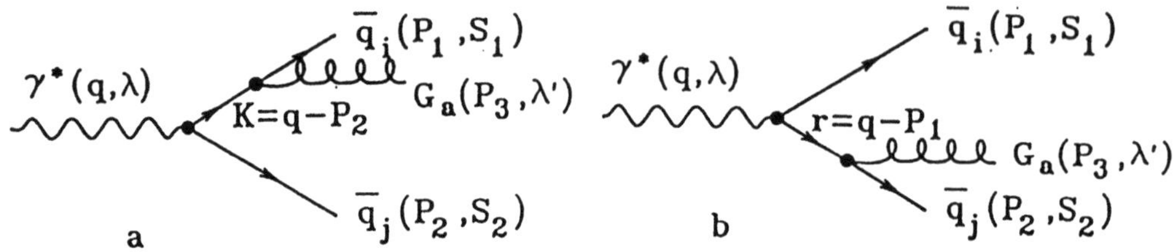

Figure 10: Gluon radiation.

$$\int_0^{\pi} (\sin\theta_i)^n d\theta_i = \sqrt{\pi}\frac{\Gamma(\frac{n+1}{2})}{\Gamma(\frac{n+2}{2})} \tag{102}$$

and see that the gamma functions cancel in pairs from the product over all angles to leave a simple expression. The $(D-1)$ - dimensional integral for particle 2 is performed as usual with the D- dimensional delta function, leaving a single delta function in energy to perform the remaining energy integral for particle 1. The final result is

$$\int d\phi_2 = \frac{Q^{D-4}}{2^{D-1}\pi^{\frac{D-2}{2}}}\frac{\Gamma(\frac{D}{2}-1)}{\Gamma(D-2)} \tag{103}$$

which gives for the rate in D dimensions

$$W(\gamma^* \to q\bar{q}) = 3\alpha e_i^2 Q(\frac{Q^2}{4\pi M^2})^{\frac{D}{2}-2}\frac{\Gamma(\frac{D}{2})}{\Gamma(D-2)} \tag{104}$$

This reduces to our previous expression for $D=4$, and we will use the expression as $\epsilon \to 0$ for $D = 4+\epsilon$:

$$W(\gamma^* \to q\bar{q}) = 3\alpha e_i^2 Q\left[\frac{Q^2}{4\pi M^2}\right]^{\frac{\epsilon}{2}}\frac{\Gamma(2+\frac{\epsilon}{2})}{\Gamma(2+\epsilon)} \tag{105}$$

The next order corrections are given by real gluon emission from the quark lines, as in the diagrams in Figure 10, with amplitudes M_1 and M_2.

$$M_1 = \bar{u}_j(p_2,s_2)(-ie_ie_D\gamma_\mu)(\frac{-i\not{k}\delta_{jk}}{k^2})(-ig_D\gamma_\nu(T^a)_{ki})v_i(p_1,s_1)\epsilon^\mu(\lambda)\epsilon^{*\nu}(\lambda') \tag{106}$$

The spin-summed square of this amplitude is

$$|M_1|^2 = \frac{e_D^2 g_D^2 e_i^2}{k^4}(-g_{\mu\mu'})(-g_{\nu\nu'}) \text{ Trace } [\gamma_\mu\not{k}\gamma_\nu\not{p}_1\gamma_{\nu'}\not{k}\gamma_{\mu'}\not{p}_2] \tag{107}$$

$$* \sum_{ij}\sum_a (T^a)_{ij}(T^a)_{ji} \tag{108}$$

The color factor trace is

$$\sum_a \text{Trace}(T^a)^2 = 8\cdot\frac{1}{2} = 4 \tag{109}$$

and the dirac matrix algebra gives

$$4(2-D)^2[8(k\cdot p_1)(k\cdot p_2) - 4k^2 p_1\cdot p_2] \tag{110}$$

where we have used the D-dimensional form

$$\gamma_\mu \gamma_\alpha \gamma^\mu = (2-D)\gamma_\alpha \tag{111}$$

For massless particles, it is convenient to use the energy fraction parameterization

$$x_i = \frac{2E_i}{Q} \tag{112}$$

with $i = 1$ (antiquark), $i = 2$ (quark), $i = 3$ (gluon).

Then $x_1 + x_2 + x_3 = 2$ and all scalar products are written in terms of Q and x_i as

$$2p_i \cdot p_j = Q^2(1 - x_k), i \neq j \neq k. \tag{113}$$

The amplitude squared term is then $|M_1|^2 = 8(2-D)^2 e_i^2 e_D^2 g_D^2 \frac{(1-x_1)}{(1-x_2)}$.

The next amplitude is

$$M_2 = \bar{u}_j(p_2, s_2)(-ig_D\gamma_\nu(T^a)_{jk})(\frac{i\not{r}\delta_{ki}}{r^2})(-ie_D\gamma_\mu)v_i(p_1, s_1)\epsilon^\mu(\lambda)\epsilon^{*\nu}(\lambda') \tag{114}$$

and its square is obtained by the substitution $x_1 \leftrightarrow x_2$ in the previous case, giving

$$|M_2|^2 = 8(2-D)^2 e_i^2 e_D^2 g_D^2 \frac{(1-x_2)}{(1-x_1)} \tag{115}$$

The interference terms are

$$2ReM_1^* M_2 = \frac{-2e_i^2 g_D^2 e_D^2}{k^2 r^2}(-g_{\mu\mu'})(-g_{\nu\nu'}) \sum_a \text{ Trace } (T^a)^2 \cdot \text{ Trace } [\gamma_{\nu'}\not{k}\gamma_{\mu'}\not{p}_2\gamma_\nu\not{r}\gamma_\mu\not{p}_1] \tag{116}$$

The trace terms are somewhat more complicated, with the result

$$\text{Trace}[\quad] = 4(2-D)Q^4(x_1x_2 - (1-x_1)(1-x_2)) + 2(D-4)(2-D)Q^4(1-x_1)(1-x_2) \tag{117}$$

where we have used the additional relation

$$\gamma_\mu \not{a}\not{b}\gamma^\mu = 4a \cdot bI_D + (D-4)\not{a}\not{b} \tag{118}$$

The final result for total amplitude squared is then

$$\begin{aligned} |M_1 + M_2|^2 &= 8e_i^2 e_D^2 g_D^2 \{(2-D)^2 \frac{x_1^2 + x_2^2}{(1-x_1)(1-x_2)} \\ &+ 2(D-4)(D-2)(1 - \frac{x_1 + x_2 - 1}{(1-x_1)(1-x_2)})\} \end{aligned} \tag{119}$$

or in terms of the ϵ parameter

$$\begin{aligned} |M_1 + M_2|^2 &= 32e_i^2 e_D^2 g_D^2 \{(1 + \frac{\epsilon}{2})^2 \frac{x_1^2 + x_2^2}{(1-x_1)(1-x_2)} \\ &+ \epsilon(1 + \frac{\epsilon}{2}) \frac{2(1 - x_1 - x_2) + x_1 x_2}{(1-x_1)(1-x_2)}\} \\ &\equiv 32e_i^2 e_D^2 g_D^2 f(x_1, x_2) \end{aligned} \tag{120}$$

Here we see the first evidence of an infrared divergence, since we will integrate over the range $0 \leq x_1, x_2 \leq 1$. The vanishing of the $(1 - x_i)$ factors at the end points corresponds to the vanishing of the internal quark line propagators in two limits. The general expression is of the form

$$E_{Quark}E_{Gluon}(1-\cos\theta_{13}). \tag{121}$$

Thus the propagator denominator vanishes when either $E_{Gluon}\to 0$, the soft gluon divergence,or when $\theta_{13}=0$, the so-called collinear divergence for massless particles. It turns out that the dimensional regularization scheme will take care of both of these effects without differentiating between them, as we shall see.

To get the correction to the rate, we also need the 3-body phase space factor in D dimensions. One uses the delta function in the expression

$$\frac{d^{D-1}p_1}{2E_1(2\pi)^{D-1}}\frac{d^{D-1}p_2}{2E_2(2\pi)^{D-1}}\frac{d^{D-1}p_3}{2E_3(2\pi)^{D-1}}(2\pi)^D\delta^{(D)}(q-p_1-p_2-p_3) \tag{122}$$

to perform D-1 integrals for particle 3, to get

$$\frac{d^{D-1}p_1d^{D-1}p_2}{8E_1E_2E_3(2\pi)^{2D-3}}\delta(Q-E_1-E_2-E_3)|_{E_3=|\vec{p}_3|=|-\vec{p}_1-\vec{p}_2|} \tag{123}$$

Since the matrix element only depends on x_i(or E_i), we can integrate over all angles in phase space first at fixed E_1 and E_2. The remaining delta function depends on the angle between $\vec{p_1}$ and $\vec{p_2}$ through the constraint on E_3. Thus we choose the $(D-1)$'st spatial axis to be along $\vec{p_1}$. The delta function will depend on θ_{12} which appears in the first angular integral for $d^{N-1}p_2$ as

$$\int_0^\pi(\sin\theta_{12})^{D-3}d\theta_{12}\equiv\int_{-1}^1dz(1-z^2)^{\frac{D-4}{2}} \tag{124}$$

The remaning $D-3$ angles in $\vec{p_2}$ and all $D-2$ angles in $\vec{p_1}$ can then be integrated directly, using the formulae given previously. Finally the z–integration can be performed using the remaining delta function, and we are left with

$$\int x_1^\epsilon dx_1\int x_2^\epsilon dx^2\frac{Q^2}{16(2\pi)^3}(\frac{1-z_0^2}{4})^{\frac{\epsilon}{2}}\frac{1}{\Gamma(2+\epsilon)}(\frac{Q^2}{4\pi})^\epsilon \tag{125}$$

where $z_0\equiv 1-\frac{2(x_1+x_2-1)}{x_1x_2}$, and the remaining integration over x_1 and x_2 must include the amplitude.

We take out regularization mass factors again to make both QED and QCD couplings dimensionless, with

$$g_D^2\equiv\frac{4\pi\alpha_s}{(M^2)^{\frac{\epsilon}{2}}} \tag{126}$$

The rate is then written

$$W(\gamma^*\to q\bar{q}g)=\frac{2\alpha\alpha_s}{\pi}e_i^2Q(\frac{Q^2}{4\pi M^2})^{\frac{\epsilon}{2}}\frac{1}{\Gamma(2+\epsilon)}\int x_1^\epsilon dx_1\int x_2^\epsilon dx_2(\frac{1-z_0^2}{4})^{\frac{\epsilon}{2}}f(x_1,x_2) \tag{127}$$

where $f(x_1,x_2)$ was defined previously. It is convenient to normalize by the lowest order rate $W_0\equiv W(\gamma^*\to q\bar{q})$, to get

$$W(\gamma^* \to q\bar{q}g) = \frac{2\alpha_s}{3\pi} W_0 (\frac{Q^2}{4\pi M^2})^{\frac{\epsilon}{2}} \frac{1}{\Gamma(2+\frac{\epsilon}{2})} \int x_1^{\epsilon} dx_1 \int x_2^{\epsilon} dx_2 (\frac{1-z_0^2}{4})^{\frac{\epsilon}{2}} f(x_1, x_2) \quad (128)$$

The integration regions $0 \leq x_1, x_2, x_3 \leq 1$ with $x_1 + x_2 + x_3 = 2$ implies a lower limit $1 - x_1$ for x_2. One usually decouples the integrals by defining a new variable

$$\eta \equiv \frac{1-x_2}{x_1} \quad (129)$$

such that

$$\int_0^1 dx_1 \int_{1-x_1}^1 dx_2 = \int_0^1 x_1 dx_1 \int_0^1 d\eta \quad (130)$$

In terms of this variable,

$$\frac{1-z_0^2}{4} = \frac{\eta(1-\eta)(1-x_1)}{(x_2)^2} \quad (131)$$

so that the integration for the rate

$$\begin{aligned} W(\gamma^* \to q\bar{q}g) &= \frac{2}{3}\frac{\alpha_s}{\pi} W_o (\frac{Q^2}{4\pi M^2})^{\frac{\epsilon}{2}} \frac{1}{\Gamma(2+\epsilon)} \\ & \int_0^1 dx_1 x_1^{1+\epsilon} (1-x_1)^{\frac{\epsilon}{2}} \int_0^1 d\eta \eta^{\frac{\epsilon}{2}} (1-\eta)^{\frac{\epsilon}{2}} f(x_1, 1-\eta x_1) \quad (132) \end{aligned}$$

can easily be performed in terms of Γ functions according to

$$\int_0^1 dz z^{\alpha} (1-z)^{\beta} = \frac{\Gamma(\alpha+1)\Gamma(\beta+1)}{\Gamma(\alpha+\beta+2)} \quad (133)$$

These integrals will develop poles of first and second order in ϵ, and hence one must keep order ϵ and ϵ^2 terms in the amplitudes. The integration gives

$$\frac{\Gamma(2+\epsilon)\Gamma^2(1+\frac{\epsilon}{2})}{\Gamma(1+\frac{3}{2}\epsilon)} (\frac{8}{\epsilon^2} - \frac{6}{\epsilon} + \frac{19}{2} + 0(\epsilon)) \quad (134)$$

One then expands the factor

$$(\frac{Q^2}{4\pi M^2})^{\frac{\epsilon}{2}} = 1 + \frac{\epsilon}{2} \log(\frac{Q^2}{4\pi M^2}) + \frac{\epsilon^2}{8} \log^2(\frac{Q^2}{4\pi M^2}) + 0(\epsilon^3) \quad (135)$$

and

$$\Gamma(1+g\epsilon) = \Gamma(1) + g\epsilon\Gamma'(1) + \frac{g^2\epsilon^2}{2}\Gamma''(1) + 0(\epsilon^3) \quad (136)$$

where

$$\Gamma'(1) = -\gamma_E = -.5772\ldots, \quad (137)$$

the Euler constant, and

$$\Gamma''(1) = \gamma_E^2 + \frac{\pi^2}{6} \quad (138)$$

to get the final result

$$W(\gamma^* \to q\bar{q}g) = \frac{2}{3}\frac{\alpha_s}{\pi} W_o \{ \frac{8}{\epsilon^2} + \frac{1}{\epsilon}[4\log(\frac{Q^2}{4\pi M^2}) + 4\gamma_E - 6]$$

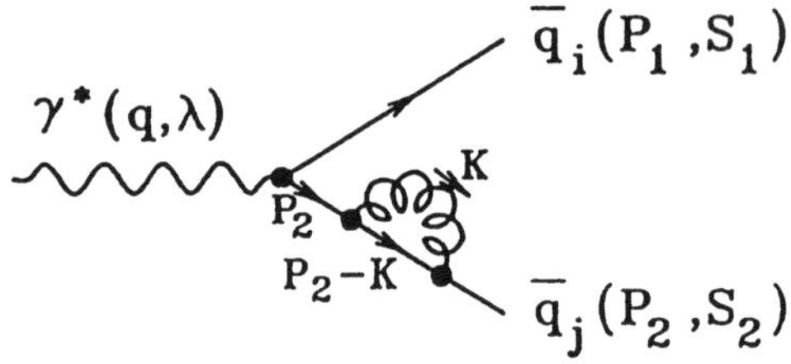

Figure 11: Correction on external line.

$$
\begin{aligned}
&+ \log^2(\frac{Q^2}{4\pi M^2}) \\
&+ (2\gamma_E - 3)\log(\frac{Q^2}{4\pi M^2}) \\
&+ \gamma_E^2 - 3\gamma_E - \frac{7\pi^2}{6} + \frac{57}{6} + 0(\epsilon)\}
\end{aligned}
\tag{139}
$$

The poles in ϵ must be cancelled by contributions from gluon loop diagrams in the same order. These will come from single loop amplitudes inserted in the lowest order diagram $\sim e\alpha_s$ interfering with that diagram itself $\sim e$ to give a total $\sim \alpha\alpha_s$.

First, consider the correction to the quark propagators from gluon loops, as in Figure 11.

The effect of the gluon loop is a multiplicative correction factor to the lowest order amplitude M_0:

$$M_3 = M_0 \frac{i\not{p}_2}{(p_2)^2} \sum(p_2) \tag{140}$$

with

$$
\begin{aligned}
\sum(p) &= (-ig_D)^2 T^a_{\alpha\beta} T^a_{\beta\alpha} \\
&\int \frac{d^Dk}{(2\pi)^D} \frac{\gamma_\mu(\not{p} - \not{k})\gamma_\nu}{(p-k)^2 + i\epsilon} (\frac{-i}{k^2})(g_{\mu\nu} + \frac{(\xi - 1)k_\mu k_\nu}{k^2})
\end{aligned}
\tag{141}
$$

where a and β are internal gluon and quark colors and are summed over, but α is the fixed external q and $\bar{q}$ color index for the final quark and antiquark. Thus the color factor is

$$\sum_a \sum_\beta T^a_{\alpha\beta} T^a_{\beta\alpha} = \frac{4}{3} \tag{142}$$

We will show that $\sum(p) = -i\not{p}\sum$ so that $M_3 = M_0 \sum$, is just a multiplicative constant. To evaluate Σ, use the Feynman parameterization of propagators:

$$\frac{1}{ab} = \int_0^1 dx \frac{1}{[ax + b(1-x)]^2} \tag{143}$$

and its derivative

$$\frac{1}{ab^2} = \int_0^1 dx \frac{2(1-x)}{[ax + b(1-x)]^3} \tag{144}$$

for $a = (p-k)^2$ and $b = k^2$. One then shifts the $k-$integrals by a constant $k = r + xp$ to make the denominators $= r^2 - C$, with $C = -x(1-x)p^2$. The numerators become

$$\gamma_\mu(\not{p} - \not{k})\gamma^\mu = (2 - D)(\not{p}(1-x) - \not{r}) \tag{145}$$

and

$$\not k(\not p - \not k)\not k = -k^2(\not p(1+x) + \not r) + (2p\cdot r + xp^2)(x\not p + \not r) \tag{146}$$

and we can drop all terms linear in r, since they will give zero under the symmetric integration $\int \frac{d^D r}{(2\pi)^D}$, and replace all quadratic terms $r_\mu r_\nu$ by $\frac{r^2}{D} g_{\mu\nu}$. Thus

$$\gamma_\mu(\not p - \not k)\gamma^\mu \to (2-D)(1-x)\not p \tag{147}$$

and

$$\not k(\not p - \not k) \to -k^2(1+x)\not p + 2\not p(x^2p^2 + \frac{r^2}{D}) \tag{148}$$

where we have kept the k^2 factor in the first term to cancel part of the $1/k^4$. After taking out a factor of $-i\not p$, we have

$$\begin{aligned}\bar{\Sigma} &= -\frac{4i}{3}g_D^2\Big[\int_0^1 dx \int \frac{d^D r}{(2\pi)^D}\frac{((2-D)(1-x) - (\xi-1)(1+x))}{(r^2-C)^2} \\ &+ \int_0^1 dx \int \frac{d^D r}{(2\pi)^D} 4(\xi-1)(1-x)\frac{(x^2p^2 + \frac{r^2}{D})}{(R^2-C)^3}\Big]\end{aligned} \tag{149}$$

Note that there are divergences in the gluon loop integrals for $D=4$, which appear gauge-dependent. To evaluate the r–integrals in D dimensions, use the general formula

$$\int \frac{d^D r}{(2\pi)^D}\frac{(r^2)^\alpha}{(r^2-C)^\beta} = \frac{i(-1)^{\alpha-\beta}}{(16\pi^2)^{\frac{D}{4}}} C^{\alpha-\beta+\frac{D}{2}}\frac{\Gamma(\alpha+\frac{D}{2})\Gamma(\beta-\alpha-\frac{D}{2})}{\Gamma(\frac{D}{2})\Gamma(\beta)} \tag{150}$$

and the remaining x–integrals are just powers of x and $1-x$, which just produces products of more Γ functions. The final result is

$$\bar{\Sigma} = \frac{\xi\alpha_s}{3\pi}(\frac{-p^2}{4\pi M^2})^{\frac{\epsilon}{2}}\frac{2+\epsilon}{\epsilon}\frac{\Gamma(1-\frac{\epsilon}{2})\Gamma^2(1+\frac{\epsilon}{2})}{\Gamma(2+\epsilon)} \tag{151}$$

One can then work in Landau gauge ($\xi = 0$) and avoid these factors. In fact, when the fermions are on-shell ($p^2 = 0$), and $\epsilon > 0$, one can see that they are zero in any gauge. Then the only loop correction we must worry about is the vertex correction (Figure 12).

$$\begin{aligned}M_4 &= \int \frac{d^D k}{(2\pi)^D}\bar{u}(p_2,s_2)(-ig_D\gamma_\nu T^a_{\eta j})\frac{i(\not p_2 + \not k)}{(p_2+k)^2}(-ie_D e_i\gamma_\alpha) \\ &(\frac{i(\not p_1 - \not k)}{(p_1-k)^2})(-ig_D\gamma_\mu T^a_{i\eta})(\frac{-i}{k^2})(g_{\mu\nu} + (\xi-1)\frac{k_\mu k_\nu}{k^2})v(p_1,s_1)\epsilon^\alpha(\lambda)\end{aligned} \tag{152}$$

We need the interference term of this amplitude with M_0:

$$\begin{aligned}2\mathrm{Re}M_0^* M_4 &= 2ig_D^2 e_D^2 e_i^2(-g_{\alpha\alpha'}) \\ &\int \frac{d^D k}{(2\pi)^D}\frac{1}{(p_2+k)^2(p_1-k)^2k^2} \\ &\mathrm{Trace}(T^aT^a) \\ &(g_{\mu\nu} + (\xi-1)\frac{k_\mu k_\nu}{k^2}) \\ &\mathrm{Trace}[\not p_2\gamma_\nu(\not p_2 + \not k)\gamma_\alpha(\not p_1 - \not k)\gamma_\mu \not p_1\gamma_{\alpha'}]\end{aligned} \tag{153}$$

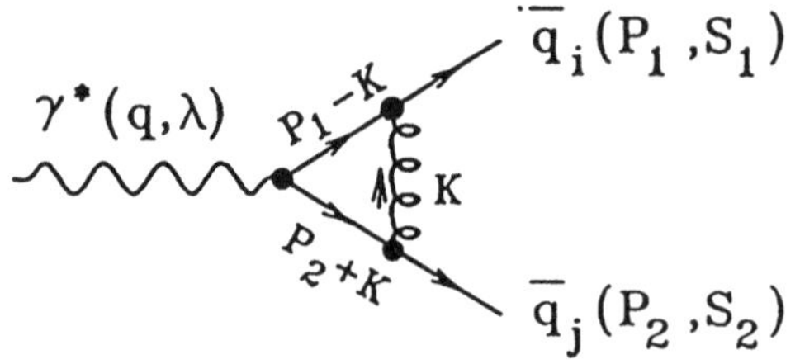

Figure 12: Vertex correction.

The color factor trace = 4 as usual, and the dirac algebra trace gives (after some algebra)

$$
\begin{aligned}
4q^2(1 \;+\; &\frac{\epsilon}{2})[-2q^2 + 4k\cdot(p_2 - p_1) + \frac{8(k\cdot p_1)(k\cdot p_2)}{q^2} \\
+\epsilon k^2 \;+\; &(\xi - 1)(-\frac{4(k\cdot p_1)(k\cdot p_2)}{k^2} + 2k\cdot(p_2 - p_1) + k^2)]
\end{aligned} \tag{154}
$$

where we have made the usual substitution $D = 4 + \epsilon$.

The $D-$dimensional loop integrals are again done with Feynman parameterization

$$
\frac{1}{abc} = 2\int_0^1 dx \int_0^1 dy \int_0^1 dz \frac{\delta(1-x-y-z)}{(ax+by+cz)^3} \tag{155}
$$

and

$$
\frac{1}{abc^2} = 2\int_0^1 dx \int_0^1 dy \int_0^1 dz \frac{3z\delta(1-x-y-z)}{(ax+by+cz)^4} \tag{156}
$$

with

$$
\begin{aligned}
a &= (p_1 - k)^2, \\
b &= (p_2 + k)^2, \\
c &= k^2
\end{aligned} \tag{157}
$$

The shift in the $k-$variable is $k = r - yp_2 + xp_1$ to place the denominator in the usual $r^2 - C$ form, with $C = (yp_2 - xp_1)^2 = -xyq^2$. The $r-$ integrals are then done with the usual general formula, but we must keep q^2 spacelike in order to avoid passing through singularities in the region of integration. The $x-$ and $y-$ integrals then again give more $\Gamma-$ functions, from which we extract the $\epsilon-$ poles. (For details, see the book by R. Field).

One must be careful in using the Feynman parameterization for the propagators to note that the $\delta(1-x-y-z)$ term wh en used to perform the z-integrals then requires an upper limit of $1-x$ for the y-integral. Then one must again define a new variable $\eta \equiv \frac{y}{1-x}$ to separate the nested x- and y- integrals.

$$
\int_0^1 dx \int_0^{1-x} dy f(x,y) = \int_0^1 dx\,(1-x)\int_0^1 d\eta f(x,\eta(1-x)) \tag{158}
$$

For this reason, some treatments first combine two of the propagators using a single Feynman parameter, and then incorporate the third denominator factor with a subsequent shift of variable and an independent second Feynman parameter.

The final result is most useful when expressed as a correction to the zeroth order rate W_o:

$$
W_{vertex} \;=\; \frac{2\alpha_s}{3\pi}(\frac{-q^2}{4\pi M^2})^{\frac{\epsilon}{2}}\frac{\Gamma(1-\frac{\epsilon}{2})\Gamma^2(1+\frac{\epsilon}{2})}{\Gamma(1+\epsilon)}
$$

$$\times \quad (\frac{-8}{\epsilon^2} + \frac{1}{2\epsilon(1+\epsilon)(2+\epsilon)}[-\epsilon(8+4\epsilon+4\xi)$$
$$- \quad 8\xi + 8(2+\xi-1)(2+\epsilon) - 4(\xi-1)(2+\epsilon)]) \qquad (159)$$

One then expands this expression in powers of ϵ to combine with the real gluon emission correction. One sees that the gauge parameter ξ-dependence drops out, as it must since this is the only place it appears to this order in α_s.

The final rate formula has terms proportional to $\log(\frac{-q^2}{4\pi M^2})$ and $\log^2(\frac{-q^2}{4\pi M^2})$, and to continue to the timelike region one uses

$$\log(\frac{-q^2}{4\pi M^2}) = \log(\frac{Q^2}{4\pi M^2}) - i\pi \qquad (160)$$

and omits the imaginary terms. This gives some additional finite terms, with the final result

$$\begin{aligned} W_{Vertex}(\gamma^* \to q\bar{q}) &= \frac{2\alpha_s}{3\pi} W_0[-\frac{8}{\epsilon^2} + \frac{1}{\epsilon}(-4\log(\frac{Q^2}{4\pi M^2}) - 4\gamma_E + 6) \\ &- \log^2(\frac{Q^2}{4\pi M^2}) - (2\gamma_E - 3)\log(\frac{Q^2}{4\pi M^2}) \\ &- \gamma_E^2 + 3\gamma_E + \frac{\pi^2}{8} - 8 + \pi^2 + 0(\epsilon)] \qquad (161) \end{aligned}$$

When combined with the real gluon emission rate, the expected cancellation of all divergences occurs, and the total first order QCD correction to the process is

$$\begin{aligned} W_{Gluon} + W_{Vertex} &= \frac{2\alpha_s}{3\pi} W_0(\frac{57}{6} - 8) \\ &= \frac{\alpha_s}{\pi} W_0 \qquad (162) \end{aligned}$$

The dimensional regularization scheme has produced a result independent of regulaization mass and also infra-red finite. This procedure works because one never gets a product of infrared and ultraviolet divergences in the same term.

The final result is then the first order QCD correction for the parton model R ratio.

$$R = 3\sum_i e_i^2(1 + \frac{\alpha_s}{\pi}) \qquad (163)$$

One can extend this procedure to higher order corrections, including multiple real gluon emission graphs. These graphs of course have soft and collinear divergences which must be cancelled by interference terms of loops in lower order diagrams, just as we have seen here. Additional complications arise, however, since the definition of higher order terms depends on the renormalization point and subtraction scheme for defining α_s. The next order calculation has been known for some time, and yields a term in $(\frac{\alpha_s}{\pi})^2$ with coefficient ≈ 1.4 in the energy region where 5 quark flavors contribute. The next order $(\frac{\alpha_s}{\pi})^3$ terms have only recently been calculated, with a somewhat surprisingly large coefficient ≈ 64. Although the values of α_s extracted from experiment agree within errors using either the second- or third-order perturbative expressions, one may still feel uneasy about this large coefficient. Perhaps one of you will extend this calculation to fourth order!

Phase Transitions in Nuclei at Low Temperatures

A.L. Goodman

Physics Department, Tulane University, New Orleans, LO70118, USA

Contents

Phase Structure of Strongly Interacting Matter
Editor: J. Cleymans

1. Introduction

The subject of these lectures is phase transitions in nuclei at low temperatures. For our colleagues in other branches of physics, this title might appear to contain a contradiction. We all learned that temperature and phase transition are concepts which apply to macroscopic systems. In what sense can they be used for a microscopic system, such as an atomic nucleus? I will try to address this question during the course of these lectures.

Traditional nuclear structure physics is concerned with nuclei at low excitation energies, where it is possible to measure and calculate discrete energy levels. However in recent years we have been interested in nuclear structure at higher energies. The nuclear level density grows exponentially with the excitation energy. At higher energies it is not feasible to calculate or measure individual energy states. It is then reasonable to use statistical techniques to calculate the average properties of neighbouring energy levels. This is accomplished by introducing a temperature and applying the methods of quantum statistical mechanics.

How do we define the temperature of a single nucleus? Let ρ be the level density, where $\rho(E,I)dE$ is the number of nuclear energy levels with spin I and energy between E and E+dE. Then the entropy is

$$S(E,I) = k \ln\rho(E,I), \tag{1}$$

where k is Boltzmann's constant. The temperature is

$$T(E,I) = \left[\frac{\partial S}{\partial E}\right]_I^{-1} . \tag{2}$$

So the level density determines the temperature for each combination of spin and energy.

We exercise special care when we apply statistical mechanics to finite systems with $N \sim 100$. For macroscopic systems the equilibrium state is given by the most probable state. For a system with a given energy, this is the state with the largest entropy. Less probable states are ignored. Fluctuations away from the most probable state are negligible, except at critical points and phase transitions. For example, in the canonical ensemble, the fractional fluctuation in the energy is proportional to $1/\sqrt{N}$. In the grand canonical ensemble the fractional fluctuation in the density is proportional to $1/\sqrt{N}$. In the limit $N \to \infty$, these fluctuations vanish. However for atomic nuclei, where $N \sim 100$, these statistical fluctuations can not be ignored. The fluctuations around the most probable state must be included to obtain the equilibrium distribution. We will see that these fluctuations can radically alter the properties of nuclei, even when we are not in the vicinity of a phase transition.

The finite number effects can also be studied by comparing the different statistical ensembles: microcanonical, canonical, and grand canonical. For large N these ensembles are equivalent. For finite nuclei these ensembles are not equivalent, but little is known about their differences. Hot nuclei provide an excellent laboratory for testing the range of validity for quantum statistical mechanics.

Phase transitions are common in macroscopic systems. A minute change in some intensive variable (temperature, pressure, etc.) produces a large alteration in an order parameter, which measures a collective property of the system. Phase transitions also occur in finite systems when they are described by a mean field theory. However this is an artifact of the mean field approximation. We will see that when statistical fluctuations in the order parameter are included, then the phase transitions are often smoothed out or even eliminated.

For a many-body system, such as atomic nuclei, our first approximation is the mean field approximation. The average effect of the nucleon-nucleon interaction is described by a one-body potential. We will be using the finite-temperature Hartree-Fock-Bogoliubov (FTHFB) theory, which is a microscopic mean field approximation. The mean field is calculated self-consistently from a N - N interaction. Nuclear properties are determined as functions of angular momentum and temperature (excitation energy). There are other mean field approximations, such as the macroscopic Landau theory of phase transitions.

We want a mean field theory which is general enough to include the following characteristics of nuclei:

1. Independent particle motion. The nucleons occupy orbitals created by the mean field. If the nuclear shape is spherically symmetric, then the nucleon orbitals have good angular momentum.

2. Deformation. For nuclei which are several nucleons away from a closed shell configuration, the equilibrium shape of the nucleus is not spherically symmetric. Then the mean field is not spherically symmetric, and the nucleon orbitals don't have a spin quantum number.

3. Pair correlations. The nucleon-nucleon interaction forms correlated pairs of nucleons. In the simplest version, the two nucleons in each pair occupy orbitals related by time-reversal. The nucleus is then described as superfluid or superconducting.

4. Rotations. Nuclei with deformed equilibrium shapes have quantized rotational spectra. In the coordinate system of the rotating nucleus, the mean field includes the Coriolis force and the centrifugal force.

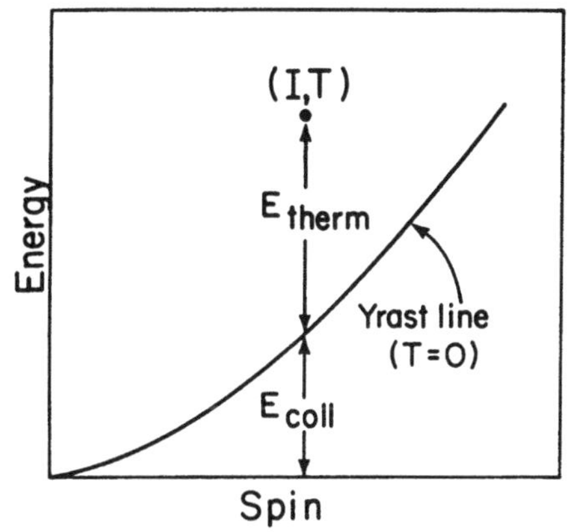

Fig. 1. Schematic diagram of energy E versus spin I. The temperature is T.

5. Thermal excitations. At finite-temperature particles (or quasi-particles) are thermally excited. The mean field is then temperature dependent. Fig. 1 shows how the nuclear excitation energy can be divided into two components: thermal and collective. By introducing a temperature we dramatically extend the domain of the mean field approximation.

6. Density. The mean field will depend upon the nucleon density.

The FTHFB theory includes all of these effects in a completely self-consistent manner. A variation in any one of these properties will create adjustments in all the other properties. The theory describes several phase transitions in nuclei:

1. Superfluid-normal fluid. This transition is induced by increasing the temperature or the rotational frequency, or some combination of them.

2. Shape transition. Heating or rotation can change the kind of shape that the nucleus has, i.e., prolate to spherical, or prolate to oblate.

3. Rotation mode. Heating or rotation can change collective rotation to noncollective rotation. In collective rotation the shape symmetry axis is perpendicular to the rotation axis. In noncollective rotation the symmetry axis coincides with the rotation axis.

4. Liquid-gas transition. The equation of state for nuclear matter predicts a transition between a high density liquid phase and a low density gas phase.

The mean field theories ignore statistical fluctuations. We will see that when fluctuations are included, these phase transitions are sometimes smoothed out or eliminated.

In Sect. 2 the finite-temperature HFB theory will be derived. Sect. 3 describes the superfluid-normal fluid phase transition. Sect. 4 discusses shape transitions and transitions in the rotation mode. Sect. 5 demonstrates the importance of statistical fluctuations in order parameters for finite systems. Sect. 6 describes the nuclear liquid-gas phase transition.

2. Finite-Temperature Hartree-Fock-Bogoliubov-Theory

We will now derive the FTHFB theory [1,2].

2.1. Thermodynamics

For an unconstrained system at zero temperature, the equilibrium state is that state which minimizes the energy E. However for a system with a given temperature T, chemical potential μ, and rotational frequency ω, the equilibrium state minimizes the grand potential in a rotating frame

$$\Omega' = E - TS - \mu N - \omega J, \tag{3}$$

where S is the entropy, N is the particle number, and J is the angular momentum. The equilibrium condition is

$$\delta\Omega' = 0. \tag{4}$$

2.2. Statistical Mechanics

The grand partition function is

$$Z = \mathrm{Tr}\left[e^{-\beta(H-\mu\hat{N}-\omega J_x)}\right] , \tag{5}$$

where $\beta=1/kT$, H is the many-body Hamiltonian, $\hat{N}$ is the particle number operator

$$\hat{N} = \sum_i C_i^\dagger C_i , \tag{6}$$

and J_x is the spin operator for a nucleus rotating about the x-axis

$$J_x = \sum_{ij} \langle i|J_x|j\rangle\, C_i^\dagger C_j . \tag{7}$$

The trace in eq. (5) implies a sum over all states with any number of particles (or quasiparticles), but with a given angular momentum. The density operator is

$$D = Z^{-1}\, e^{-\beta(H-\mu\hat{N}-\omega J_x)} . \tag{8}$$

The expectation value of any operator O is given by the average in the grand canonical ensemble

$$\langle O\rangle = \mathrm{Tr}(DO) . \tag{9}$$

The energy, entropy, particle number, and spin are evaluated by

$$E = \langle H\rangle = \mathrm{Tr}(DH) , \tag{10}$$

$$S = \langle -k \ln D\rangle = -k\, \mathrm{Tr}(D \ln D), \tag{11}$$

$$N = \langle\hat{N}\rangle = \mathrm{Tr}(D\hat{N}) , \tag{12}$$

$$J = [I(I+1)]^{1/2} = \langle J_x\rangle = \mathrm{Tr}\,(DJ_x) . \tag{13}$$

2.3. Independent Quasiparticle Model

The Hamiltonian

$$H = \sum_{i=1}^{N} K_i + \sum_{i<j=1}^{N} v_{ij} , \tag{14}$$

includes the kinetic energy of each nucleon and the nucleon-nucleon interactions. We are confronted with a many-body problem, which we cannot solve exactly. So we must make approximations. To evaluate the grand partition function and the density operator in the mean field approximation, the interactions are replaced by one-body potentials.

The mean fields can be defined by means of quasiparticles. A quasiparticle is an elementary excitation of the system. The idea is to find a transformation from particle coordinates to quasiparticle coordinates, such that the quasiparticles are weakly interacting. That is, we want to write the Hamiltonian as

$$H - \mu\hat{N} - \omega J_x = E_o + H_{qp} + H_{qp\text{-}int} , \tag{15}$$

where E_o is the energy of the quasiparticle vacuum, H_{qp} describes the quasiparticle excitations, and $H_{qp\text{-}int}$ is the weak quasiparticle interaction. At zero-temperature this interaction is not included in mean field theories. However at finite-temperature certain components of the quasiparticle interactions are included in the mean fields.

In the HFB theory the quasiparticles are defined by the general Bogoliubov transformation

$$a_i^{\dagger} = \sum_j (U_{ij} C_j^{\dagger} + V_{ij} C_j) . \tag{16}$$

Each quasiparticle is a linear combination of all particle creation and annihilation operators. If there are n single-particle basis states, then U and V are $n \times n$ complex matrices. Since the quasiparticles must satisfy the fermion commutation relations, it follows that the transformation (16) must be unitary

$$UU^{\dagger} + VV^{\dagger} = U^{\dagger}U + \tilde{V}V^{*} = I, \tag{17}$$

$$U\tilde{V} + V\tilde{U} = U^{\dagger}V + \tilde{V}U^{*} = 0 . \tag{18}$$

To obtain the HFB density operator, the exact Hamiltonian is approximated by the independent quasiparticle Hamiltonian

$$H - \mu\hat{N} - \omega J_x \approx H_{HFB} = E_o + \sum_i E_i a_i^{\dagger} a_i , \tag{19}$$

where E_i is a quasiparticle energy. Substitute approximation (19) into eq. (5) for the grand partition function. The trace is evaluated by summing over all n quasiparticle states, where n is any integer from 0 to ∞. The result is the grand partition function for independent quasiparticles

$$Z_{HFB} = \prod_i (1+e^{-\beta E_i}) \ . \tag{20}$$

Using approximation (19) to evaluate the density operator (8), we find

$$D_{HFB} = \prod_i \left[f_i \hat{n}_i + (1-f_i)(1-\hat{n}_i)\right] , \tag{21}$$

where $\hat{n}_i$ is the quasiparticle number operator

$$\hat{n}_i = a_i^\dagger a_i , \tag{22}$$

and f_i is the Fermi-Dirac function

$$f_i = \frac{1}{1 + e^{\beta E_i}} \ . \tag{23}$$

The quasiparticle occupation probability is

$$<\hat{n}_i> = Tr(D_{HFB} \hat{n}_i) = f_i \ . \tag{24}$$

Normally the quasiparticle energies are all positive. Then at zero-temperature all f_i equal zero, and the HFB state is the quasiparticle vacuum. However at finite-temperature $0<f_i<1$. There is a statistical mixture of all n quasiparticle excitations, where n is any integer from 0 to ∞.

The single-particle density matrix ρ and the pairing tensor t are defined by

$$\rho_{ij} = <C_j^\dagger C_i> = Tr(DC_j^\dagger C_i) , \tag{25}$$

$$t_{ij} = <C_j C_i> = Tr(DC_j C_i) \ . \tag{26}$$

If D is approximated by D_{HFB}, then

$$\rho = \tilde{U}fU^* + V^\dagger(1-f)V , \tag{27}$$

$$t = \tilde{U}fV^* + V^\dagger(1-f)U , \tag{28}$$

where $f_{ij} = f_i \delta_{ij}$. Remember that for zero-temperature f = 0.

The energy (10) is the expectation value of the Hamiltonian

$$H = \sum_{ij} K_{ij} C_i^\dagger C_j + \frac{1}{4} \sum_{ijkl} v_{ijkl} C_i^\dagger C_j^\dagger C_l C_k , \tag{29}$$

where the matrix elements of v are anti-symmetrized. The finite-temperature Wick's theorem states that the ensemble average of an operator is equal to the sum of all possible fully contracted terms. For example

$$\begin{aligned} <C_i^\dagger C_j^\dagger C_l C_k> &= <C_i^\dagger C_k> <C_j^\dagger C_l> \\ &- <C_i^\dagger C_l> <C_j^\dagger C_k> + <C_i^\dagger C_j^\dagger> <C_l C_k> \ . \end{aligned} \tag{30}$$

These contractions are given by the densities ρ and t. So the energy is

$$E = \sum_{ij} K_{ij}\, \rho_{ji} + \frac{1}{2} \sum_{ijkl} v_{ijkl}\, \rho_{lj}\, \rho_{ki} + \frac{1}{4} \sum_{ijkl} v_{ijkl}\, t^*_{ij}\, t_{kl} \tag{31}$$

The entropy (11) is evaluated with D_{HFB}

$$S = -k \sum_i [f_i \ln f_i + (1-f_i) \ln (1-f_i)] \; . \tag{32}$$

The particle number (12) and the angular momentum (13) are

$$N = \mathrm{Tr}\, \rho \; , \tag{33}$$

$$J = \mathrm{Tr}(\rho\, J_x) \; . \tag{34}$$

The grand potential in a rotating frame (3) is

$$\Omega' = \sum_{ij} (K - \mu - \omega J_x)_{ij}\, \rho_{ji} + \frac{1}{2} \sum_{ijkl} v_{ijkl}\, \rho_{lj}\, \rho_{ki} + \frac{1}{4} \sum_{ijkl} v_{ijkl}\, t^*_{ij}\, t_{kl} + kT \sum_i [f_i \ln f_i + (1-f_i) \ln (1-f_i)] \; . \tag{35}$$

2.4. Variational Principle

The potential Ω' is minimized

$$\delta\Omega' = 0 \; . \tag{36}$$

The quantities which are varied are U, V and f. The variation (36) demonstrates that the equilibrium value of f is the Fermi-Dirac distribution. Because of the unitarity constraints (17) and (18), the variations δU and δV are not independent. You can verify that [1]

$$\delta U = \varepsilon_1 U + \varepsilon_2 V^* \; , \tag{37}$$

$$\delta V = \varepsilon_1 V + \varepsilon_2 U^* \; , \tag{38}$$

where ε_1 and ε_2 are infinitesimal matrices which satisfy

$$\varepsilon_1^\dagger = -\varepsilon_1, \quad \tilde{\varepsilon}_2 = -\varepsilon_2 \; . \tag{39}$$

You can also evaluate $\delta\Omega'$ and show that the variational principle (36) gives the FTHFB equation [1]

$$\begin{pmatrix} \mathscr{H} & \Delta \\ -\Delta^* & -\mathscr{H}^* \end{pmatrix} \begin{pmatrix} U_i \\ V_i \end{pmatrix} = E_i \begin{pmatrix} U_i \\ V_i \end{pmatrix} \; . \tag{40}$$

The energy matrix contains two mean fields. The Hartree-Fock (HF) Hamiltonian in a rotating frame is

$$\mathscr{H} = K - \mu - \omega J_x + \Gamma \; , \tag{41}$$

where the HF potential is

$$\Gamma_{ij} = \sum_{kl} v_{ijkl}\, \rho_{lk}. \tag{42}$$

The pair potential is

$$\Delta_{ij} = \frac{1}{2} \sum_{kl} v_{ijkl}\, t_{kl} \; . \tag{43}$$

The FTHFB equation is an eigenvalue equation. The eigenvectors define the quasiparticles (16), where U_i indicates the vector $(U_{i1}, U_{i2} \ldots)$ and similarly for V_i. The eigenvalues are the quasiparticle energies E_i .

2.5. Properties of the HFB Equation

The mean fields Γ and Δ depend upon the densities ρ and t. But the densities are functions of the eigenvectors (U,V) of the mean fields. Similarly the quasiparticle occupation probability f_i depends on the quasiparticle energy E_i, and vice-versa. The theory is self-consistent: the densities create mean fields which in turn lead to the original densities.

The FTHFB equation is non-linear, and it is solved by iteration. Begin with an educated guess for the fields or densities, and iterate until convergence is achieved. At each iteration the chemical potential μ is adjusted to produce a specified value of the particle number N, the rotational frequency ω is varied to give a specified value of the spin I, and the temperature T can be varied to give a specified energy E.

The nucleon orbitals are given by the eigenfunctions of the density ρ. The orbital occupation probabilities are given by the eigenvalues of ρ. It should be emphasized that all quantities (mean fields, densities, quasiparticles, nucleon orbitals, etc.) are self-consistent functions of N, I and E (or T).

The FTHFB theory has several limiting cases. The finite-temperature HF theory is obtained by setting the pair field Δ and the pair tensor t equal to zero. The zero-temperature HFB and HF theories are recovered by setting the quasiparticle occupations f equal to zero. If H is the pairing Hamiltonian

$$H = \sum_i (\varepsilon_i - \mu)\, C_i^\dagger C_i - \sum_{ij>0} G_{ij}\, C_i^\dagger C_{\bar{i}}^\dagger C_{\bar{j}} C_j \; , \tag{44}$$

where $|\bar{i}>$ is the time-reverse of $|i>$, and $\varepsilon_{\bar{i}} = \varepsilon_i$, and if the spin I is zero, then the FTHFB equation simplifies to the finite-temperature BCS equation [1]

$$\Delta_i = \frac{1}{2} \sum_{j>0} G_{ij} \frac{\Delta_j}{E_j} \tanh\left(\tfrac{1}{2}\beta E_j\right) , \tag{45}$$

where the quasiparticle energy is

$$E_i = [(\varepsilon_i - \mu)^2 + \Delta_i^2]^{1/2} \, . \tag{46}$$

For the simple pair force, $G_{ij} = G$, the pair gap has the same value for all orbitals, $\Delta_i = \Delta$.

2.6. Hamiltonian

For most of our examples we will use the pairing plus quadrupole (PPQ) Hamiltonian of Kumar and Baranger (KB) [3,4]. It has the form

$$H = \sum_i \varepsilon_i C_i^\dagger C_i - \frac{1}{2} \chi \sum_{\mu=-2}^{2} Q_{2\mu}^\dagger Q_{2\mu} - GP^\dagger P , \tag{47}$$

where the quadrupole operator is

$$Q_{2\mu} = \sum_{ij} <i|r^2 Y_{2\mu}|j> C_i^\dagger C_j , \tag{48}$$

and the pair creation operator is

$$P^\dagger = \sum_{i>0} C_i^\dagger C_{\bar{i}}^\dagger . \tag{49}$$

This is a phenomenological Hamiltonian. The two parameters are χ, the strength of the quadrupole interaction, and G, the strength of the pair interaction. KB fit these parameters by requiring that H reproduce the systematic features of ground state deformations and pair gaps in rare earth nuclei. So H is completely determined by the nuclear ground states. We use the same Hamiltonian to determine how temperature and spin affect the shapes and pair gaps. So there are no parameters available to fit the thermal and rotational responses of nuclei.

3. Pairing Phase Transitions

The residual components of the nucleon-nucleon interaction which are not included in the HF potential can create correlated pairs of nucleons. The pair field Δ measures the strength of these correlations.

3.1. Neutron-Neutron and Proton-Proton Pair Correlations

In the conventional pairing theory each pair consists of two neutrons or two protons in orbitals related by time-reversal. These Cooper pairs exist in the ground states of many nuclei. These correlations reduce the moment of inertia by a factor of two. For an even nucleus, the ground state is the quasiparticle vacuum, and the two quasiparticle states have excitation energies $E_1 + E_2$. Eq. (46) shows that the minimum excitation energy is 2Δ. So the pair field creates a large energy gap between the ground state and the lowest two quasiparticle state.

Heating a nucleus creates statistical quasiparticle excitations, which break the correlated pairs. Therefore raising the temperature weakens the pair gap. Fig. 2 shows the proton and neutron pair gaps in ^{170}Er [5]. At a critical temperature T_c, the pair correlations vanish. This is a second order phase transition from a superfluid to a normal fluid. Notice that $kT_c \sim \frac{1}{2} \Delta(T=0)$. For a typical rare earth nucleus, $kT_c \sim 0.5$ MeV.

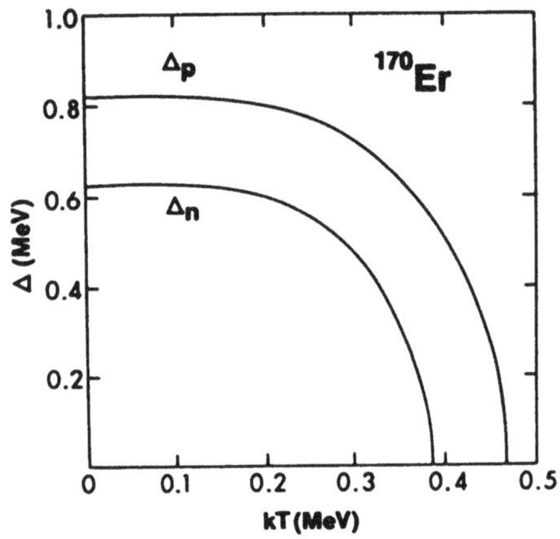

Fig. 2. The proton pair gap and the neutron pair gap versus the temperature for ^{170}Er.

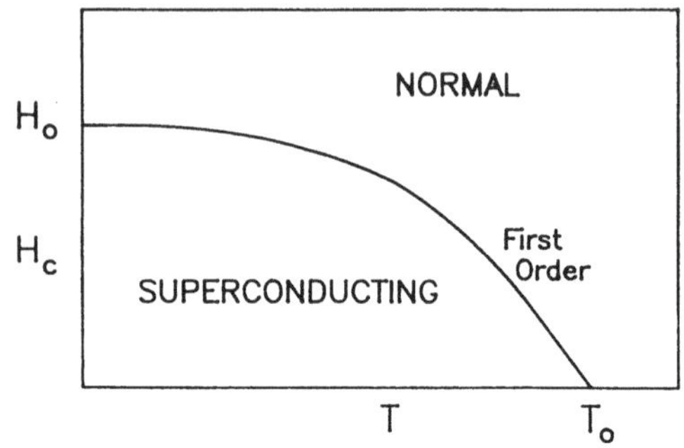

Fig. 3. Critical field curve and phase diagram for a superconductor: critical magnetic field H versus temperature T.

Nuclear rotation also weakens the pair correlations. In 1960 Mottelson and Valatin [6] drew an analogy between nuclei and superconductors. A sufficiently large magnetic field destroys the superconductivity in a metal. Similarly, for a nucleus rotating with a sufficiently large rotational frequency, the Coriolis force should break the Cooper pairs.

This analogy can be extended to finite-temperature [7]. Fig. 3 shows the critical field curve, or phase diagram, for a superconductor. A second order phase transition from superconductor to normal is caused by a magnetic field H_0 at zero temperature, or a temperature T_0 with no magnetic field. For temperatures between 0 and T_0 a first order transition is caused by a temperature-dependent critical field H_c .

I have solved the FTHFB equation for the two-level model [7]. The critical frequency curve is given in Fig. 4. The transition from superfluid to normal is obtained by either rotating or heating the system, or more generally by applying a temperature-dependent critical rotational frequency. There is a tricritical point which separates first order from second order transitions. The phase diagram for superconductors does not have a tricritical point.

We next consider the proton pair gap in ^{166}Er. Fig. 5 shows how angular momentum and temperature conspire to eliminate the pair correlations [8]. Each curve is for a different angular momentum. The phase diagram for the protons is given by the spin-dependent critical temperature in Fig. 6 [8]. There are corresponding figures for the neutrons.

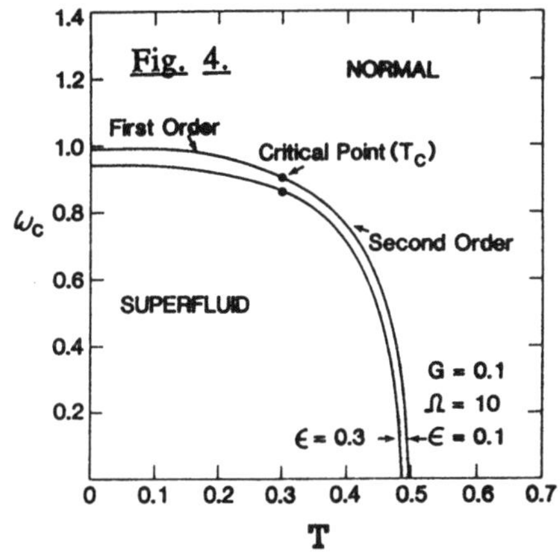

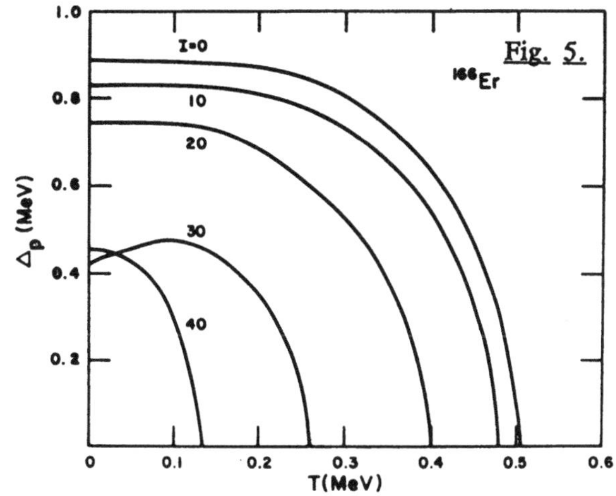

Fig. 4. Critical frequency curve and phase diagram for the two-level model: critical rotational frequency ω_c versus temperature T. The quantities $\hbar\omega_c$, kT, G and ε have units of MeV. The single-particle splitting is ε.

Fig. 5. The proton pair gap Δ_p vs the temperature T for ^{166}Er. The spin is I.

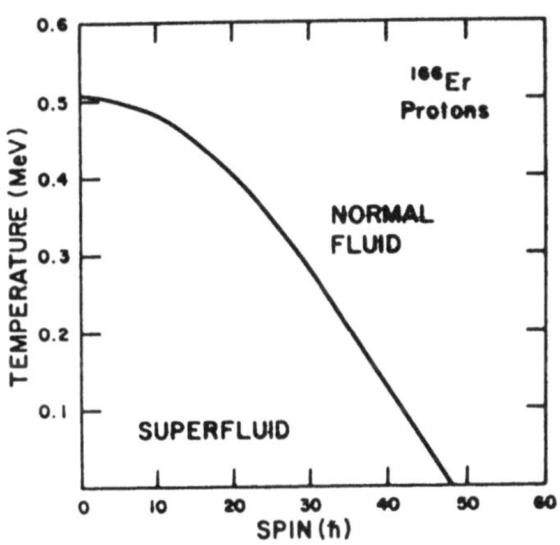

Fig. 6. Phase diagram for the proton pair correlations in ^{166}Er. The critical temperature T_c vs the spin I.

3.2. Neutron-Proton Pair Correlations

Since nuclei contain protons and neutrons, there are other types of Cooper pairs than the usual ones, i.e. two protons or two neutrons in time-reversed orbitals For example a Cooper pair could consist of a proton and a neutron in time-reversed orbitals ($p\bar{n}$, T=0 or 1), or a proton and a neutron in identical spatial orbitals (pn, T=0). This increases the possible number of superfluid phases and phase transitions.

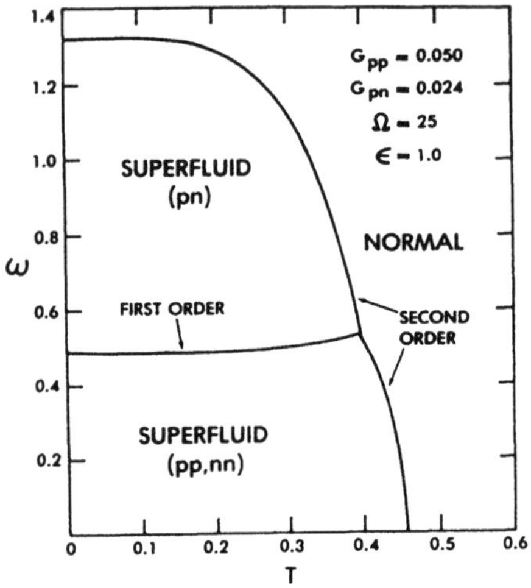

Fig. 7. Phase diagram for the two-level model: rotational frequency ω versus temperature T.

Suppose we consider the two-level model and permit the $p\bar{p}$, $n\bar{n}$ superfluid phase and the pn superfluid phase. The phase diagram is shown in Fig. 7 [9]. At low spins and low temperatures the system is a $p\bar{p}$, $n\bar{n}$ superfluid. Raising the rotational frequency causes a first order phase transition to a pn superfluid. This is because it is easier for the Coriolis force to break a $p\bar{p}$ or $n\bar{n}$ Cooper pair than a pn Cooper pair. Increasing the temperature of either superfluid phase causes a second order transition to the normal phase. This illustrates the kind of additional phase transitions which might occur in nuclei, because they contain protons and neutrons.

4. Shape Phase Transitions

Many nuclear ground states have intrinsic shapes which are deformed, i.e., not spherically symmetric. A common shape is a prolate spheroid. These deformations are caused by shell effects. When the temperature is increased, there are statistical excitations of particles (quasiparticles), and the shell effects are weakened.

4.1. Hot Nonrotating Nuclei

Let's first consider the effect of temperature on nuclei which are not rotating. For an axially symmetric shape, the quadrupole deformation is characterized by β. For a prolate spheroid β is positive, for an oblate spheroid β is negative, and for a sphere β is zero. If the ratio of the long axis to a short axis is 1.4:1, then $\beta \sim 0.3$.

If a nucleus has a given temperature, then the equilibrium shape is the shape which minimizes the free energy

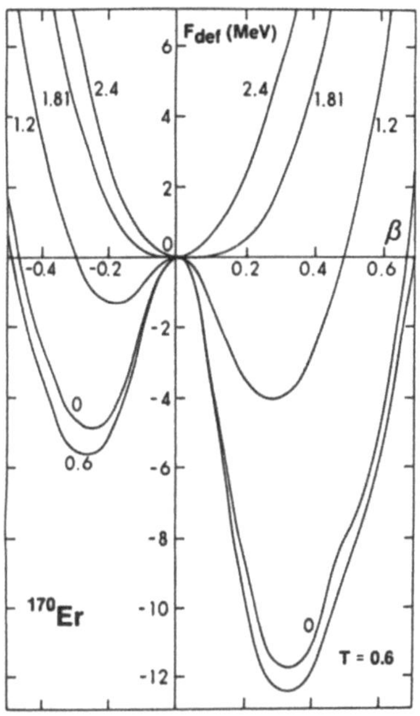

Fig. 8. The deformation free energy versus the quadrupole deformation. Each curve corresponds to a different temperature, which has units of MeV. The nucleus is ^{170}Er.

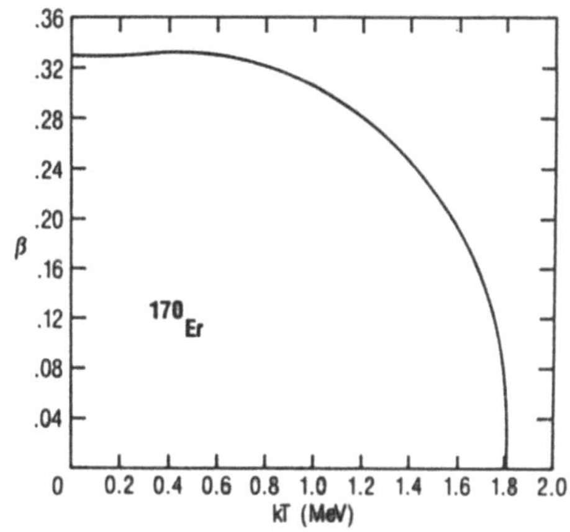

Fig. 9. The quadrupole deformation vs the temperature for ^{170}Er.

$$F = E - TS. \tag{50}$$

The deformation free energy is

$$F_{def}\,(\beta,T) = F(\beta,T) - F(0,T). \tag{51}$$

Fig. 8 shows F_{def} for ^{170}Er [5] . This nucleus has a strongly deformed prolate ground state (β = 0.33). When the temperature is increased to 1.2 MeV, the prolate minimum is much shallower, and the equilibrium shape is less deformed. At the critical temperature of 1.81 MeV the prolate and oblate minima and the spherical maximum coalesce into a single minimum at β = 0. The equilibrium shape is spherical. Characteristically the minimum is very flat at the critical temperature. Fig. 9 shows how the equilibrium shape changes with temperature [10]. This is a second order phase transition.

Does the shape remain axially symmetric when the temperature is increased? A quadrupole shape is characterized by its moments $<Q_{2\mu}>$, or alternatively by β and γ, where

$$\beta^2 = \sum_{\mu} \left[\frac{4\pi}{5} \; \frac{<Q_{2\mu}>}{<r^2>} \right]^2 , \tag{52}$$

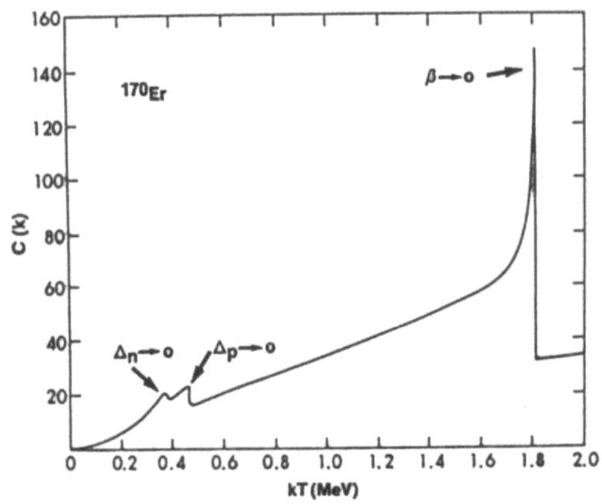

Fig. 10. The specific heat vs the temperature for ^{170}Er.

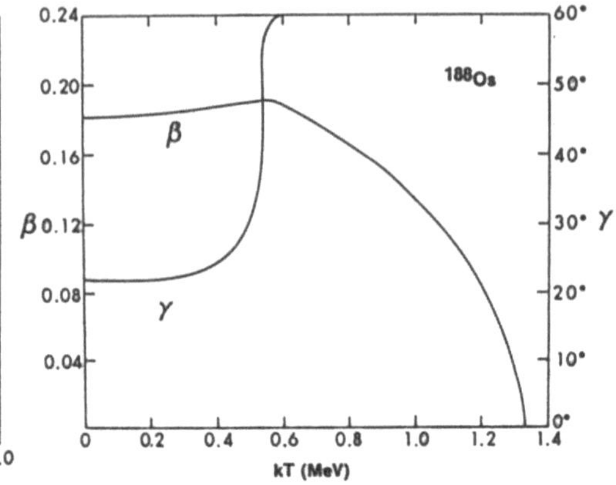

Fig. 11. The quadrupole deformation parameters β and γ vs the temperature for ^{188}Os.

$$\gamma = \tan^{-1} (2^{1/2} <Q_{22}> / <Q_{20}>) \ . \tag{53}$$

For a prolate spheroid $\gamma = 0^\circ$. An oblate spheroid has $\gamma = 60^\circ$. Intermediate values of γ describe triaxial ellipsoids, with three different axis lengths. The free energy surfaces $F(\beta,\gamma;T)$ for ^{170}Er show that the equilibrium shape remains axially symmetric and prolate at each temperature. Furthermore the oblate shape is a saddle point, not a relative minimum.

A signature of a second order phase transition is a peak in the specific heat

$$C = \partial E/\partial T \ . \tag{54}$$

Fig. 10 shows C(T) for ^{170}Er [5]. There is a large peak when the deformation vanishes, and two smaller peaks when the neutron and proton pair gaps vanish.

Our next example is ^{188}Os. This is a transitional nucleus. The ground state shape is not axially symmetric. The shape is very soft in the γ coordinate. The response of the deformation to heating is shown in Fig. 11 [5]. When the temperature is increased to 0.60 MeV, the triaxial shape suddenly becomes an oblate spheroid. When the temperature reaches 1.33 MeV, the oblate shape collapses to a sphere. There are two shape transitions in this nucleus.

Our last example is the transitional nucleus ^{148}Sm. Fig. 12 shows the equilibrium shape as a function of temperature [10]. At T=0 the shape is spherical. But when the temperature is raised to 0.40 MeV, the shape suddenly becomes prolate. Raising the temperature actually creates a deformation. This effect is the opposite of what we have seen in other nuclei. When the

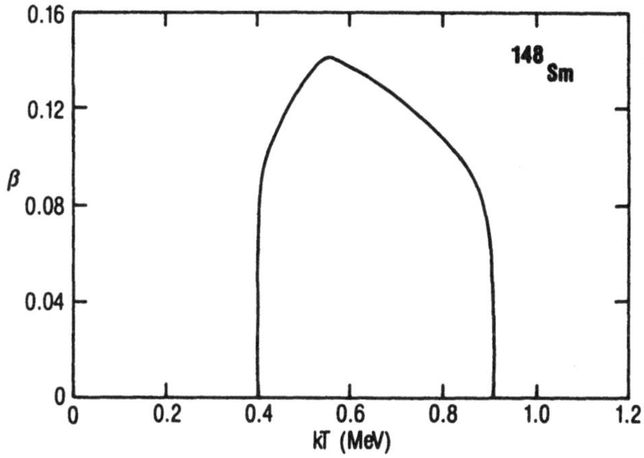

Fig. 12. The quadrupole deformation vs the temperature for ^{148}Sm.

temperature is increased to 0.91 MeV, the deformation suddenly disappears, and the shape reverts to a sphere.

What is the explanation for this temperature induced deformation? The nucleus ^{148}Sm has a spherical ground state because the pairing interaction (which favors spherical shapes) dominates over the quadrupole interaction (which favors deformed shapes). Raising the temperature destroys both the pair gap Δ and the quadrupole deformation β. However the neutron pair gap Δ_n starts to decline at T ~ 0.40 MeV, and Δ_n disappears at T_c = 0.56 MeV. The critical temperature for the shape transition is 0.91 MeV. Since the neutron pair gap disappears at a lower temperature, the quadrupole interaction is able to create a deformation for an intermediate range of temperatures. The nucleus ^{148}Sm exhibits a delicate balance between two competing tendencies, and the balance is shifted simply by changing the temperature.

4.2. Hot Rotating Nuclei

For nonrotating nuclei the range on γ of 0° to 60° includes all quadrupole shapes. However for rotating nuclei, the range on γ is extended to -60° to 120° to include all rotation modes. At γ = 0° a prolate shape rotates collectively, i.e. the symmetry axis is perpendicular to the rotation axis. For γ = 60° an oblate shape rotates collectively. At γ = - 60° an oblate shape rotates noncollectively, i.e. the rotation axis coincides with the symmetry axis. In this mode the nucleon orbitals have their spins aligned along the rotation axis, and the nuclear spin is generated by individual nucleons, instead of a collective rotation. For γ = 120° a prolate shape rotates noncollectively. At intermediate values of γ the shape is triaxial.

Consider ^{166}Er, which is a strongly deformed nucleus. The ground state is prolate. Fig. 13 shows how γ changes with temperature and spin [8,11]. When this nucleus is cold (T = 0), the angular momentum is generated by a

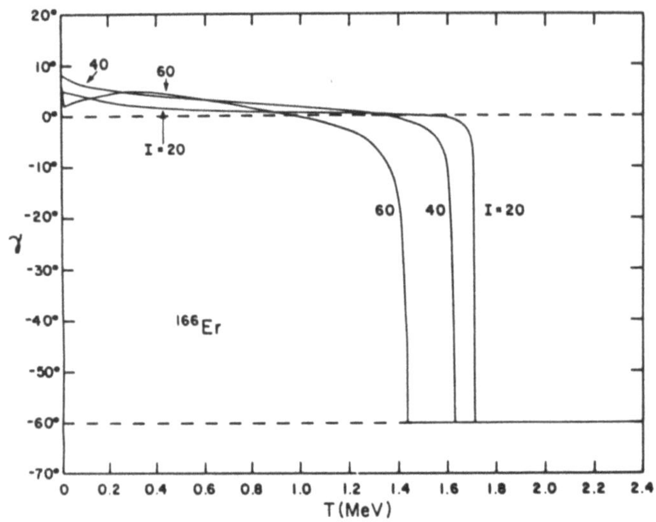

Fig. 13. The quadrupole deformation γ vs the temperature T for ^{166}Er. The spin is I.

prolate collective rotation. Since this nucleus has a stiff shape, the deviation from axial symmetry is small for all yrast states.

If this nucleus rotates with a fixed spin I, which is between 0 and 60ħ, then raising the temperature to a moderate value (T ~ 1.0 MeV) has little effect upon the shape. However when the temperature is increased to a spin-dependent critical value $T_c(I)$, then there is a rapid transition from a nearly prolate shape which rotates collectively ($\gamma \sim 0°$) to an oblate shape which rotates noncollectively ($\gamma = -60°$). The oblate shape has a small deformation β. The critical temperature decreases from 1.74 MeV at I = 0 to 1.44 MeV at I = 60ħ. The FTHFB phase diagram is shown in Fig. 14 [8, 11].

Alhassid et al. found this shape transition in ^{166}Er using the Landau theory [12, 13]. The FTHFB and the Landau theories give phase diagrams which are qualitatively similar.

Experiments on giant dipole resonances (GDR) in hot rotating nuclei may show how the deformation changes with temperature and spin. This resonance is a collective vibration of protons against neutrons. For axially symmetric deformations, the resonance has two components: a high-energy part coming from vibrations along the short axis, and a low-energy part from vibrations along the long axis. The high-energy/low-energy strength ratio S_2/S_1 is 2:1 for a prolate shape and 1:2 for an oblate shape. Since the GDR can be excited at different spins and temperatures, variations in shape can be studied.

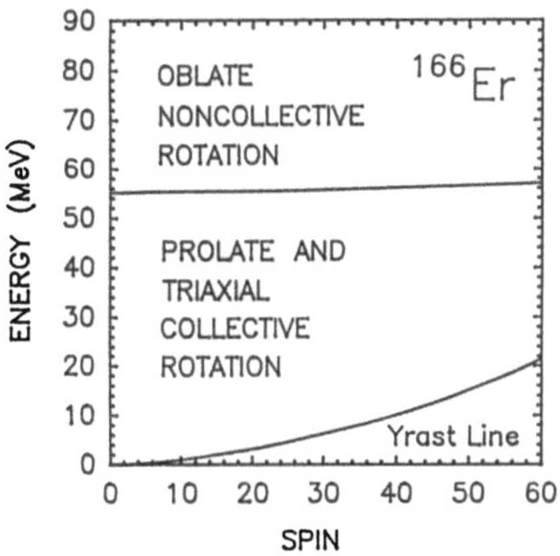

Fig. 14. Phase diagram for the shape and rotation mechanism of ^{166}Er.

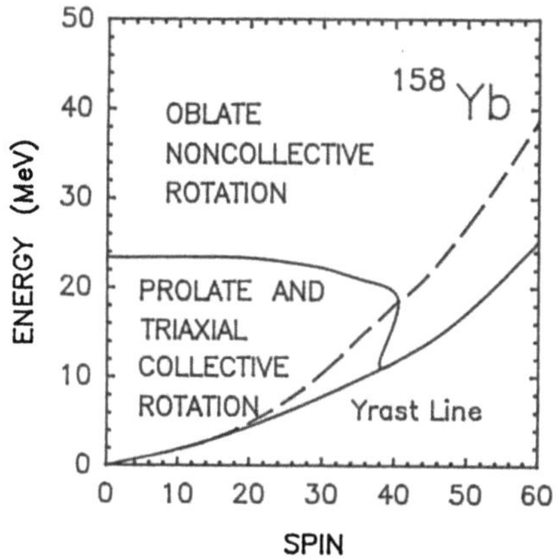

Fig. 15. Phase diagram for the shape and rotation mechanism of ^{158}Yb.

A Seattle experiment on ^{166}Er indicates that the prolate ground state deformation persists for temperatures up to 1.0 MeV, if I ~ 25ħ [14]. A Copenhagen experiment suggests a transition to an oblate shape at T ~ 1.6 MeV, if I ~ 40ħ [15]. These observations agree with the calculations. However a Stony Brook experiment claims that this shape transition is not seen [16]. Calculations of the GDR in hot rotating nuclei have been done by M. Gallardo et al. [17] and Alhassid et al. [18, 19] .

The nucleus ^{158}Yb is transitional. It has a soft shape, which is easily changed by heating or rotation. The phase diagram for ^{158}Yb is shown in Fig. 15 [20, 21] . The ground state has a weak prolate deformation. If the nucleus is not rotating, then the relativly low critical temperature of 1.0 MeV causes a transition from prolate to spherical shape. If the nucleus is cold, then rotating to spin 40ħ causes a transition from nearly prolate collective rotation to oblate noncollective rotation. For a constant spin between 0 and 40ħ, raising the temperature causes the same transition. A comparison of Figs. 14 and 15 shows how a strongly deformed nucleus and a transitional nucleus have different thermal and rotational responses.

Experiments on ^{158}Yb at Oak Ridge National Laboratory (ORNL) show that for spins 40-50ħ, the noncollective yrast structures give way to collective structures as the temperature increases [22]. Experiments on the isotone ^{154}Dy at Argonne National Laboratory (ANL) also suggest collective structure at moderate temperatures for spins above 40ħ [23]. Earlier measurements at Lawrence Berkeley Laboratory (LBL) showed that transitions in the E2 bump are strongly collective [24] . For ^{158}Yb, the FTHFB theory predicts the absence of collective structure for all temperatures when I > 40ħ . (See Fig. 15.)

This apparent discrepancy between theory and experiment is partially resolved in the following section.

5. Statistical Fluctuations in Order Parameters for Finite Systems

For the thermodynamic limit $N \to \infty$, the equilibrium state is usually given by the most probable configuration. In this limit, statistical fluctuations are negligible, except at critical points and phase transitions. However for a finite nucleus, the statistical fluctuations can create large deviations from the most probable configuration. These fluctuations must be included to obtain the equilibrium distribution for a hot nucleus.

Mean field theories omit statistical fluctuations in order parameters. For example, the FTHFB theory predicts a specific equilibrium value for the pairs gaps and deformation at each combination of spin and temperature. How do we include statistical fluctuations away from these equilibrium values? The fundamental postulate of statistical mechanics is that all accessible states are equally probable. Therefore the probability that a system will have a particular value of an order parameter Δ is proportional to the number of accessible states which give that value of Δ. For a system with given temperature, this is the partition function Z, so that

$$P(\Delta) \propto Z(\Delta) = e^{-F(\Delta)/T} , \qquad (55)$$

where F is the free energy (50). For any operator $O(\Delta)$, the average value is

$$\bar{O} = \langle O \rangle = \frac{\int O(\Delta)\, P(\Delta)\, d\Delta}{\int P(\Delta)\, d\Delta} . \qquad (56)$$

The standard deviation in O is

$$\Delta O = [\langle O^2 \rangle - \langle O \rangle^2]^{1/2} . \qquad (57)$$

5.1. Pair Fluctuations

There are thermal fluctuations in the pair gap Δ . This was first considered by Moretto, for the nonrotating uniform model [25]. The average pair gap is

$$\bar{\Delta} = \frac{\int_0^\infty \Delta\, P(\Delta)\, d\Delta}{\int_0^\infty P(\Delta)\, d\Delta} . \qquad (58)$$

As an illustration we adopt the $i_{13/2}$ model [26, 27]. First consider a nonrotating system. The FTHFB equation is used to evaluate the function $F(\Delta;T)$ for all values of Δ, not merely the equilibrium value. Then the average pair gap $\bar{\Delta}(T)$ is determined by eq. (58). The BCS mean field theory gives a second order phase transition from superfluid to normal fluid, as shown in Fig. 16. However when the fluctuations are included, then the

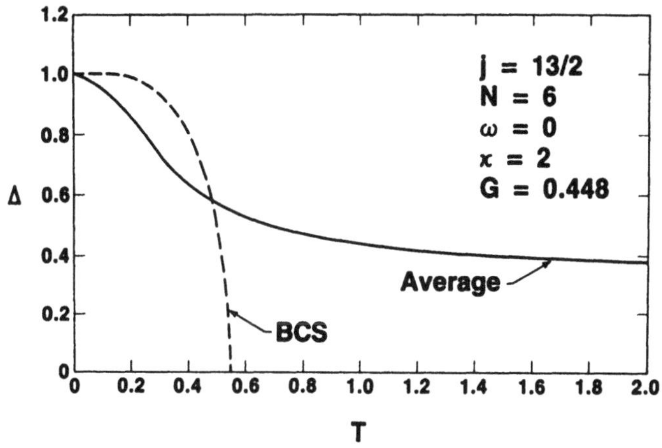

Fig. 16. The pair gap Δ vs the temperature T for $\omega = 0$ in the $i_{13/2}$ model. The dashed curve is the BCS or most probable Δ. The solid curve is the average Δ.

average gap $\bar{\Delta}$ does not vanish at high temperatures. The fluctuations eliminate the phase transition. Egido et al. find similar results for ^{168}Yb [28].

Next let the $i_{13/2}$ system rotate with frequency ω. Then the free energy in a rotating frame F' is minimized, where

$$F' = E - TS - \omega J. \tag{59}$$

Therefore F' replaces F in eq. (55). The function $F'(\Delta;T,\omega)$ is calculated with the FTHFB equation. The average gap $\bar{\Delta}(T,\omega)$ is determined by eq. (56). In the mean field approximation, increasing the rotational frequency at constant temperature causes a first order phase transition from superfluid to normal fluid, as long as the temperature is below a critical value T_c. That is, Δ jumps discontinuously from a positive value to zero at a critical frequency $\omega_c(T)$. There is a tricritical point at $\omega_c(T_c)$. However when fluctuations in Δ are included, this first order transition is smoothed out for temperatures between ~ $T_c/2$ and T_c.

5.2. Shape Fluctuations

There are also statistical fluctuations in the nuclear shape. For quadrupole deformations there are two order parameters β and γ. The FTHFB theory is used to evaluate the free energy $F(\beta,\gamma;I,T)$. For each combination of angular momentum and temperature, we consider an ensemble of nuclei with the shape probability distribution

$$P(\beta,\gamma) \propto \varepsilon^{-F(\beta,\gamma)/T} . \tag{60}$$

For any operator $O(\beta,\gamma)$ the average value is

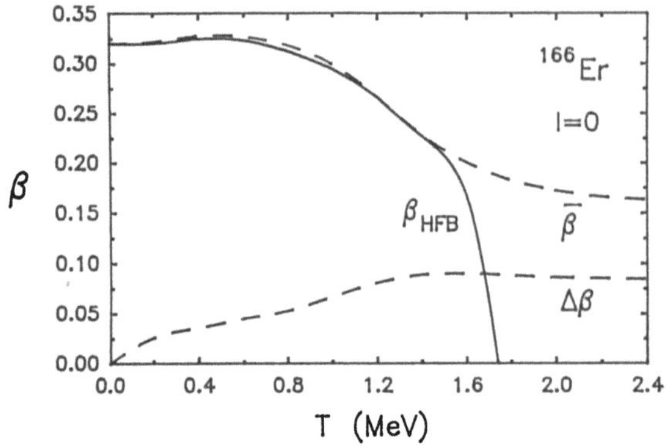

Fig. 17. The quadrupole deformation β vs the temperature T for ^{166}Er at spin zero.

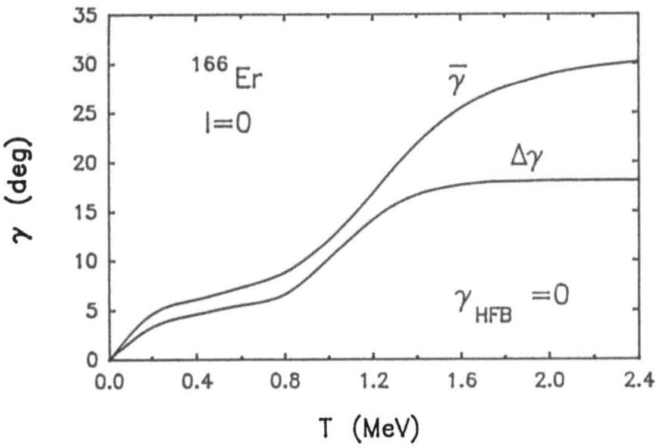

Fig. 18. The quadrupole deformation γ vs the temperature T for ^{166}Er at spin zero.

$$\bar{O} = \langle O \rangle = \frac{\int O(\beta,\gamma)\, P(\beta,\gamma)\, d\tau}{\int P(\beta,\gamma)\, d\tau} \quad . \tag{61}$$

There are two metrics $d\tau$ in current use: $\beta d\beta d\gamma$ and $\beta^4 |\sin 3\gamma| d\beta d\gamma$.

Our first example is the strongly deformed nucleus ^{166}Er [29]. Let the spin equal zero. Figs. 17 and 18 compare the average values of β and γ to the most probable, or mean field values. (These figures use the first metric.) The mean field theory predicts a shape transition from prolate to spherical at a critical temperature T_c. However for temperatures above T_c, the shape fluctuations produce an average shape which has nearly maximum triaxiality ($\bar{\gamma} \sim 30^\circ$) and a deformation $\bar{\beta}$ which is ~1/2 the ground state value. The fluctuations eliminate the shape transition. Egido et al. [30] found similar results in ^{158}Er.

If the nucleus ^{166}Er is rotating and $T > T_c(I)$, the mean field phase is oblate noncollective rotation. (See Fig. 14.) However the shape fluctuations

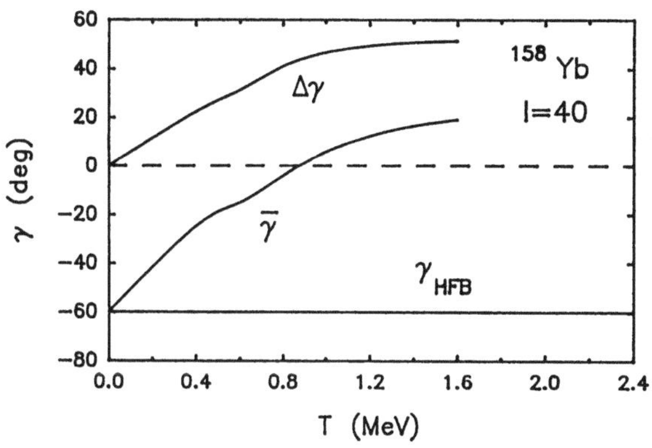

Fig. 19. The quadrupole deformation γ vs the temperature T for ^{158}Yb. The spin is 40ħ.

create a high probability for populating prolate collective, oblate collective, and prolate noncollective rotations. For T > T_c(I) the fluctuations produce an average shape which is triaxial and the rotation is collective. The average shape does not experience the transition from prolate collective to oblate noncollective which was predicted by the mean field theories. This may explain the ambiguity in interpreting the GDR experiments.

Now we return to the puzzle presented by ^{158}Yb. Recall that the FTHFB theory gives no collective structure for any excitation energy when I > 40ħ. (See Fig. 15.) However the ORNL experiment [22] shows that although the yrast line is noncollective for spins 40-50ħ, when the thermal excitation energy E^* is raised to 5 - 10 MeV, there is significant collective structure.

The discrepancy is resolved by the thermal shape fluctuations, which populate collective structures at moderate temperatures [20, 21]. For example, Fig. 19 shows that when I = 40ħ and T = 0.9 MeV (E^* = 12.7 MeV), the average phase is prolate collective rotation, even though the mean field phase is oblate noncollective rotation.

The collective strength is measured by the reduced probability for an E2 transition, or B(E2) value. For I >> 1, this is

$$B(E2,\ I \rightarrow I \pm 2) \approx \frac{1}{2}\left[\frac{3}{4\pi}\right]^2 Z^2\, e^2\, R^4\, \beta^2 \cos^2(\gamma - 30^\circ)\ , \qquad (62)$$

where R is the nuclear radius. The B(E2) value for I = 40ħ is given in Fig. 20. The mean field value is zero at all temperatures. When shape fluctuations are included, the B(E2) value increases with excitation energy. For T ~ 0.6 MeV (E^* ~ 6.5 MeV), metric 1 gives B(E2) ~ 50 W.u. and metric 2 gives B(E2) ~ 100 W.u. Inclusion of the orbitals $\nu j_{15/2}$ and $\pi i_{13/2}$ should substantially increase these values. The significant conclusion is that thermal shape fluctuations provide a simple mechanism for generating

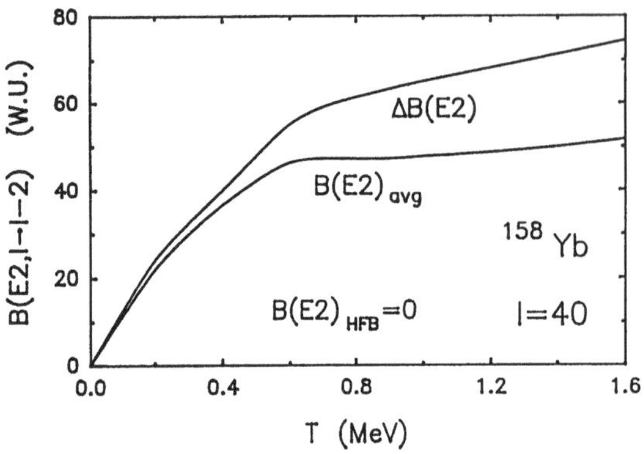

Fig. 20. The B(E2) value vs the temperature T for ^{158}Yb. The spin is 40ħ.

collective B(E2) strength at moderate temperatures, even when the equilibrium mean field is noncollective.

The thermal shape fluctuations also create fluctuations in the moment of inertia and the rotational frequency [31,32].

In the examples above, statistical fluctuations smooth out the structural transitions predicted by the mean field theory. However this is not always the case. The mean field theory predicts that ^{158}Yb has a transition from quasirotational structure ($E_\gamma \propto I$) at high spins and moderate temperatures to quasivibrational structures ($E_\gamma \sim$ constant) at low spins and low temperatures [31, 32]. Thermal shape fluctuations do not alter this transition. Experiments at ANL found this transition in the isotone ^{154}Dy [23].

6. Nuclear Liquid-Gas Transition

The equation of state for nuclear matter contains a critical point at a temperature T_c and density ρ_c. For temperature below T_c, a liquid-gas phase transition is predicted.

6.1. Equation of State

This can be illustrated by deriving a simplified equation of state. The energy contains a thermal term and an interaction term, which includes the T=0 kinetic energy

$$E = E_{int} + E_{therm} . \tag{63}$$

The interaction between the nucleons has a short range repulsion and a long range attraction. Consequently at T = 0 the system saturates at an equilibrium density ρ_0 and an equilibrium energy per nucleon E_0. In the vicinity of this equilibrium state, the interaction energy can be approximated as a quadratic in the density

$$E_{int} = E_0 + \frac{K}{18}\left[\frac{\rho-\rho_0}{\rho_0}\right]^2 , \tag{64}$$

where K is the compressibility. As an example, we use the ideal classical gas limit of the thermal energy

$$E_{th} = \frac{3}{2} T . \tag{65}$$

The classical gas describes the high temperature low density limit. The degenerate Fermi gas gives the low temperature high density limit. Nuclear matter at T_c, ρ_c lies between these two limits. The virial expansion provides the correct E_{th} for the critical point. All of these limits lead to a critical point and a liquid-gas phase transition. The entropy of a classical gas is

$$S = \ln (\rho_Q/\rho) + 5/2 , \tag{66}$$

where the quantum density is

$$\rho_Q = (MT/2\pi\hbar^2)^{3/2} , \tag{67}$$

and M is the nucleon mass. The free energy is

$$F = E - TS = E_o + \frac{K}{18} \left[\frac{\rho - \rho_o}{\rho_o}\right]^2 + T \left[\ln \left[\frac{\rho}{\rho_Q}\right] -1\right] . \tag{68}$$

The pressure is

$$P = - \left[\frac{\partial F}{\partial V}\right]_T = \rho^2 \left[\frac{\partial F}{\partial \rho}\right]_T . \tag{69}$$

This gives a simple equation of state for nuclear matter

$$P = \frac{K}{9} \left[\frac{\rho}{\rho_o}\right]^2 (\rho - \rho_o) + \rho T . \tag{70}$$

The first term is the interaction pressure and the second is the thermal pressure, as given by the ideal gas law. The conditions for a critical point are

$$(\partial P/\partial \rho)_T = (\partial^2 P/\partial \rho^2)_T = 0 . \tag{71}$$

The equation of state (70) has a critical point at $T_c = K/27$ and $\rho_c = \rho_o/3$. At lower temperatures there is a liquid-gas phase transition.

An improved equation of state is shown in Fig. 21 [33]. The critical point has a temperature $T_c = 15.3$ MeV and a density $\rho_c = 0.415\rho_o$. For $T < T_c$ the function $P(\rho)$ has the van der Waals behavior which is characteristic of a liquid-gas phase transiton.

The phase diagram for nuclear matter is shown in Fig. 22 [34]. For temperatures below T_c there are regions of stable liquid and gas phases. There is a mechanically unstable region, where $\partial P/\partial \rho < 0$. The unstable region on the left is the supersaturated vapor.

6.2. Statistical Density Fluctuations

Finally we consider the statistical density fluctuations in finite nuclei [34]. For given temperature and pressure, the equation of state defines the

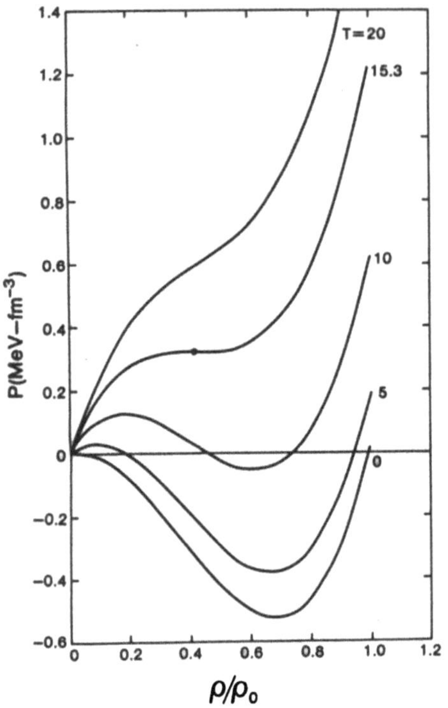

Fig. 21. Equation of state found from the zero-range Skyrme-type interaction. The temperatures are shown in MeV for each isotherm.

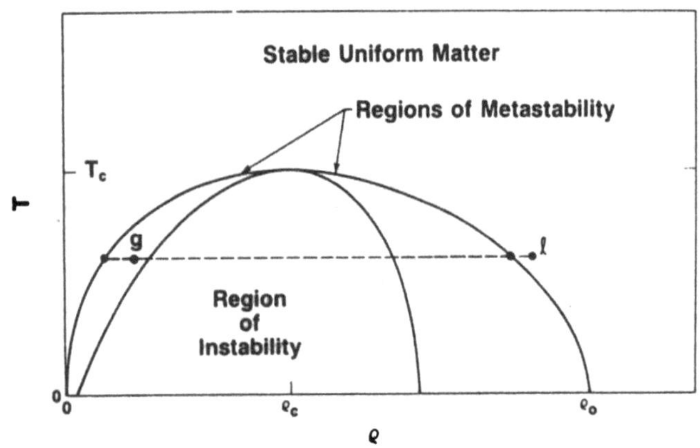

Fig. 22. Regions of stability, metastability, and instability for uniform nuclear matter.

equilibrium densities. The maximum number of these densities is three (see Fig. 21). However for fixed T and P, the statistical fluctuations can transport the system to any density ρ. The relative probability for a given value of ρ is

$$P(\rho) \alpha e^{-G(\rho)/T}, \tag{72}$$

where G is the Gibbs free energy

$$G = E - TS + PV. \tag{73}$$

The Landau theory provides the function $G(\rho;\ T,\ P)$, where ρ is treated as an independent variable, not constrained by P and T.

The conclusion is that for a finite system statistical density fluctuations are important not only at the critical point, but also in some neighborhood of the critical point. For N = 100 and the temperature range $0.9 < T/T_c < 1.0$, there are large density fluctuations which transport the system to every density in the unstable and metastable regions. It is then improper to speak of a first order phase transition between liquid and gas phases which have unique densities. At lower temperatures the density fluctuations diminish in amplitude, and the first order transition gradually emerges.

Acknowledgement

This work was supported in part by the National Science Foundation.

References

1. A. L. Goodman, Nucl. Phys. A352, 30 (1981).
2. A. L. Goodman, Advances in Nuclear Physics (Plenum Press, New York, 1979) Vol. 11.
3. M. Baranger and K. Kumar, Nucl. Phys. A110, 490 (1968).
4. K. Kumar and M. Baranger, Nucl. Phys. A110, 529 (1968).
5. A. L. Goodman, Phys. Rev. C34, 1942 (1986).
6. B. Mottelson and J. Valatin, Phys. Rev. Lett. 5, 511 (1960).
7. A. L. Goodman, Nucl. Phys. A369, 365 (1981).
8. A. L. Goodman, Phys. Rev. C38, 977 (1988).
9. A. L. Goodman, Nucl. Phys. A402, 189 (1983).
10. A. L. Goodman, Phys. Rev. C33, 2212 (1986).
11. A. L. Goodman, Phys. Rev. C35, 2338 (1987).
12. Y. Alhassid, S. Levit, and J. Zingman, Phys. Rev. Lett. 57, 539 (1986).
13. Y. Alhassid, J. Zingman, and S. Levit, Nucl. Phys. A469, 205 (1986).
14. C. A. Gossett et al., Phys. Rev. Lett. 54, 1486 (1985).
15. J. J. Gaardhoje, C. Ellegaard, and B. Herskind, Phys. Rev. Lett. 53, 148 (1984).
16. D. R. Chakrabarty et al., Phys. Rev. C37, 1437 (1988).
17. M. Gallardo, M. Diebel, T. Dossing, and R. A. Broglia, Nucl. Phys. A443, 415 (1985).
18. Y. Alhassid, B. Bush, and S. Levit, Phys. Rev. Lett. 61, 1926 (1988).
19. Y. Alhassid and B. Bush, Phys. Rev. Lett. 63, 2452 (1989).
20. A. L. Goodman, Phys. Rev. C38, 1092 (1988).
21. A. L. Goodman, Phys Rev. C39, 2008 (1989).

22. C. Baktash, in Proc. of the Workshop on Nuclear Structure at Moderate and High Spin, Berkeley, 1986.
23. R. Holzmann et al., Phys. Rev. Lett. 62, 520 (1989).
24. H. Hubel et al., Phys. Rev. Lett. 41, 791 (1978).
25. L. G. Moretto, Phys. Lett. 40B, 1 (1972).
26. A. L. Goodman, Phys. Lett. 131B, 5 (1983).
27. A. L. Goodman, Phys. Rev. C29, 1887 (1984).
28. J. L. Egido, P. Ring, S. Iwasaki, and H. J. Mang, Phys. Lett. 154B, 1 (1985).
29. A. L. Goodman, Phys. Rev. C37, 2162 (1988).
30. J. L. Egido, C. Dorso, J. O. Rasmussen, and P. Ring, Phys. Lett. B178, 139 (1986).
31. A. L. Goodman, Phys. Rev. C39, 2478 (1989).
32. A. L. Goodman, Nucl. Phys. A504, 413 (1989).
33. D. H. Boal and A. L. Goodman, Phys. Rev. C33, 1690 (1986).
34. A. L. Goodman, J. I. Kapusta, and A. Z. Mekjian, Phys. Rev. C30, 851 (1984).

Chiral Symmetry

M.D. Scadron

Department of Physics, University of Arizona, Tucson, AZ 85721, USA

Abstract: We present many varied chiral symmetry models at the quark level which consistently describe strong interaction hadron dynamics. The pattern that emerges is a nonstrange current quark mass scale $\hat{m}_{cur} \sim$ (34–69) MeV and a current quark mass ratio $(m_s/\hat{m})_{cur} \sim$ 5–6 along with no strange quark content in nucleons.

1. Introduction

The notion of "spontaneous symmetry breakdown" appears to be significant in most branches of modern physics. In nonrelativistic BCS theories of superconductivity and also valence nuclear pairing, two fundamental equations emerge:

$$\Delta = 2\omega_D\, e^{-1/\lambda} \quad , \quad \Delta/T_c = \pi e^{-\gamma_E} \approx 1.76 \quad , \tag{1.1}$$

where Δ is the gap energy, T_c the critical temperature and ω_D the Debye energy. Charge symmetry breakdown for superconductivity and isospin symmetry breakdown for valence nuclear pairing are thought to be characterized by the "order parameters" $\Delta \sim 3^\circ$K, $T_c \sim 2^\circ$K for superconductivity and $\Delta \sim 1$ MeV, $T_c \sim 0.6$ MeV for valence nuclear pairing. For relativistic strong interactions, however, it is chiral (handedness) symmetry which is broken and is characterized by the quark mass (energy gap) $\Delta = m_{qk} \sim 300$ MeV and critical temperature $T_c \sim 200$ MeV. More specifically, for low energy nonperturbative quantum chromodynamics (QCD), there are in fact five (equivalent) order parameters measuring chiral symmetry breakdown: the quark condensate $\langle -\bar{q}q \rangle \sim (250\text{ MeV})^3$, the dynamically generated quark mass $m_{dyn} \sim 320$ MeV (a relativistic version of the nonrelativistic quark mass $m_{qk} = M_N/3 \approx 313$ MeV), the pion decay constant $f_\pi \sim 90$ MeV, the scalar sigma mass $m_\sigma = 2m_{dyn} \sim 630$ MeV and the chiral symmetry restoration temperature, $T_c = 2f_\pi \sim 180$ MeV.

In these lectures we shall focus on the predictions of chiral symmetry as they pertain to strong interactions at the quark level. Many years ago such reviews were made at the hadron level [1,2] in a model-independent manner based on Gell-Mann's SU(3) equal-time charge algebra

$$[Q^i, Q^j] = if^{ijk}Q^k \quad , \quad [Q^i, Q_5^j] = if^{ijk}Q_5^k \quad , \quad [Q_5^i, Q_5^j] = -if^{ijk}Q^k \tag{1.2}$$

which has the chiral form for $Q_\pm = Q \pm Q_5$,

$$[Q_\pm^i \,,\, Q_\pm^j] = if^{ijk}Q_\pm^k \quad , \quad [Q_\pm^i \,,\, Q_\mp^j] = 0 \quad , \tag{1.3}$$

where $i,j = 1$–8, $k = 0$–8. At the quark level, however, the charge algebra (1.2) is complemented by the $(3,3^*) + (3^*,3)$ quark model algebra

$$[Q^i, \bar{q}\,\lambda^j q] = if^{ijk}\bar{q}\,\lambda^k q \quad , \quad [Q^i, \bar{q}\,\lambda^j\gamma_5 q] = if^{ijk}\bar{q}\,\lambda^k\gamma_5 q \tag{1.4a}$$

$$[Q_5^i, \bar{q}\,\lambda^j q] = -d^{ijk}\bar{q}\,\lambda^k\gamma_5 q \quad , \quad [Q_5^i, \bar{q}\,\lambda^j\gamma_5 q] = -d^{ijk}\bar{q}\,\lambda^k q \quad . \tag{1.4b}$$

Since we now believe the quark model is of physical significance, the algebra of quark densities (1.4) can also be taken as model independent.

Apart from the combined algebra (1.2-1.4) (which we shall refer to as current algebra), the predictive power of the theory [1,2] becomes greatly enhanced when the notion of "partial conservation of axial currents" (PCAC) is introduced. For a hadron transition $h_i \to \pi_j + h_f$, strong interaction PCAC is via the operator divergence equation $\partial A^j = f_\pi m_\pi^2 \phi_\pi^j$ or given by the "soft pion reduction"

$$\langle \pi_j h_f | H | h_i \rangle \underset{q_\pi \to 0}{\to} (-i/f_\pi) \langle h_f | [Q_5^j, H] | h_i \rangle \tag{1.5}$$

where H represents a semistrong, electromagnetic or weak hamiltonian density. In spite of the importance of PCAC as a phenomenological tool, it nonetheless seems to be somewhat ad hoc at the hadron level. However, the deeper meaning of PCAC is based on the breakdown of chiral symmetry at the quark level. In 1961, Goldstone enumerated the general criteria for "spontaneous" symmetry breakdown involving massless bosons [3]. As Nambu [4] had already understood in 1960, the light isovector pion (and later the isospinor K and isoscalar η_8) were such "Nambu-Goldstone bosons" (NGB) for strong interactions. At the quark level, chiral symmetry is dynamically (or spontaneously) broken when quarks acquire (nonperturbative) mass $m_{dyn} \neq 0$. Then the NGB massless pion is seen as a pole in the axial-vector current satisfying the axial Ward identity (AWI)

$$q^\mu \Gamma_{\mu 5}(p', p) = \gamma_5 S^{-1}(p') + S^{-1}(p)\gamma_5 \quad , \tag{1.6}$$

such that for low $q^2 = (p'-p)^2$, the AWI can be solved for the axial-vector current in the Nambu form [5]

$$\Gamma_{\mu 5}(p', p) \propto \bar{q}\, [\gamma_\mu \gamma_5 + \frac{2\, \Sigma(p^2)}{q^2} q_\mu i\gamma_5] q + \mathcal{O}(q) \quad . \tag{1.7}$$

In these lectures we will break chiral symmetry explicitly via nonstrange and strange "current" quark masses $\hat{m}_{cur}$ and $m_{s,cur}$ in order to give the NGB π and K their physical masses. To do so one must go beyond the model-independent equations (1.2)-(1.7) by studying various model field theories. We begin in section 2 with the SU(2) chiral field theories of the linear sigma model form and the four-fermion Nambu-Jona-Lasinio model. In section 3 we look at more model-independent current divergences obtained from Heisenberg equations of motion. Alternatively in section 4 we study the model-dependent approach of nonperturbative quantum chromodynamics (QCD) and associated "constituent" quark models. Then in section 5 we examine the more kinematical infinite momentum frame (IMF) scheme and its related scaling predictions. The valence quark model with no strange quarks in nucleons is found to hold. In section 6 we draw our conclusions.

2. SU(2) Chiral Field Theories

Thirty years ago our first understanding of chiral symmetry began with apparently independent chiral field theories: the SU(2) linear σ model [6,7] (LσM) and the four-fermion Nambu-Jona-Lasinio (NJL) model [8]. We link these field theories together by requiring the otherwise general σ mass in the LσM to be fixed at the NJL value [8] $m_\sigma = 2m_{qk}$, where m_{qk} is the quark mass.

2.1 Chiral-limiting LσM

We begin with the SU(2) LσM, but first shift the σ field vacuum expectation value from the spontaneously broken value $\langle\sigma_{old}\rangle = -f_\pi \neq 0$ to $\sigma = \sigma_{old} + f_\pi$. Then $\langle\sigma\rangle = 0$ signals the true vacuum and the new lagrangian density has the interacting part [6,7]

$$\mathscr{L}_{int} = g'\sigma(\sigma^2+\vec{\pi}^2) - (g'/4f_\pi)(\sigma^2+\vec{\pi}^2)^2 + g\,\bar{\psi}(\sigma+i\gamma_5\vec{\tau}\cdot\vec{\pi})\psi - gf_\pi\bar{\psi}\psi \quad . \tag{2.1}$$

In (2.1) we take the fermion fields as quark operators. At this quark level the chiral-limiting (CL) couplings in (2.1) are

$$g = m_{qk}/f_\pi \approx 3.5 \quad , \quad g' = m_\sigma^2/2f_\pi \approx 2.2 \text{ GeV} \quad , \tag{2.2}$$

where $f_\pi \approx 90$ MeV, $m_{qk} \approx M_N/3 \approx 313$ MeV and we invoke the NJL value [8] $m_\sigma = 2m_{qk} \approx 626$ MeV.

In the CL one requires $m_\pi = 0$, not only to tree order in the lagrangian, but in higher loop orders as well. Since fermion loops depend on color number N_c, but meson loops do not, this null result must hold for quark loops and independently for meson loops. More specifically, for quark loops in one-loop order, the "vacuum polarization (VP) and tadpole (qktad) pion self-energy graphs of figure 1 generate the CL amplitudes as $q \to 0$,

$$M^o_{VP} = -i4N_cN_fg^2 \int \frac{đ^4p}{p^2-m_{qk}^2} \tag{2.3a}$$

$$M^o_{qktad} = \frac{i4N_cN_f2g'g}{m_\sigma^2} \int \frac{đ^4p\; m_{qk}}{p^2-m_{qk}^2} \quad , \tag{2.3b}$$

where $đ^4p = d^4p/(2\pi)^4$. With the lagrangian couplings (2.2), it is indeed clear from (2.3) that the pion remains massless in the CL: $M^o_{VP} + M^o_{qktad} = 0$. Likewise the meson loop graphs of figure 2 give the CL amplitude as $q \to 0$,

$$M^o_{meson\ loops} = 4g'^2i \int \frac{đ^4p}{p^2(p^2-m_\sigma^2)} + \left[\frac{5g'}{f_\pi} - \frac{6g'^2}{m_\sigma^2}\right] i \int \frac{đ^4p}{p^2} + \left[\frac{g'}{f_\pi} - \frac{6g'^2}{m_\sigma^2}\right] i \int \frac{đ^4p}{p^2-m_\sigma^2} \quad . \tag{2.4a}$$

$$= 0 \tag{2.4b}$$

so that the pion continues to be massless in the CL [9].

Fig. 1. Quark loops contributing to m_π in LσM (a) vacuum polarization, (b) σ tadpole.

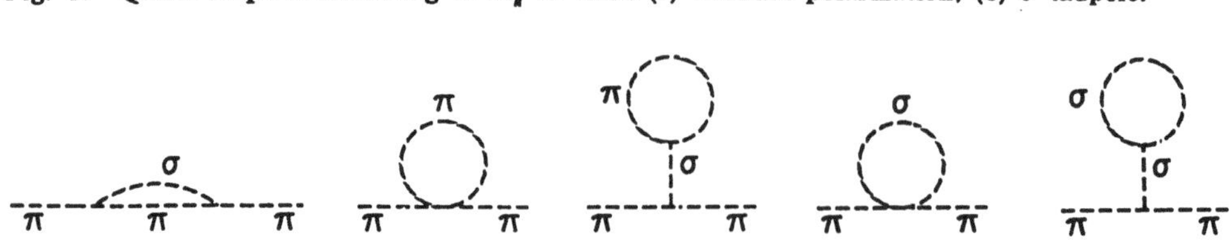

Fig. 2. Meson loops contributing to m_π in LσM.

For future reference we note the CL exact "gap equation"

$$1 = -i4N_c g^2 \int \frac{d^4p}{(p^2-m_{qk}^2)^2} \quad , \tag{2.5a}$$

and the approximate LσM gap equation for flavor number $N_f = 2$,

$$m_\sigma^2 \approx i4N_c g^2\, 2 \int \frac{d^4p}{p^2-m_{qk}^2} \quad , \tag{2.5b}$$

with both (2.5a, b) holding numerically for ultraviolet cutoff $\Lambda \sim 700$ MeV.

2.2 LσM away from chiral limit

When the pion gets mass $q^2 = m_\pi^2 \neq 0$, the LσM self-energy quark loop amplitudes (2.3) respectively become

$$M_{VP} = -i4N_c N_f g^2 \int \frac{d^4p[p^2-\hat{m}^2-m_\pi^2/4]}{\left[\left(p+\frac{1}{2}q\right)^2-\hat{m}^2\right]\left[\left(p-\frac{1}{2}q\right)^2-\hat{m}^2\right]} \tag{2.6a}$$

$$M_{qktad} = \frac{i4N_c N_f}{m_\sigma^2} 2g'g \int \frac{d^4p\ \hat{m}}{p^2-\hat{m}^2} \quad . \tag{2.6b}$$

Here [6,7] $2f_\pi g' = m_\sigma^2 - m_\pi^2$ away from the CL and the chiral-broken nonstrange quark mass is $\hat{m} = m_{qk} + \hat{m}_{cur}$, where $\hat{m}_{cur}$ is the nonstrange current quark mass. Then subtracting (2.3a) from (2.6a) and (2.3b) from (2.6b), making a Taylor series expansion $\hat{m}^2 \approx m_{qk}^2 + 2\hat{m}_{cur}m_{qk}$ for $\hat{m}_{cur}/m_{qk} << 1$ and applying the gap equations (2.5), it is straightforward to compute the incremental self-energy pion mass shifts [9]

$$\delta M_{VP} = M_{VP} - M_{VP}^o \approx -(5/4)m_\pi^2 + 4\hat{m}_{cur}m_{qk} \quad , \tag{2.7a}$$

$$\delta M_{qktad} = (M-M^o)_{qktad} \approx -4\hat{m}_{cur}m_{qk} + i4N_c N_f \int \frac{d^4p}{p^2-m_{qk}^2}\left[\frac{\hat{m}_{cur}}{m_{qk}} - \frac{m_\pi^2}{m_\sigma^2}\right] \approx -m_\pi^2 \quad , \tag{2.7b}$$

$$\delta M_{meson\ loops} = (M - M^o)_{meson\ loops} \approx m_\pi^2 \quad . \tag{2.7c}$$

Adding together the amplitudes in (2.7) we obtain the net incremental pion mass shift away from the CL in the LσM,

$$\delta M_{L\sigma M} = \delta M_{VP} + \delta M_{qktad} + \delta M_{meson\ loops} \approx -(5/4)m_\pi^2 + 4\hat{m}_{cur}m_{qk} \quad . \tag{2.8}$$

This must be identified with the total pion mass squared because the mass shift is the mass in this chiral picture, $\delta M_{L\sigma M} = m_\pi^2$. Then we find from (2.8) that [9]

$$\hat{m}_{cur} \approx (9/16)m_\pi^2/m_{qk} \approx 34\ \text{MeV} \quad . \tag{2.9}$$

Apart from generating an unambiguous current quark mass (2.9), the above LσM analysis emphasizes the importance of the σ tadpole graphs in fig. 1 and fig. 2 - without them the pion gets a mass in the CL in violation of chiral symmetry.

2.3 Chiral-limiting NJL model

In order to support the above LσM results, we now turn to the four-fermion NJL chiral model with lagrangian density [8]

$$\mathscr{L} = i\bar{\psi}\not{\partial}\psi + \mathscr{g}[(\bar{\psi}\psi)^2 + (\bar{\psi}i\gamma_5\vec{\tau}\psi)^2] \quad . \tag{2.10}$$

The π and σ mesons are then bound states with $m_\pi = 0$ and $m_\sigma = 2m_{qk}$ in the CL with the exact gap equation (2.5a) still valid in the NJL model but with (2.5b) replaced by the gap equation

$$1 = 2\mathscr{g}\ i4N_cN_f \int \frac{d^4p}{p^2-m_{qk}^2} \quad . \tag{2.11}$$

To link this NJL theory to the LσM, we note that the tree-order LσM couplings of (2.2) become the one-loop NJL couplings for cutoff $\Lambda \approx 700$ MeV,

$$g \approx 2\pi/\sqrt{N_c} \approx 3.63 \quad . \tag{2.12a}$$

$$g' = -i4N_cN_f g^3 m_{qk} \int \frac{d^4p}{(p^2-m_{qk}^2)^2} = 2g^2f_\pi \approx 2.3 \text{ GeV} \quad , \tag{2.12b}$$

where (2.12a) follows from the exact CL gap equation (2.5a) and (2.12b) represents the $\sigma\pi\pi$ quark triangle. The near identity of the couplings (2.12) with (2.2) when $m_\sigma = 2m_{qk}$ suggests the compatibility of these chiral symmetric theories.

Pion mass generation is another link between the LσM and NJL model. Following NJL one "sums bubbles" to all orders to obtain the pion propagator [8]

$$-g^2D_\pi(q^2) = \frac{2\mathscr{g}}{1+2\mathscr{g}J(q^2)} \quad , \tag{2.13}$$

where the quark loop representing $J(q^2)$ is depicted in fig. 3. In the CL with $q \to 0$ one has

$$J(0) = -i4N_cN_f \int \frac{d^4p}{p^2-m_{qk}^2} \quad . \tag{2.14}$$

Comparing (2.14) with the gap equation (2.11) one finds $-2\mathscr{g}J(0) = 1$. Then the pion propagator (2.13) has a pole as $q \to 0$, signaling $m_\pi = 0$ in this (chiral) limit.

Fig. 3. Quark bubble coupling to pseudoscalar pions in NJL model.

2.4 NJL model away from chiral limit

Now we compute the pion propagator away from the CL. In that case the quark bubble J in fig. 3 becomes

$$J(q^2) = -iN_cN_f \int d^4p \ \mathrm{Tr}\ \{i\gamma_5(\not{p} + \not{q}/2 - \hat{m})^{-1}\ i\gamma_5(\not{p} - \not{q}/2 - \hat{m})^{-1}\} \tag{2.15}$$

for $q^2 = m_\pi^2$. Carrying out the trace in (2.15) and comparing with the LσM VP amplitude (2.6a), we see that

$$M_{VP} = g^2J(m_\pi^2) \quad . \tag{2.16}$$

Also inverting (2.13) we find

$$-D_\pi^{-1}(q^2) = (g^2/2\mathcal{g})[1+2\mathcal{g}J(q^2)] = (g^2/2\mathcal{g}) + g^2J(q^2) \quad . \tag{2.17}$$

Just as in the LσM with $\delta M_{L\sigma M} = m_\pi^2$, so in the NJL model the pion mass shift being the entire (chiral-breaking) mass requires $-\delta D_\pi^{-1} = D_\pi^{-1}(q^2{=}0) - D_\pi^{-1}(q^2{=}m_\pi^2){=}m_\pi^2$. Since g and $\mathcal{g}$ are constant in the NJL model, the ratio $g^2/2\mathcal{g}$ in (2.17) is eliminated in the incremental inverse pion propagator, which then implies from (2.7a) and (2.16),

$$m_\pi^2 = g^2\delta J(m_\pi^2) = \delta M_{VP} \approx -(5/4)m_\pi^2 + 4\hat{m}_{cur}m_{qk} \quad , \tag{2.18a}$$

$$\hat{m}_{cur} \approx (9/16)\, m_\pi^2/m_{qk} \approx 34 \text{ MeV} \quad . \tag{2.18b}$$

Not surprisingly, once the LσM generates the cancelling factors $\delta M_{qktad} \approx -m_\pi^2$ and $\delta M_{meson\ loops} \approx +m_\pi^2$ (of opposite sign from the Feynman rule for fermion vs. boson closed loops), the resulting LσM net incremental self-energy shift δM_{VP} becomes equivalent to the NJL inverse pion propagator shift by virtue of (2.16). The identical pion mass expressions in the LσM and NJL model, (2.9) and (2.18) are then manifest.

2.5. Chiral symmetry restoration temperature

At finite temperature the σ mass decreases from its zero temperature value of $m_\sigma = 2m_{dyn}$ until finally at the chiral symmetry restoration temperature the σ mass "melts" with $m_\sigma(T_c) = 0$, thereafter merging with the pion mass to restore chiral symmetry. Again this picture is quantitatively quite similar in both the LσM [10] and in the NJL model [10,11]. In particular, the chiral-limiting relation between the "order parameters" $T_c = 2f_\pi$ for $N_f = 2$ appears to hold in various models.

In the case of the LσM, one has at finite temperature T [12]

$$m_\sigma^2(T) = -m_\sigma^2/2 + C\lambda T^2 \quad , \tag{2.19}$$

where $\lambda = m_\sigma^2/2f_\pi^2$ is the quartic meson coupling in (2.1) and C is a combinatorial factor. At $T = T_c$, the left-hand side (LHS) of (2.19) vanishes, as do the σ tadpole graphs analogous to fig. 1 and fig. 2. Then C is determined from the $\vec{\pi}$ and σ quartic loops and quark (VP) loop self-energy graphs for the σ meson [10]: $C = (N_f^2 - 1 + 3 - N_c)/12$. For $N_f = 2$ and also $N_c = 3$ one then finds that $C = 1/4$. Therefore from (2.19) and $m_\sigma(T_c) = 0$ one has

$$0 = -(1/2) + CT_c^2/2f_\pi^2 \quad \text{or} \quad T_c = 2f_\pi \quad . \tag{2.20}$$

One can also show in the NJL model [10,11] that $T_c = 2f_\pi$ likewise holds.

2.6. Compatibility with physical data

Experimental data underscores in many ways the significance of this combined LσM-NJL chiral picture. In particular, the pion-quark pseudoscalar coupling constant $g \approx 3.5 - 3.6$ from (2.2) and (2.12a) is related to the πN coupling [13] $g_{\pi NN} = 13.4 \pm 0.1$ as $g_{\pi NN} \approx 3\, g_A g \approx 13.4$. The latter factor of 3 enhances the "bad" quark density $\bar{q}\,\tau^3\gamma_5 q$ up to the hadron level $\bar{N}\tau^3\gamma_5 N$ since there are 3 quarks in the nucleon and $\langle x^{-1}\rangle \approx 3$. The cubic or $\sigma\pi\pi$ coupling constant $g' \approx 2.2$ GeV in (2.2b) and (2.12b) is also seen in $\delta \to \eta\pi$ decay with observed rate [14] $\Gamma_{\delta\eta\pi} \approx 57$ MeV. Converting η to its nonstrange (NS) component $|\eta\rangle \to$

$\cos\phi\ |\eta_{NS}>$ with [15] $\phi \sim 40°$ (corresponding to a singlet-octet mixing angle of $\theta \sim -15°$), one obtains $g' = g_{\delta\eta_{NS}\pi}$ with

$$|g'| = m_\delta \cos^{-1}\phi\ [8\pi\ \Gamma_{\delta\eta\pi}/p]^{1/2} \sim 2.7\ \text{GeV} \quad , \tag{2.21}$$

for momentum $p \approx 319$ MeV. Note again that (2.21) is close to the theoretical values $g' \approx 2.2$ GeV in (2.2b) and 2.3 GeV in (2.12b).

The scalar σ particle with mass $m_\sigma = 2m_{qk} \sim 626$ MeV used throughout this section with above $\sigma\pi\pi$ coupling g' corresponds to an extremely large decay width $\Gamma_\sigma \sim m_\sigma \sim 600$ MeV. Although this precludes the PDG tables from classifying such a scalar σ meson as a (narrow width) "particle," it nonetheless has been seen in many independent experiments over the past decade [16] with $m_\sigma \sim$ (600-700) MeV and $\Gamma_\sigma \sim$ (400-800) MeV.

As for the chiral symmetry restoration temperature, the CL value $f_\pi \approx 90$ MeV predicts $T_c = 2f_\pi \approx 180$ MeV for $N_f = 2$. Lattice simulations in fact infer a critical temperature [17,18] near 200 MeV, but it is not clear if it is a chiral symmetry restoration or deconfinement temperature T_D although T_c and T_D are expected to be close by.

Finally we comment on the physical pion mass now converted into the nonstrange current quark mass scale $\hat{m}_{cur} \approx 34$ MeV in (2.9) and in (2.18). It has long been appreciated that current quark mass scales are model dependent (in our case LσM-NJL model dependent). While the strange to nonstrange current quark mass ratio is thought to be more model independent (to be considered in subsequent sections), we can still test this $\hat{m}_{cur}$ scale by constructing the nonstrange "constituent" quark mass (con), which is known to be $\hat{m}_{con} \sim 340$ MeV in any model (but more about this constituent mass scale in section 4). Here we simply note that for our constant (non-running) current quark mass of 34 MeV,

$$\hat{m}_{con} = m_{qk} + \hat{m}_{cur} \approx 313\ \text{MeV} + 34\ \text{MeV} \approx 347\ \text{MeV} \quad , \tag{2.22}$$

which is close to the anticipated nonstrange constituent quark mass $\hat{m}_{con} \approx 340$ MeV.

3. Current Divergences and Current Quark Masses

In the 1960's, Gell-Mann envisioned a perturbation theory expansion in the world hamiltonian density H as given by

$$H = H_o + H_{ss} + H_{em} + H_w \quad , \tag{3.1}$$

where H_o is the chiral-symmetric U(3) hamiltonian (now interpreted as the pure glue H_{QCD}) satisfying $[Q, H_o] = [Q_5, H_o] = 0$, H_{ss} is the "semistrong" SU(3)-breaking, H_{em} the effective electromagnetic and H_w the nonleptonic weak hamiltonian density, respectively.

In this section we shall focus on the semistrong hamiltonian which can be expressed in terms of the nonstrange and strange current quark masses as

$$H_{ss} = \hat{m}_{cur}\ (\bar{u}u + \bar{d}d) + m_{s,cur}\bar{s}s \quad . \tag{3.2}$$

Here $\hat{m}_{cur}$ explicitly breaks SU(2) × SU(2) chiral symmetry, while $m_{s,cur}$ explicitly breaks SU(3) × SU(3) chiral symmetry. Both $\hat{m}_{cur}$, $m_{s,cur}$ vanish in the chiral limit (CL) as do the

π and K masses and the associated current divergences, which otherwise satisfy the Heisenberg equations of motion

$$i\partial V^i = [Q^i, H_{ss}] \quad , \quad i\partial A^i = [Q_5^i, H_{ss}] \tag{3.3}$$

for flavors i = 3, 4+i5, 6+i7 and vector (axial-vector) current divergences ∂V^i,(∂A^i).

3.1. Nonstrange current quark mass and PCAC

By analogy with the pion self energies of section 2, the pion matrix elements of H_{ss} must be $\langle \pi | H_{ss} | \pi \rangle = m_\pi^2$. As noted by Gell-Mann-Oakes and Renner (GMOR) [19], as one pion becomes soft, the PCAC relation (1.5) becomes

$$\langle \pi^o | H_{ss} | \pi^o \rangle \underset{\vec{q}\to 0}{\to} (-i/f_\pi)\langle 0| [Q_5^3, H_{ss}]| \pi^o \rangle = f_\pi^{-1} \langle 0 | \partial A^3 | \pi^o \rangle = m_\pi^2 \quad , \tag{3.4}$$

due to the Heisenberg equation of motion (3.3) and $\langle 0 | \partial A^3 | \pi^o \rangle = f_\pi m_\pi^2$. Then (3.4) is just a consistency condition recovering the on-shell relation $\langle \pi | H_{ss} | \pi \rangle = m_\pi^2$.

However, when two pions are taken soft in $\langle \pi | H_{ss} | \pi \rangle$, double pion PCAC does more than act as the square of (3.4). As we have seen in section 2 for the LσM and NJL models, chiral symmetry requires that two-pion matrix elements be supplemented by σ tadpole graphs (otherwise $m_\pi \neq 0$ in the CL). In the case of $\langle \pi | H_{ss} | \pi \rangle$, double pion PCAC must account for the σ tadpole graph of fig. 4, leading to the null result [20]

$$m_\pi^2 = \langle \pi | H_{ss} | \pi \rangle \to (-i/f_\pi)^2 \langle 0 | [Q_5^3, [Q_5^3, H_{ss}]] | 0 \rangle + 2 g_{\sigma\pi\pi} \langle \sigma | H_{ss} | 0 \rangle m_\sigma^{-2} \tag{3.5a}$$

$$= (\hat{m}_{cur}/f_\pi^2) \left[-\langle \bar{u}u + \bar{d}d \rangle_0 + \langle \bar{u}u + \bar{d}d \rangle_0 \right] + \mathcal{O}(\hat{m}_{cur}^2) \tag{3.5b}$$

$$= 0 \; \hat{m}_{cur} + \mathcal{O}(\hat{m}_{cur}^2) \quad , \tag{3.5c}$$

by virtue of the current algebra (1.4b). In short, the lead (GMOR) quark condensate term in (3.5a) is completely cancelled by the rapidly varying σ tadpole contribution. Such double-pion PCAC meson tadpoles were long ago shown to be needed in $K \to \pi\pi e\nu$ weak decays [21], in weak $K \to 2\pi$ decays and in electromagnetic $\eta \to 3\pi$ decays [2,22]. Thus it is not surprising that σ tadpoles also supplement the $\langle \pi | H_{ss} | \pi \rangle$ transition.

The solution of (3.5c) is $\hat{m}_{cur} = \mathcal{O}(m_\pi)$ compatible with our LσM-NJL results $\hat{m}_{cur} \sim 34$ MeV, but neither are compatible with the presently accepted tiny scale $\hat{m}_{cur} \sim 5$ MeV. The physical content of (3.5) is that m_π cannot set both the hadron scale in the problem and simultaneously suppress the PCAC extrapolation in pion momentum. Evidently, pion PCAC can only do the latter as indicated by the vanishing lead $\hat{m}_{cur}$ term in (3.5c).

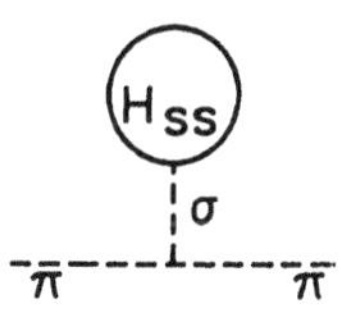

Fig. 4. Semistrong hamiltonian coupling to σ tadpole contributing to pion PCAC.

3.2 Mass ratio $(m_s/\hat{m})_{cur}$ from $\langle\pi|\partial V|K\rangle$ using PCAC

For both the pion and kaon on mass shell we invoke the nonrenormalization theorem [23] $f_+(0) = 1 - \mathcal{O}(\epsilon^2)$ to write at zero invariant momentum transfer $t = 0$:

$$\sqrt{2}\ \langle\pi^0|\, i\partial V^{6+i7}\,|\bar{K}^0\rangle = f_+(0)\,(m_K^2-m_\pi^2) \approx m_K^2-m_\pi^2 \quad . \tag{3.6}$$

On the other hand, reducing in only the π^0 on the LHS of (3.6), using the divergence relation $i\partial V^{6+i7} = (m_s-\hat{m})_{cur}\bar{d}s$ from (3.3) and the current algebra (1.4), along with $\langle 0|\partial A|K\rangle = f_K m_K^2$, we are led to

$$\sqrt{2}\ \langle\pi^0|\, i\partial V^{6-i7}\,|\bar{K}^0\rangle \to (-i/f_\pi)\ \langle 0|\,[Q_5^3, i\partial V^{6+i7}]\,|\bar{K}^0\rangle = \left[\frac{f_K}{f_\pi}\right] m_K^2 \left[\frac{m_s-\hat{m}}{m_s+\hat{m}}\right]_{cur} \quad . \tag{3.7}$$

Pion PCAC is in part the first soft pion step in (3.7), and in part that the extrapolation length from (3.6) to (3.7) is in fact small, $\mathcal{O}(m_\pi^2/m_K^2)$. Then equating (3.6) to (3.7) gives the current quark mass ratio [20]

$$\left[\frac{m_s}{\hat{m}}\right]_{cur} \approx \frac{f_K/f_\pi + 1 - m_\pi^2/m_K^2}{f_K/f_\pi - 1 + m_\pi^2/m_K^2} \approx 6.6 \quad . \tag{3.8}$$

This first-order SU(3)-breaking ratio $(m_s/\hat{m})_{cur} \approx 6.6$ follows from both the observed first-order SU(3)-breaking ratio $f_K/f_\pi \approx 1.25$ and the second-order SU(3)-breaking ratio $m_\pi^2/m_K^2 \approx 0.078$.

Had we instead assumed, as did GMOR [19], that $f_K/f_\pi = 1$ even in the SU(3)-breaking relation (3.8), then the latter becomes in this artificial limit $(m_s/\hat{m})_{cur} \to 2(m_K^2/m_\pi^2)-1 \approx 25$, the original GMOR value. While this demonstrates that (3.8) encompasses GMOR, the extremely sensitive $f_K \to f_\pi$ limit should be avoided in (3.8) because the first-order denominator term $f_K/f_\pi-1 \approx 0.25$ is over three times the second-order denominator term $m_\pi^2/m_K^2 \approx 0.078$.

In fact Glashow and Weinberg (GW) [24] have shown that $f_K-f_\pi \approx f_\kappa$ (assuming $f_+(0) \approx 1$), so the GMOR assumption $f_K=f_\pi$ requires $f_\kappa=0$ and the absence of scalar kappa mesons, which they ignored anyway. Of course, we also have avoided discussion of κ mesons and f_κ in the derivation of $(m_s/\hat{m})_{cur}$ in (3.8), but that was because we exploited the nonrenormalization theorem at $t=0$. Had we instead worked at $t\neq 0$, then a kappa tadpole term would contribute to (3.6). However, using the GW relation $f_\kappa \approx f_K - f_\pi$ along with the CTMOP relation [25] $f_+ + f_- \approx f_K/f_\pi$ at $t=m_\pi^2$ and the pion PCAC equation (3.7), we are again led to $(m_s/\hat{m})_{cur} \approx 6.6$ from (3.8) [20]. This demonstrates the internal consistency of (3.8); the consequent ratio $(m_s/\hat{m})_{cur} \sim 6$ must <u>not</u> be ignored.

3.3 Nonstrange current quark mass from $\langle N|\partial A|N\rangle$

Although the pion PCAC result (3.5) only mildly sets the scale for $\hat{m}_{cur}$, the (nonzero) spin-flip transition of nucleons sandwiched around the axial divergence at zero momentum transfer gives a more precise relation. First we write [20,26]

$$\langle\uparrow N|\, i\partial A^3|\downarrow N\rangle = -m_N g_A\ \langle\uparrow N|\gamma_5\tau^3|\downarrow N\rangle = -f_\pi g_{\pi NN}\langle\uparrow N|\gamma_5\tau^3|\downarrow N\rangle - \hat{m}_{cur}\ \langle\uparrow N|\bar{q}\gamma_5\lambda^3 q|\downarrow N\rangle_{NP} \ . \tag{3.9}$$

Here the large first term on the RHS of (3.9) is due to the pion tadpole graph of fig. 5,

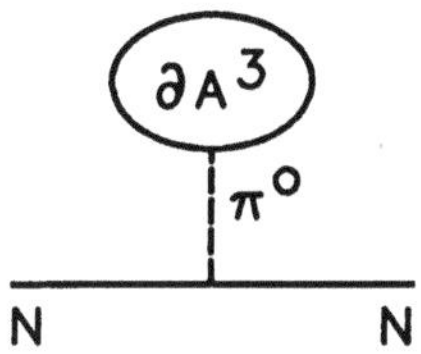

Fig. 5. Pion tadpole contribution to $\langle N|\partial A|N\rangle$.

while the small second nonpole (NP) term is generated by the quark operator equation $i\partial A^3 = -\hat{m}_{cur}\bar{q}\gamma_5\tau^3 q$ obtained from (3.3).

Applying the SU(6) qqq valence wave functions for nucleons [27], the latter NP quark transition satisfies [26]

$$\langle\uparrow N|\bar{q}\gamma_5\tau^3 q|\downarrow N\rangle_{NP} = -\langle\uparrow N|\gamma_5\tau^3|\downarrow N\rangle \quad . \tag{3.10}$$

Alternatively, the sign change in (3.10) can be understood in this valence picture (where the nucleon spin is due to the spectator quark) with a spin zero diquark in an antisymmetric quark spin configuration. Since the spin-flip nucleon requires all three (valence) quarks also to flip their spins, the minus sign in (3.10) follows [20].

Finally, substituting (3.10) back into the NP term of (3.9), the overall nonvanishing spin-flip transition $\langle\uparrow N|\gamma_5\tau^3|\downarrow N\rangle$ factors out and we are left with the SU(2) × SU(2) chiral breaking equation

$$\hat{m}_{cur} = f_\pi g_{\pi NN} - m_N g_A = (66 \pm 9)\ \text{MeV} \quad , \tag{3.11}$$

for the present values [13,14] $f_\pi = (93.1 \pm 0.1)$ MeV, $g_{\pi NN} = 13.4 \pm 0.1$, $m_N = 938.9$ MeV, $g_A = 1.259 \pm 0.004$ and an observed $(5.5 \pm 0.7)\%$ Goldberger-Treiman (GT) discrepancy. Were it not for the valence quark model sign change in (3.10), the quark mass deduced from (3.11) would be negative. Note too the importance of the π^0 tadpole graph of fig. 5. Only then is the GT relation valid in the CL, $f_\pi g_{\pi NN} = m_N g_A$. The scale of $\hat{m}_{cur}$ in (3.11) is loosely compatible with our earlier results, (2.9), (2.18), and (3.5), but more about this later.

3.4 Baryon spin-flip ΔS=1 transitions ⟨B'|∂A|B⟩

For baryon spin-flip strangeness-changing transitions $\langle N|\partial A|\Lambda\rangle$, $\langle N|\partial A|\Sigma\rangle$, the SU(3) × SU(3)-breaking analogs of (3.11) are [20,26],

$$\tfrac{1}{2}(m_s+\hat{m})_{cur} = f_K g_{K\Lambda N} - \tfrac{1}{2}(m_\Lambda+m_N)\, g_A^{\Lambda N} \quad , \tag{3.12a}$$

$$\tfrac{1}{2}(m_s+\hat{m})_{cur} = f_K g_{K\Sigma N} - \tfrac{1}{2}(m_\Sigma+m_N)\, g_A^{\Sigma N} \quad , \tag{3.12b}$$

where $f_K/f_\pi \approx 1.25$ and $f_\pi \approx 93$ MeV requires $f_K \approx 116$ MeV. To find the pseudoscalar (P) and axial (A) couplings on the RHS of (3.12), we invoke the respective d/f ratios [13,28] $(d/f)_P \approx 2.0$ and $(d/f)_A \approx 1.74$. Then we obtain

$$g_{KN\Lambda} = g_{\pi NN}\sqrt{3}\left[\frac{f+{}^1/_3 d}{f+d}\right]_P \approx 12.9 \quad , \quad g_{K\Sigma N} = g_{\pi NN}\sqrt{2}\left[\frac{d-f}{d+f}\right]_P \approx 6.3 \quad , \tag{3.13a}$$

$$g_A^{\Lambda N} = g_A\sqrt{3}\left[\frac{f+{}^1/_3 d}{f+d}\right]_A \approx 1.26 \quad , \quad g_A^{\Sigma N} = g_A\sqrt{2}\left[\frac{d-f}{d+f}\right]_A \approx 0.48 \quad . \tag{3.13b}$$

Substituting eqs. (3.13) into (3.12), we obtain

$$\tfrac{1}{2}(m_s+\hat{m})_{cur} \approx (116\ \text{MeV})\,(12.9) - (1027\ \text{MeV})\,(1.26) \approx 205\ \text{MeV} \quad , \qquad (3.14a)$$

$$\tfrac{1}{2}(m_s+\hat{m})_{cur} \approx (116\ \text{MeV})\,(6.3) - (1067\ \text{MeV})\,(0.48) \approx 219\ \text{MeV} \quad . \qquad (3.14b)$$

These results are reasonably compatible. If we then input $\hat{m}_{cur} \approx 68$ MeV from (3.11) into (3.14), we respectively find $m_{s,cur} \sim 332$ MeV, 370 MeV. Although the latter strange mass is slightly at variance, the current quark mass ratio from (3.14) and (3.11) is not [29],

$$(m_s/\hat{m})_{cur} \sim 5.0 \text{ to } 5.4 \quad , \qquad (3.15)$$

but is close to the ratio obtained from $\langle\pi|\partial V|K\rangle$, namely 6.6. In fact similar results follow from $\langle B'|\partial V|B\rangle$ $\Delta S = 1$ transitions [20], but they involve detailed dynamical calculations of kappa tadpole graphs and we avoid such complications here.

3.5. Up and down current masses

Thus far we have focused on the average nonstrange current quark mass $\hat{m}_{cur} = \frac{1}{2}(m_u + m_d)_{cur}$ occurring in the semistrong hamiltonian (3.2). The latter can be expressed in the SU(3) λ_8 breaking form $u_0 + cu_8$, or more specifically

$$H_{ss} = \epsilon_0\bar{q}\lambda^0 q + \epsilon_8\bar{q}\lambda^8 q \quad \text{with} \quad \epsilon_8 = (\hat{m}-m_s)_{cur}/\sqrt{3} \quad . \qquad (3.16a)$$

Likewise the SU(2) λ_3 breaking of the u and d quark masses occurs in the electromagnetic hamiltonian $H_{em} = H^3_{tad} + H_{JJ}$ in (3.1) where H^3_{tad} is the "tadpole" hamiltonian density [30]

$$H^3_{tad} = c'H_3 = \epsilon_3\bar{q}\lambda^3 q = \tfrac{1}{2}(m_u-m_d)_{cur}\,(\bar{u}u - \bar{d}d) \quad , \qquad (3.16b)$$

with $\epsilon_3 = \frac{1}{2}(m_u-m_d)_{cur}$. Then the complete quark mass matrix is $H_{ss} + H^3_{tad} = (m_u\bar{u}u + m_d\bar{d}d + m_s\bar{s}s)_{cur}$ through SU(2)-breaking em order.

To extract ϵ_3 from the observed em mass splittings, we must also subtract off the current-current component H_{JJ}. This leads to the universal em splitting ratios [31]

$$(c'/c)_P \approx (c'/c)_V \approx 0.022 \quad , \quad (c'/c)_B \approx 0.0172 \quad , \quad (c'/c)_D \approx 0.0179 \quad , \qquad (3.17)$$

where P,V,B,D respectively refer to pseudoscalar, vector, octet baryon and decuplet baryon em mass splittings. Apart from being essentially the same value for mesons and baryons, the Coleman-Glashow ratio $c'/c \approx 0.02$ in (3.17) can be transformed into the SU(2)-breaking current quark mass ratio

$$\frac{c'}{c} = \frac{\epsilon_3}{\epsilon_8} = \left[\frac{\sqrt{3}}{2}\right]\left[\frac{m_d-m_u}{m_s-\hat{m}}\right]_{cur} \approx 0.02 \quad . \qquad (3.18)$$

Just as single pion PCAC applied to $\langle\pi|H_{ss}|\pi\rangle$ only leads to the GMOR consistency relation (3.4), single kaon PCAC applied to $\langle K^+|H_{em}|K^+\rangle - \langle K^o|H_{em}|K^o\rangle$ or to the isospin-rotated transition $\langle K^+|\partial V|K^o\rangle$ just recovers the Coleman-Glashow ratio (3.18). That is, not only is the SU(3) quark mass <u>ratio</u> (3.8) or (3.15) reasonably model independent, so too the SU(2) quark mass <u>ratio</u> (3.18) is model independent. To infer an SU(2)-breaking quark mass scale $(m_u-m_d)_{cur}$ from (3.18), one must assume an SU(3)-breaking quark mass scale $(m_s -$

$\hat{m})_{cur}$. In the approach of section 3, this latter scale is [20,26] $(m_s - \hat{m})_{cur} \sim$ 300-350 MeV, so that from (3.18) one obtains

$$(m_d - m_u)_{cur} \sim (7\text{-}8)\ \text{MeV} \quad . \tag{3.19a}$$

Since we continue to assume $m_{con} = m_{qk} + m_{cur}$ for each quark flavor and because [32] $(m_d - m_u)_{con} \sim 6$ MeV, it should not be too surprising that the flavor-independent quark mass cancels out of the difference

$$(m_d - m_u)_{cur} \sim (m_d - m_u)_{con} \sim 6\ \text{MeV} \quad . \tag{3.19b}$$

Furthermore the SU(2) ratio deduced from (3.11) and (3.19), $(m_d/m_u)_{cur} \sim 1.1$, can be translated to a uud proton or ddu neutron to suggest a small SU(2) isospin-breaking at the hadron level,

$$(\Delta I)_N / I_N \sim (m_d/m_u)_{cur}/3 \sim 0.03 \quad , \tag{3.20}$$

close to observed 2% isospin-breaking in nucleons. This should be compared with the conventional choice [33] $(m_d/m_u)_{cur} \sim 1.8$, which suggests that the measured 2% isospin-breaking in hadrons is an accident.

4. Nonperturbative QCD and Constituent Quark Models

We start with the chiral-invariant lagrangian density of QCD

$$\mathscr{L}_{QCD} = -\tfrac{1}{4}\, G^a_{\mu\nu}\, G^{\mu\nu,a} + \bar{\psi}\gamma_\mu \left[i\partial^\mu - \tfrac{1}{2}\, g_s \lambda^a B^{\mu,a} \right] \psi \quad . \tag{4.1}$$

The absence of a mass term in (4.1) suggests that quarks acquire nonperturbative dynamical mass m_{dyn} via the dynamical breakdown of chiral symmetry. This latter mass m_{dyn} (equivalent to the constant quark mass m_{qk} in the LσM and NJL models) runs with momentum and is the residue of the q^{-2} massless pion pole in the axial current (1.5). In QCD the nonvanishing nonperturbative quark condensate $<-\bar{q}q> \sim (250\ \text{MeV})^3$ at cutoff M ~ 1 GeV (obtained from QCD sum rules for hadron masses) also breaks chiral symmetry. A dynamical model is then needed to link the chiral-broken order parameters m_{dyn} and $\langle \bar{q}q \rangle$.

4.1. Dynamical quark mass in nonperturbative QCD

To connect the quark condensate $\langle \bar{q}q \rangle$ to the nonperturbative mass m_{dyn}, one studies the quark propagator self energy $\Sigma(p)$ as a function of p^2 as depicted in fig. 6. For spacelike $p^2 << 0$, the operator product expansion [OPE] shows that $\Sigma(p)$ "runs" [34] as $\Sigma(p) \sim \langle \bar{q}q \rangle / p^2$. But for timelike p^2, the quark self energy obtained from the Schwinger-Dyson equation [SDE] monotonically increases [35], thus simulating quark confinement also suggested in fig. 6. We shall continue down the OPE channel for $p^2 > 0$, however, as we are focusing here on quark mass generation, not on SDE confinement. Then for both $p^2 < 0$ and $p^2 > 0$ we shall represent the quark propagator self-energy by the nonperturbative graph of fig. 7, which gives [36,37]

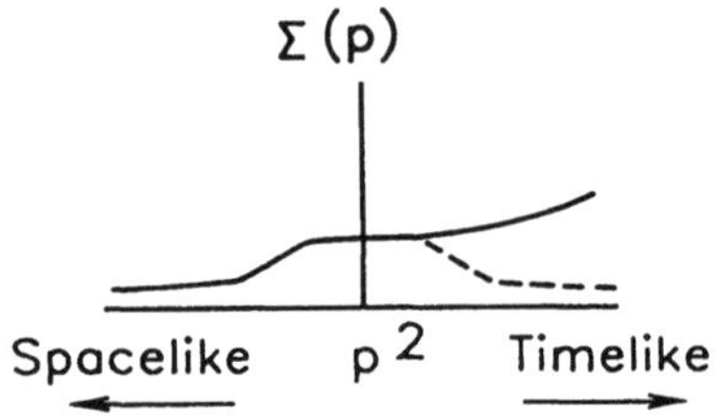

Fig. 6. Quark self energy Σ(p) in spacelike and timelike regions.

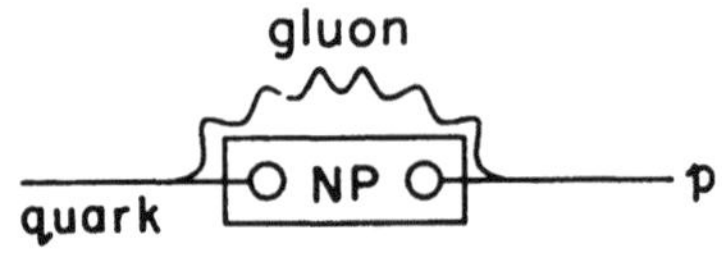

Fig. 7. Nonperturbative (NP) quark condensate component of quark self energy.

$$\Sigma(p) = \frac{g_s^2 \langle \bar{q}q \rangle}{9p^2} \left[3 + a - a \frac{\not{p}\, m_{dyn}}{p^2} \right] \quad . \tag{4.2}$$

where a is the gauge parameter of the perturbative gluon propagator.

For low p^2, $\Sigma(p)$ in (4.2) reaches a plateau and "freezes out" [38], becoming gauge-parameter-independent on the timelike "mass shell" $\not{p} = m_{dyn}$, $p^2 = m_{dyn}^2$ with gauge-invariant dynamical mass in the chiral limit [36,37]

$$m_{dyn} = \left[\frac{4\pi\alpha_s}{3} \langle -\bar{q}q \rangle \right]^{1/3} \approx 320 \text{ MeV} \quad \text{for} \quad \alpha_s(1 \text{ GeV}) \sim 0.5 \quad . \tag{4.3}$$

It is satisfying and we believe significant that m_{dyn} in (4.3) is close to the usual nonrelativistic quark model value $m_{qk} = M_N/3 \approx 313$ MeV, only now m_{dyn} is a fully relativistic quark mass.

Lastly, the bound CL $\bar{q}q$ scalar mass m_σ can be inferred in nonperturbative QCD from the σ tadpole coupling to the quark condensate analogous to fig. 1b, giving the dynamical quark mass via a NJL-type gap equation [39]

$$m_{dyn} = \frac{\langle \bar{q}q \rangle_\sigma}{-m_\sigma^2} g_{\sigma qq}^2 \quad . \tag{4.4}$$

The latter in turn leads to the dimensionless ratio $m_\sigma/m_{dyn} \approx [\pi/\alpha_s(m_\sigma)]^{1/2}$, close indeed to the NJL four fermion value [8] $m_\sigma/m_{dyn} = 2$ for $\alpha_s(m_\sigma) \approx 0.75$ as follows from asymptotic freedom and [40] α_s (1 GeV) ≈ 0.50.

4.2. Nonstrange and strange constituent quark masses

Away from the CL, the dynamical mass m_{dyn} in (4.3) becomes the nonstrange constituent quark mass $\hat{m}_{con}$ in the nonstrange sector and the strange constituent quark mass $m_{s,con}$ in the strange sector. The first estimate of constituent quark masses comes from assuming baryon total magnetic moments are generated by the valence quark components [40,41]. The SU(6) valence wave functions [27] not only recover the SU(6) ratio $\mu_n/\mu_p = -2/3$, but also predict that the uud proton magnetic moment is due to one of the constituent up quarks:

$$\mu_p = e/2m_{u,con} \approx 2.79\ e/2m_p \quad \text{or} \quad m_{u,con} \approx m_p/2.79 \approx 336 \text{ MeV} \quad . \tag{4.5}$$

Since one expects [32] the down constituent mass to be about 7 MeV greater than the up mass, we conclude from (4.5) that the average nonstrange constituent quark mass is

$$\hat{m}_{con} \approx 340 \text{ MeV} \quad . \tag{4.6}$$

A similar SU(6) analysis for the Λ s[u,d] baryon magnetic moment leads to cancelling u and d parts, enhancing the constituent strange quark contribution to

$$\mu_\Lambda \approx -e/6m_{s,con} \approx -0.61\ e/2m_p \quad \text{or} \quad m_{s,con} \approx m_p/3(0.61) \approx 510 \text{ MeV} \quad . \tag{4.7}$$

This suggests that nonperturbative SU(3)-breaking is characterized by the constituent quark mass ratio [40]

$$(m_s/\hat{m})_{con} \approx 510 \text{ MeV}/340 \text{ MeV} = 1.5 \quad . \tag{4.8}$$

To support these nonrelativistic constituent quark mass estimates (4.5–4.8), we turn to relativistic chiral symmetry considerations described by Goldberger-Treiman (GT) relations at the quark level. The chiral-limiting (CL) GT relation is $f_\pi^{CL} g_{\pi qq} = m_{dyn}$, where [2,42] $f_\pi^{CL} \approx$ 90 MeV and $m_{dyn} \approx 320$ MeV from (4.3), consistent with $g_{\pi qq} \approx 3.63$ from (2.12a). Away from the CL the (flavor-blind) $g_{\pi qq}$ should remain fixed at 3.63, but the observed $f_\pi \approx 93.3$ MeV means that the GT relation becomes in the nonstrange sector,

$$\hat{m}_{con} = f_\pi\ g_{\pi qq} \approx (93.3 \text{ MeV})\ (3.63) \approx 339 \text{ MeV} \quad , \tag{4.9}$$

close indeed to (4.6). To estimate the strange constituent mass, we extend the GT relation at the quark level to the $\bar{q}q$ kaon:

$$\tfrac{1}{2}\,(m_s + \hat{m})_{con} = f_K\, g_{Kqq} \quad . \tag{4.10}$$

The ratio of (4.10) to (4.9) for flavor blind $g_{Kqq} \approx g_{\pi qq}$ given the observed decay constant ratio $f_K/f_\pi \approx 1.25$ means that [5,43]

$$(m_s/\hat{m})_{con} = 2f_K/f_\pi - 1 \approx 1.50 \quad , \tag{4.11}$$

again similar to (4.8).

4.3 Constituent quark triangle loops

A third measure of nonperturbative SU(3) breaking stems from the PVV decay graphs of fig. 8, normalized to the ABJ $\pi^o \to 2\gamma$ amplitude [44] $F_{\pi\gamma\gamma} = \alpha/\pi f_\pi \approx 0.025\ \text{GeV}^{-1}$. This follows from nonstrange constituent quarks traversing the triangle of fig. 8. However for $\eta \to 2\gamma$ and $\eta' \to 2\gamma$ decays, both nonstrange and strange quarks traverse the triangle graphs of fig. 8, giving the amplitude ratio in the latter case [15]

$$F_{\eta'\gamma\gamma}/F_{\pi\gamma\gamma} = \frac{5}{3}\sin\phi_P + \frac{\sqrt{2}}{3}\cos\phi_P \left[\frac{\hat{m}}{m_s}\right]_{con} \approx 1.35 \quad , \tag{4.12}$$

for the pseudoscalar mixing angle $\phi_P \approx 42^o$ relative to the quark basis and $(m_s/\hat{m})_{con} \approx 1.5$.

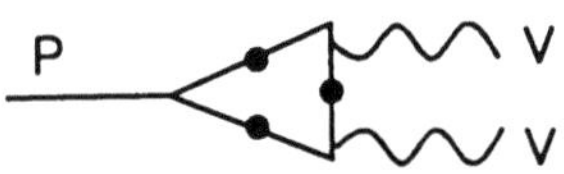

Fig. 8. PVV quark triangle. Dots indicate constituent quark masses.

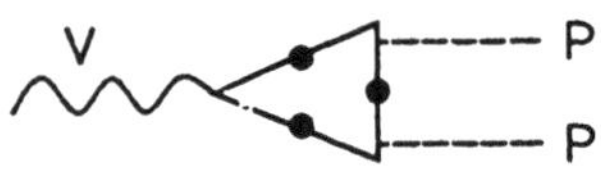

Fig. 9. VPP quark triangle. Dots indicate constituent quark masses.

Then the $\eta' \to 2\gamma$ decay rate is $\Gamma_{\eta'\gamma\gamma} = m_{\eta'}^3 M_{\eta'\gamma\gamma}^2/64\pi \approx 4.87$ keV, close to experiment [14] $\Gamma_{\eta'\gamma\gamma} = (4.54 \pm 0.55)$ keV.

A more accurate determination of this constituent quark mass ratio from PVV transitions is for $K^* \to K\gamma$ radiative decays, both neutral and charged. The amplitude ratio in the vector-dominance model (VDM) fit to the measured ratio [45].

$$F_{K^{*o}K^o\gamma}/F_{K^{*+}K^+\gamma} = -2\left[1 - \frac{1}{2}\delta\right] = -1.53 \pm 0.11 \tag{4.13a}$$

requires $\delta \approx 0.47$ and the corresponding constituent quark mass ratio to be

$$(m_s/\hat{m})_{con} = 1 + \delta \approx 1.5 \quad , \tag{4.13b}$$

again consistent with (4.8) and (4.11). This result is valid beyond the VDM [45].

A fourth measure of this ratio comes from VPP triangle graphs of fig. 9 and meson charge radii normalized to [46] $r_{\pi^+} = \sqrt{N_c}/2\pi f_\pi \sim 0.6$ fm also valid in the VDM. Then the kaon charge radii are [47]

$$\langle r^2_{K^+}\rangle/\langle r^2_{\pi^+}\rangle = 1 - {}^5/_6\delta + {}^3/_5\delta^2 + \ldots \approx 0.70 \tag{4.14a}$$

$$\langle r^2_{K^o}\rangle/\langle r^2_{\pi^+}\rangle = -{}^1/_3\delta + {}^1/_2\delta^2 + \ldots = -0.12 \pm 0.06 \tag{4.14b}$$

$$\langle r^2_{K\pi}\rangle/\langle r^2_{\pi^+}\rangle = 1 - {}^3/_4\delta + {}^3/_{10}\delta^2 + \ldots \approx 0.81 \tag{4.14c}$$

for $\delta = (m_s/\hat{m})_{con} - 1$. From the experimental ratios on the RHS of (4.14) one deduces that

$$(m_s/\hat{m})_{con} \sim 1.3 \text{ to } 1.6 \quad , \tag{4.15}$$

reasonably compatible with (4.8), (4.11), and (4.13b).

4.4 Gauge invariant constituent quark masses from QCD

Now that we have identified the nonstrange and strange constituent quark masses to be $\hat{m}_{con} \approx 340$ MeV and $m_{s,con} \approx 510$ MeV, we recall the gauge-parameter-independent value of m_{dyn} in the CL to be $m_{dyn} \approx 320$ MeV - but only on the "mass shell" $p^2 = m^2_{dyn}$, where the gluon gauge parameter cancels out of (4.2), leading to (4.3). We express the running structure of QCD quark masses away from the CL as

$$m(p^2) = m_{dyn}(p^2) + m_{cur} \quad , \tag{4.16}$$

with m_{cur} approximately constant for any quark flavor because it varies slowly like a logarithm. Since $m_{dyn}(p^2)$ can be written as m^3_{dyn}/p^2 so that it is gauge invariant when $p^2 = m^2_{dyn}$, so too the constituent quark masses correspond to $m(p^2)$ in (4.16) on the appropriate mass shell [2,26,36]:

$$m(p^2 = \hat{m}^2_{con}) = m_{dyn}(p^2 = \hat{m}^2_{con}) + \hat{m}_{cur} \; , \quad \hat{m}_{cur} \approx (340\text{-}280) \text{ MeV} \approx 60 \text{ MeV} \quad , \tag{4.17a}$$

$$m(p^2 = m^2_{s,con}) = m_{dyn}(p^2 = m^2_{s,con}) + m_{s,cur} \;, \quad m_{s,cur} \sim (510-120)\text{ MeV} \sim 390\text{ MeV} \;, \tag{4.17b}$$

$$(m_s/\hat{m})_{cur} \sim 390\text{ MeV}/60\text{ MeV} \sim 6.5 \;. \tag{4.17c}$$

We find it significant that $\hat{m}_{cur} \sim 60$ **MeV** in (4.17a) is near (3.11), but believe that the ratio $(m_s/\hat{m})_{cur} \sim 6.5$ in (4.17c) is more model independent, being in approximate agreement with (3.8) and (3.15).

Such model independence of the current quark mass ratio also holds for the (nonrunning) LσM-NJL models even though their nonstrange current mass scale of (2.9) or (2.18) is a factor of two smaller than (4.17a). More specifically, anchoring the nonstrange constituent quark mass as in (2.22), we generalize the LσM-NJL model to SU(3) and estimate the ratio

$$\left[\frac{m_s}{\hat{m}}\right]_{cur} \sim \frac{m_{s,con} - m_{qk}}{\hat{m}_{con} - m_{qk}} \sim \frac{510-313}{347-313} \sim 6 \;, \tag{4.18}$$

which is indeed the model-independent current quark mass ratio found throughout these lectures.

From the perspective of dispersion theory, the above $\hat{m}_{cur}$ scale and ratio $(m_s/\hat{m})_{cur}$ do not generate anomalously large unphysical effects in the theory. In fact the observed GT discrepancies,

$$\Delta_{\pi NN} = 1 - \frac{m_N g_A}{f_\pi g_{\pi NN}} \approx 5.5\% \;, \quad \Delta_{K\Sigma N} = 1 - \frac{(m_\Sigma + m_N) g_A^{\Sigma N}}{2 f_K g_{K\Sigma N}} \approx 33\% \tag{4.19}$$

discussed in sections 3.3 and 3.4 are respectively the order of the current to constituent quark mass ratios (squared, due to dispersion relations)

$$\Delta_{\pi NN} \sim (\hat{m}_{cur}/\hat{m}_{con})^2 \approx (70/340)^2 \approx 4\% \;, \tag{4.20a}$$

$$\Delta_{KN\Sigma} \sim [(m_s+\hat{m})_{cur}/(m_s+\hat{m})_{con}]^2 \approx (460/850)^2 \sim 30\% \;. \tag{4.20b}$$

4.5 Current verses QCD lagrangian masses

Although the original definition of current quark mass was related to a current divergence (both vanish in the CL), our present emphasis on QCD has focused on the lagrangian masses in QCD (m_L) as currentlike, because they too vanish in the CL. However, as stressed in ref. [48] m_{cur} is not identical with m_L even though both vanish in the CL. Rather, in connection with the gauge invariant relation associated with (4.16) and (4.17) for each quark flavor,

$$m_{con} = m_{cur} + m^3_{dyn}/m^2_{con} \;, \tag{4.21}$$

the lagrangian mass m_L becomes enhanced by the gluon condensate $\alpha_s\langle G^2\rangle \approx (440\text{ MeV})^4$ to the current mass scale

$$m_{cur} = m_L\left[1 + \frac{\pi\alpha_s\langle G^2\rangle m_{con}}{3(m_{con}-m_L)^2(m_{con}+m_L)^3}\right] \;. \tag{4.22}$$

This gluon enhancement in (4.22) is about a factor of five for the nonstrange sector, converting a $\hat{m}_L \sim 10$ MeV lagrangian mass as found [17] in lattice simulations to $\hat{m}_{cur} \sim 50$ MeV. The latter is near the current quark mass scales of (2.9), (2.18), (3.11), (4.17a) and not incompatible with the nonstrange PCAC scale of (3.5).

4.6 Hyperfine splitting constituent quark model

The nonrelativistic (valence) quark model approach to qqq baryon spectroscopy [41,49] predicts the ground state baryon (Ba) masses to be the sum of all nonstrange and strange constituent quark masses along with hyperfine splitting terms:

$$M_{Ba} = (m_1 + m_2 + m_3)_{con} + B\left[\frac{\vec{S}_1\cdot\vec{S}_2}{(m_1m_2)_{con}} + \frac{\vec{S}_1\cdot\vec{S}_3}{(m_1m_3)_{con}} + \frac{\vec{S}_2\cdot\vec{S}_3}{(m_2m_3)_{con}}\right] \quad . \tag{4.23}$$

All ground state octet and decuplet hadron masses are then fit to within 5% if

$$\hat{m}_{con} \approx 363 \text{ MeV} \quad , \quad m_{s,con} \approx 538 \text{ MeV} \quad , \quad B \approx (298 \text{ MeV})^3 \quad . \tag{4.24}$$

Surprisingly, even the pseudoscalar and vector meson (Me) masses obey an analog hyperfine splitting mass formula

$$M_{Me} = (m_q + m_{\bar{q}})_{con} + M\left[\frac{\vec{S}_q\cdot\vec{S}_{\bar{q}}}{(m_q m_{\bar{q}})_{con}}\right] \quad , \tag{4.25}$$

but then most of the meson masses are fit if the parameters of (4.24) are shifted to

$$\hat{m}_{con} \approx 306 \text{ MeV} \quad , \quad m_{s,con} \approx 488 \text{ MeV} \quad , \quad M \approx (390 \text{ MeV})^3 \quad . \tag{4.26}$$

The uniform constituent quark mass shift of about 50 MeV between (4.24) and (4.26) is discussed elsewhere [50]. In fact it was recently noted [51] that a similar shift of 40 MeV occurs when one applies (4.25) to massless pions in the CL:

$$m_\pi \to 0 = 2\, m_{dyn} - 3\, M/4m_{dyn}^2 \quad , \tag{4.27}$$

which suggests that $m_{dyn} \approx 281$ MeV. But then away from the CL, (4.27) and the gauge invariance condition (4.21) convert the mass formula (4.25) for pions to [51]

$$m_\pi = 2\, \hat{m}_{con} - 3\, M/4\hat{m}_{con}^2 = 2\, \hat{m}_{cur} \quad . \tag{4.28}$$

Because $m_{dyn}(p^2) \sim 1/m_{con}^2$ at $p^2 = m_{con}^2$ for QCD and the hyperfine splitting term in the nonrelativistic <u>linear</u> mass formula (4.25) also behaves as M/m_{con}^2, it is not surprising that m_π and $\hat{m}_{cur}$ in (4.28) are related in a <u>linear</u> fashion. What is surprising, however, is that the numerical solution of (4.28) is

$$\hat{m}_{cur} = m_\pi/2 \approx 69 \text{ MeV} \quad , \tag{4.29}$$

extremely near to our other determinations of the nonstrange current quark mass: (3.11), (4.17a), and even (4.22) and (3.5). The same conclusion holds for strange quarks [51].

5. Infinite Momentum Frame and Quark-Parton Scaling

The relativistic dynamical models such as the LσM and NJL in section 2 require $m_\pi^2 \propto \hat{m}_{cur}$, while the PCAC and spin-flip prescriptions of section 3 find $m_\pi \propto \hat{m}_{cur}$ as does the nonrelativistic hyperfine splitting model of section 4. Interestingly enough, the apparently linear relation between $\hat{m}_{cur}$ and $\hat{m}_L$ in (4.22) dynamically requires [51] $m_\pi^2 \propto \hat{m}_{cur}$. In the infinite momentum frame (IMF), however, these various dynamical relations are replaced by a quadratic approximation between hadron masses, which is of a kinematical nature because $E^2 = p^2 + m^2$ becomes $E \approx p(1 + m^2/2p^2)$ in the IMF, regardless of the detailed dynamical model employed. The kinematic simplicity of the IMF or equivalently the "light plane" is that troublesome dynamical tadpole graphs then vanish because propagator denominators become infinite when $(\vec{p}' - \vec{p})^2 \to \infty$ as used to derive the nonrenormalization theorem [23] $f_+(0) = 1 - O(\epsilon^2)$. But when dynamical questions do arise in this frame, we will rely on the usual quark-parton (valence) scaling model.

5.1 Hadron mass equal splitting laws in IMF

To demonstrate the simplicity of the IMF and hadron quadratic mass formulae, we first note that the quadratic Gell-Mann-Okubo (GMO) octet baryon mass formula $m_\Sigma^2 + 3m_\Lambda^2 = 2(m_N^2+m_\Xi^2)$ is valid to within 3% (as is the linear version). However, the quadratic GMO meson mass formula $m_\pi^2 + 3m_{\eta 8}^2 = 4m_K^2$ is definitely preferable to a linear formula. But it is the equal splitting laws (ESL) measuring SU(3) breaking which really decide in favor of quadratic mass formulae and the IMF.

More specifically, for $\bar{q}q$ pseudoscalar and vector mesons we recall the (SU(6)) quadratic mass ESL

$$m_K^2 - m_\pi^2 \approx m_{K^*}^2 - m_\rho^2 \approx m_\phi^2 - m_{K^*}^2 \quad , \tag{5.1}$$

which are numerically accurate. Since only one quark spin is flipped in vectors (V) relative to pseudoscalars (P), one expects the V mass splittings to be close to the P mass splittings and they are. In the same spirit, one also anticipates that the ESL for spin $^1/_2$ octet qqq baryons (B) to be the same as for spin $^3/_2$ decuplet baryons (D). Although there are approximate Okubo ESL for linear mass decuplets, $m_{\Sigma^*}-m_\Delta \sim m_{\Xi^*} - m_{\Sigma^*} \sim m_\Omega - m_{\Xi^*} \sim 145$ MeV, there are also ESL for quadratic mass decuplets:

$$m_{\Sigma^*}^2 - m_\Delta^2 \approx m_{\Xi^*}^2 - m_{\Sigma^*}^2 \approx m_\Omega^2 - m_{\Xi^*}^2 \approx 0.43 \text{ GeV}^2 \quad . \tag{5.2a}$$

The important point is that the corresponding ESL for linear mass octet baryons, $m_{\Sigma\Lambda} - m_N \sim m_\Xi - m_{\Sigma\Lambda} \sim 190$ MeV (where $4m_{\Sigma\Lambda} = m_\Sigma + 3m_\Lambda$ in order that the GMO formula identically holds), does not have the same 145 MeV scale of the linear mass decuplets. However, the quadratic mass ESL for octets [2,52]

$$m_{\Sigma\Lambda}^2 - m_N^2 \approx m_\Xi^2 - m_{\Sigma\Lambda}^2 \approx 0.43 \text{ GeV}^2 \tag{5.2b}$$

does have the same scale as for decuplets in (5.2a).

It is these universal quadratic mass ESL for P and V mesons in (5.1) and for D and B baryons in (5.2) which are as close to model independent as possible, presumably because of kinematical suppression of tadpole graphs in the IMF.

5.2. Meson-quark mass formulae in IMF

The meson ESL (5.1) can be generalized to the $\bar{q}q$ quark model in the IMF by employing structure function scaling integrals $\tilde{h}$ [53,54],

$$m_\pi^2 = 2\hat{m}_{cur}^2 \tilde{h} \quad , \quad m_K^2 = (m_s^2+\hat{m}^2)_{cur}\tilde{h} \tag{5.3a}$$

for Nambu-Goldstone pseudoscalar mesons and

$$m_\rho^2 = m_{0V}^2 + 2\hat{m}_{cur}^2 \tilde{h} \quad , \quad m_{K^*}^2 = m_{0V}^2 + (m_s^2+\hat{m}^2)_{cur} \tilde{h} \tag{5.3b}$$

for vector mesons. The quark structure function integral $\tilde{h}$ is the same for both P and V mesons by the valence SU(6) picture or by the meson ESL (5.1). While we shall discuss the scale of $\tilde{h}$ later, here we divide it out of (5.3a) to find [2,52-54]

$$(m_s/\hat{m})_{cur}^{P} = \sqrt{2m_K^2/m_\pi^2-1} \approx 5 \quad . \tag{5.4a}$$

Equivalently the ESL for vector mesons or (5.3b) predicts

$$(m_s/\hat{m})_{cur}^{V} = \sqrt{2\left[m_{K^*}^2-m_\rho^2\right]/m_\pi^2+1} \approx 5 \quad . \tag{5.4b}$$

Needless to say, these values for $(m_s/\hat{m})_{cur}$ are reasonably near the ratio $(m_s/\hat{m})_{cur} \sim 6$ found in both sections 3 and 4.

In fact, one might expect that the quadratic structure $m_\pi^2 \sim \hat{m}_{cur}^2$ in (5.3) could also be reexpressed in pure linear language $m_\pi \sim \hat{m}_{cur}$. On the light plane it was originally shown [55] that

$$(m_s/\hat{m})_{cur}^{P} \approx m_K/m_\pi - 1 \approx 6 \quad . \tag{5.5}$$

In the sense of perturbation theory, (5.5) is equivalent to (5.4a).

5.3. Baryon-quark mass formulae in IMF and nucleon σ term

The analogue quadratic mass IMF formula for qqq nucleons is [56,57]

$$m_N^2 = m_{0B}^2 + \hat{m}_{cur}^2 (\tilde{f}_1 + \tilde{f}_2) \quad . \tag{5.6}$$

Here the structure function integrals $\tilde{f}_{1,2}$ refer to the one-quark and two-quark distributions, respectively for the uud proton or ddu neutron. To isolate the chiral symmetry-breaking content in (5.6), we must first fold in the nucleon "σ term" measured in low energy πN scattering of magnitude [58]

$$\sigma_N \equiv \langle N|\,[Q_5^3, i\partial A^3]\,|N\rangle \approx 60 \text{ MeV} \quad . \tag{5.7}$$

In the quark model, the operator $[Q_5^3, i\partial A^3] = \hat{m}_{cur}(\bar{u}u + \bar{d}d)$ vanishes in the SU(2) $\times$ SU(2) chiral limit $\hat{m}_{cur} \to 0$. Expressing (5.7) in the IMF language of (5.6) one writes [56]

$$m_N \sigma_N = \hat{m}_{cur}^2 (\tilde{f}_1+\tilde{f}_2) \quad . \tag{5.8}$$

To proceed in the structure function framework of the meson formulae (5.3), we search here for another combination of octet baryon masses which has the $\tilde{f}_1 + \tilde{f}_2$ dependence of (5.8). The SU(6) valence extension of (5.6) is

$$m_\Sigma^2 = m_{0B}^2 + \hat{m}_{cur}^2 \tilde{f}_2 + m_{s,cur}^2 \tilde{f}_1 \quad , \quad m_\Xi^2 = m_{0B}^2 + \hat{m}_{cur}^2 \tilde{f}_1 + m_{s,cur}^2 \tilde{f}_2 \quad , \tag{5.9a}$$

giving the desired combination

$$m_\Xi^2 + m_\Sigma^2 - 2m_N^2 = (m_s^2-\hat{m}^2)_{cur} (\tilde{f}_1+\tilde{f}_2) \approx 1.40 \text{ GeV}^2 \quad . \tag{5.9b}$$

Dividing (5.9b) by (5.8) so as to eliminate $\tilde{f}_1+\tilde{f}_2$, we arrive at the current quark mass ratio

[2,52-54]

$$\left[\frac{m_s}{\hat{m}}\right]^B_{cur} = \sqrt{\frac{m_{\Xi}^2 + m_{\Sigma}^2 - 2m_N^2}{m_N \sigma_N} + 1} \approx 5 \tag{5.10}$$

for $\sigma_N \approx 60$ MeV.

For decuplet states, SU(6) symmetry requires the Δ analogue of the nucleon IMF expression (5.6) to be

$$m_{\Delta}^2 = m_{0D}^2 + \hat{m}_{cur}^2 \tilde{f}_3 \quad , \tag{5.11}$$

where the three-quark structure function integral $\tilde{f}_3$ is equivalent to $\tilde{f}_1+\tilde{f}_2$ (by SU(6)). Then the ESL in (5.2a) translate under SU(6) to

$$m_{\Sigma^*}^2 - m_{\Delta}^2 = m_{\Xi^*}^2 - m_{\Sigma^*}^2 = m_{\Omega}^2 - m_{\Xi^*}^2 = \tfrac{1}{3}\,(m_s^2-\hat{m}^2)_{cur}\,\tilde{f}_3 \quad . \tag{5.12}$$

So dividing (5.12) by (5.8) with $\tilde{f}_3 = \tilde{f}_1+\tilde{f}_2$ leads to [2,52]

$$\left[\frac{m_s}{\hat{m}}\right]^D_{cur} = \sqrt{\frac{3\left[m_{\Sigma^*}^2-m_{\Delta}^2\right]}{m_N \sigma_N} + 1} \approx 5 \quad . \tag{5.13}$$

We note that both B and D current quark mass ratios (5.10) and (5.13) are essentially equivalent due to B and D ESL and in fact they are numerically the same as $(m_s/\hat{m})_{cur}$ obtained from P and V mesons in (5.4). This overall simplicity and consistency pattern is one of the advantages of working in the IMF.

Instead of always solving for the current quark mass ratio in the case of baryons, we can also solve directly for the nucleon σ term to predict in this IMF valence picture with $(\bar{s}s)_N = 0$,

$$\sigma_N = [m_{\Xi}^2+m_{\Sigma}^2-2m_N^2]m_{\pi}^2/2m_N[m_K^2-m_{\pi}^2] \approx 62 \text{ MeV} \quad . \tag{5.14}$$

Note that the bracketed terms in (5.14) are baryon SU(3)-breaking divided by meson SU(3)-breaking, while $m_{\pi}^2/2m_N$ measures the SU(2) $\times$ SU(2) chiral breaking of σ_N. The fact that (5.14) is perfectly compatible with the measured value $\sigma_N \approx 60$ MeV speaks clearly for the utility of the IMF.

5.4. Exclusive structure functions and current mass scale

So far in this section we have exploited SU(6) symmetry and the equal splitting laws for mesons and baryons to obtain the (model-independent) quark mass ratio $(m_s/\hat{m})_{cur}$ while dividing out the structure function integrals $\tilde{h}$ for mesons and $\tilde{f}_1+\tilde{f}_2$, $\tilde{f}_3$ for baryons. Now we focus on the nonstrange scale $\hat{m}_{cur}$ by computing $\tilde{h}$ and $\tilde{f}_3$ in the (model-dependent) context of quark-parton scaling.

Consider first the pion mass described by the structure function integral $\tilde{h}$ in (5.3a). In the valence quark parton model with the IMF Feynman scaling variable $x = (p_0+p_3)_{qk}/(p_0+p_3)_{had}$, the exclusive $\bar{q}q$ structure function of the pion (called "bare" in ref. [54]) h(x) behaves like $(1-x)^2$ as $x\to 1$ in vector gluon theories [59] or approximately like $x^2(1-x)^2$ over the whole x region 0 to 1. Normalizing the valence integral to unity we have [2,54]

$$\int_0^1 dx\ h(x) = 1 \quad , \quad h(x) = 30x^2\,(1-x)^2 \quad , \qquad (5.15a)$$

so that the "bad" quark density $\bar{q}q$ has the associated structure function integral

$$\tilde{h} = \int_0^1 dx\ h(x)/x = 30 \int_0^1 dx\ x(1-x)^2 = \frac{5}{2} \quad . \qquad (5.15b)$$

Then (5.3a) predicts the nonstrange current quark mass to be

$$\hat{m}_{cur} = m_\pi \sqrt{1/2\tilde{h}} = m_\pi/\sqrt{5} \approx 62 \text{ MeV} \quad . \qquad (5.16)$$

Alternatively we may study the baryon sector and (5.8) where $\tilde{f}_1 + \tilde{f}_2$ is replaced by $\tilde{f}_3$ describing qqq decuplets. The associated exclusive structure function behaves as $(1-x)^5$ as $x \to 1$ or approximately as $x^2(1-x)^5$ over the whole x region. This can be verified by applying the quark spectator-helicity rule (also giving $x^2(1-x)^2$ for $\bar{q}q$ pions), leading to [2,54]

$$\int_0^1 dx\ f_3(x) = 3 \quad , \quad f_3(x) = 504x^2(1-x)^5 \quad , \quad \tilde{f}_3 = \int_0^1 dx\ f_3(x)/x = 12 \quad . \qquad (5.17)$$

Furthermore $\tilde{f}_1$ and $\tilde{f}_2$ are computed in ref. [54], with the result that $\tilde{f}_1 \approx 4$ and $\tilde{f}_2 \approx 8$ so that $\tilde{f}_1 + \tilde{f}_2 \approx \tilde{f}_3$ as expected from SU(6) symmetry. Returning to (5.8) we then predict the nonstrange current quark mass scale given $\tilde{f}_3 = 12$, $\sigma_N \approx 60$ MeV:

$$\hat{m}_{cur} \approx \sqrt{m_N \sigma_N/\tilde{f}_3} \approx 69 \text{ MeV} \quad . \qquad (5.18)$$

We note the approximate consistency between $\hat{m}_{cur}$ in (5.16) due to m_π and (5.18) due to m_N, as well as the many similar estimates in sections 3 and 4.

5.5. Quark flavor and spin in proton

In the standard nonrelativistic (valence) quark model, it has long been assumed that the proton is primarily composed of uud valence quarks. Twenty-five years of successful quark model physics supports in large part this valence picture or associated quark line (OZI) rule. Indeed, we have continued to exploit this valence scheme throughout these relativistic quark model lectures.

Granted that the OZI rule need not be exact, but characteristic OZI breaking, such as in quark $\bar{q}q$ density models tends to be at most 20% and usually much less [60]. Defining the measure of nonvalence $\bar{s}s$ content in the proton as

$$y = 2\langle\bar{s}s\rangle_N/\langle\bar{u}u+\bar{d}d\rangle_N \quad , \qquad (5.19)$$

the IMF relation between the nucleon σ term and the current quark mass ratio $(m_s/\hat{m})_{cur}$ in (5.10) is modified to

$$\left(\frac{m_s}{\hat{m}}\right)^B_{cur} \approx \sqrt{\frac{m_\Xi^2+m_\Sigma^2-2m_N^2}{m_N\sigma_N(1-y)}} + 1 \to 5.7 \qquad (5.20)$$

for the maximal moderate OZI breaking of $y \sim 0.2$. Even if y takes on the extreme value [61] of 0.6, then $(m_s/\hat{m})_{cur}$ in (5.20) increases only to ~ 8. <u>If</u> instead of (5.20) a linear relation $(m_s/\hat{m})_{cur} \sim (m_\Xi+m_\Sigma-2m_N)/\sigma_N(1-y)$ held, then a large OZI breaking $y \sim 0.6$ would increase $(m_s/\hat{m})_{cur}$ to the much larger GMOR value of 25. Such a possibility, however, not only destroys the valence quark model, but it is inconsistent with the observed quadratic mass equal splitting laws (5.1) and (5.2) which instead suggest the square root structure of (5.4), (5.10), or (5.20) because of IMF considerations.

The recent EMC measurements [62] have generated much interest in the quark spin in the proton. Unfortunately this has become linked with the strange quark flavor content in the proton for a large quark-parton axial-spin vector $\Delta s = (\bar{s}\gamma_\mu\gamma_5 s)_N$. This also suggests [63] a large $y \sim 0.6$ and 60% nucleon $(\bar{s}s)_N$ strange quark density in (5.19) when an incorrect hyperon semileptonic d/f ratio is used in the spin analysis [64].

From our perspective, the gluon-dependent EMC measurement involving the λ^o isoscalar axial current $\Delta u + \Delta d + \Delta s$ should be postponed until after the gluon-independent λ^3 and λ^8 axial currents are properly analyzed. More specifically, the λ^3 current measured in neutron β decay is [14]

$$\Delta u - \Delta d = g_A \approx 1.26 \quad , \tag{5.21a}$$

while the remaining semileptonic hyperon decays predict a λ^8 nucleon spin

$$\Delta u + \Delta d - 2\Delta s = g_A \left[\frac{3f - d}{f + d}\right]_A \approx 0.58 \tag{5.21b}$$

for the axial $(d/f)_A$ ratio [28] 1.74 already used in (3.13). Subtracting the two measurements (5.21) we eliminate the Δu part that is the dominant component in the EMC analysis to find,

$$\Delta d - \Delta s \approx \tfrac{1}{2}(-1.26 + 0.58) \approx -0.34 \quad . \tag{5.22}$$

Since the valence (V) spin numbers in nucleons are

$$\Delta u_V = {}^4/_3 \quad , \quad \Delta d_V = -{}^1/_3 \quad , \quad \Delta s_V = 0 \quad , \tag{5.23}$$

it is reasonable to anticipate from (5.22) that $\Delta d \approx -0.33$ and $\Delta s \approx 0.01$ are both extremely near their valence values.

But the final confirmation is the new EMC measurement [62], which combined with the quark-parton model and recently realized gluon components [65] requires

$$\int_0^1 dx\, g_1^P(x) = \tfrac{1}{2}\left[\tfrac{4}{9}\Delta u + \tfrac{1}{9}\Delta d + \tfrac{1}{9}\Delta s - \tfrac{2}{3}\frac{\alpha_s \Delta g}{2\pi}\right] \approx 0.126 \quad . \tag{5.24}$$

When (5.24) is solved with (5.21a) and (5.21b), and a Δg estimated from angular momentum conservation to be [52] $\Delta g \sim 1.7$ at cutoff 1 GeV, the resulting spin flavor components are

$$\Delta u \approx 0.93 \quad , \quad \Delta d \approx -0.33 \quad , \quad \Delta s \approx 0.01 \quad , \tag{5.25}$$

again consistent with the valence values and (5.22) for Δd and Δs, which are therefore model independent.

More model dependence enters the total spin of the quarks, which from (5.25) $\Delta u + \Delta d + \Delta s \approx 0.61$ suggests that almost 40% of the proton's spin comes from gluons and orbital angular momentum. But this conclusion does not contradict the valence quark model.

6. Conclusion

In this review we have surveyed low-energy strong interaction quark-hadron dynamics, always focusing on chiral symmetry. We have studied the LσM and NJL model lagrangians in section 2, the more model-independent current algebra and Heisenberg equations of motion in section 3, nonperturbative QCD and constituent quark models in section 4, and finally the (kinematical) infinite momentum frame and quark-parton scaling in section 5. Always we have arrived at a consistent physical picture with current quark masses

$$\hat{m}_{cur} \sim (34\text{–}69)\ \text{MeV} \tag{6.1}$$

depending if the masses are constant or run with momentum and ratio

$$(m_s/\hat{m})_{cur} \sim 5\text{–}6 \quad , \tag{6.2}$$

along with the quark-valence scheme finding

$$(\bar{s}s)_N \approx 0 \quad , \quad (\bar{s}\gamma_\mu \gamma_5 s)_N \approx 0 \quad . \tag{6.3}$$

With hindsight, if we had begun by assuming the current quark mass ratio (6.2) and scale (6.1), then all past and present theories of chiral symmetry breaking would appear to be unified: the SU(2) chiral field theories of the LσM and NJL, the current algebra-PCAC program, nonperturbative QCD and constituent quark models and finally the infinite momentum frame quark-parton scaling approach. It is this chiral symmetry unification (rather than the resulting values of the current quark masses) that we find appealing and quite exciting.

Acknowledgements

The author wishes to thank V. Elias, N. H. Fuchs, and A. Kocić for many useful discussions during preparation of this manuscript and J. Cleymans, S. A. Coon, R. Delbourgo, J. F. Gunion, H. F. Jones, B. H. J. McKellar, P. C. McNamee, and N. Paver for past collaborations. He also appreciates the word processing assistance of R. Miller. Finally, he is grateful for partial support from the U.S. Department of Energy.

References

[1] J. D. Bjorken and M. Nauenberg, Ann. Rev. Nucl. Sci. 18 (1968) 229; W. I. Weisberger, Brandeis Summer School Lectures 1967, eds. Chretian and S. Schweber (Gordon and Breach, 1968).

[2] M. D. Scadron, Rep. Prog. Phys. 44 (1981) 213.

[3] J. Goldstone, Nuovo Cim. 19 (1961) 154.

[4] Y. Nambu, Phys. Rev. Lett. 4 (1960) 380.

[5] R. Delbourgo and M. D. Scadron, J. Phys. G5 (1979) 1621.

[6] M. Gell-Mann and M. Lévy, Nuovo Cim. 16 (1960) 205.

[7] V. de Alfaro, S. Fubini, G. Furlan, and C. Rossettii, "Currents in Hadron Physics" (Amsterdam, North-Holland, 1973), Chapter 5; B. W. Lee "Chiral Dynamics" (Gordon and Breach, NY, 1972).

[8] Y. Nambu and G. Jona-Lasinio, Phys. Rev. 122 (1961) 345.

[9] T. Hakioğlu and M. D. Scadron, "Field Theory Calculations of the Pion Mass to One-Loop Order," submitted for publication (1989).

[10] D. Bailin, J. Cleymans, and M. D. Scadron, Phys. Rev. D31 (1985) 164; J. Cleymans, A. Kocic and M. D. Scadron, ibid, D39 (1989) 3323.

[11] T. Hatsuda and T. Kunihiro, Prog. Theor. Phys. 91 (1987) 284.

[12] L. Dolan and R. Jackiw, Phys. Rev. D9 (1974) 3320.

[13] See e.g., M. Nagels et al., Nucl. Phys. B109 (1979) 1.

[14] Particle Data Group, Phys. Lett. 204B (1988) 1.

[15] H. F. Jones and M. D. Scadron, Nucl. Phys. B155 (1979) 409; M. D. Scadron, Phys. Rev. D 29 (1984) 2076.

[16] See e.g., P. Estabrooks, Phys. Rev. D19 (1979) 2678; J. Donoghue and Y. Leyorer, Nucl. Phys. B158 (1979) 123; N. Biswas et al., Phys Rev. Lett 47 (1981) 1378; T. Akesson et al., Phys. Lett. 133B (1983) 241; A. Courau et al., Nucl. Phys. B271 (1986) 1; J. Augustin et al., Nucl Phys. B320 (1989) 1.

[17] See e.g. S. Gottlieb, W. Liu, R. L. Renken, R. L. Sugar and D. Toussaint, Phys. Rev. D38 (1988) 2245.

[18] J. Kogut et al., Phys. Rev. Lett. 50 (1983) 392.

[19] M. Gell-Mann, R. Oakes, and B. Renner (GMOR), Phys. Rev. 175 (1968) 2195.

[20] M. D. Scadron and N. H. Fuchs, Journ. Phys. G15 (1989) 943.

[21] S. Weinberg, Phys. Rev. Lett. 17 (1966) 336; W. I. Weisberger in Ref. 1, p. 421.

[22] P. C. McNamee and M. D. Scadron, Phys. Rev. D10 (1974) 2280; D11 (1975) 226.

[23] M. Ademollo and R. Gatto, Phys. Rev. Lett. 13 (1964) 264; S. Fubini and G. Furlan, Physics 1 (1965) 229.

[24] S. Glashow and S. Weinberg (GW), Phys. Rev. Lett. 20 (1968) 224; P. Auvil and N. Deshpande, Phys. Rev. 183 (1969) 1463; 185 (1969) 2043.

[25] C. G. Callan and S. B. Treiman, Phys. Rev. Lett. 16 (1966) 197; V. Mathur, S. Okubo, and L. K. Pandit, ibid 16 (1966) 371 (CTMOP).

[26] V. Elias and M. D. Scadron, Jour. Phys. G 14 (1988) 1175.

[27] W. Thirring, Acta Phys. Austriaca, Supp. II (1966) 205; J. Kokkedee, The Quark Model (W. A. Benjamin Publ., 1969), p. 187.

[28] Z. Dziembowski and J. Franklin, Temple U. preprint, Jan. 1989; F. E. Close, Phys. Rev. Lett. 64 (1990) 361.

[29] M. D. Scadron, Jour. Phys. G7 (1981) 1325.

[30] S. Coleman and S. Glashow, Phys. Rev. 134 (1964) B671.

[31] M. D. Scadron and N. H. Fuchs, Jour. Phys. G15 (1989) 957.

[32] N. Isgur, Phys. Rev. D 21 (1980) 779.

[33] J. Gasser and H. Leutwyler, Phys. Reps. 87 (1982) 77; and references therein.

[34] H. D. Politzer, Nucl. Phys. B117 (1976) 397; also see K. Lane, Phys. Rev. D 10 (1974) 2605.

[35] R. Fukuda and T. Kugo, Nucl. Phys. B 117 (1976) 250.

[36] V. Elias and M. D. Scadron: Phys. Rev. D 30 (1984) 647.

[37] V. Elias, M. D. Scadron, and R. Tarrach: Phys. Lett. 162B (1985) 176; V. Elias, T. Steele, M. D. Scadron, and R. Tarrach, Phys. Rev. D 34 (1986) 3537; V. Elias, T. Steele and M. D. Scadron, Phys. Rev. D 38 (1988) 1584.

[38] L. Reinders and K. Stam, Phys. Lett. 180B (1986) 125.

[39] V. Elias and M. D. Scadron, Phys. Rev. Lett. 53 (1984) 1129.

[40] A. De Rújula, H. Georgi, and S. Glashow, Phys. Rev. D 12 (1975) 147.

[41] O. W. Greenberg, Phys. Rev. Lett. 13 (1964) 598; R. Dalitz in High Energy Physics, ed. M. Jacob and C. Derviett, (Gordon and Breach, 1965); H. Lipkin, Phys. Repts. 8 (1973) 173.

[42] S. A. Coon and M. D. Scadron, Phys. Rev. C 23 (1981) 1150.

[43] L. N. Chan, Phys. Rev. Lett. 39 (1977) 1125.

[44] S. L. Adler, Phys. Rev. 177 (1969) 2426; J. Bell and R. Jackiw, Nuovo Cim. 60 (1969) 47 (ABJ).

[45] Ll. Ametller and A. Bramon, Ann. Phys. (NY) 15 (1984) 308; A. Bramon and M. D. Scadron, Phys. Rev. D40 (1989) 3779.

[46] S. Gerasimov, Sov. J. Nucl. Phys. 29 (1979) 259; R. Tarrach, Z. Phys. C1 (1979) 221.

[47] C. Ayala and A. Bramon, Europhys. Lett. 4 (1987) 777; also see Ll. Ametller, C. Ayala and A. Bramon, Phys. Rev. D 24 (1981) 233; D 29 (1984) 916.

[48] V. Elias and T. Steele, Phys. Lett. 199B (1987) 547.

[49] N. Isgur and G. Karl, Phys. Rev. D20 (1979) 1191.

[50] H. J. Lipkin in Baryon 1980; Proceedings of the IVth International Conference on Baryon Resonance, ed. N. Isgur, p. 461.

[51] V. Elias, M. Tong, and M. D. Scadron, Phys. Rev. D, in press.

[52] M. Anselmino and M. D. Scadron, Phys. Lett. 229B (1989) 117.

[53] J. F. Gunion, P. C. McNamee, and M. D. Scadron, Phys. Lett. 63B (1976) 81; Nucl. Phys. B123 (1977) 445.

[54] N. H. Fuchs and M. D. Scadron, Phys. Rev. D 20 (1979) 2421.

[55] H. Sazdjian and J. Stern, Nucl. Phys. B94 (1975) 163.

[56] R. L. Jaffe and C. H. Llewellyn Smith, Phys. Rev. D7 (1973) 2506.

[57] S. J. Brodsky, F. E. Close and J. F. Gunion, Phys. Rev. D5 (1972) 1384; 6 (1972) 177; 8 (1973) 3678.

[58] H. Nielsen and G. Oades, Nucl. Phys. B72 (1974) 310; G. Hite and R. Jacob, Phys. Lett. 53B (1974) 200; W. Langbein, Nuovo Cimento 51A (1979) 219; M. Olsson, J. Phys. G6 (1980) 431; R. Koch, Z. Phys. C15 (1982) 161; G. Höhler, Pion-Nucleon Scattering, Vol. 1/9 B (Springer-Verlag, 1983), Table 2.4.7.1.

[59] G. R. Farrar and D. R. Jackson, Phys. Rev. Lett. 35 (1975) 1416; A. I. Vainshtein and V. I. Zakharov, Phys. Lett. 72B (1978) 368.

[60] T. Kunihiro, Prog. Theor. Phys. 80 (1988) 34; V. Bernard, R. L. Jaffe and U. G. Meissner, Nucl. Phys. B 308 (1988) 753; T. Hatsuda, Phys. Lett. 213b (1988) 361;

H. Yabu, Phys. Lett. B 218 (1989) 124; M. Anselmino and M. D. Scadron, submitted for publication (1989).

[61] J. F. Donoghue and C. R. Nappi, Phys. Lett. 168B (1986) 105.

[62] J. Ashman et al., Phys. Lett. 206B (1988) 364, Cern preprint (1989).

[63] See, e.g. S. J. Brodsky, J. Ellis and M. Karliner, Phys. Lett. 206B (1988) 309.

[64] F. E. Close, Phys. Rev. Lett. 64 (1989) 361.

[65] A. V. Efremov and O. V. Teryaev, JNR Preprint (1988); G. Altarelli and G. C. Ross, Phys. Lett. 212B (1988) 391; E. Leader and M. Anselmino, Santa Barbara preprint (1988); R. D. Carlitz, J. C. Collins, and A. H. Mueller, Phys. Lett. 214B (1988) 229.

[66] S. Weinberg, in A Festschrift for I. I. Rabi, ed. L. M. Ots (NY, 1977), p. 185.

[67] J. Jersak and J. Stern, Nucl. Phys. B7 (1968) 413; H. Leutwyler, Springer Tracts Mod. Phys. 50 (1969) 29; H. Fritzsch, M. Gell-Mann and H. Leutwyler, Caltech Report 68-456 (1974) unpublished.

[68] U. Wiedner et al., Phys. Rev. Lett. 58 (1987) 648; Phys. Rev. D, in press (1989).

[69] J. Gasser, Ann. Phys. (NY) 136 (1981) 62.

[70] H. F. Jones and M. D. Scadron, Phys. Rev. D11 (1975) 174; A. Cass and B. McKellar, Phys. Rev. D18 (1978) 3269.

Dynamical Evolution and Particle Production in Relativistic Nuclear Collisions

U. Heinz[1], *P. Koch*[1], *K.S. Lee*[2], *E. Schnedermann*[1], *and H. Weigert*[1]

[1]Institut für Theoretische Physik, Universität Regensburg, Postfach 397, D-8400 Regensburg, Fed. Rep. of Germany

[2]Department of Physics, Chonnam National University, Kwangju 500, South Korea

Abstract: We present an overview of the different stages in the dynamical evolution of relativistic nuclear collisions, from the entropy generating initial pre-equilibrium stage through the hydrodynamic expansion phase to particle freeze-out. We discuss the various theoretical models which have been applied to these different stages, and compare their predictions with the recent results from heavy-ion experiments at CERN and Brookhaven. Particular attention is given to the possible formation and subsequent hadronization of a quark-gluon plasma in these collisions. The observed strange particle abundancies are interpreted as an indication for chemical equilibration in nuclear collisions, and the particle momentum spectra are analyzed for signs of collective expansion flow.

Prologue

In the last three years a new field in experimental nuclear physics has begun to florish: the investigation of ultra-relativistic heavy-ion collisions. The presently available beam energies range from 14.5 A GeV, provided by the Brookhaven AGS, to 200 A GeV at the CERN SPS, corresponding to total energies in the nucleon-nucleon center of mass of 3 to 10 GeV per nucleon. Up to now only rather light ions up to ^{32}S can be accelerated by these machines, but upgrades to allow the acceleration of even the heaviest natural nuclei are under construction or planned for the next years. Furthermore, there are very good chances that in the second half of this decade a dedicated relativistic heavy-ion collider (RHIC) will be available at Brookhaven, providing center of mass energies of up to 100 GeV/nucleon and beams as heavy as Uranium.

The reasons for this activity can be traced back to theoretical discussions of strongly interacting matter at high temperatures and densities which started a little over ten years ago. These discussions centered around the possibility for a phase transition from a hadronic resonance gas to a quark-gluon plasma (QGP) in regions of very high energy density $\epsilon \gtrsim 1-2$ GeV/fm^3. This is about 10-20 times the energy density of cold nuclear matter in its ground state. Short of recreating the early universe or probing the interior core of large neutron stars, the only hope to reach such high energy densities in the laboratory is in relativistic nuclear collisions, by converting large amounts of directed beam energy into particle excitation and randomly oriented internal motion within a compact collision zone. After extensive theoretical work during the first half of the eighties the physics community became convinced that this is not a hopeless task. In 1986/87 a first round of experiments began with the aim of checking the global features of nuclear collisions at relativistic energies, to provide evidence for sufficiently large energy densities in the collision zone, and to take a preliminary look at a few of the so-called "quark-gluon plasma signatures".

Phase Structure of Strongly Interacting Matter
Editor: J. Cleymans

Given the fact that hardly anything was known experimentally or even theoretically on the dynamical evolution of ultrarelativistic nuclear collisions, there was obviously first a need for a rough global view of what was going on, and more dedicated experimental setups with focus on precision measurements of particular observables on an event-by-event basis had to be postponed for the second round of experiments presently under design. While (at least in our opinion) the first experiments did not really have a chance to "prove" QGP formation in nuclear collisions, the experimental results are as positive as could be hoped for and certainly not in contradiction with the theoretical expectations. We will show in these lectures that some very encouraging discoveries have been made which, although not proving QGP formation, demonstrate quite convincingly the creation of a very dense and highly excited, quasi-equilibrated fireball in the reaction zone, *i.e.* a state of matter never before observed anywhere else.

Lecture 1: The Space-Time Evolution of a Heavy-Ion Collision

1.1 What is a quark-gluon plasma?

Fig. 1 shows in a cartoon the transition from cold nuclear matter to a quark-gluon plasma. In cold nuclear matter a useful set of effective degrees of freedom are protons and neutrons, plus virtual mesons as exchange particles to describe their interactions. The average distance between nucleons (about 2 fm) is large compared to their size (0.86 fm) thus that they can usually be considered as pointlike particles. The fact that each nucleon consists of three valence quarks and a whole sea of gluons and $q\bar{q}$ pairs is of little practical importance in this energy region, since the effective coupling constant between these constituents is large ($\alpha_s \gtrsim 1$), leading to confinement of all colored quanta inside the hadrons and no direct evidence for individual quark-gluon consituents at larger distances. The space between the nucleons is the vacuum of Quantum ChromoDynamics (QCD) which has a rather complicated structure and contains gluon and $q\bar{q}$ condensates. While the former expels color electric flux and is thus responsible for color confinement into small bags called hadrons, the $q\bar{q}$ condensate signals spontaneous breakdown of chiral symmetry and is responsible for giving the quarks (and thus the nucleons) large effective masses. Cold nuclear matter is a *color insulator*, since color cannot propagate.

As the cold nucleus shown on the left in Fig. 1 is compressed and heated up, the density of hadrons increases. The excitation energy manifests itself in increased thermal

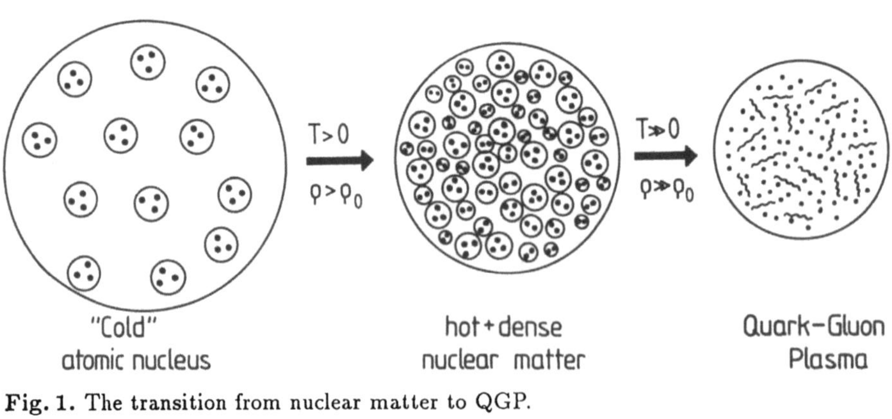

Fig. 1. The transition from nuclear matter to QGP.

motion as well as creation of real (*i.e.* thermally distributed) mesons and baryon resonances. For the description of such a hot and dense "hadron resonance gas" the set of effective degrees of freedom has to be enlarged to include more and more baryon and meson resonances.

At some point a situation will be reached where the space becomes closely packed with these resonances, and they begin to overlap. At this point the internal structure of the hadrons becomes essential, and the quark and gluon constituents manifest themselves and begin to percolate from hadron to hadron. Compressing the nucleus a little further, individual hadrons loose their identity. Clearly now it does not make sense anymore to use hadrons as effective degrees of freedom: although an infinite set of hadronic resonance states may form a complete Fock-space basis, it is impossible to truncate this basis and still have a reliable description. It is much more economical to switch to quarks and gluons as effective degrees of freedom. Furthermore, since the QCD coupling is energy dependent and decreases for large energies, the coupling between these constituents becomes weaker and weaker as the thermal and Fermi energies increase. In the situation shown on the right in Fig. 1, color confinement can no longer be upheld, and quarks and gluons can move around freely throughout the dense and hot region. Inside this quark-gluon plasma the vacuum is free of quark and gluon condensates (which have been "evaporated") and is called the "perturbative" vacuum, on which the quarks and gluons are constructed as (weakly interacting) thermal excitations. The QGP is a *color conductor*.

1.2 How to make QGP in nuclear collisions

The ideas leading to the expectation that a QGP might be created in relativistic nuclear collisions, thereby recreating in the laboratory conditions similar to those in the first microsecond of our universe, are summarized in Fig. 2. Based on experience [1] from nuclear collisions at the BEVALAC, where beam energies up to 1 or 2 GeV per nucleon were available, one might expect also at higher energies a situation like the one shown in the upper half of Fig. 2: the two nuclei approach each other with relativistic velocity and stop each other completely in the collision, thereby fully converting the kinetic energy of the directed relative motion into randomly oriented internal excitation energy. After a short time of rescattering between the constituents of the collision zone a locally equilibrated fireball of high temperature is formed, which has also a very large baryon density, because all the baryons from the two colliding nuclei have been stopped and compressed into the collision region. In thermodynamic language, the fireball is described by a temperature T and a chemical potential μ_B which controls the baryon number content, and we expect large values for both μ_B and T, with $\mu_B/T \gg 1$ due to the high baryon density. Such a situation is encountered at BEVALAC energies [1,2].

Unfortunately, this picture has to break down at very high energies, due to the onset of nuclear transparency. This feature, which is expected to show its first effects at c.m. energies above 5 GeV/nucleon or so [3], is due to the relativistic kinematics: in each nucleon-nucleon collision the collision partners lose only about 50% of their longitudinal energy, and thus the total energy loss due to N-N-collisions is limited by the longitudinal dimension of the target and projectile nuclei. Additional scatterings with the secondary particles created in the collision, which could cause further energy loss, are kinematically forbidden: until these secondary particles are formed in decays of the strongly excited struck nucleons, a certain time (of order of the nucleon radius divided by c, the maximum velocity with which the decay meson can leave the excited nucleon) has passed in the rest

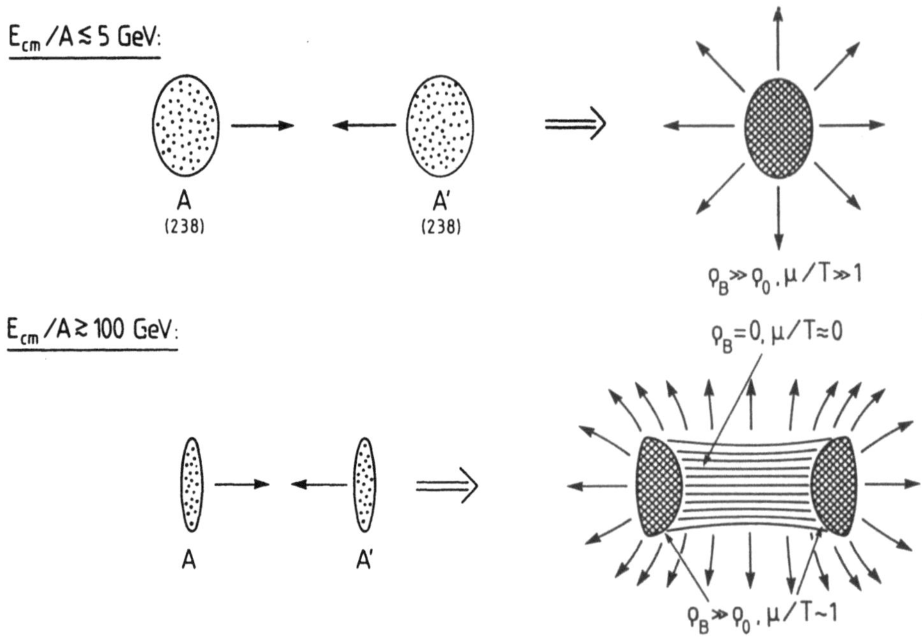

Fig. 2. The spatial evolution of a heavy-ion collision in the stopping and in the transparency regime.

frame of the struck nucleon. Since this, say, projectile nucleon moves with a very large velocity relative to the target nucleons, the latter see this decay time strongly Lorentz-dilated, and by the time the secondary particle formation is completed, the projectile nucleon has long passed all target nucleons, leaving the latter ones no chance to rescatter with the decay secondaries.

As a consequence, at extremely high energies a complementary picture is expected, as shown in the lower half of Fig. 2: The two nuclei, now much more Lorentz contracted due to the larger relative momentum, pass essentially through each other, thereby getting considerably decelerated and excited, but not completely stopped. Thus the baryon number from the projectile and target nuclei flows strongly into the backward and forward directions, rather than getting stopped in the c.m. system. The region between these "target and projectile fragmentation regions", which is called the "central region", contains nearly no net baryon number, *i.e.* it is symmetric in its quark and antiquark content. It is fed by decay particles left behind by the decaying excited nucleons from the rapidly moving projectile and target. In thermodynamic language it would be described by high temperature and a nearly vanishing baryonic chemical potential, very much like the early universe. The fragmentation regions, instead, would contain the baryon number and thus have large values for μ_B, although probably not as large as in the stopping region ($\mu_B/T \sim 1$).

Scanning the energy region from a few GeV/nucleon to several hundred GeV/nucleon in the center of mass we thus expect to create matter of varying temperature and baryon density, allowing us to map a considerable fraction of the $T-\mu_B$ phase diagram. Whether the energy densities deposited in the collision zone are sufficient to create the QGP depends on the effectiveness of the stopping process. Before the experiments were perfor-

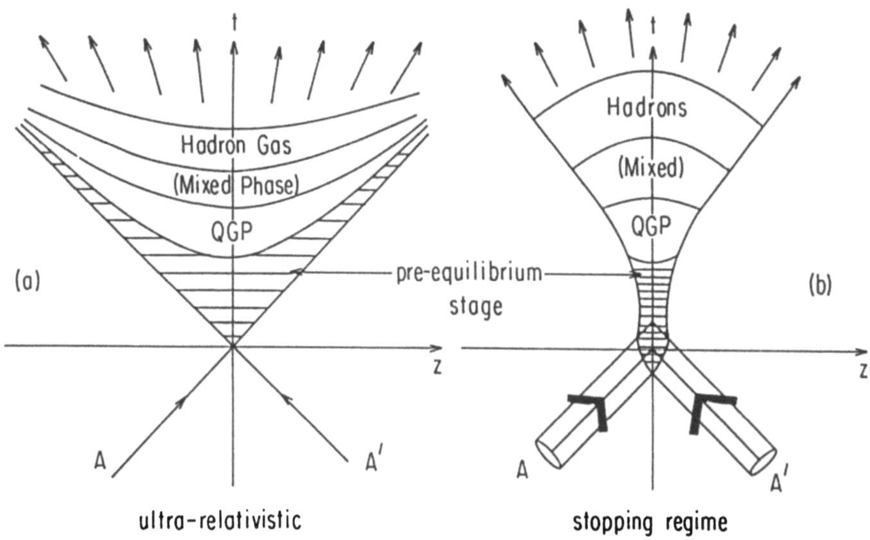

Fig. 3. Space-time evolution along the beam direction for relativistic nuclear collisions in (a) the ultrarelativistic full-transparency limit, and (b) in the complete stopping limit.

med, theoretical estimates of the nuclear stopping power were not very reliable. Still, it appeared possible to exceed the critical energy density inspite of the onset of nuclear transparency [4], with a continuing increase proportional to the logarithm of the c.m. energy, $\epsilon \sim \ln \sqrt{s}$.

In Fig. 3 we show the space-time evolution for these two extreme scenarios. The right picture refers to the stopping regime: it shows the two nuclei approaching each other along the z-axis with the velocity of light, colliding at $t = 0$ in the center of mass $z = 0$, stopping each other and staying together at $z = 0$ as a compact object for some time, and then eventually exploding into all directions (in particular along the z-axis) due to its high pressure. Immediately after the collision a pre-equilibrium stage follows during which secondary particles are produced and rescatter, eventually leading to local equilibrium. If equilibration occurs sufficiently fast and the energy density at this point is still large enough, a QGP will be formed. Expansion will lead to dilution and cooling, such that after a short time the system reaches the critical conditions for rehadronization. If the phase transition back to hadronic matter is of first order (as indicated by lattice QCD results [5]), a mixed phase will ensue, followed by an interacting hadron gas stage, until the system eventually becomes so dilute that the hadrons stop interacting with each other and freeze out.

In the regime of nuclear transparency (left figure) the two nuclei (now very Lorentz-contracted) pass each other with nearly the velocity of light, filling the central region in between with baryon-free matter of secondaries. Again a pre-equilibrium region of particle formation and rescattering is followed after a certain proper time by local equilibrium. For sufficiently high initial energy densities a QGP will form, expanding now due to the ongoing longitudinal motion mostly into the z-direction, cooling, rehadronizing, and finally decoupling into free-streaming hadrons. In this case there will be a large asymmetry in longitudinal and transverse expansion, because the latter is entirely created by the pressure of the initial hot stage, while the former has a strong primordial component due to the partial transparency.

Lecture 2: Pre-Equilibrium Dynamics – Quark-Gluon Kinetic Theory

A crucial time scale in these scenarios is the one for equilibration. If equilibration occurs too slowly, the system will fall apart into individual hadrons before local equilibrium has been reached. In this case it will never be possible to create from the many produced partons a collective system like the quark-gluon plasma which is described by just a few statistical parameters like temperature, chemical potential, and an average, temperature- and density-dependent strong coupling constant. In such a case it would probably be impossible to distinguish the outcome from a superposition of hadron-hadron interactions, perhaps slightly modified by some rescattering.

It thus seems very important to derive some estimates for the equilibration time scale in a system of partons. A proper framework for this would be a kinetic theory of quarks and gluons, based on the gauge theory QCD. This has only recently been developed (see [6] for an up-to-date review) and is not yet sufficiently well understood to serve as a practical tool for extensive non-equilibrium studies in dynamical quark-gluon systems. Despite its still rather academic character, we will now shortly introduce the basic concepts of quark-gluon kinetic theory, hoping to convey its interesting intrinsic structure and to stimulate the readers' fantasy with the many open problems on the way to turning this approach into the basis for a numerical description of heavy-ion dynamics.

2.1 Classical Kinetic Theory for Colored Particles

2.1.1 Classical Equations of Motion for Particles with Color and Spin

Let us consider the quark color generators $\hat{Q}_a \equiv -t_a = -\frac{\hbar}{2}\lambda_a$ as Heisenberg operators satisfying Heisenberg's equations of motion,

$$\frac{d\hat{Q}_a}{d\tau} = \frac{i}{\hbar}[\mathcal{H}, \hat{Q}_a] \, . \tag{2.1}$$

If the theory is to be formulated in a relativistically invariant way, the time derivative in (2.1) should be with respect to proper time, and the "Hamiltonian" $\mathcal{H}$ (a Lorentz scalar) generating the proper-time evolution is to be taken as the "quadratic Dirac Hamiltonian" [7],

$$\mathcal{H} \equiv -\frac{1}{2m}\left[\left(i\hbar\partial_\mu + \frac{g}{c}A_\mu\right)^2 + \frac{g}{2c}S^{\mu\nu}F_{\mu\nu} - m^2c^2\right] \, . \tag{2.2}$$

Here $S^{\mu\nu} = (i\hbar/4)[\gamma_\mu, \gamma_\nu]$ is the quark spin tensor, and $(1/2)S^{\mu\nu}F_{\mu\nu}$, which reduces to $\boldsymbol{s}\cdot\boldsymbol{B}$ in the particle's rest frame, is the coupling of the spin to the magnetic field. In the classical c-number limit (2.1) leads to the equation of motion [8]

$$m\frac{dQ_a}{d\tau} = -\frac{g}{c}f_{abc}\left(p^\mu A^b_\mu - \frac{1}{mc}p^\alpha \tilde{F}^b_{\alpha\beta}S^\beta\right)Q^c \, , \tag{2.3}$$

where now Q_a are the c-number (i.e. commuting) components of an 8-component classical color vector $\boldsymbol{Q}$ describing the coupling of a classical colored particle to the eight color potentials A^a_μ. In (2.3) $\tilde{F}_{\alpha\beta} = \frac{1}{2}\epsilon_{\alpha\beta\mu\nu}F^{\mu\nu}$ is the dual field strength tensor, and

$$S^\beta = -\frac{1}{2mc}\epsilon^{\beta\nu\lambda\rho}p_\nu S_{\lambda\rho} \tag{2.4}$$

defines a normalized ($p_\mu S^\mu = 0$, $S^\mu S_\mu = -\boldsymbol{s}^{\,2}$) classical spin 4-vector in terms of the classical limit of the spin tensor. Equation (2.3) conserves the SU(3) Casimir invariants, i.e. the length Q^aQ_a of Q and the cubic invariant $d_{abc}Q^aQ^bQ^c$ (where d_{abc} are the symmetric structure constants of SU(3)). Thus, the equation describes precession of the classical color vector of the particle due to two effects: direct interaction (color exchange) with an external color field A^a_μ, and coupling of the particle's spin to the color magnetic field. In the case of vanishing particle spin this equation was derived by Wong [9].

If the spin couples to the color magnetic field, it will similarly start to precess, and its equation of motion is given by the c-number limit of $\dot{S}^\mu = \frac{i}{\hbar}[\mathcal{H}, S^\mu]$, namely [8]

$$m\frac{dS^\mu}{d\tau} = \frac{g}{c}Q^a\left[\underset{a}{F}^{\mu\nu}S_\nu + \frac{1}{(mc)^3}(p^\mu S^\nu - p^\nu S^\mu)(D_\nu \tilde{F}_{\alpha\beta})_a p^\alpha S^\beta\right]. \tag{2.5}$$

In absence of inhomogeneities of the external field and of non-Abelian effects, this equation is reduces to the BMT equation [10] for a spinning particle with Landé g-factor 2. Had we started from the Yang-Mills Hamiltonian rather than the Dirac Hamiltonian, we would have obtained for the spin-1 gluons an analogous equation, but for a g-factor of 1. So with respect to color, we cannot distinguish in the structure of the equations between quarks and gluons: by going to the classical limit (i.e. effectively to very high-dimensional representations of the color generators) *we have lost the difference in color between quarks and gluons.* However, *in their spin aspects they still remain different*: The Landé factor distinguishes quarks from gluons in their coupling to the magnetic-field. Since we are not going to work out the spin aspects of the resulting classical kinetic theory in any detail, we will not further elaborate on this point.

Finally, we need an equation of motion for the momentum of the particle. It is given by

$$m\frac{dp^\mu}{d\tau} = \frac{g}{c}Q^a\left[\underset{a}{F}^{\mu\nu}p_\nu - \frac{1}{mc}(D^\mu \tilde{F}_{\alpha\beta})_a p^\alpha S^\beta\right] = \frac{g}{c}Q^a\left[\underset{a}{F}^{\mu\nu}p_\nu + \frac{1}{2}D^\mu(S^{\alpha\beta}F_{\alpha\beta})_a\right]. \tag{2.6}$$

In the first term we recognize the (colored version of) the relativistic Lorentz force law. The second term describes the possible gain in energy-momentum due to the space-time variation of the spin magnetic interaction energy in an inhomogeneous color magnetic field.

2.1.2 Classical Kinetic Equations for the 1-Particle Distribution Functions in a Quark-Gluon Plasma

We can now write down a classical kinetic equation of the Vlasov type for the single-particle phase space distribution for colored and spinning particles. Since the particles' momentum p^μ, the color Q_a, and the spin S_μ all are dynamical variables (i.e. evolve in time under the influence of an external or intrinsic mean color field), phase space has to be spanned by the 20 coordinates x^μ, p^μ, Q^a and S^μ. Only in the absence of classical color and spin it reduces to the conventional 8-dimensional phase space (x^μ, p^μ) which, by using the mass-shell constraint between p^0 and $\boldsymbol{p}$ for the classical particles, is further reduced to the well-known dimensions $(\boldsymbol{x}, \boldsymbol{p}, t)$. In our larger 20-dimensional phase space the integration measure is given by $d\Sigma_\mu\, dP\, dQ\, dS$, where $d\Sigma_\mu$ is the surface element for some space-like hypersurface Σ, and

$$\begin{aligned} dP &= 2\,\theta(p_0)\,\delta(p^2 - m^2)\frac{d^4p}{(2\pi\hbar)^3} = \left.\frac{d^3p}{(2\pi\hbar)^3 E}\right|_{p_0 = E = \sqrt{\boldsymbol{p}^2 + m^2}} \\ dQ &= \delta(Q^aQ_a - q^2)\,\delta(d_{abc}Q^aQ^bQ^c - \tilde{q}^3)\,d^8Q \\ dS &= \delta(p_\mu S^\mu)\,\delta(S^\mu S_\mu + s^2)\,d^4S\,, \end{aligned} \tag{2.7}$$

with δ-functions fixing the mass-shell and normalization constraints for p^μ, Q^a, and S^μ.

The probability to find a classical particle at a given point in this phase space is given by the 1-particle distribution function $f(x,p,Q,S)$. It has to be a Lorentz scalar and gauge invariant. A gauge and Lorentz invariant [6] expression for the time evolution of $f(x,p,Q,S)$ is given by

$$m\frac{df}{d\tau} \equiv \left[p^\mu \frac{\partial f}{\partial x^\mu} + m\dot{p}^\mu \frac{\partial f}{\partial p^\mu} + m\dot{Q}^a \frac{\partial f}{\partial Q^a} + m\dot{S}^\mu \frac{\partial f}{\partial S^\mu}\right] = C(x,p,Q,S) \,, \tag{2.8}$$

where C is a collision term describing short-range 2-body collisions. Inserting (2.3) and (2.6) and leaving all spin effect aside (to keep the expressions manageable), we obtain the following equation for the 1-particle distribution function of a plasma of classical colored particles [11]:

$$p^\mu \left[\partial_\mu - \frac{g}{c} Q_a F^a_{\mu\nu}(x) \partial^\nu_p - \frac{g}{c} f_{abc} A^b_\mu(x) Q^c \partial^a_Q\right] f(x,p,Q) = C(x,p,Q) \,. \tag{2.9}$$

If there are antiparticles involved, their distribution function $\bar{f}(x,p,Q)$ obeys a similar equation, with Q^a replaced by $-Q^a$ (i.e. the second term changes sign). These equations have to be closed by an equation for the mean color field A_μ which, of course, is the Yang-Mills equation

$$(D_\mu F^{\mu\nu})_a(x) = -\frac{g}{c} j^\nu_a(x) = \frac{g}{c} \int p^\nu Q_a \left[f_q(x,p,Q) - \bar{f}_{\bar{q}}(x,p,Q) + f_g(x,p,Q)\right] dPdQ \,, \tag{2.10}$$

where $f_q, \bar{f}_{\bar{q}}$, and f_g are the distribution functions for quarks, antiquarks, and gluons respectively.

Equations (2.9/10) together form the basis of a relativistic kinetic description for a plasma of colored particles. The mean field terms on the left hand side of (2.9) generalize those known from the usual Vlasov equation for electromagnetic plasmas; however, in addition to the drift in momentum induced by the electric and magnetic fields (non-relativistically the combination $\boldsymbol{E} + \boldsymbol{v}/c \times \boldsymbol{B}$ occurs as the coefficient of the momentum derivative of f), there are now also drift terms in the color sector of phase space, due to the non-Abelian interaction between the colored particles and the potentials A_μ.

The collision term on the right hand side of (2.9) couples the 1-particle distribution function to two-body correlations. So actually (2.9/10) generally do not close; closure can be obtained, however, by factorizing the two-body correlations into products of single-particle distribution functions (Boltzmann approximation). Without this approximation further kinetic equations are needed describing the evolution of the 2-body distribution function which then again couples to 3-body correlations, and so on. This BBGKY hierarchy of coupled equations has not been constructed yet for the non-Abelian case; in principle it should emerge from the quantum mechanical formulation of Sect. 2.2 in the classical limit, but this has so far not been shown explicitly.

2.1.3 Color Moment Equations

From the kinetic equations (2.9) one can construct several infinite hierarchies of moment equations, by forming moments involving powers of the color vector Q^a, of the momentum vector p^μ, or both. The color moment equations prove useful later when comparing with the quantum mechanical formulation, since it turns out that the lowest color moments of $f(x,p,Q)$ can be identified with the classical limit of the color components of the Wigner function. The two lowest moments of the momentum operator formed with these color moments then lead to equations of motion for macroscopic entities, namely the space-time densities of energy-momentum, baryon number and color current, i.e. they yield a chromohydrodynamic description of the plasma [11].

For later reference we will now shortly review the color moment equations. We define the color singlet, octet, etc. distribution functions as the following moments of $f(x,p,Q)$:

$$
\begin{aligned}
f(x,p) &= \int f(x,p,Q)\, dQ \,, \\
f_a(x,p) &= \int Q_a\, f(x,p,Q)\, dQ \,, \\
f_{ab}(x,p) &= \int Q_a Q_b\, f(x,p,Q)\, dQ \,, \quad \text{etc.}
\end{aligned}
\tag{2.11}
$$

For these one obtains from (2.9), by taking appropriate color moments,

$$
p^\mu \partial_\mu f(x,p) = \frac{g}{c} p^\mu F^a_{\mu\nu}(x)\, \partial^\nu_p f_a(x,p) + \int C(x,p,Q)\, dQ \,, \tag{2.12a}
$$

$$
\begin{aligned}
&p^\mu \left[\partial_\mu \delta_{ac} + \frac{g}{c} f_{amc} A^m_\mu(x)\right] f_c(x,p) \\
&\quad = \frac{g}{c} p^\mu F^b_{\mu\nu}(x)\, \partial^\nu_p f_{ab}(x,p) + \int Q_a C(x,p,Q)\, dQ \,,
\end{aligned}
\tag{2.12b}
$$

$$
\begin{aligned}
&p^\mu \left[\partial_\mu \delta_{ac}\delta_{bd} + \frac{g}{c}(\delta_{ac} f_{bmd} + \delta_{bd} f_{amc}) A^m_\mu(x)\right] f_{cd}(x,p) \\
&\quad = \frac{g}{c} p^\mu F^c_{\mu\nu}(x)\, \partial^\nu_p f_{abc}(x,p) + \int Q_a Q_b C(x,p,Q)\, dQ \,, \quad \text{etc.}
\end{aligned}
\tag{2.12c}
$$

Classically, all these moments are independent. Quantum mechanically, the color charges Q_a do not commute, and the color algebra between them allows in the case of quarks (where $Q_a \leftrightarrow -\lambda_a/2$) to express the second color moment f_{ab} in terms of f and f_a [11]:

$$
f_{ab} = \frac{\delta_{ab}}{6} f - \frac{1}{2} d_{abc}\, f_c \,. \tag{2.13}
$$

Hence for quarks the color hierarchy can be truncated by hand by imposing (2.13) even on the classical level and rewriting (2.12b) as

$$
\begin{aligned}
&p^\mu \left[\delta_{ac}\partial_\mu + \frac{g}{c}\, f_{amc} A^m_\mu(x) + \frac{g}{2c} d_{amc} F^m_{\mu\nu}(x)\, \partial^\nu_p\right] f_c(x,p) \\
&\quad = \frac{g}{6c}\, p^\mu F^a_{\mu\nu}(x)\, \partial^\nu_p f(x,p) + \int Q_a C(x,p,Q)\, dQ \,.
\end{aligned}
\tag{2.14}
$$

In Sect. 2.2 we will show that (2.12a) and (2.14) are exactly reproduced in the semiclassical limit of the Wigner function formulation.

For gluons which are in the adjoint representation ($Q_a \leftrightarrow T_a$ with $(T_a)_{bc} = i f_{bac}$) (2.13) is not applicable, and the color hierarchy cannot be closed so easily. In fact, the gluon Wigner function contains also color moments which are not symmetric in the color indices a, b, which have no classical analogue. Thus the classical limit for gluons is more complicated.

2.2 Quantum Kinetic Theory for Quarks and Gluons

2.2.1 Quark and Gluon Wigner Functions

The quantum mechanical analogue of the 1-particle distribution function in phase space is the Wigner function [12,13], which is therefore the basic object of interest in quantum kinetic theory. In a non-relativistic theory it is essentially the Fourier transform of the density matrix. In a relativistic gauge theory like QCD or QED, the Wigner *function* is defined as the ensemble expectation value of the Wigner *operator*, where the latter is again related by Fourier transformation to an object which looks like a density operator, except for some modifications which are required by the gauge invariance of the theory. Specifically, the Wigner operator for fermions (quarks) is defined as [11,14] (the minus sign is convention and has to do with the fermion statistics of the quark fields ψ):

$$\hat{W}^{AB}_{\alpha\beta}(x,p) = -\int \frac{d^4y}{(2\pi\hbar)^4} e^{-\frac{i}{\hbar}p\cdot y} \left[e^{-\frac{y}{2}\cdot D_x}\psi(x)\right]^A_\alpha \left[\overline{\psi}(x)e^{\frac{y}{2}\cdot D^\dagger_x}\right]^B_\beta , \qquad (2.15)$$

where $D_\mu(x) = \partial_\mu - (ig/\hbar c)A_\mu(x)$ with $A_\mu(x) \equiv A^a_\mu(x)t_a$, $t_a = (\hbar\lambda_a/2)$ is the covariant derivative for the quark field operators, and A, B=1,2,3 and α,β=1,...,4 denote color and spinor indices. One sees that the partial derivative in the usual shift operator $\exp[\pm(y/2)\cdot\partial_x]$, which occurs in the density matrix $\psi(x-y/2)\,\overline{\psi}(x+y/2)=$ $[\exp[-(y/2)\cdot\partial_x]\psi(x)]\,[\overline{\psi}(x)\exp[(y/2)\cdot\partial^\dagger_x)]$, here is replaced by the covariant derivative operator. This guarantees gauge invariance (which the usual density matrix does not possess) and ensures through Weyl's correspondence principle that the momentum argument p of the Wigner operator is the *kinetic* (rather than the canonical) momentum of the particles which classically obeys the mass–shell condition $p^2 = m^2$.

In [14] it was shown that (2.15) can be rewritten in terms of so-called link-operators

$$U(x',x) = P\exp\left\{\frac{ig}{\hbar c}\int_x^{x'} dz^\mu A_\mu(z)\right\} , \qquad (2.16)$$

where the path from x to x' is the *straight line* $z(s) = x + s(x'-x)$, $0 \le s \le 1$, and P is the path–ordering operator:

$$\hat{W}(x,p) = -\int \frac{d^4y}{(2\pi\hbar)^4} e^{-\frac{i}{\hbar}p\cdot y}\, U(x,x-\tfrac{y}{2})\psi(x-\tfrac{y}{2})\,\overline{\psi}(x+\tfrac{y}{2})U(x+\tfrac{y}{2},x) . \qquad (2.17)$$

There is no path ambiguity in (2.17) because the identity of (2.15) with (2.17) holds only for the straight path in the definition of (2.16).

The underlying relationship to the quark density operator is best reflected in the expression for the quark (baryon number) current

$$b_\mu(x) = \int d^4p\,\langle \mathrm{Tr}\gamma_\mu \hat{W}(x,p)\rangle . \qquad (2.18)$$

The quark color vector current is given by

$$j^a_\mu(x) = \int d^4p\,\langle \mathrm{Tr}t_a\gamma_\mu \hat{W}(x,p)\rangle . \qquad (2.19)$$

Please note the formal similarity of (2.18) with the classical expression [11]

$$\begin{aligned} b_\mu(x) &= \int \frac{d^4p}{(2\pi\hbar)^3} 2\delta(p^2-m^2)\, p_\mu[\theta(p_0)f(x,p) + \theta(-p_0)\overline{f}(x,-p)] \\ &= \int \frac{d^4p}{(2\pi\hbar)^3} 2\delta(p^2-m^2)\, \theta(p_0)\, p_\mu[f(x,p) - \overline{f}(x,p)] , \end{aligned} \qquad (2.20)$$

where f and $\overline{f}$ are the classical distribution functions for quarks and antiquarks. The classical mass-shell condition (which in general does not apply to the quantum mechanical quark fields described by $W(x,p)$) has been extracted explicitly; it can be combined with the momentum space integration measure to yield the more familiar expression ($E_p = \sqrt{m^2 + \boldsymbol{p}^2}$)

$$b_\mu(x) = \int \frac{d^3p}{(2\pi\hbar)^3} \frac{p_\mu}{E_p} \left[f(x, E_p, \boldsymbol{p}) - \bar{f}(x, E_p, \boldsymbol{p})\right] . \tag{2.21}$$

Indeed, if (2.18) is evaluated in local thermal equilibrium in the absence of any color background fields (this means that all spin effects due to couplings of the kind $\boldsymbol{\sigma} \cdot \boldsymbol{B}$ vanish), then, as $g \to 0$, (2.17) exactly reduces to (2.21).

For the gluons (or more generally the gauge field quanta) the starting point is not the gluon density operator, but an object more related to gluon energy density [6,15]

$$\hat{\Gamma}^{ab}_{\mu\nu}(x,p) = \int \frac{d^4p}{(2\pi\hbar)^4} e^{-\frac{i}{\hbar}p\cdot y} \left[e^{-\frac{y}{2}\cdot\mathcal{D}(x)} \boldsymbol{F}_\mu{}^\lambda(x)\right]^a \left[e^{\frac{y}{2}\cdot\mathcal{D}(x)} \boldsymbol{F}_{\lambda\nu}(x)\right]^{\dagger b} . \tag{2.22}$$

Now the color indices a, b run from 1 to 8, and $\boldsymbol{F}_{\mu\nu}$ denotes the 8-component color vector of the gluon field strength tensor. The covariant derivatives occurring in the shift operators are now in the adjoint representation, *i.e.* $\mathcal{D}_\mu(x) = \partial_\mu(x) - \frac{ig}{\hbar c}\mathcal{A}_\mu(x)$ with $\mathcal{A}_\mu = A^a_\mu T_a$, $(T_a)_{bc} = -i\hbar f_{abc}$. (2.22) can be reexpressed in a form analogous to (2.17) by using link operators (2.16) with $\mathcal{A}$ in the adjoint representation.

The choice of the field strength tensor as a starting point is due to its gauge covariant transformation properties, in contrast to the vector potentials. This variant of the gluon Wigner function is most directly related to the gluon energy-momentum tensor

$$T_{\mu\nu}(x) = \int d^4p \,\mathrm{Tr}\langle\hat{\Gamma}_{\mu\nu}(x,p) - \frac{1}{4} g_{\mu\nu}\hat{\Gamma}^\lambda_\lambda(x,p)\rangle . \tag{2.23}$$

Now the similarity to the classical expression

$$T_{\mu\nu}(x) = \int \frac{d^4p}{(2\pi\hbar)^3} 2\delta(p^2 - m^2)\theta(p_0)p_\mu p_\nu f(x,p) \tag{2.24}$$

is less obvious. However, near thermal equilibrium and in the weak coupling limit $g \to 0$ (2.23) can be shown [15] to reduce to (2.24) (with $m = 0$ for gluons). In the classical limit $\Gamma_{\mu\nu}$ has the Lorentz-tensor structure

$$\Gamma_{\mu\nu}(x,p) = (p_\mu p_\nu - p^2 g_{\mu\nu})\mathcal{G}(x,p) - ip^2 \mathcal{S}_{\mu\nu}(x,p) . \tag{2.25}$$

Here $\mathcal{G}$ (still an 8×8 color matrix) describes the Lorentz-scalar gluon density, in terms of which the gluon color vector current is defined as [16]

$$j^\nu_a(x) = \frac{g}{\hbar c} \int d^4p\, p^\nu \,\mathrm{Tr}\langle(T_a\mathcal{G}(x,p)\rangle = \frac{ig}{c} \int d^4p\, p^\nu f_{abc}\mathcal{G}^{bc}(x,p) . \tag{2.26}$$

$\mathcal{S}_{\mu\nu}$ (which is the antisymmetric part of the Lorentz-tensor Γ) describes the gluon spin tensor density (for ensembles which are locally spin polarized) and generates an axial vector (spin) current

$$j^{5\nu}_a(x) = -\frac{\hbar}{2}\, \epsilon^{\nu\rho\sigma\tau} \int d^4p\, p_\rho \,\mathrm{Tr}\langle T_a \mathcal{S}_{\sigma\tau}(x,p)\rangle . \tag{2.27}$$

The analogous object for quarks is obtained from (2.19) by replacing γ_μ by $\gamma_\mu\gamma_5$ [17].

Recently it has been attempted to formulate gluon Wigner functions directly in terms of the gluon potentials. This requires the choice of a specific gauge, since the potentials do not transform covariantly under gauge changes. Elze [18] has suggested using covariant background gauge; below we will discuss our own approach [19] using Fock-Schwinger gauge.

2.2.2 Kinetic Equations of Motion for the Quark and Gluon Wigner Operators

Using the quark and gluon field equations given by QCD, one may derive equations of motion for the Wigner functions which resemble the Vlasov-Boltzmann equations for the 1-particle distribution function in classical kinetic theory. The derivation is lengthy due to the occurrence in the definition of the Wigner operator of the link operators (2.16), and the full result can be found in [6]. Below we will only give the leading terms in an $\hbar$-expansion.

A semiclassical expansion of these equations is not without problems. They arise from the presence of the link operators (2.16) in the definition (2.17) of the Wigner function or, alternatively, through the covariant derivatives in (2.15). Formally the analogue of the Wigner-Kirkwood $\hbar$-expansion is generated by substituting in (2.15) $y \to i\hbar\partial_p$ in the shift operators and expanding the resulting expressions $\exp[(i\hbar/2)D(x)\cdot\partial_p]$ into a Taylor series. However, the term $-(ig/\hbar c)A(x)$ in the covariant derivative spoils the resulting ordering in terms of powers of $\hbar$. (For Abelian gauge theories this problem does not occur [17], since it turns out that the combination of covariant derivatives acting from the left and from the right leads to the A-fields occuring only inside commutators which vanish in the case of U(1).) Thus a well-defined semiclassical expansion generally requires strong constraints on the properties of the non-Abelian gluon potentials.

It turns out that this problem can be circumvented by chosing a particular gauge. The idea is to interpret the link operators (2.16) in (2.17) as a particular gauge transformation and introducing new, gauge transformed fields according to

$$\psi_Z(x') := U(Z,x')\psi(x') \,, \tag{2.28a}$$

$$\overline{\psi}_Z(x') := \overline{\psi}(x')U(x',Z) \,, \tag{2.28b}$$

$$\begin{aligned} A_Z^\nu(x') &:= U(Z,x')A^\nu(x')U(x',Z) - \tfrac{\hbar c}{ig}U(Z,x')\left(\partial_{x'}^\nu U(x',Z)\right) \\ &= \int\limits_{z(s)=Z+(x'-Z)s} ds\, s\left[F_Z^{\mu\nu}\left(z(s)\right)(x'-Z)_\mu\right] \,, \end{aligned} \tag{2.28c}$$

$$F_Z^{\mu\nu}(x') := U(Z,x')F^{\mu\nu}(x')U(x',Z) \,. \tag{2.28d}$$

Antisymmetry of $F_{\mu\nu}$ leads immediately to

$$(x'-Z)_\mu A_Z^\mu(x') \equiv 0 \,, \tag{2.29}$$

the gauge condition of Fock-Schwinger (FS) gauge [20], supplemented by the boundary condition (see (2.28c))

$$A_Z^\mu(Z) = 0. \tag{2.30}$$

Obviously this gauge allows to reconstruct the gluon potentials from the field strength tensor (see (2.28c)). The idea is to identify the fixed point Z to which this FS-gauge is hooked up with the spatial argument x of the Wigner function. In this gauge all link operators along any radial direction away from the fixed point Z reduce to the identity operator since for arbitrary a, b due to the gauge condition (2.29) we have

$$\begin{aligned} U(Z+by, Z+ay) &= P\exp\Big[\frac{ig}{\hbar c}\int_{Z+ay}^{Z+by} dz_\mu A_Z^\mu(z)\Big] \\ &= P\exp\Big[\frac{ig}{\hbar c}(b-a)\int\limits_{z(s)=Z+ay+sy(b-a)} ds\, s\, y_\mu A_Z^\mu\left(z(s)\right)\Big] = 1 \,. \end{aligned} \tag{2.31}$$

We can now define a new quark Wigner operator

$$\tilde{W}_Z(x,p) = -\int \frac{d^4y}{(2\pi\hbar)^4} e^{-\frac{i}{\hbar}p\cdot y}\ \psi_Z(x-\tfrac{y}{2})\ \overline{\psi}_Z(x+\tfrac{y}{2})\ , \tag{2.32}$$

which obviously reduces to (2.17) for $Z = x$. As a candidate for the gluon Wigner operator we may try [19,21]

$$\tilde{\Gamma}_Z^{\mu\nu,ab}(x,p) = \int \frac{d^4y}{(2\pi\hbar)^4} e^{-\frac{i}{\hbar}p\cdot y}\ A_Z^{\mu,a}(x-\tfrac{y}{2})\ A_Z^{\nu,b}(x+\tfrac{y}{2})\ . \tag{2.33}$$

This is different from (2.22) in that it contains two derivatives less and thus has immediately the correct dimension to serve as a gluon number density (rather than energy-momentum density) in phase space. Its Lorentz-tensor decomposition will thus not contain the additional factors p^2 seen in (2.25), leading to a more well-behaved classical limit (in which $p^2 \to 0$).

If we now let the covariant kinetic drift operator $p\cdot[D(x),\ .\ ]$ act on the Wigner operator (2.17), we can write this in terms of $\tilde{W}_Z$ as

$$p\cdot[D(x),\hat{W}(x,p)] = p\cdot\left[\partial_x\tilde{W}_Z(x,p) + [D(Z), W_Z(x,p)]\right]_{|Z=x}\ . \tag{2.34}$$

The first term turns out to be rather simple: defining

$$F_{Z=x}^{\mu\nu}(x-\tfrac{i\hbar}{2}\partial_p) := \exp\left[-\tfrac{i\hbar}{2}\partial_p\cdot\partial_x\right]F_Z^{\mu\nu}(x)_{|Z=x}\ , \tag{2.35a}$$

$$A_{Z=x}^{\nu}(x-\tfrac{i\hbar}{2}\partial_p) := \left[\int_0^1 ds\, sF_{Z=x}^{\mu\nu}(x-s\tfrac{i\hbar}{2}\partial_p)\right]\left(-\tfrac{i\hbar}{2}\partial_{p,\mu}\right)\ , \tag{2.35b}$$

one finds [19]

$$\begin{aligned}
p\cdot\hbar\partial_x\tilde{W}_Z(x,p) = &\\
&+\tfrac{ig}{c}p_\mu\left\{A_Z^\mu(x-\tfrac{i\hbar}{2}\partial_p)\tilde{W}_Z(x,p) - \tilde{W}_Z(x,p)A_Z^\mu(x+\tfrac{i\hbar}{2}\partial_p^\dagger)\right\}\\
&-\tfrac{g}{2c}\left\{A_Z^\mu(x-\tfrac{i\hbar}{2}\partial_p)(\hbar\partial_\mu\tilde{W}_Z(x,p)) + (\hbar\partial_\mu\tilde{W}_Z(x,p))A_Z^\mu(x+\tfrac{i\hbar}{2}\partial_p^\dagger)\right\}\\
&+\tfrac{ig^2}{2c^2}\left\{A_Z^2(x-\tfrac{i\hbar}{2}\partial_p)\,\tilde{W}_Z(x,p) - \tilde{W}_Z(x,p)\,A_Z^2(x+\tfrac{i\hbar}{2}\partial_p^\dagger)\right\}\\
&+\tfrac{ig}{2c}\left\{\tfrac{\hbar}{2}\sigma_{\mu\nu}F_Z^{\mu\nu}(x-\tfrac{i\hbar}{2}\partial_p)\,\tilde{W}_Z(x,p) - \tilde{W}_Z(x,p)\,\tfrac{\hbar}{2}\sigma_{\mu\nu}F_Z^{\mu\nu}(x+\tfrac{i\hbar}{2}\partial_p^\dagger)\right\}.
\end{aligned} \tag{2.36a}$$

In a similar way one derives the generalized mass-shell condition

$$\begin{aligned}
(p^2-m^2c^2-\tfrac{1}{4}\hbar^2\partial_x^2)\tilde{W}_Z(x,p) = &\\
&-\tfrac{g}{c}p_\mu\left\{A_Z^\mu(x-\tfrac{i\hbar}{2}\partial_p)\tilde{W}_Z(x,p) + \tilde{W}_Z(x,p)A_Z^\mu(x+\tfrac{i\hbar}{2}\partial_p^\dagger)\right\}\\
&-\tfrac{ig}{2c}\left\{A_Z^\mu(x-\tfrac{i\hbar}{2}\partial_p)(\hbar\partial_\mu\tilde{W}_Z(x,p)) - (\hbar\partial_\mu\tilde{W}_Z(x,p))A_Z^\mu(x+\tfrac{i\hbar}{2}\partial_p^\dagger)\right\}\\
&-\tfrac{g^2}{2c^2}\left\{A_Z^2(x-\tfrac{i\hbar}{2}\partial_p)\,\tilde{W}_Z(x,p) + \tilde{W}_Z(x,p)\,A_Z^2(x+\tfrac{i\hbar}{2}\partial_p^\dagger)\right\}\\
&-\tfrac{g}{2c}\left\{\tfrac{\hbar}{2}\sigma_{\mu\nu}F_Z^{\mu\nu}(x-\tfrac{i\hbar}{2}\partial_p)\,\tilde{W}_Z(x,p) + \tilde{W}_Z(x,p)\,\tfrac{\hbar}{2}\sigma_{\mu\nu}F_Z^{\mu\nu}(x+\tfrac{i\hbar}{2}\partial_p^\dagger)\right\}
\end{aligned} \tag{2.36b}$$

The second term in (2.34), containing the covariant derivative with respect to the "gauge parameter" Z, describes the effects due to the gauge change that occurs when comparing the Wigner function (2.32) at neighboring points. It is given by [19]

$$
\begin{aligned}
[D(Z), W_Z(x,p)] = &+ \tfrac{ig}{\hbar c} \int_0^1 ds(1-s)\left[e^{-s\frac{i}{2}\hbar\partial_p\cdot\partial_x} F_Z^{\mu\nu}(x)\right]\left(-\tfrac{i}{2}\hbar\partial_{p,\mu} W_Z(x,p)\right) \\
&- \tfrac{ig}{\hbar c}\left(\tfrac{i}{2}\hbar\partial_{p,\mu} W_Z(x,p)\right) \int_0^1 ds(1-s)\left[e^{s\frac{i}{2}\hbar\partial_p^\dagger\cdot\partial_x} F_Z^{\mu\nu}(x)\right] .
\end{aligned}
\tag{2.37}
$$

Unlike (2.36), (2.37) does not admit a more economical rewriting in terms of gauge potentials. The somewhat unusual derivatives with respect to a "gauge parameter" are due to the fact that the physical Wigner operator, which is gauge covariant and in the classical limit has an interpretation as a probability density in phase space, is given *at each point* by (2.32) with $Z=x$. Thus, if we insist on expressing at each point the Wigner function by the simple bilinear expression (2.32), we have to readjust the gauge correspondingly when evolving the Wigner-function from one point to another. If we did not do so and kept the gauge (*i.e.* Z) fixed, we would still obtain an object which is mathematically well-defined, which contains the same amount of information as the physical Wigner function, and whose solution is completely determined by only the terms contained in equations (2.36); however, this object would have no intuitive physical interpretation at any point other than $x = Z$.

2.2.3 Semiclassical Expansion and Mean-Field Approximation

The major advantage of choosing FS gauge to represent the right hand side of (2.34) results from the fact that in this gauge all covariant spatial derivatives degenerate into *partial* ones, and that thus a straightforward Taylor expansion of the exponentials in (2.35) yields a systematic $\hbar$-expansion (which at the same time is a derivative expansion in the FS-gauge gluon fields). This exactly parallels the Wigner-Kirkwood expansion in non-gauge theories, and no additional restrictions on the gauge field configuration are required. On the other hand, the equations (2.36/37) are still gauge covariant, because the index Z reminds us that we can easily go over to any other gauge by reinserting explicitly the U-operators from the definitions (2.28).

In order to see the connection between the classical and quantum kinetic equations more clearly, the gluon fields should be separated into a mean-field and a fluctuating part. Then a *mean-field approximation* can be defined by replacing in the quark kinetic equation all gluon fields by their mean-field values; for the gluon kinetic equation, the gluon Wigner operator is first split into a mean-field part $\overline{\Gamma}_{\mu\nu}$ and a fluctuating part $\hat{G}_{\mu\nu}$, and then again in the equation of motion for $\hat{G}_{\mu\nu}$ all gluon fields are replaced by their mean-field values [15]. It is important to note that this procedure breaks the gauge invariance of the formalism: Elitzur's theorem forbids the appearance of a gauge-invariant vacuum expectation value $\langle A_\mu \rangle$, hence the separation $\hat{A}_\mu = \langle A_\mu \rangle + \hat{a}_\mu$ has to be gauge dependent. Since transformations of $\langle A_\mu \rangle$ from one gauge to another involve the coupling constant g, results in the lowest order of g (linear response approximation) can be expected to be still gauge invariant. However, at higher orders in g, keeping gauge invariance becomes a non-trivial problem also in the kinetic formulation. We will discuss below how this will affect the issue of the damping rate for QCD plasmons.

Writing $\hat{A}_\mu = \overline{A}_\mu + \hat{a}_\mu$ (where $A_\mu = \langle \overline{A}_\mu \rangle$) and denoting by $\overline{F}_{\mu\nu}$ and $\overline{\Gamma}_{\mu\nu}$ the contributions to the field strength tensor and to the gluon Wigner function coming entirely from the mean field $\overline{A}_\mu$ (i.e., all $\hat{A}$'s are replaced by $\overline{A}$'s), we can define the fluctuating part of the gluon Wigner operator $\hat{G}_{\mu\nu}$ by

$$
\hat{G}_{\mu\nu} = \hat{\Gamma}_{\mu\nu} - \overline{\Gamma}_{\mu\nu} . \tag{2.38}
$$

This has a non-vanishing expectation value $\langle \hat{G}_{\mu\nu} \rangle$ (because $\langle \hat{\Gamma}_{\mu\nu} \rangle \neq \overline{\Gamma}_{\mu\nu}$) which we call $G_{\mu\nu}$, the gluon Wigner *function*.

The definition of the mean field reads

$$\overline{D}_\mu \overline{F}^{\mu\nu} = \frac{g}{\hbar c}\left(\langle j^\nu \rangle_{\mathrm{q}} + \langle j^\nu \rangle_{\mathrm{g}}\right) , \tag{2.39}$$

with the average color currents generated by the distribution functions for the quarks and fluctuating gluons. For simplicity we will from now on leave off the bars from the mean fields.

In the lowest order of a semiclassical expansion of the exact equations of motion for the Wigner functions [6] we find

$$\begin{aligned} p^\mu \mathcal{D}_\mu(x) W(x,p) + \frac{g}{2c}\left\{p^\mu F_{\mu\nu}(x), \partial_p^\nu W(x,p)\right\} - \frac{ig}{2\hbar c}\left[S^{\mu\nu} F_{\mu\nu}(x), W(x,p)\right] \\ - \frac{g}{4c}\left\{D_\sigma(x)(S^{\mu\nu} F_{\mu\nu}(x)), \partial_p^\sigma W(x,p)\right\} = C(x,p) \end{aligned} \tag{2.40}$$

for the quarks, and

$$\begin{aligned} p^\sigma \tilde{\mathcal{D}}_\sigma G_{\mu\nu}(x,p) + \frac{g}{2c}\left\{p^\sigma \mathcal{F}_{\sigma\tau}, \partial_p^\tau G_{\mu\nu}(x,p)\right\} + \frac{g}{c}\left(\mathcal{F}_{\mu\alpha} G^\alpha{}_\nu(x,p) - G_\mu{}^\alpha(x,p)\mathcal{F}_{\alpha\nu}\right) \\ - \frac{i\hbar g}{2c^2}\left((\mathcal{D}_\sigma \mathcal{F}_\mu{}^\lambda)\partial_p^\sigma G_{\lambda\nu}(x,p) + \partial_p^\sigma G_{\mu\lambda}(x,p)(\mathcal{D}_\sigma \mathcal{F}^\lambda{}_\mu)\right) = C_{\mathrm{g}}(x,p) \end{aligned} \tag{2.41}$$

for the gluons. C and C_{g} are quark and gluon collision terms, defined by the contributions from the fluctuating parts of the gluon fields and by corrections of higher order in $\hbar$.

If one projects these equations onto their various color channels (by taking traces with the appropriate generators for color singlet, octet, etc.), and separates the various spin components (scalar, vector, axial vector, etc.) by either taking traces with the appropriate combinations of γ-matrices (quarks [17]) or by inserting the decomposition (2.25) (gluons), one can convince oneself [22] that one obtains an exact mapping onto the classical color moment equations of Sect. 2.1.3 (even including the spin coupling terms of Sect. 2.1.1 which were not taken into account in Sect. 2.1.3 to keep things simple).

2.2.4 Linear Response Theory and Plasma Oscillations

Equations (2.40/41) can be solved in linear approximation around global thermal equilibrium [23]. Splitting the quark Wigner function into its quark and antiquark contributions, f and $\tilde{f}$, we can write

$$\begin{aligned} f(x,p) &= n(p)\,\mathbf{1} + \delta f(x,p) \; ; \\ \tilde{f}(x,p) &= \tilde{n}(p)\,\mathbf{1} + \delta \tilde{f}(x,p) \; ; \\ G(x,p) &= n_{\mathrm{g}}(p)\,\mathbf{1} + \delta G(x,p) \; ; \end{aligned} \tag{2.42}$$

where $n(p)$, $\tilde{n}(p)$, $n_{\mathrm{g}}(p)$ are the usual Fermi-Dirac and Bose-Einstein distributions for quarks, antiquarks, and gluons. Since in global thermal equilibrium there are no color fields and thus a gauge can be chosen such that also the vector *potentials* vanish in equilibrium, $F_{\mu\nu}$ and A_μ can be considered as small quantities, too. The procedure is outlined in detail in [23,24].

Perturbing the equilibrium by a small external $F_{\mu\nu}$ and leaving out the collision terms in the kinetic equations, one finds response functions for both the longitudinal and transverse components [23] which are completely real in the region of timelike momenta $\omega^2 > k^2$ and have an imaginary part for $\omega^2 < k^2$ which is due to Landau (mean-field) damping. Both response functions have a pole in the timelike region, namely at

$\omega_L^2 = \omega_0^2 + \frac{3}{5}k^2 + \mathcal{O}(k^4)$ and $\omega_T^2 = \omega_0^2 + \frac{6}{5}k^2 + \mathcal{O}(k^4)$, where $\omega_0^2 = g^2T^2(N_f + 2N)/18$ is the QCD plasma frequency known from one-loop calculations in finite temperature QCD. These poles correspond to collective longitudinal and transverse plasma oscillations, and in the absence of collision terms they are completely undamped. The result is independent of the gauge (due to using a gauge invariant kinetic framework as a starting point and using an approximation scheme which to this order in the perturbation preserves the gauge invariance — see comments in the previous section). The above dispersion relations agree with the leading term from a high-temperature expansion for the poles of the one-loop finite temperature gluon propagator, which is also known to be gauge invariant.

Of course, there exist physical mechanisms which can contribute an imaginary part to the dispersion relation, like pair decay of the perturbation into massless quark-antiquark or gluon pairs, or higher order scattering processes. Depending upon the sign of this imaginary part, the plasm oscillations will be either damped or unstable. The underlying mechanisms are contained in the (so far neglected) collision terms which unfortunately have not yet been sufficiently analyzed. In the framework of finite temperature QCD, the imaginary part resides in the non-leading terms of a high-temperature expansion for the poles of the response function, which have recently been analyzed in a series of papers [25]. Unfortunately, the response function defining plasma oscillation has so far not been written down in a generally accepted, gauge invariant way, and as a result its imaginary part (and thus the plasmon damping rate) is plagued by gauge dependence. (This gauge dependence does not show up in the leading term of a high-temperature expansion, but crucially affects the next-to-leading term.) The different results obtained up to now do not even agree in sign!

Clearly this shows that there is a severe problem in our understanding the nature of QCD plasma oscillations within finite temperature QCD. The different approaches tried so far [25] include both old-fashioned linear response theory and the method of the

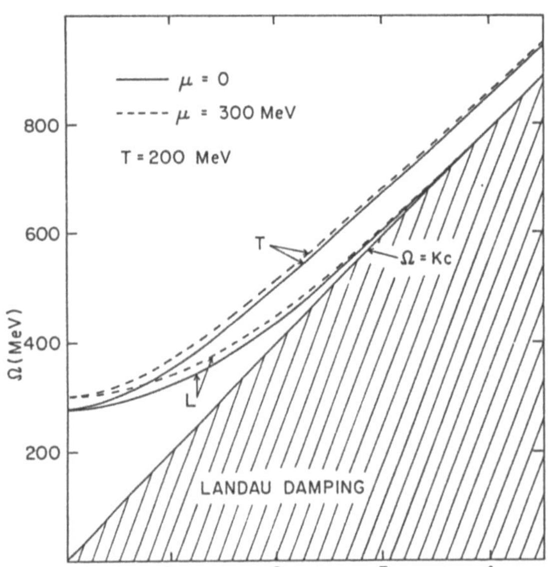

Fig. 4. The dispersion relation for longitudinal (L) and transverse (T) colored plasma modes. The calculation was done for finite quark masses ($m_{u,d}$=10 MeV, m_s=150 MeV) and α_s=0.3.

thermal partition function with sources ("effective action"). Within finite temperature QCD these methods have not reached mutual consistency, *i.e.* it is not known how they translate into each other (which might shed some light onto the origin of the discrepancies). We believe that starting from the gauge invariant kinetic framework outlined above, the discrepancies may be eventually resolved (see also [18]). For this to happen, the kinetic equations have to be analyzed beyond the classical limit; in particular, the effect of correlations has to be investigated, because they will be responsible for entropy generation and thus also for plasmon damping. The major stumbling block to preserving gauge invariance in this approach is the separation of gluon fluctuations from the mean field, and more work is needed to more clearly understand this step. However, we are am confident that these problems can be overcome, thereby clarifying the unsatisfactory present situation.

As affected party in the question of plasmon damping [25], we have, of course, a prejudice as to the most reliable answer for the damping rate. In the first paper of Ref. [25] we found a positive damping rate for color perturbations in the plasma, which is due to pair decay into quark-antiquark and gluon pairs and which translates into an equilibration time scale of order 1–2 fm/c. This is not very short, but perhaps short enough on the total time scale of the nuclear collision to justify the use of local equilibrium concepts during its later stages. Until more practical versions of the kinetic approach are available, we will thus take this as preliminary justification for a thermo- and hydrodynamic description of the heavy-ion dynamics.

Lecture 3: The Hadronization Phase Transition

3.1 Equation of State and Phase Diagram

Our analysis of the further evolution of the collision zone will be very rough and phenomenological; in particular, since it is much too early to worry about details of the nuclear equation of state in the temperature and density regions probed by the new experiments, we will confine ourselves to the discussion of just one such EOS, whose only interesting feature is that it provides a phase transition to quark matter at a reasonable critical temperature and for reasonable baryon density. Much more elaborate (and realistic) nuclear equations of state, which, although differing in detail, do the same thing, are on the market, and anyone who is not happy with the version used here is welcome to pick his own. Our conclusions will not in a qualitative way be changed by that.

We describe [26] the hadronic phase as a hadron resonance gas with hard core repulsion (implemented through a "finite proper volume correction" [27]), taking into account all baryons and mesons from the SU(3) ground state octets plus the Δ-resonance, the Ω^- and $\overline{\Omega}^+$ baryons, the η' and the ϕ mesons. Above the transition, our model uses a free gas of massless up- and down-quarks, massive (m_s = 150 MeV) strange quarks, and gluons, subject to a vacuum ("bag-") pressure B.

This equation of state implements both thermal and chemical equilibrium. For the discussion in Lecture 5 of the shape of the p_T-spectra we need only local thermal equilibrium. Chemical equilibrium becomes a serious issue only when discussing particle abundancies within this EOS. We will later present arguments that chemical equilibrium can be assumed to exist in the QGP phase, but not *a priori* in a hadronic fireball. The question of chemical equilibration will be discussed in detail in Lecture 4. Until then, we

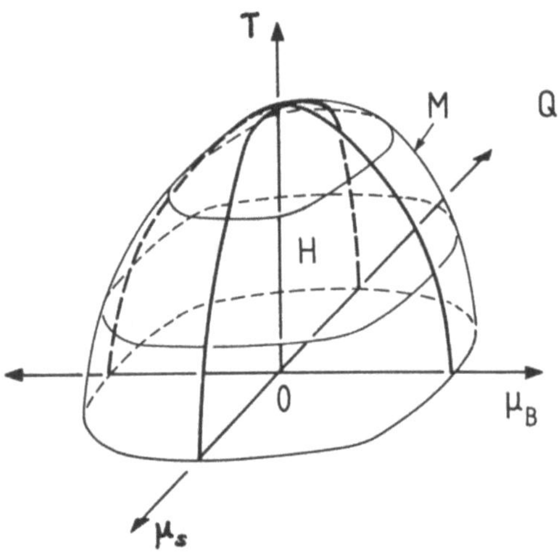

Fig. 5. Generic phase diagram in the space spanned by T, μ_B and μ_s for the quark-matter to hadron-gas phase transition. The inside of the igloo contains the hadronic phase, the igloo surface represents the mixed phase, and the outside is the quark-gluon plasma phase.

should be very careful to avoid unjustified assumptions about the existence of chemical equilibrium when talking about particle production rates.

In thermodynamic equilibrium the total system is described by three parameters, the temperature T, the baryon chemical potential $\mu_B(= 3\mu_q)$, and the strange chemical potential $\tilde{\mu}_s$, which control its total energy, baryon number, and strangeness. In the hadron gas phase the chemical potentials of the differrent hadrons are expressed in terms of μ_B and $\tilde{\mu}_s$ as $\mu_i = b_i\,\mu_B^H + s_i\,\tilde{\mu}_s^H$ for a particle with baryon number b_i and strangeness s_i.

The thermodynamic conditions identifying the equilibrium phase coexistence surface are $T_H = T_Q$ (thermal equilibrium), $P_H = P_Q$ (mechanical or pressure equilibrium), and $\mu_B^H = \mu_B^Q = 3\mu_q^Q$ as well as $\tilde{\mu}_s^H = \tilde{\mu}_s^Q$ (chemical equilibrium). These conditions yield a 2-dimensional phase transition surface in the 3-dimensional $(T, \mu_B, \tilde{\mu}_s)$ space which generically has the form of an igloo (Fig. 5). (We often prefer to use instead of $\tilde{\mu}_s$ the total chemical potential $\mu_s = \mu_q - \tilde{\mu}_s$ of the strange quarks which obviously carry both strangeness and baryon number.)

Fixing the net strangeness per baryon s/A to a certain value, e.g. zero as in the initial state of a nuclear collision, determines another 2-dimensional surface in this space. This is shown in Fig. 6; here H, Q, and M denote the hadronic, quark-gluon plasma, and mixed phase for a system with zero strangeness, and the latter is obtained from the intersection of the $s/A = 0$ surface with the phase coexistence igloo of Fig. 5. Due to the presence of free strange quarks in the plasma, zero strangeness (*i.e.* $\rho_s = \rho_{\bar{s}}$) implies $\mu_s^Q = 0$; on the other hand, in the hadron gas strange quarks come always together with light quarks (e.g. in kaons or hyperons), and the condition of zero strangeness involves both μ_s^Q and μ_q^Q as shown in Fig. 6 by the H-surface [28].

By construction, the phase transition is of first order: as is seen in Fig. 7, the baryon density, energy density, and entropy per baryon show strong decreases as the system passes from the plasma to the hadron gas side.

The discontinuity of the specific entropy S/A is particularly important: it is due to the liberation of gluons which in the plasma phase contribute a large fraction (of order 50%) to the entropy while not contributing to A. Since dynamically during expansion the entropy per baryon cannot decrease, the system cannot hadronize at constant T and μ_B

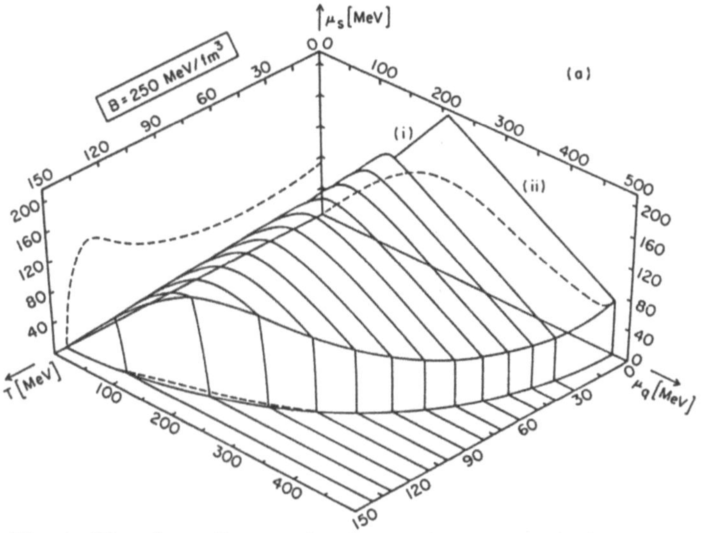

Fig. 6. The phase diagram for the quark-matter to hadron-gas phase transition for a system with zero net strangeness, for B=250 MeV/fm^3 [28].

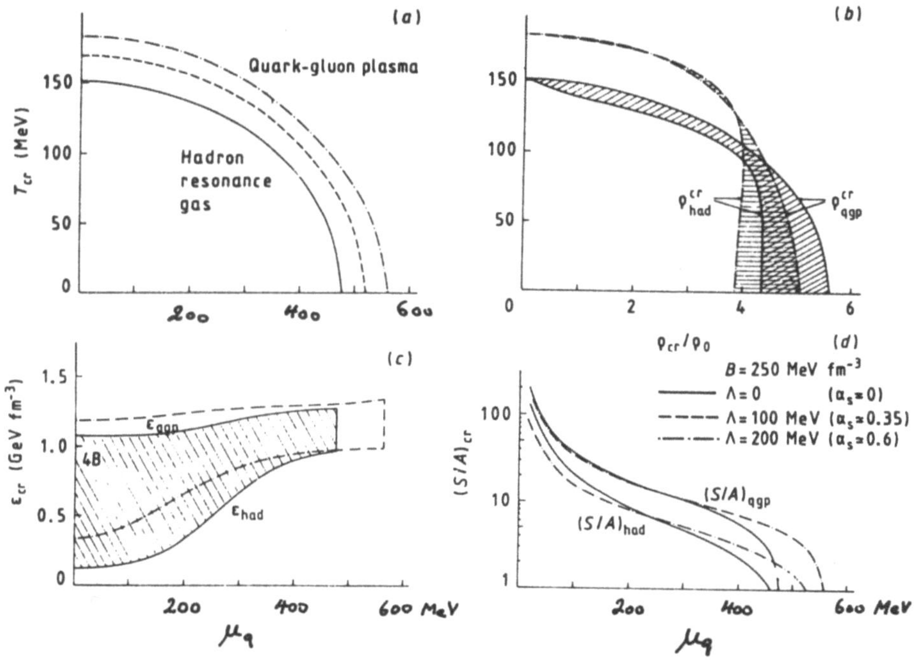

Fig. 7. Typical curves for the critical values for (b) baryon density, (c) energy density, and (d) entropy per baryon along the critical line in the T-μ_q diagram (a). (For simplicity of representation, in (a) only the projection of the phase coexistence surface of Fig. 6 on the T-μ_q plane is shown, and the small change of μ_q across the transition has been neglected.) Different curves indicate various choices for the strong coupling constant by which perturbative interactions between quarks and gluons in the plasma phase are taken into account. Most of the changes induced by varying α_s could be reabsorbed by changing the vacuum pressure B which here has been chosen as B=250 MeV/fm^3; therefore in all other calculations presented here α_s has been set to zero.

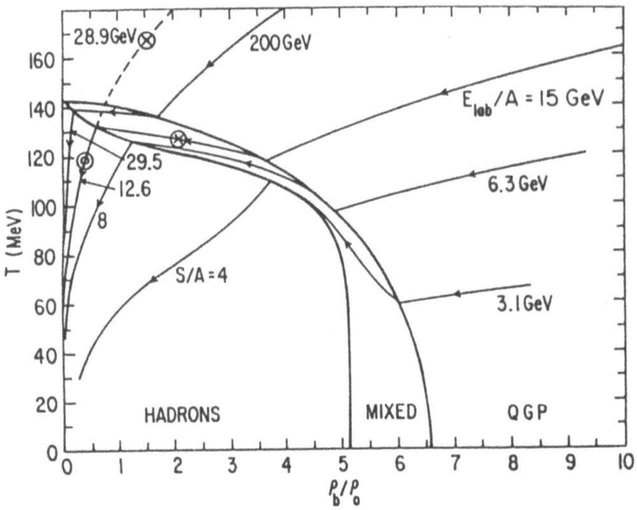

Fig. 8. Isentropic expansion trajectories for a hadronizing quark-gluon plasma, for several values of S/A. The quoted beam energies stem from a 1-dimensional shock calculation. The dashed line describes an expanding hadron gas (no phase transition). The crosses and circles indicate freeze-out points for kaons and nucleons (see text). (From [31].)

(under these conditions S/A is always lower in the hadron gas phase than in the plasma phase [26], by factors 2 to 5), but has to reheat and expand towards smaller baryon densities while crossing the mixed phase [29,30]. This is shown in Fig. 8 where lines of constant S/A are plotted in the T-ρ_B plane.

3.2 Chemical Composition and Strangeness Abundance Near the Phase Transition Surface

From Fig. 8 one concludes that a comparison of the composition of the two phases at constant T and μ makes little sense [30,32] since these states cannot be dynamically connected and cannot both be the dynamical origin of a given (say, experimentally determined) final state. It is very important to keep this fact in mind when discussing hadronic abundancy ratios: since most of the interesting hadronic abundancies, in particular those of all antibaryons, increase exponentially with increasing temperature and decreasing baryon chemical potential, they are underestimated in the hadronic phase by orders of magnitude if the hadron gas is taken to be at the same T and μ_B as the plasma at the beginning of hadronization, rather than at the same S/A.

As an example we show in Fig. 9 the strange quark content per unit entropy along the phase coexistence surface, comparing hadron gas and quark-gluon plasma at equal values of S/A [30]. It is seen that the strangeness content is never more than about 20% higher in the plasma phase (see also [32]); indeed, in the high baryon density (small S/A) section of the phase transition surface, an equilibrated hadron gas contains relatively more strangeness than the corresponding quark-gluon plasma.

It was argued [30] and later verified by a kinetic calculation [33] that the same processes that guarantee entropy conservation during hadronization also drive the system continuously towards hadronic chemical equilibrium. These processes essentially convert gluons into $q\bar{q}$ pairs [34], thereby transferring the gluonic part of the entropy to the quark and thus hadron sector. As a result, the level of strangeness in general and abundancy

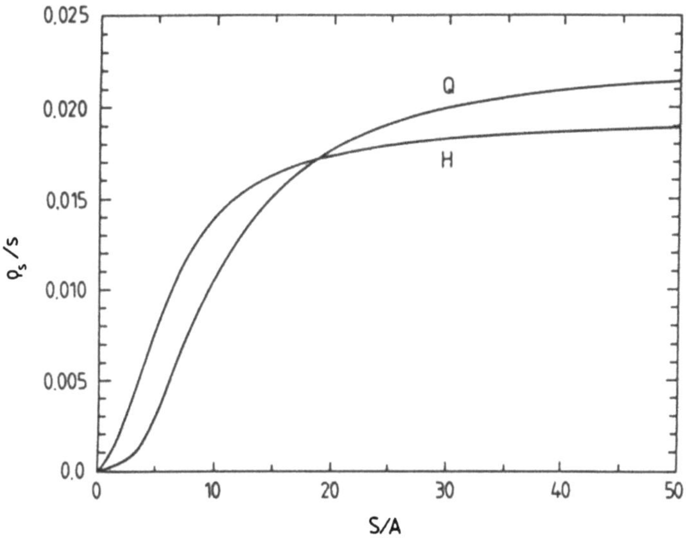

Fig. 9. Strangeness contents in a hadron gas (H) and a quark-gluon plasma (Q) at the critical surface, as a function of S/A [30].

ratios of strange to non-strange particles in particular from a hadronized quark-gluon plasma are likely to be rather close to those from a chemically equilibrated hadron gas with the same specific entropy. While the first part of this statement seems unavoidable, it is possible that strong non-equilibrium aspects of the hadronization process on a microscopic level lead to a different distribution of the produced strange quarks among the hadrons in the two scenarios, thereby giving unusual abundancy ratios [34]. One such scenario has recently been investigated [35] in connection with antibaryon production and will be discussed in the next Lecture. The final resolution of this issue will have to come from experiment.

Thus a high level of strangeness from nuclear collisions would primarily indicate chemical equilibrium in the fireball. To use this as an argument in favor of QGP formation, one must further argue that chemical equilibrium is not expected in a purely hadronic environment but easily achieved in a (hadronizing) quark-gluon plasma. Such arguments exist [34], but they rely on thermal momentum distributions, which implies that in a hadron gas many inelastic strangeness producing channels are effectively closed by mass thresholds. It is necessary to investigate the possible influence on hadronic strangeness production of large, not yet thermalized relative momenta in the initial pre-equilibrium stages of the collision [36], before dismissing the possibility of chemical equilibration in a purely hadronic phase and trusting chemical equilibrium as indirect evidence for QGP.

3.3 Equilibrium Hadronization and $s\bar{s}$ Separation in the Mixed Phase

As we see from Fig. 6, the strange chemical potential changes nearly discontinuously [28,37,38] during hadronization. As a consequence, the values of μ_s in the mixed phase (which smoothly interpolate between the plasma and hadron gas limits) cannot correspond to zero strangeness in either subphase: indeed, in the mixed phase the partial volume still occupied by quark matter contains more s than $\bar{s}$ quarks, while in the hadron subphase hadrons containing strange antiquarks are more abundant than their antiparticles [28,38]. (The system as a whole remains strangeness neutral.) For systems with a

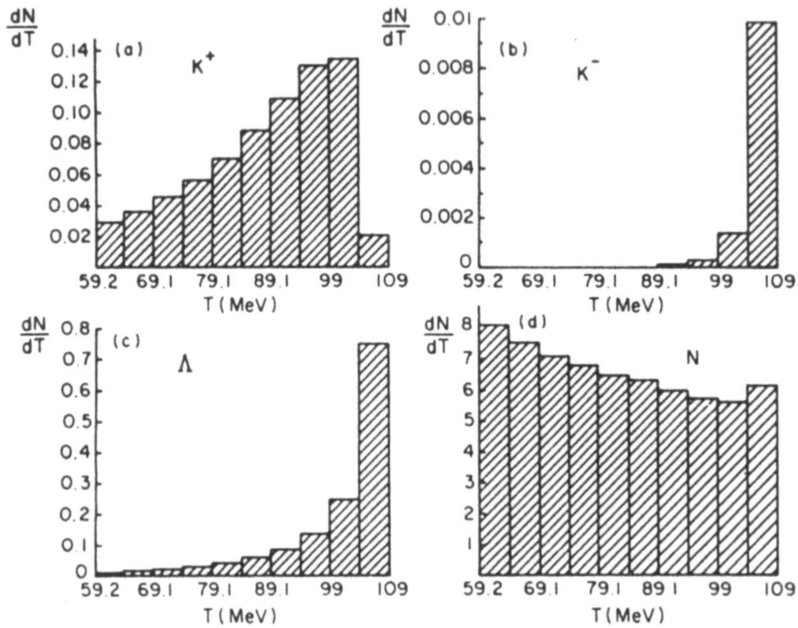

Fig. 10. Hadronization rates along a constant entropy contour with $S/A=4$, for a system with $A=500$ [31].

finite net strangeness per baryon s/A this s-$\bar{s}$ separation occurs in qualitatively the same way [39], and an analogous effect is well known from the process of distilling alcohol.

There is a very simple intuitive explanation for this [28,31]: while in the plasma just before hadronization there are as many s as $\bar{s}$ quarks, light quarks are much more abundant than light antiquarks, due to the non-zero baryon density. Thus it is (at least initially) easier to form a kaon ($K^{+,0} \equiv (q\bar{s})$) than an antikaon ($\overline{K}^{-,0} \equiv (s\bar{q})$), and since this is the dominant channel to hadronize strangeness (the hyperons being suppressed by their large mass), the above separation effect ensues.

One can study this in more detail by following the system during hadronization along a constant entropy line and asking at each point how many particles of a given species have already hadronized. In Fig. 10 one sees that K^+ start to hadronize immediately and keep doing so throughout the mixed phase (with a drop towards the end because the $\bar{s}$ quarks are eventually exhausted), while K^- and hyperons only hadronize towards the end of the mixed phase (or perhaps don't hadronize at all, but form little blobs of strange matter as suggested in [39]).

Fig. 10d also shows that the system is trying to hadronize as much baryon number as possible right from the start. This is quite different from the hadronization phase transition in the early universe where it was argued [40] that baryon number tends to stay behind in the quark subphase. The origin for the difference in the two cases is the different entropy balance: as we had shown, in our case S/A is larger in the quark-gluon plasma than in the hadron gas. This situation is inverted in the case of the early universe, where also the leptons and photons are in chemical equilibrium with the strongly interacting particles and thus contribute to the entropy balance, which in turn determines the hadronization trajectory. As a consequence, in the early universe the system actually *cools* and moves towards *larger* values of μ_B (however, not to larger baryon density!) during equilibrium hadronization, quite in contrast to the behaviour shown in Fig. 8.

Lecture 4: Chemical Equilibration in Relativistic Nuclear Collisions

4.1 K/π Ratios

In [41] it was suggested that enhanced production of strangeness as a consequence of QGP formation should reflect itself in an increase of the K/π ratio in nuclear collisions when compared with pp collisions. We have argued in the previous lecture that for such a phenomenon to happen it is sufficient that the system approaches chemical equilibrium, and it is not necessary that this equilibration has its origin in QGP formation. Since in pp collisions strangeness is known to be produced considerably below chemical saturation levels, a tendency towards chemical equilibration would in any case signal *new physics*, be it because of plasma formation or as a result of other new processes occuring inside the dense and hot collision zone.

From detailed studies at the ISR [42] we know that in the relevant energy region at central rapidity the $K\pi$ ratios for both charge channels are equal and close to 10%. (Further into the target and projectile fragmentation regions the K^+/π^+ ratio decreases very slightly, while the K^-/π^- ratio drops by about a factor of 3 – see Fig. 7.10 in [42].) In contrast, the E802 collaboration at Brookhaven [43] found in central Si+Au collisions at 14.5 A GeV beam energy at central rapidity $K^+/\pi^+ = 0.192 \pm 0.03$ (*i.e.* twice as big as in pp collisions) and $K^-/\pi^- = 0.036 \pm 0.008$ (*i.e.* nearly a factor 3 smaller than in pp collisions).

This goes exactly in the right direction, if one supposes complete stopping of the Si-projectile by the Au-target (which was demonstrated in the experiment [44]) and the formation of a baryon-rich ($\mu_B \gg 0$) fireball in chemical equilibrium: since at finite baryon density light antiquarks are suppressed relative to light quarks (while strange quarks and antiquarks are equal in number due to strangeness conservation), ($\bar{s}q$) pairs (kaons) should be more abundant than ($s\bar{q}$) pairs (antikaons), with no difference between π^+ and π^-. Thus with increasing light quark chemical potential the kaon abundance rises relative to the pions, while the antikaon abundance drops.

In Fig. 11 we show, for a thermally and chemically equilibrated hadron resonance gas as discussed before, those regions in the phase diagram that correspond to fixed values for the K/π ratios. We also include the π^+/p ratio which was also measured by E802 and found to be close to 50% at midrapidity [43]. We see that the situation $(K^+/\pi^+) = (K^-/\pi^-) = 0.1$ observed in pp collisions at midrapidity corresponds to $\mu_B = 0$, *i.e.* a baryon-free central region, and a temperature of about 110 MeV. This temperature is much lower than the typical temperature of about 160 MeV extracted from the transverse momentum spectra in the Hagedorn fireball picture [45], which is nothing but a rephrasing of the statement made before that strangeness is produced below chemical saturation in those collisions. As μ_B increases, the lines for constant K^+/π^+ move to smaller temperatures (implying a rising K^+/π^+ ratio at fixed T), while the lines for constant K^-/π^- move towards higher temperature (consistent with a decreasing K^-/π^- ratio at constant T). The lines corresponding to the measured values from Si+Au collisions cross at $T \simeq 100$ MeV and $\mu_B \simeq 600$ MeV.

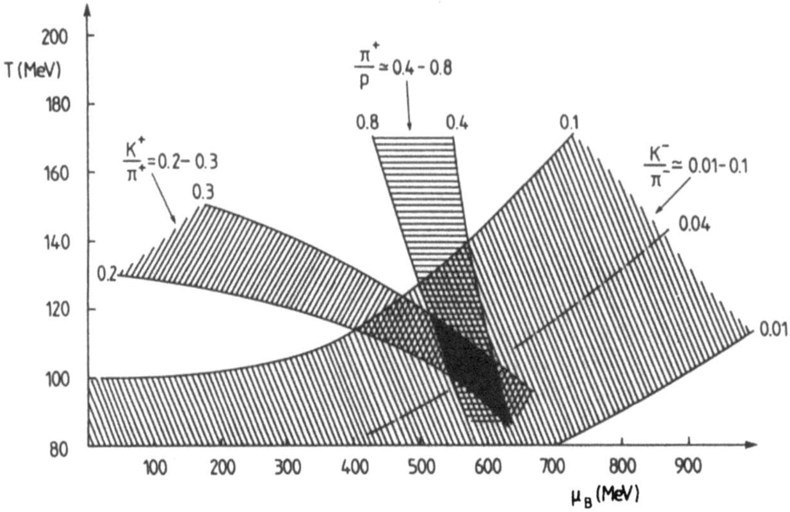

Fig. 11. Lines of constant K^+/π^+, K^-/π^-, and π^+/p ratios in the phase diagram for a hadron resonance gas in thermodynamic equilibrium.

The surprising confirmation of this interpretation in terms of a chemically equilibrated hadron gas comes from the fact that the line corresponding to the measured π^+/p ratio of $\simeq 0.5$ also crosses these lines at nearly the same point. Thus the particle ratios measured in this experiment seem to be perfectly consistent with a thermally and chemically equilibrated baryon-rich hadronic fireball freezing out at a temperature of around 100 MeV and a baryon density of $\simeq 0.3\rho_0$ ($\mu_B \simeq 600$ MeV)!

Such an approach to chemical equilibrium is a very interesting effect and was quite unexpected on the basis of what many people thought they knew about final state interactions in a hadron gas. In the next subsection we will give further evidence for this effect from other experiments. In light of the possibility discussed above that this might be an indirect signature for an intermediate QGP phase in these collisions it is certainly necessary to work towards a more detailed understanding of the dynamics of the equilibration process itself. Until then these experimental findings are at least very promising.

4.2 ϕ-Meson Production in Relativistic Nuclear Collisions

Asher Shor [46] first suggested that a strong enhancement of the ϕ/ω ratio above the value of $\approx \frac{1}{20}$ observed in pp collisions from $\sqrt{s} = 7$ to $\sqrt{s} = 53$ GeV should be a rather clean signature of QGP formation. Let me give a short and improved presentation of his arguments: In Ref. 10 the ϕ/ω ratio was estimated by calculating the density ratio between $s\bar{s}$ pairs (determining the coalescence rate into ϕ) and light $q\bar{q}$ pairs (determining the coalescense rate into ρ^0 and ω), using the thermal and chemical equilibrium values in the QGP:

$$\frac{\langle\phi\rangle}{\langle\omega\rangle} \approx \left(\frac{n_s n_{\bar{s}}}{n_q n_{\bar{q}}}\right)_{\text{QGP}} . \tag{4.1}$$

However, recent progress in our understanding of the chemistry of the hadronization process make this estimate look somewhat too high, by a factor 2–3. The reason is (see Sec. 3.2) that during hadronization more quark-antiquark pairs are created from gluons, and that most ($\sim 85\%$) of these $q\bar{q}$ pairs are light ones. Obviously, by this process the denominator in (4.1) is more strongly affected than the numerator, and therefore entropy conservation during hadronization necessarily leads to a dilution of the ratio (4.1). As argued before, a better estimate of the result for this ratio at the end of hadronization is a value corresponding to a thermalized hadron resonance gas close to chemical equilibrium, such that from a QGP we expect

$$\frac{\langle\phi\rangle}{\langle\omega\rangle} \approx \frac{\int_0^\infty \frac{p^2 dp}{\exp[\sqrt{p^2+m_\phi^2}/T_c]-1}}{\int_0^\infty \frac{p^2 dp}{\exp[\sqrt{p^2+m_\omega^2}/T_c]-1}} , \tag{4.2}$$

where T_c is the temperature of the hadronization phase transition. Values for this ratio range from 1/3 to 1/2 for T_c between 150 and 250 MeV. These ratios may be slightly enhanced by final state absorption effects in the hadron gas before chemical freeze-out: while the absorption of ϕ is hindered by a small absorption cross section (about 8 mb on nucleons and even less on mesons) and it is thus reasonable to assume that the number of ϕ's produced from the plasma freezes in at the point of hadronization, the interaction of the ρ and ω with the surrounding resonance gas is much stronger. They may remain in chemical equilibrium until the hadron gas has cooled down to a lower chemical freeze-out temperature T_f, leading to a reduction of their abundance. However, chemical kinetic simulations of the hadronization process [33,34] indicate that for most particle species chemical freeze-out occurs soon after completion of the hadronization, such that these final state effects may be quite small.

We thus conclude that an enhancement of $\langle\phi\rangle/\langle\omega\rangle$ by a factor 6–10 above the *pp*-value of $\frac{1}{20}$ is the most optimistic scenario we can expect even in the case of QGP formation in every single collision. In the present experiments it is unlikely that this highest possible enhancement value could be obtained since most probably only a mixed phase is reached (leading to incomplete chemical equilibration of the strange quarks in the plasma subphase), and even that only in a fraction of all events.

In the light of these arguments the recent observation by the NA38 collaboration [47] of an increase of the $\phi/(\omega+\rho^0)$ ratio in central 200 A GeV O+U and S+U collisions by about a factor 3 over the value measured in the same experiment in 200 A GeV p+U collisions appears as a very interesting effect. This is particularly true since the same collaboration found a quite opposite effect for the J/ψ meson (namely a suppression by a factor of two under the same kinematic conditions [48]), and it was shown [47] that ρ^0 and ω mesons show neither of these effects.

It is easy to "explain" all of these effects with the QGP hypothesis: ρ and ω mesons are always produced with chemical equilibrium abundancies, irrespective of whether a QGP or only a hadron gas is formed. ϕ mesons are enhanced in central nuclear collisions, because the probability to create QGP blobs increases with centrality, and QGP blobs lead to chemical equilibration of strangeness. The J/ψ is suppressed because it cannot bind inside the QGP, due to screening of the color potential [49]. The crucial question is therefore: is this explanation unique, or can the observed features also be understood in a hadronic scenario?

In [50] we analyzed the effects of secondary vector meson production and absorption caused by the rescattering of hadrons in the central rapidity region which were produced in the primary nucleon-nucleon collisions. We set up rate equations for the secondary production processes of the type $i+j \to V+X$ and absorption processes of the type $V+l \to X$,

$$\frac{dN_V}{d^4x} = \sum_{i,j} \langle\sigma v\rangle_{i,j}^{V,X}\, \rho_i(x)\, \rho_j(x) - \sum_l \langle\sigma v\rangle_{l,V}^{X}\, \rho_l(x)\, \rho_V(x) \, , \tag{4.3}$$

where the ρ_i are particle densities for all the species occurring in the fireball, and the σ_{ij}'s are the relevant inelastic cross sections suitably averaged over the particle momentum distributions. In a cylindrically symmetric environment with boost-invariant longitudinal expansion, the densities can be reexpressed in terms of the measured rapidity densities dN_i/dy, and, with a few not very restrictive simplifying assumptions, a simple analytical expression for the asymptotic rapidity density for vector mesons can be found [50]:

$$\frac{dN_V(b,\tau)/dy}{dN_V^{NN}/dy} = N_{part}(b)\left[e^{-A_V(b,\tau)} + R_V(1 - e^{-A_V(b,\tau)})\right] . \tag{4.4}$$

Here dN_V^{NN}/dy is the vector meson density produced in primary NN collisions, τ is the total proper time available for rescattering (*i.e.* until freeze-out), and N_{part} is the impact parameter dependent number of participant nucleons in the collision. The absorption factor

$$A_V(b,\tau) = \frac{dN_{ch}}{dyS_{eff}}(b)\ln\left(\frac{\tau}{\tau_0}\right)\sum_l \langle\sigma v\rangle_{lV}^X\, \alpha_l \tag{4.5}$$

(where S_{eff} is the effective transverse overlap area in the collision, and α_l is the fraction of the total charged multiplicity contributed by particle species l) depends on impact parameter through the multiplicity density of charged particles created by the primary collisions, and contains information on the particle density and the total proper lifetime τ of the fireball. It is hard to calculate reliably (due to the many unknown absorption cross sections under the sum) and will be estimated using the experimental results on J/ψ suppression. R_V is the ratio of secondary production and absorption rates:

$$R_V = \frac{\sum_{i,j}\langle\sigma v\rangle_{ij}^{VX}\, \rho_i\, \rho_j}{\sum_l \langle\sigma v\rangle_{lV}^X\, \rho_l\, \rho_V} , \tag{4.6}$$

where all densities are those produced in the primary NN collisions. In [50] we estimated that $R_\psi \simeq 0$ (due to the high mass threshold secondary J/ψ production is suppressed), $R_\rho \simeq R_\omega \simeq 1$ (light non-strange vector mesons are produced at chemical equilibrium abundancies even in NN collisions), and $R_\phi \sim 3-4$ (which is essentially the ratio between hadronic absorption cross sections for non-strange vector mesons and the ϕ whose absorption in the dominant channels is OZI suppressed).

With $R_\psi = 0$ the absorption factor A_ψ can be expressed for a given value of produced transverse energy E_T (*i.e.* for a given impact parameter b) via the experimental [48] J/ψ suppression in this E_T window. The absorption factors for other vector mesons are given as a power of the one for the J/ψ, the exponent being just the ratio of absorption cross sections. Using experimental data, the ϕ absorption cross section is found to be about 3–4 times bigger than the J/ψ absorption cross section, which fixes the remaining parameters in (4.4). We thus find [50]

$$\left(\frac{dN_\phi/dy}{dN_{\omega+\rho^0}/dy}\right)[E_T]\Big/\left(\frac{dN_\phi^{NN}/dy}{dN_{\omega+\rho^0}^{NN}/dy}\right) = R_\phi + \left(S[E_T]\right)^{\gamma_\phi}(1 - R_\phi) , \tag{4.7}$$

with R_ϕ and γ_ϕ both being in the range 3–4, and $S[E_T]$ being the experimental J/ψ suppression as a function of E_T. Inserting the fit for $S[E_T]$ shown in Fig. 12a into (4.7) we get the curves in Fig. 12b for the $\phi/(\omega+\rho^0)$ ratio as a function of E_T. One sees that rather good qualitative agreement is obtained. Since the ϕ is produced at a very low level in the primary collisions (much below chemical equilibrium), secondary production (which depends quadratically on total multiplicity [50]) wins over absorption (which rises

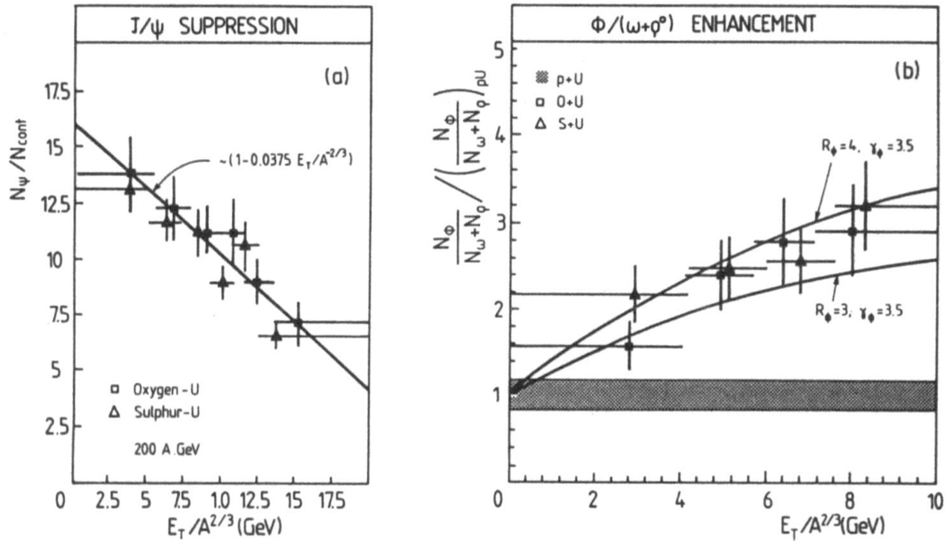

Fig. 12. (a) Linear fit to the J/ψ suppression as a function of transverse energy measured by NA38. (b) $\phi/(\omega+\rho)$ enhancement as a function of E_{T} from O+U and S+U collisions, normalized to p+U events. Data are from NA38, the calculation uses eq. (6) with the fit shown on the left. (From [50])

only linearly with multiplicity), and a sizeable enhancement of ϕ mesons by secondary processes appears possible.

This analysis depends crucially on the postulated link between J/ψ absorption and ϕ enhancement: If the former is due to final state rescattering, we are also able to explain the latter in this way. Whether the first part is really true, is still under debate (for example, it requires very high particle densities in the fireball where a hadronic picture begins to become doubtful). If it is, then the new ϕ data appear not to require QGP formation either, but indicate a well-developed tendency towards chemical equilibration of strangeness via hadronic processes.

4.3 Anti-Baryon Production in Relativistic Nuclear Collisions

There are two alternative approaches to baryon structure and dynamics. One is the naive quark model, according to which baryons are made out of 3 nonrelativistic constituent quarks: $m_{\mathrm{B}} \simeq 3m_q$, $m_q \simeq 300$ MeV. The other is the Skyrme model [51], based on approximate chiral symmetry with current quark masses $m_{u,d} \ll \Lambda_{QCD}$, which views baryons as topological solitons in an effective low-energy theory of pseudoscalar mesons. The NQM is very successful for calculating the static properties and spectrum of hadrons, and has inspired most models of baryon production. Its philosophy was implicit in the suggestion several years ago [26,52] to study antibaryon and antinuclei production in heavy-ion collisions as a signature for the large antiquark densities implied by quark-gluon plasma (QGP) formation. However, the NQM has never been derived from QCD for light quarks, and it does not explain naturally the spin content of the proton [53] as revealed by the EMC [54]. The Skyrme model has had some successes with the spectrum and static properties of baryons [55], provides a good description of their internal structure [56], explains naturally the spin content of the proton [57], and has the cardinal merit of being justified in QCD for light quarks [51]. We will now discuss [34] a model of baryon production inspired by the Skyrme model, which is analogous to the Kibble mechanism [58] for monopole or string formation in the Early Universe.

The basic idea is to view the hadronization of an initial quark-gluon state like a phase transition, with the formation of independent domains in which local order parameters adopt vacuum expectation values independently. One of these order parameters, the quark condensate $\langle 0|\bar{q}q|0\rangle$, can take a topologically non-trivial set of values, and defects may form when orientations in adjacent domains have a topological mismatch. We interpret these topological defects as baryons, and topologically trivial inhomogeneities may correspond to pseudoscalar mesons (cf. spin waves in ferromagnets). We apply this approach to the hadronization of a thermalized quark-gluon plasma (QGP), concentrating for simplicity on the initially baryon-free central rapidity region.

The most attractive feature of this approach, compared with previous models for QGP hadronization [26,33,34], is that it in principle allows for a strong violation of chemical equilibrium during the hadronization process, *i.e.* (anti-)baryon production might exceed the limits set by a thermally and chemically equilibrated hadron resonance gas. All existing models are based on the notion that (anti-)baryons arise via coalescence of 3 (anti-)quarks and are thus more or less directly sensitive to the thermal (anti-)quark distributions and chemically equilibrated (anti-)quark abundances in the initial QGP state. The optimistic original suggestions, based on the combinatoric recombination model [34,59], that the high antiquark densities in the QGP might lead to antibaryon [34] and antinucleus [52,26] abundances greatly exceeding the values in a chemically equilibrated hadron gas, were already corrected in Sect. 4.2 where it was pointed out that entropy conservation, and the implied production of additional $q\bar{q}$ pairs during hadronization, leads within any combinatorical model to abundancy levels very close to chemical equilibrium in a hadron resonance gas.

In contrast, the topological baryons of the Skyrme model do not "count" quarks, and since (according to the model assumptions in [35,60]) all partons (quarks, antiquarks and gluons) are equally effective as seeds for the development of the chiral $\bar{q}q$ condensate, the number of created topological defects is essentially independent of the chemical composition of the QGP. The only constraints on baryon-antibaryon formation are of global or topological nature (*i.e.* that the chiral condensate outside the collision zone has a topological winding number corresponding to the total baryon number contained inside, and that the total energy of the collision zone is conserved while its entropy can only increase), or arise from the explicit breaking of $SU_f(3)$ chiral symmetry by the finite (~ 150 MeV) strange quark mass, which will affect the ratio of strange to nonstrange baryons subject to the constraints imposed by the strange to light quark ratio in the original QGP phase. In this section we study whether these features might lead to a breaking of chemical equilibrium between (anti-)baryons and mesons by the hadronization process itself, and argue that an experimental study of the systematics of antibaryon production in nuclear collisions, by checking chemical equilibration also in the antibaryon sector, might provide indirect information on the validity of our topological approach to QGP hadronization.

To set the scene, let us perform a first rough estimate of the baryon production rate disregarding any energy and entropy constraints as well as quark mass effects. We follow the QGP to the beginning of the confinement phase transition (*i.e.* down to a temperature T_c somewhere between 160 and 200 MeV) and consider the thermally distributed quarks, antiquarks and gluons there as seeds for a developing chiral condensate. For a completely equilibrated QGP we checked in [35] that it is allowed to assume that the orientation in flavor space of $\langle\bar{q}q\rangle$ around each seed is completely random.

Since at T_c the density of partons is large ($\sim 2-4$ partons/fm^3 for a free quark-gluon gas), already for a collision between medium-sized nuclei the total number of partons in the fireball is several hundreds. Using a geometric algorithm to identify pointlike defects in the developing chiral condensate [60], we found in the limit of a large number of seed partons a probability of $\frac{24\pi^2}{35\cdot 16} \simeq 0.42$ defects with unit winding number (*i.e.* either a baryon or an antibaryon) per parton.

To normalize properly the rate in a way which does not depend on the actual fireball size we use again the entropy. Due to the approximate adiabaticity of the hydrodynamic expansion of the fireball, quantities which are normalized to the initial QGP entropy are hardly affected by the strong expansion during hadronization and allow for a sensible comparison with alternative scenarios, for example that of a much more dilute equilibrated hadron gas which results from most other approaches. Of course, in both scenarios $B\bar{B}$ pair annihilation in the hadron gas stage is a matter of concern; it was discussed in [35,61] with somewhat different conclusions. For very high energy collisions it appears that freeze-out occurs sufficiently fast after the hadronization phase transition that annihilation in the hadron phase is not a serious problem.

Since the QGP partons can be taken as massless (even for the strange quarks with $m_s \simeq T_c$ the error is less than 10%), each one of them carries $\simeq 4$ units of entropy. Our above naive estimate thus yields a baryon-antibaryon production rate of

$$\alpha \equiv \frac{B+\bar{B}}{S} = \frac{0.42}{4} \simeq 0.1\,, \tag{4.8}$$

i.e. 0.1 topological defects per unit of entropy.

When contrasting this with the corresponding ratio in a chemically equilibrated hadron resonance gas with vanishing baryon chemical potential, one finds [35] $0.018 \leq \alpha \leq 0.039$ for the range 160 MeV $\leq T_c \leq$ 200 MeV for the phase transition temperature. Thus the topological baryon production mechanism appears to be able to generate a factor 2.5 to 6 more baryons than would be present in chemical equilibrium at T_c. For a plasma with the size of a sulphur nucleus ($\sim$ 130 fm^3) at hadronization a total of about 130 fm^3 $\times$ (2 – 4 partons/fm^3) $\times$ ($\frac{0.42}{2}$ antibaryons/parton) = 50 – 100 antibaryons might thus be created (compared to only $\sim$ 8 in an equilibrium hadron gas at T=160 MeV). Reviving the ideas of [26,52], one might thus speculate that this would open reasonable chances for creating small antinuclei by antibaryon coalescence at freeze-out. In this process the enhancement factor above hadronic chemical equilibrium would enter with a power $\bar{A}$ (*i.e.* the mass number of the antinucleus formed), and already for $\bar{\alpha}$ nuclei it could be of the order 40 – 1000.

Given such a large (anti-)baryon production rate, the obvious question arises whether the above estimate is consistent with energy and entropy conservation. The original QGP has a certain well-defined energy and entropy content. Since entropy cannot decrease during hadronization, it has to reappear in the hadronic phase, mostly in terms of pions which are light and thus carry away the entropy most economically. A large fraction of the available thermal energy is already spent this way, and the question arises whether there is enough fireball energy left to also create the large amount of massive baryon-antibaryon pairs estimated above. Indeed, starting from an ideal gas model for the QGP phase, and using a noninteracting hadron resonance gas with the usual free masses for the hadrons, one finds that even if no additional entropy is produced and no thermal energy is lost to collective expansion, energy conservation limits the $B\bar{B}$ enhancement factor to values between 1.4 (T_c=200 MeV) and 2.5 (T_c=160 MeV)!

Fortunately, this is largely an artifact of the oversimplified equation of state employed. QCD lattice calculations show [5] that above the phase transition strong nonperturbative effects appear which, although not very prominent in the energy density, lead to severe deviations for the pressure and entropy density from the ideal gas behaviour: already at several times T_c the entropy density begins to drop below the ideal gas curve, with only 30% of the ideal gas value left at T_c. On the other hand, the energy density is hardly affected until $T < 1.2T_c$, and at T_c is only suppressed to $\sim$ 45% of the ideal gas value. This raises the ratio of ϵ/s, *i.e.* the energy available to spend when transforming the entropy into hadronic degrees of freedom, by a factor 1.5 above the ideal gas estimate used above.

A further softening of these constraints arises from another consequence of the topological baryon model: baryon masses are reduced as T approaches the chiral restoration phase transition, as a consequence of a decreasing chiral condensate [62]. Using numerical results for $\langle\bar{q}q\rangle$ near the phase transition [63] and a parabolic extrapolation to T=0 motivated by chiral perturbation theory [64] we found in [35] that at T_c the chiral condensate has decreased to about 1/4 of its value at zero temperature (before it suddenly drops to zero in the phase transition), and that in parallel baryon masses have been lowered [65] to about 45% of their rest masses:

$$m_B^*(T_c) = m_B \left(\frac{\langle\bar{q}q\rangle_{T_c}}{\langle\bar{q}q\rangle_0}\right)^{9/16} \simeq 0.24^{\frac{9}{16}} m_B = 0.45\ m_B\ . \qquad (4.9)$$

This implies that we do not have to spend immediately the energy of a full baryon rest mass when creating a topological defect. A quantitative analysis [35] shows that indeed the rough estimate (4.8) is fully consistent with energy and entropy conservation (at T_c and below) once this strong effective mass effect is taken into account.

On the other hand, the prediction of such small effective masses near T_c has also serious consequences for (anti-)baryon abundancies in an equilibrated hadron gas. Obviously a strong baryon mass suppression will lead to much higher saturation densities than given in the naive estimate just below (4.8). Indeed, one finds that the (anti-)baryon production level implied by (4.8) is hardly distinguishable from an equilibrated hadron gas with effective baryon masses given by (4.9).

This means that, after all, chemical equilibrium is *not* drastically broken in the topological baryon production model, but that the strong baryon-antibaryon enhancement is mostly a consequence of the strong effective mass effect for the baryons. This can be viewed as a precursor of chiral symmetry restoration at the deconfinement transition, which can leave traces even in a non-QGP environment. In any case, the observation of enhanced $B\bar{B}$ production in the central region of relativistic nuclear collisions would be most exciting: it would not only again indicate an approach towards chemical equilibrium, but the observation of antibaryon levels above the naive hadron gas equilibrium estimates (without effective baryon masses) would lend in one way or the other support to the picture of baryons as topological solitons in a chiral condensate. Similar to the case of strangeness, it would directly demonstrate, however, only that *(i)* baryons have a small effective mass at high T and that *(ii)* a system of such low-mass baryons close to chemical equilibrium had been formed. While this would clearly be a new effect of highly excited nuclear matter, it should not be mistaken as a direct proof of a QGP origin of these baryons.

Lecture 5: Momentum Spectra of Hadrons from Nuclear Collisions

If approximate thermal and chemical equilibrium is reached in the collision, the thermodynamic ideas of the preceding chapter can be applied to the expansion phase until the point of freeze-out. Actually, freeze-out will in general not occur simultaneously all over the expanding fireball, but rather affect the more dilute surface regions first and subsequently proceed towards the inner regions as they also cool down and expand. Thus freeze-out occurs along a three-dimensional hypersurface in the four-dimensional space-time history of the expanding system. In this picture the measured momentum distributions of the emitted particles will be determined by two basic ingredients: (a) the intrinsic thermodynamic conditions on the freeze-out hypersurface and (b) the collective flow of the expanding matter along this surface. The hadronic energy spectra provide us thus with a snap-shot of the system at the "point" of freeze-out, rather than with the whole expansion history. But if this is so, how can we expect to obtain any information on the early phase of the collision, in particular on whether or not a quark-gluon plasma had been formed?

The answer is that different kinds of particles freeze out at different points of the expansion trajectory. It is intuitively obvious that weakly interacting particles decouple earlier from the collective flow (*i.e.* cease to be in local equilibrium with the surrounding matter) than strongly interacting ones. As two extreme examples one can compare on the one hand photons and leptons, which interact only electromagnetically with the quarks or hadrons and thus, once created, in practice escape without any further interactions, with pions on the other hand, which (depending on their energy) can have very large cross sections of up to hundreds of millibarns with other hadrons (e.g. with nucleons), and therefore after formation are likely to rescatter several times before escaping ("freezing out"). Thus *direct* photons and lepton-pairs (*i.e.* photons and leptons which do not originate from decay of secondary hadrons) probe the early, very hot and dense stage of the collision, while different hadrons with varying interaction cross sections probe (with some differentiation) the late, colder and more dilute hadronic freeze-out stage.

5.1 The Freeze-out Condition for Different Kinds of Hadrons

Particle freeze-out is an intrinsically non-equilibrium process. Particles of species i will drop out of equilibrium and freeze out from the collective flow, if the following condition is satisfied [66]:

$$\tau^i_{\text{scattering}} > \min\left(\tau_{\text{expansion}},\ \tau^i_{\text{escape}}\right) . \tag{5.1}$$

The nature of the three time scales occurring in this inequality is the following:

(*i*) τ^i_{scatt} is the average time between (elastic or inelastic) scattering events of particle i; it is determined by the local densities of all other particle species with which it can scatter, the average (thermal) relative velocity between the scattering partners, and by the interaction cross section:

$$\frac{1}{\tau^i_{\text{scatt}}} = \sum_j \langle v_{ij}\sigma_{ij}\rangle \rho_j . \tag{5.2}$$

Our (approximate) evaluation of this expression in a thermodynamic ensemble with given T and μ is described in detail in [31,67]. The scattering time scale is fully determined by the local thermodynamic conditions and is completely independent of the global aspects like size and collective dynamics of the system. Its most important feature is that it is (via the cross sections) ***particle-specific.***

(*ii*) Dynamical aspects enter on the right hand side of the criterium (5.1): intuitively, particles will decouple from the flow when by the time they have travelled one mean free path by thermal motion, the potential scattering partner has receded due to collective expansion by more than a mean free path. This picture is implemented in (5.1) through the expansion time scale $\tau_{\rm exp} \equiv -\rho/(\partial\rho/\partial t)$ which is defined locally through the time derivative of the density and by the continuity equation (for details see [68,69]). It involves the collective expansion velocity profile which for a spherically expanding fireball can be parametrized as

$$\beta(t,r) = \beta_s(t)\left(\frac{r}{R(t)}\right)^n , \tag{5.3}$$

where $R(t) = R_0 + \int_0^t \beta_s(t')\,dt'$ is the fireball radius at time t. Most of our results have been obtained for $n = 2$.

(*iii*) The "escape" time scale $\tau_{\rm esc}$

$$\tau^i_{\rm esc} = \frac{r_{\rm f}(t)}{\langle v_i \rangle} \tag{5.4}$$

takes into account the finite size of the system: $r_{\rm f}(t)$ is the size of the still equilibrated inner part of the fireball at time t; this is smaller than the fireball radius $R(t)$, because the outer parts of the fireball have in general already decoupled, due to $\tau_{\rm scatt}$ there being larger than the dynamical time scale $\tau_{\rm exp}$. Equation (5.1) implies that as soon as the "dynamical" freeze-out radius $r_{\rm f}(t)$ becomes smaller than one mean free path, the whole remaining inner part of the fireball also freezes out.

5.2 Isentropic Expansion of the Fireball

Let us assume that at the beginning of expansion the fireball formed in the nuclear collision has a certain initial energy density ϵ_0, baryon density $\rho_{\rm B,0}$, and zero net strangeness $\rho_s = 0$. We estimate the number of baryons contained in this fireball from a geometric picture, e.g. for central O + Au collisions we set $A = 16 + 52 = 68$ where $A_{\rm T} = 52$ is the number of target nucleons contained in a central tube with the same transverse area as that of the oxygen projectile. This fixes the initial fireball volume and its total thermal energy $E_{\rm FB}$.

Since we know very little about the initial pre-equilibrium stage, it is hard to theoretically estimate ϵ_0 and $\rho_{\rm B,0}$. Upper estimates can be obtained from 1-dimensional relativistic shock calculations, but this assumes a vanishing mean free path of the colliding nucleons and does not properly account for the onset of nuclear transparency e.g. at CERN energies. Therefore, in practical applications we have to take a more pragmatic point of view and try to extract that information by back-extrapolating the expansion model directly from the data. This will be the philosophy followed below.

Assuming equilibration, the initial parameters can be converted via the equation of state into a temperature T_0 and chemical potentials $\mu_{\rm B,0}$, $\mu_{s,0}$, and the specific entropy S/A can be determined. The expansion is assumed to occur isentropically. Without any explicit dynamics we can calculate from the equation of state a line of constant S/A in the $\rho_{\rm B}$-T plane, which yields a trajectory $\rho_{\rm B}(T)$ along which the fireball expands (see Fig. 8). Following this trajectory, for each value of T all thermodynamic parameters are determined by the equation of state.

As the temperature decreases, some part of the initial energy E_{FB} is converted into collective motion. To properly deal with this we should solve the hydrodynamical equations. As a result we would obtain the collective velocity field $\beta(t,r)$ for each fireball volume element as it develops in time. However, this requires knowledge of the initial density and velocity profile which again we don't have. Therefore, we instead *assume* the form of the velocity profile $\beta(r)$ at each point of the expansion trajectory $\rho_B(T)$, thereby losing the information of the global time evolution of the fireball. Since in the end we only need the velocity profile *at freeze-out*, this freedom of choice is exactly equivalent to a choice of the initial profiles, and a reasonable guess here is as good or bad as it is for the initial conditions.

For each value of T along the expansion trajectory, we thus have to fix two parameters: the fireball radius $R(T)$ and its expansion velocity on the surface $\beta_s(T)$. This is done [67,69,70] employing conservation of total energy E_{FB} and baryon number A. Once these are determined the three time-scales for the freeze-out criterium (5.1) are calculated and freeze-out is checked. As soon as freeze-out occurs, the corresponding values T_f, $\beta_s(T_f)$, and R_f are stored and the spectrum is calculated (following sextion). Should it not fit the data, the procedure is repeated with new initial conditions until convergence is reached.

Obviously, this is a very rough procedure which completely neglects the fact that in reality freeze-out proceeds successively from outer to inner regions of the fireball and thus occurs on a rather complicated freeze-out *surface*. In the meantime we know a little better how to deal with this more general problem [68]. A realistic freeze-out hypersurface is shown in Fig. 13. Nevertheless, all results presented in this lecture are still based on the rough procedure discussed above.

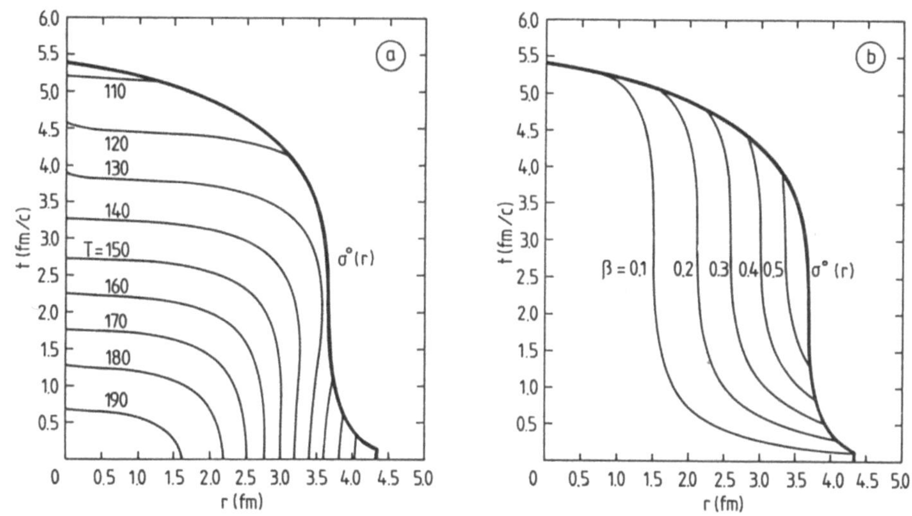

Fig. 13. The freeze-out surface $\sigma^0(r)$ for a spherically expanding fireball with $A = 68$, assuming an initial Fermi-distribution for the baryon density with mean radius of 2.6 fm [68]. The thick line separates the still equilibrated matter (left) from the region which has already frozen out (right). (a) shows isotherms inside the equilibrated region and demonstrates that freeze-out occurs mostly near temperatures between 110 and 130 MeV. (b) shows lines of constant expansion velocity and demonstrates that most of the matter freezing out does so with a flow velocity between 0.4 and 0.6 c.

5.3 The Calculation of Momentum Spectra

The spectrum of hadrons emitted from the fireball at freeze-out is in our model given by a local thermal distribution with the freeze-out temperature T_f, boosted by the local collective velocity field $\beta(x_f^\mu)$ along the freeze-out hypersurface. The general expression for the invariant momentum distribution is [71]

$$E\frac{d^3N}{d^3p} = \frac{g}{(2\pi)^3}\int_{\sigma_f} f(x,p)\, p^\mu d\sigma_\mu \,, \tag{5.5}$$

where g is a spin-isospin degeneracy factor, and σ_f is the freeze-out surface, with normal vector $d\sigma_\mu$. The Lorentz invariant local thermal distribution $f(x,p)$ is expressed in local momentum coordinates by

$$f(x,p) = \frac{1}{e^{(\overline{E}-\mu(x))/T(x)} \pm 1} \tag{5.6}$$

for bosons and fermions, respectively, where $\mu(x)$ and $T(x)$ are the local chemical potential and temperature, and $\overline{E} = u^\mu(x)p_\mu$ is the local energy, expressed in terms of the local flow velocity $u^\mu(x)$ relative to the observer. In most cases the Boltzmann approximation $f(x,p) \approx e^{-(\overline{E}-\mu(x))/T(x)}$ is adequate.

To obtain the cross section in, say, center-of-fireball coordinates E, p, a Lorentz-boost with the collective velocity field $\beta(x_\mu)$ has to be performed. Its evaluation is best done by choosing coordinate systems (u, v, w) which reflect the global symmetries of the expanding fireball. A detailed discussion of two special cases, spherically and cylindrically symmetric expansion, can be found in [68]. Here we only give the relevant result for the invariant spectrum for spherical symmetry:

$$\begin{aligned} E\frac{d^3N}{d^3p} &= \frac{d^3N}{dy\, p_T dp_T\, d\phi_p} \\ &= \frac{g}{2\pi^2}\sum_{n=1}^{\infty}(\pm)^{n+1}\int r^2 dr e^{n(\mu - E\cosh\rho)/T}\sqrt{\frac{\pi}{2\alpha}}\left\{EI_{\frac{1}{2}}(n\alpha) - p\frac{\partial\sigma^0}{\partial r}I_{\frac{3}{2}}(n\alpha)\right\}. \end{aligned} \tag{5.7}$$

Here $\rho(t,r) = \tanh^{-1}\beta(t,r)$ is the flow rapidity, $\alpha(r) \equiv p\sinh\rho(r)/T(r)$, and $I_{\frac{1}{2}}(\alpha)$ and $I_{\frac{3}{2}}(\alpha)$ are modified Bessel functions. The upper (lower) sign refers to bosons (fermions). The transverse momentum and rapidity distributions are obtained by integrating over $y = \frac{1}{2}\ln\frac{E+p_L}{E-p_L}$ and p_T, respectively.

The radial integral extends over the freeze-out surface in (t,r)-space defined by $\sigma^0(r)$. The term $\sim \partial\sigma^0/\partial r$ distinguishes different shapes of the freeze-out surface. For instantaneous freeze-out in the detector frame it vanishes. For freeze-out at constant local time it is given by $\partial\sigma^0/\partial r = \beta(r)$. The difference in the spectral shape between these two choices is minor [69] as we will see below in examples. Calculations with $\partial\sigma^0/\partial r$ determined from Fig. 13 are in progress.

After integrating (5.7) over rapidity, we obtain for spherical expansion the general form of the p_T-spectrum [69]

$$\frac{dN}{p_T dp_T} = \frac{dN}{m_T dm_T} = \frac{gR^3 e^{\mu/T}}{\pi}\Big(F_1(m_T) - F_2(m_T)\Big)\,, \tag{5.8}$$

where $\alpha = \beta\gamma p/T$, and (with $\overline{m}_T = \gamma m_T/T$)

$$F_1(m_T) = m_T\int dy\cosh y\int_0^1 \xi^2 d\xi e^{-\overline{m}_T\cosh y}\frac{\sinh\alpha}{\alpha}\,; \tag{5.9}$$

$$F_2(m_T) = T\int dy\int_0^1 \xi^2 d\xi\frac{\partial\sigma^0}{\partial r}\frac{e^{-\overline{m}_T\cosh y}}{\beta\gamma}\left(\cosh\alpha - \frac{\sinh\alpha}{\alpha}\right). \tag{5.10}$$

One sees that $F_1(m_T)$ is the contribution from a hypersurface of constant global time, while any local variations of the freeze-out surface due to $\partial\sigma^0/\partial r \neq 0$ contribute to the spectrum via F_2. Even for a constant expansion velocity the two contributions have a different m_T-dependence [72], due to the two different Bessel functions $\sinh\alpha/\alpha=\sqrt{\pi/2\alpha}I_{1/2}(\alpha)$ and $\cosh\alpha - \sinh\alpha/\alpha=\sqrt{\alpha\pi/2}I_{3/2}(\alpha)$ entering the respective integrands: for example, at y=0, while $F_1(m_T)$ approaches for $p_T \to 0$ a constant limit, F_2 vanishes in this limit as $(p_T/T)^2$. For $\partial\sigma^0/\partial r > 0$ (as in the case of a constant local time freeze-out surface) the difference between these two contributions leads to concavity of the combined spectrum in the region $p_T < T$. For more realistic freeze-out surfaces as in Fig. 13 which contain some timelike sections with $\partial\sigma^0/\partial r < 0$, F_2 contributes with the opposite sign and, depending on the details of the rapidity average (for example whether one integrates over all y or only selects a small window around y=0), the resulting spectra may tend to be slightly convex in the small-p_T ($p_T < T$) region [72]. Please note that this effect has practically no influence on the slope in the medium- and large-p_T region ($p_T > T$).

In Fig. 14 we analyze in some detail the various influences on the transverse mass spectra caused by collective flow. We plot the different contributions to the spectrum $(1/m_T^{3/2})\,dN/dm_T$ (with the additional factor $m_T^{-1/2}$ extracted to facilitate comparison with thermal radiation) for pions, kaons and protons for different dynamical assumptions and cuts.

Fig. 14a shows the spectra at $y = 0$ for a spherical shell of temperature $T = 110$ MeV, expanding with constant velocity $\beta = 0.45$. The contributions F_1, F_2, and $F_1 - F_2$ (see (5.7–9)) are shown separately. (While the relative normalizations for F_1, F_2, and $F_1 - F_2$ are correct, the total spectra $F_1 - F_2$ for each particle species have been arbitrarily normalized to 1 at $m_T = m_0$. This facilitates the comparison of slopes.) The (arbitrarily normalized) purely thermal radiation spectrum is integrated over y and has been calculated for an effective temperature corresponding to a blueshift of the true one by the absolute value of the flow velocity, $T_{eff} = T\sqrt{(1+\beta)/(1-\beta)}$. At first sight all contributions to the spectra seem to approach the same slope at large m_T corresponding to this blueshifted temperature. Actually, they stay a little steeper, due to contributions from volume elements with longitudinal flow components whose transversal blueshift factor is smaller than the maximum one reached by volume elements whose flow points purely in the transverse direction. This effect is somewhat stronger in Fig. 14b where the spectrum is integrated over y which gives more weight to contributions from longitudinally moving volume elements. At small m_T the global-time contribution F_1 turns slightly concave for pions, but has a convex shape for protons. Kaons lie in between. The convexity for protons is nearly gone in the local-time spectrum $F_1 - F_2$, while the concavity of the pion spectrum is enhanced, in agreement with our above discussion.

For very large constant expansion velocities ($\beta > 0.65$) we found it possible to produce a small dip in the proton spectrum near $m_T = m_p$ and thus a peak in the m_T-spectrum [73]; however, this peak weakens with integration over y and usually vanishes once the spectrum is integrated over a realistic velocity profile.

Fig. 14b shows the same spectra as Fig. 14a, but now integrated over y. The convexity of the proton spectrum becomes less pronounced, while the concavity of the pion spectrum gets slightly stronger.

In Figs. 14c,d we additionally integrate the spectrum over a parabolic radial flow velocity profile with the surface velocity chosen such that the same average flow of $\langle\beta\rangle =$

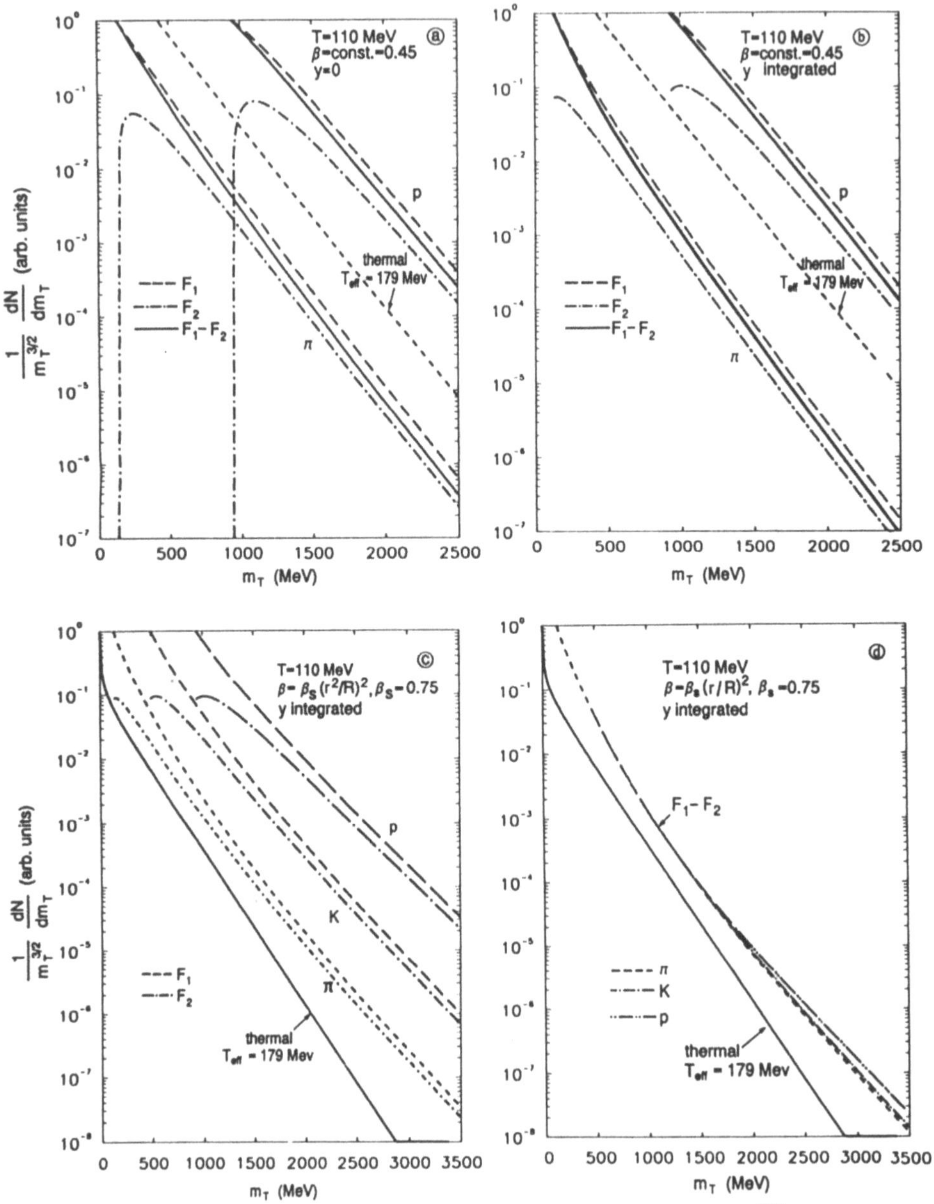

Fig. 14. The various contributions to the transverse mass spectrum $(1/m_{\mathrm{T}}^{3/2})\, dN/dm_{\mathrm{T}}$ for different dynamical assumptions. All spectra $F_1 - F_2$ have been arbitrarily normalized to 1 at $m_{\mathrm{T}} = m_0$, and the relative normalization of the F_i is exact.

0.45 is recovered. Now we observe *strong concavity* in all global-time spectra F_1, and this feature is not qualitatively affected by the F_2 contribution. The asymptotic slopes now are much flatter than expected from a blueshifted temperature calculated with the average flow velocity; they correspond more closely to a blueshift factor calculated with the larger surface velocity β_s. Thus in the large-m_{T} tail of the distribution the fast expanding outer shells of the fireball dominate. Hence the strong concavity is primarily due to the fact that different shells with different flow velocities contribute different effective temperatures. Fig. 14d shows that if the spectra from different particles are normalized onto each other, a nearly universal curve is obtained, with only small differences due to the different masses. The different average slopes obtained for different particles [31,67,70] in a fixed interval of p_T (or of transverse kinetic energy $m_T - m_0$) are just a reflection of the intrinsic concave curvature of this universal curve: small-p_T protons probe larger values of m_T than small-p_T pions and thereby a region of flatter slope.

The deviations in Figs. 14c,d from the thermal curve show for comparison are the tell-tale signs of collective transverse flow which we should look for in the particle spectra.

5.4 Kaon Slope Parameters as a Signature for Baryon-rich QGP

Since for the typical thermal energies occurring in nuclear collisions $\sigma_{KN} \ll \sigma_{\overline{K}N}$, in a baryon-rich environment kaons (K^+, K^0) have a considerably smaller effective interaction cross section than anti-kaons (K^-, $\overline{K}^0$), pions and nucleons, and are expected to freeze out earlier than these other particles. Thereby they may give us a glimpse from an earlier stage of the expanding fireball than possible through other common hadrons [31,67]. We saw in Fig. 8 that these earlier stages of the expansion proceed through quite different regions in the T-ρ_B-plane, if the reaction involves only a hadronic phase or if it also goes through an early quark matter phase: the dashed curve in Fig. 8 shows that a given final state with a certain (say, measured [74]) entropy per baryon S/A extrapolates back to a very hot, but rather dilute initial state of the fireball, if the equation of state is that of a hadron resonance gas. On the other hand, as shown by the solid lines, if the transition to quark matter is allowed for, the isentropic expansion trajectories have a quite different form and extrapolate back to a much colder but denser initial state.

This initial quark-gluon plasma state also possesses a lower internal pressure than the corresponding initial hadronic state with the same S/A, which means that it will expand at a slower rate and take a longer total time until freeze-out. However, the amount of thermal energy which is eventually converted into collective motion is rather similar in both cases, so that the expansion velocity at pion freeze-out is very close in both scenarios, and the effect of the phase transition on the pion, K^-, and proton spetra (which all freeze out about simultaneousely) will be minimal. The situation is different for the K^+-mesons (at least as long as the system contains a large net baryon number): in the case of a phase transition they freeze out as soon as they are formed by hadronization from the plasma, *i.e.* already in the mixed phase, which is comparatively cold; without the phase transition, their freeze-out point would be located at a much higher temperature in the hadron gas (see Fig. 8). Adding to this difference in freeze-out temperature the difference in the amount of collective flow at the earlier point of K^+-freeze-out (less flow) compared to the later point of K^--freeze-out (more flow), one arrives at the conclusion that the relative order of the K^+ and K^- slope parameters (*i.e.* the inverse slope of their energy spectra) may be inverted by a phase transition to quark matter (see Fig. 15).

To complete the picture, one has also to check the pion, proton and (if possible) even other hadronic slope parameters: for a given freeze-out temperature and expansion velocity, the effect of flow on the slope of the spectra is the stronger the heavier the particle is. This is shown in Fig. 15 by the pions, K^- mesons and protons which all freeze out at the same temperature, but show flatter and flatter spectra as their mass increases. Thus (as long as we are in the baryon-rich region – the asymmetry between K^+ and K^- vanishes as $\rho_B \to 0$!) a comparison of all these slope parameters should exhibit the pattern shown in Fig. 15a if no phase transition occurs, but should switch to

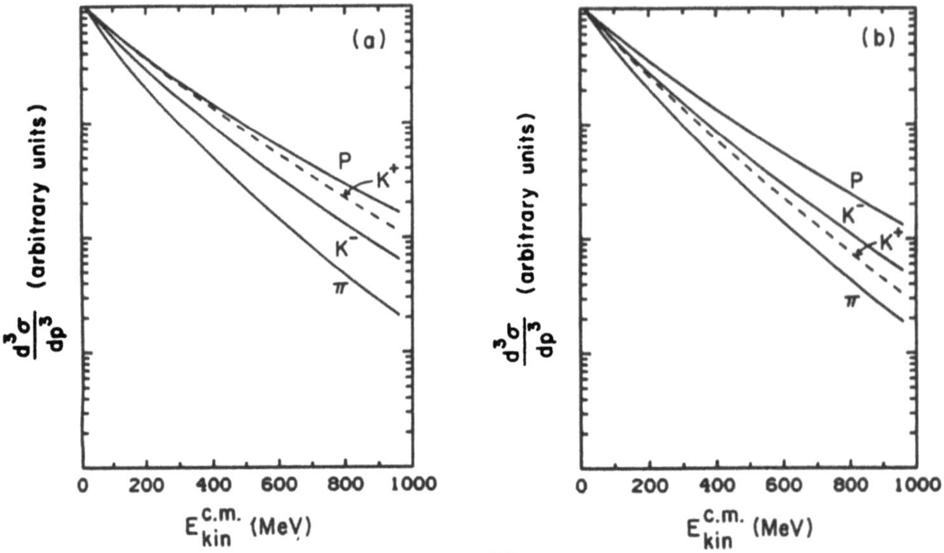

Fig. 15. Kinetic energy spectra for π, p, K, and $\overline{K}$ in the fireball c.m. Assumed was a spherically expanding fireball with A=100 and a velocity profile $\beta(r) = (r/R)^2\beta_s$, and initial conditions obtained from a shock calculation with $E_{\rm lab}/A$ = 15 GeV. The spectra were normalized at E=10 MeV to facilitate comparison of slopes. (a) no phase transition; (b) with phase transition, i.e. the system is initially in the QGP phase.

the pattern of Fig. 15b if the collision proceeds through an initial quark matter (or even mixed!) phase.

An easy way to look for this effect is to plot the ratios K^+/π, K^-/π, p/π, and in particular K^+/K^- against their transverse kinetic energy $m_{\rm T} - m = \sqrt{m^2 + p_{\rm T}^2} - m$ (to eliminate trivial kinematic effects of the different particle masses on the *thermal* component in the spectrum and to isolate the flow effects). If our picture is correct, the slope of the K^+/K^- ratio should be positive without quark matter formation, but might turn negative if the phase transition occurs. The available data [43] from the recent experiments at the Brookhaven AGS with Si-beams of 14.5 GeV/nucleon beam energy (see Fig. 23) tend to show a somewhat flatter slope for K^+ than for K^- as a function of $m_{\rm T}$, leading to a positive slope for the K^+/K^- rather than a negative one. Thus at the present stage they do not favor the hypothesis of an initial quark-gluon plasma phase (at least not for the *average* "central" collision).

5.5 Quantitative Comparison of the Flow Model with Spectra from CERN and BNL

5.5.1 CERN data

The analysis in Sect. 5.4 was based on initial conditions obtained from a shock model. As mentioned before this is unreliable and overestimates the initial compression and temperature. In this section we will follow a different philosophy, namely to try and use the experimental data on $p_{\rm T}$-spectra to obtain directly from the experiment a better insight into the initial conditions of the fireball, by extrapolating back from the observed spectra.

In [70] we showed that with the flow model one can obtain an excellent fit to the π^0 transverse momentum spectra [75] from central O+Au collisions at 200 A GeV. Since the fireball radius scales out from the spectrum and thus does not affect its shape, but only its normalization (which is anyway theoretically uncertain since it strongly depends on the degree of chemical equilibrium reached in the collision), it only enters the final results in a very weak way through the rarefaction time scale in the freeze-out criterion (5.1). Thus it essentially has to be determined independently (for example by 2-pion interferometry [76]). If we leave it as a free parameter, we obtain from our fit to the π^0 spectrum a whole set of possible initial conditions ϵ_0, $\rho_{B,0}$. It is interesting to note [70] that they all correspond roughly to the same initial thermal energy of the fireball, *i.e.* E_{TOT}/A, and only differ in its specific entropy. Thus, for a given spectral shape, fixing the fireball radius is in our model equivalent to fixing S/A.

In practice we have for concreteness assumed that the initial fireball radius is equal to the transverse (rms) radius of the (smaller) projectile; *i.e.* for O+Au, Si+Au, and S+S collisions we took R_0=2.5, 3.0, and 3.1 fm, respectively. The consistency of this choice will have to be checked in the future by an independent determination of the specific entropy (e.g. from particle ratios [74,77]) or of the radius at freeze-out from 2-particle interferometry. According to the analysis of 2-pion correlation data by the NA35 collaboration there are indications for a very large transverse freeze-out radius (of order 5 - 6 fm in S+S collisions and up to 8 fm at midrapidity in O+Au collisions). Their fit, however, does not take into account the possible effects of a strong radial flow on the correlation function.

In Figs. 15a,b we show (for the two cases of constant-local-time and constant-global-time freeze-out) our theoretical curves obtained from a fit of the initial conditions to the pion data, as well as the predicted shapes for other hadron spectra. Except for the K^+ mesons, all hadrons shown freeze out essentially together, due to their strong coupling with each other via the nucleons which here form the dominant fraction of the particle density. (The initial state of the fireball has $T \simeq 160$ MeV, $\epsilon \simeq 1$ GeV/fm^3, and $\rho_B \simeq 4\rho_0$.) Basically, since $\sigma_{\pi N}$ and $\sigma_{\overline{K}N}$ are resonance dominated in the momentum region of interest, while $\sigma_{\pi\overline{K}}$ is much smaller, the nucleons serve as a heat bath and their freeze-out temperature ($T_f \simeq 110$ MeV) determines the freeze-out temperature also for the other particles. The average flow velocity at freeze-out is $\simeq 0.44c$. Only the K^+ mesons are different, since they have a very small interaction cross section (< 10 mb) with nucleons, and their (isospin averaged) interaction with the (in our applications) less abundant pions does not exceed 60 mb even in the peak of the rather narrow $K^*(892)$ resonance [78]. Thus they freeze out considerably earlier ($T_f \simeq 145$ MeV). While the higher freeze-out temperature should lead to a flatter spectrum, but the less developed flow at this point ($\langle\beta\rangle \simeq 0.25$) counteracts such an effect. The final result is a K^+ spectrum which for small p_T is marginally flatter than the K^- one, but nearly straight (*i.e.* thermal) due the small flow velocity. In contrast, the K^--spectrum is visibly curved as an effect of the stronger flow velocity at K^- freeze-out, leading to a flatter slope than the one of the K^+ mesons at large p_T. The cross-over point between the two slopes depends somewhat on which freeze-out surface is chosen and here lies in the region between 0.5 GeV< $m_T - m_0$ <1 GeV.

While the small difference between K^+ and K^- spectra, at least in the low-p_T region, is very hard to measure and requires very high kaon statistics (first experimental data from the E802 spectrometer will be discussed below), we see in Fig. 16 that the predicted splitting between the slopes of pions, protons and Λ's due to the flow effects is much

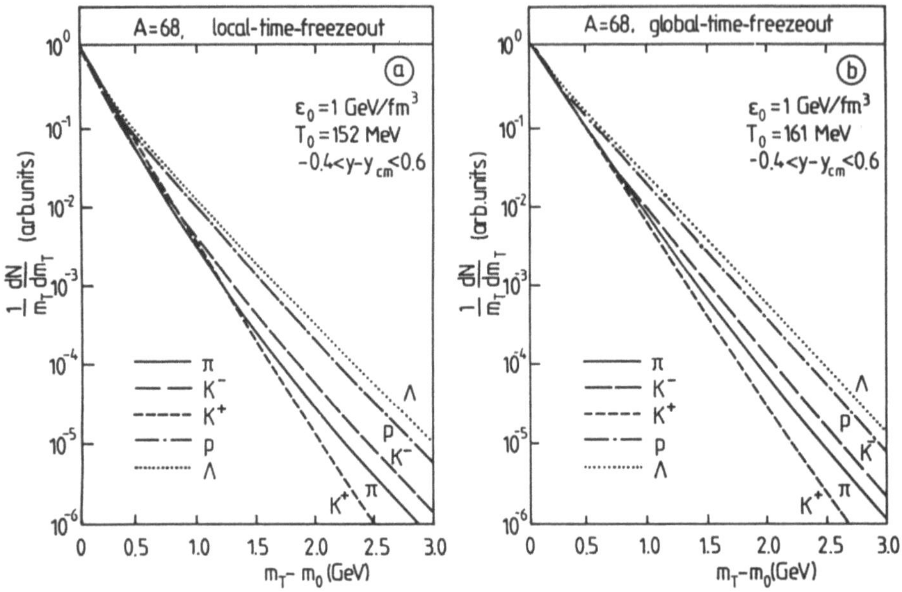

Fig. 16. (a) Transverse mass spectra for pions, kaons, protons, and Λ's in central 200 A GeV O+Au collisions, for freeze-out at constant local time. All the curves are arbitrarily normalized at $m_T = m_0$.
(b) The same for constant-global-time freeze-out. For initial state and freeze-out parameters see text.

stronger and thus easier to test. This is done in Fig. 17, using the pion, K_s^0, Λ and $\overline{\Lambda}$ m_T-spectra from 200 A GeV O+Au collisions [79,80].

In Fig. 17a we show our fit to the pion spectra together with the very high statistics π^0 data from WA80 [75] (which were the basis of our fit) and the π^- spectra from NA35 [79]. The fit with a constant-global-time freeze-out surface appears to be somewhat better than the one corresponding to constant local time. While the fit is excellent for the WA80 data which do not give information below p_T=400 MeV/c but reach out to transverse momenta of nearly 3 GeV/c, the NA35 data seem to overshoot the theoretical curve at very small transverse momenta $p_T < 200$ MeV/c. This effect has recently been confirmed by data from the NA34 external spectrometer [81]. Given the fact that more realistic freeze-out surfaces rather tend to flatten the theoretical spectra in this region (see Sect. 5.3), our model is not able to reproduce such a behaviour, which would thus require a different mechanism in the very-low p_T region. One might think of a cold target spectator contribution, but since the data are selected for a central rapidity slice where the total pion multiplicity is very large, any target spectator component leaking into this y-interval would be much smaller and make a negligible correction to the spectrum. A more likely candidate are resonance decays [82].

In Fig. 17b the predicted shape for the $K_s^0 = (K^0 + \overline{K}^0)/2$ is shown together with the data [80] (see also J. Harris [83] for a similar comparison). We show two curves for comparison: the flatter one (labelled K^-) corresponds to $\overline{K}^0$ only, the steeper one to a mix of K^0 and $\overline{K}^0$ according to the relative abundances of kaons and anti-kaons measured in the Brookhaven E802 experiment (see Sect. 5.5.2). The data appear to be even flatter than the K^--curve. A comparison with preliminary evidence from other experiments and

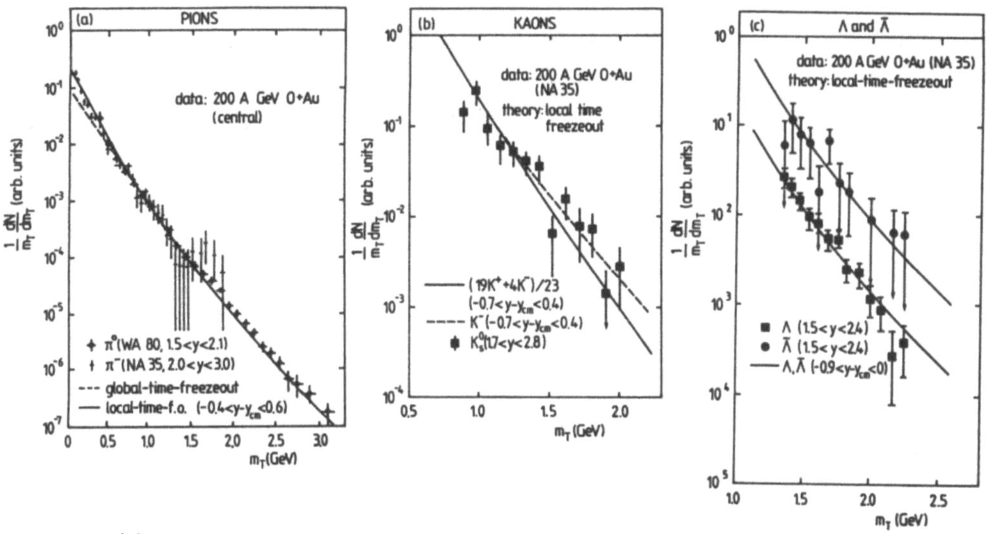

Fig. 17. (a) Transverse mass spectra for neutral and negative pions from central 200 A GeV O+Au collisions as obtained by WA80 [75] and NA35 [79]. These data were used to obtain the model parameters. The solid (dashed) curves are fits using a local-time (global-time) freeze-out surface. The relative normalization of data and theory is arbitrary. (b,c): Transverse mass spectra for K_s^0, Λ, and $\overline{\Lambda}$ [80]. The curves are those from Fig. 16a, but the corresponding ones from Fig. 16b are hardly distinguishable.

a discussion of possible origins for such an effect is given in [69]. However, the situation with kaons is far from clear, and more data with better statistics over a larger m_T region will be needed to clarify this issue.

Fig. 17c shows the Λ and $\overline{\Lambda}$ spectra from the O+Au collisions [80] together with our prediction from Fig. 16. The theoretical curves for Λ's and $\overline{\Lambda}$'s are identical, since we assumed simultaneous freeze-out for both species. The agreement with the data is quite good and demonstrates that hyperons have indeed a flatter spectrum than pions in the measured range of transverse kinetic energies $m_T - m_0$ or, in other words, that the "universal" m_T-spectrum shown in Fig. 14d is indeed curved.

It is worth noting that even when using the formulae (5.8-10) from our model to analyze the data, such an immediate success is not guaranteed if one does not properly include the freeze-out concept. In other words, doing a free fit of T_f and β_s to the spectrum of a single species of hadrons without checking the compatibility of this pair of parameters with the freeze-out criterion will in general lead to failure with other particles. In particular, if e.g. in the pion spectrum a hard scattering component at large p_T becomes visible (as is, for example, the case in peripheral nuclear [75] or proton-nucleus collisions [75,81,84]), a l.m.s. fit of the spectrum inevitably leads to a combination of low T_f and large β_s (the latter being enforced by the strong bending of the spectra at large p_T which however in this case has nothing to do with collective flow); the high β_s would in turn influence the heavier Λ's very strongly and lead to a much too flat predicted spectrum.

Several studies [83,85] have shown that in a free fit to a given spectrum the compatible values of T and β are anticorrelated, and the fit quality may allow for a rather wide variation (high-T, low-β combinations giving similar quality fits as low-T, high-β ones). In contrast, the freeze-out criterion provides a direct correlation between these parameters:

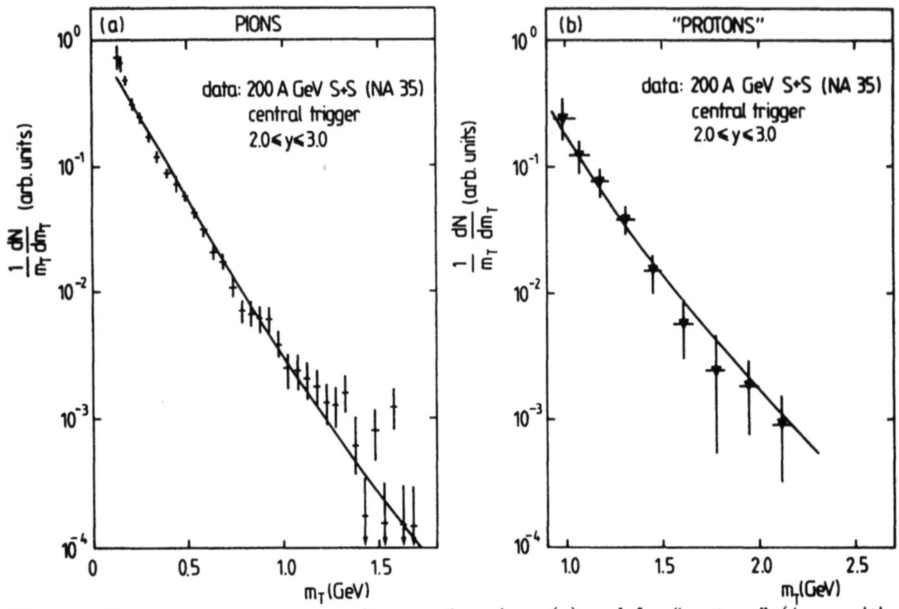

Fig. 18. Transverse mass spectra for negative pions (a) and for "protons" (*i.e.* positive minus negative tracks) from central 200 A GeV S+S collisions [83]. Theoretical curves are from Fig. 16a.

high flow values lead to earlier freeze-out (*i.e.* at high T), and vice versa. Combining both pieces of information considerably restricts even for a single particle species the allowed range of parameter pairs, and a careful comparison with other particle species finally constrains the system completely.

The available particle spectra from 200 A GeV S+S collisions [86] can be fit by the same set of initial conditions (we only changed the initial fireball radius to the one for sulfur). Since in this case in central collisions all projectile and target nucleons are involved, the baryon number of the fireball is nearly identical to the O+Au case. Our results for pions and protons (experimentally obtained by subtracting all negative from all positive tracks [83] rather than direct identification) are shown in Fig. 18. Again the predicted flatter slope of the protons is borne out in the data.

In Fig. 19 we show the same data (with slightly different kinematic cuts [86]), including also kaon and Λ data from the S+S system. We here chose to plot $(1/m_T^{3/2})dN/dm_T$ as in Fig. 14 where we compared spectra with flow to purely thermal ones. One sees that in this representation, with the different spectra normalized onto each other at the point $m_T = m_0$, both the theory and the data lie rather well on a universal curve with concave curvature. In the data the curvature is clearly visible only up to $m_T \sim 1$ GeV (*i.e.* it is mostly due to the pions), while for larger m_T the data could also be well represented by a straight line with a slope parameter of about 200 MeV. It is clear that more data at larger values of m_T are required to settle the question of continuing concave curvature in the spectra and thus to clarify whether or not there is collective flow. Theoretically, it has to be studied whether there are other mechanisms, in particular in the pion channel which is mostly responsible for the apparent curvature, which can explain the nonthermal nature of their m_T-spectra. An analysis of resonance decay contributions to the spectra is under way.

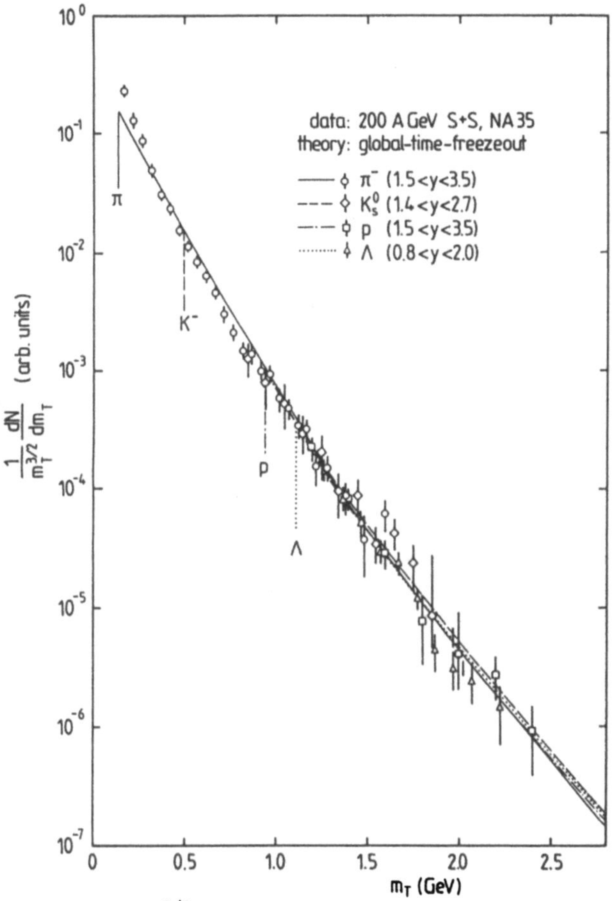

Fig. 19. $(1/m_T^{3/2})dN/dm_T$ for pions, kaons, "protons", and Λ's from central 200 A GeV S+S collisions [86]. The initial state parameters for the theoretical curves are as in Fig. 16b. Globally the fit to the pion spectrum shown here (global-time freezeout) is a little better than the one in Fig. 18 (local-time freezeout), however the low-p_T "anomaly" appears more significant.

Fig. 19 clarifies why it was found in [83] that it is also possible to fit the proton and Λ^0 spectra by a purely thermal distribution with a single temperature (this is not possible for the pions). The resulting values for T are much higher than those obtained with the flow hypothesis and are different for different particle species; calculating for example a thermal pion spectrum with the apparent Λ temperature yields a much flatter spectrum than measured. Such different temperatures for pions, protons and Λ's are hard to understand theoretically, while the collective flow hypothesis yields a much more natural interpretation of the data with one single set of parameters.

In Fig. 20 we show that our spherical fireball picture fails to describe the longitudinal momentum distributions. The rapidity distribution from our flow model is even slightly narrower than a purely thermal one with an effective temperature of 200 MeV (*i.e.* the asymptotic slope parameter in Fig. 19). Clearly the restriction to spherical symmetry is too restrictive, and the much wider experimental rapidity distribution is an indication for much stronger flow occuring along the beam axis than in the transverse directions. We

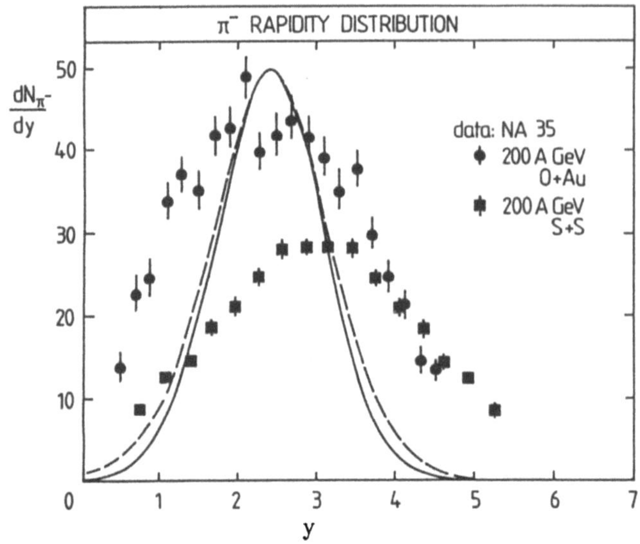

Fig. 20. Rapidity distributions for negative pions from central 200 A GeV O+Au and S+S collisions [80,83,86]. The solid line represents our flow model and uses the parameters from Fig. 16 which fit well the m_T-spectra. The dashed line correponds to purely thermal radiation of a spherical fireball with an effective temperature $T_{eff} = 200$ MeV.

hope to present a combined analysis of both the transverse and longitudinal momentum distributions within a flow model with only cylindrical symmetry in a future paper.

The failure of the spherical model with respect to the rapidity distributions renders a quantitative interpretation of the initial state parameters ϵ_0, $\rho_{B,0}$ questionable: obviously, our analysis includes in the energy balance only the kinetic energy from transverse collective flow (which had to come from thermal energy density initially). If one interprets the measured rapidity distribution as evidence for even stronger collective flow in the longitudinal direction, and furthermore assumes that also this flow energy was at one point thermalized (Fermi-Landau model [87,88]), one of course obtains much higher estimates for the initial energy densities [89].

5.5.2 The E802 Data from BNL

The E802 collaboration has measured various particle spectra in central Si + Au collisions at 14.5 A GeV. In contrast to the situation at CERN, at this energy large target nuclei can stop the projectile completely [44], leading to a very large baryon density in the collision zone and making the assumption of a spherical fireball a litle less suspicious. Indeed, the y-dependence of the slope parameter extracted from the proton m_T spectra shows a behaviour compatible with that of protons isotropically emitted from a spherical fireball [43]:

$$T_{\text{eff}}(y) = \frac{T_{\text{eff}}(y_{\text{cm}})}{\cosh(y - y_{\text{cm}})}. \tag{5.11}$$

In this section we will now analyze the spectra from this experiment in detail, following the same procedure as in the previous section, with a fireball containing A_{FB}=103 baryons and having the initial radius of the Si projectile.

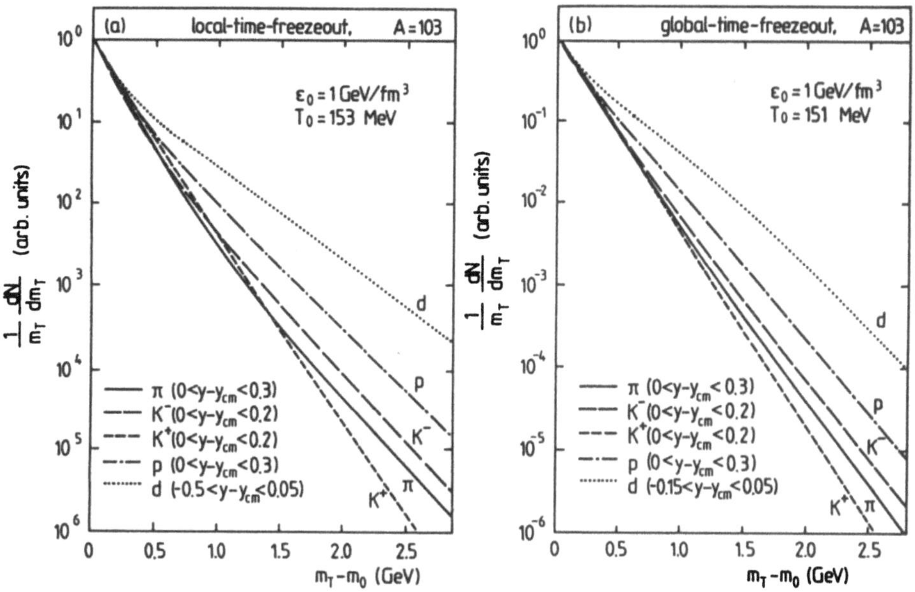

Fig. 21. (a) Transverse mass distributions for pions, kaons, protons, and deuterons in central 14.5 A GeV Si+Au collisions, for freeze-out at constant local time. All curves are arbitrarily normalized at $m_T = m_0$. The y-integration limits have been chosen to agree with the experimental acceptances for the data in Fig. 22, for a fireball centered at $y_{cm} = 1.2$. (b) The same curves for constant-global-time freeze-out.

Again we begin by determining a set of initial state parameters such that after expansion and freeze-out the measured pion spectra are reproduced. We find $\epsilon_0 \simeq 1$ GeV/fm^3, $\rho_{B,0} \simeq 4.2\rho_0$, and $T_0 \simeq 150$ MeV, while freeze-out occurs at $T_f \simeq 105$ MeV with an average flow velocity of $0.45c$ (for K^+ mesons $T_f \simeq 140$ MeV, and the flow velocity $\simeq 0.25c$). Once the initial state is thus fixed, the slopes of all other particle spectra are found to come out right without any adjustment of parameters.

The theoretical particle spectra obtained from these initial conditions are shown together in Fig. 21. Again the theoretical spectra are normalized arbitrarily, in order to concentrate on their shapes. There is a large difference between the slope parameters calculated for pions, protons and deuterons; the difference between pions and kaons is smaller, and the details also depend somewhat on the choice of freeze-out hypersurface. While for local-time freeze-out the K^+ spectrum is considerably flatter than the K^- one up to $m_T \simeq 1$ GeV, this is not the case for global-time freeze-out where the two slopes come out nearly equal in the low-m_T region while at large m_T the K^+ are steeper.

In [43] the m_T-spectra were fitted under the assumption of purely thermal radiation, with a Boltzmann distribution centered in the middle of the measured rapidity interval ($1.2 < y < 1.4$ for pions, kaons and protons) and not integrated over y. The data explore roughly the regions 0.2 GeV$< m_T - m_0 <$0.8 GeV for π^+ and π^-, $0 < m_T - m_0 <$0.5 GeV for K^+ and K^-, and $0 < m_T - m_0 <$1 GeV for protons. In these regions the following effective temperatures were found: $T_{eff}(\pi^+) = T_{eff}(\pi^-) = 126 \pm 10$ MeV, $T_{eff}(K^-) = 140 \pm 25$ MeV, $T_{eff}(K^+) = 160 \pm 15$ MeV, and $T_{eff}(p) = 187 \pm 5$ MeV, corresponding to the pattern $T_{eff}(\pi) < T_{eff}(K^-) < T_{eff}(K^+) < T_{eff}(p)$. While for the

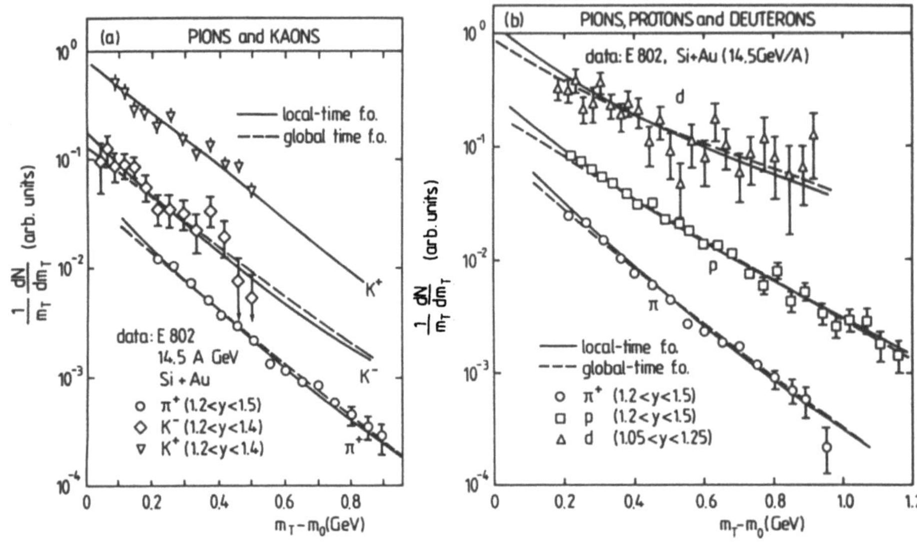

Fig. 22. Comparison of the theoretical curves in Fig. 21 with the E802 data. (a) Pions [43b] and kaons [43a]. For K^+ the results for local- and global-time freeze-out are indistinguishable. (b) Pions, protons, and deuterons [43b]. All normalizations are arbitrary.

pions the spectra for the two charge states are identical to high accuracy [43], this does not appear to be the case for the kaons. In Fig. 22 we show that, within the error bars, this pattern can be quantitatively reproduced by our model. Of course, the statistics on the kaons are not yet good enough to clearly say whether or not there is a difference in slope between the two charge states; within the errors, both our local-time and global-time freeze-out curves pass through the data. Still, there are indications for some small differences between the K^+ and K^- data, not only in abundance but also in the spectral shape; this is very interesting and should be further studied with higher accuracy.

In Fig. 22 we show our fit to the pion spectrum and the predicted spectra for the other particles together with the data. While the pion and kaon spectra are for all practical purposes parallel, and (in contrast to the pions at CERN energies) one cannot see a strong curvature within the restricted m_T-ranges of the experiment, the presence of the curvature is implicit in the different slopes seen for pions, protons and deuterons. This is very clearly demonstrated in Fig. 22b, which shows that the larger m_T, the flatter the spectra become, both in the flow model and in experiment.

That the deuterons show the flow so clearly may seem surprising at first sight, since they are only weakly bound and certainly will not exist as elementary particles in the hot fireball. Instead they will form by coalescence of protons and neutrons at freeze-out. However, the weak binding of those two nucleons in the deuteron requires a very small coalescence radius in momentum space, such that the nucleon momenta (which carry both the thermal and flow information) add up to a deuteron momentum which is exactly as it would have been for an elementary deuteron in the same temperature and flow velocity field. This feature was observed previously at the BEVALAC (for reviews see [90]), where the coalescence model proved very successful in explaining the properties of small nuclear fragments. A calculation of the deuteron spectra within a relativistic coalescence model applicable to BNL and CERN energies is under way.

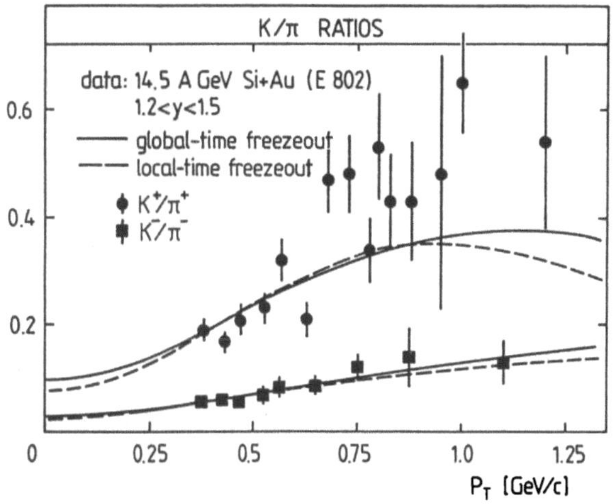

Fig. 23. K/π ratios as a function of p_T from the E802 experiment [43] and from our model. The theoretical curves are normalized to the measured integrated K/π ratios.

The kaon data of Fig. 22a have also been presented in a different way [43b] which has given rise to some excitement: Fig. 23 shows the kaon-to-pion ratios as functions of p_T. Both ratios rise with p_T, but the slope is significantly steeper for K^+/π^+ than for K^-/π^-. On the other hand, the π^+/π^- ratio is 1 independently of p_T [43]. Also shown in Fig. 11 are the predictions from our model. For this the theoretical spectra were first reevaluated as a function of p_T, then normalized relative to each other according to the measured integrated K/π ratios and finally divided by each other. The second step is necessary since our model implicitly assumes chemical equilibrium all the way until thermal freeze-out, and thus we did not a priori expect very good agreement of our integrated K/π ratios with the experimental ones. However, the deviations are surprisingly small: we obtain K^+/π^+=0.24 and K^-/π^-=0.056 for local-time freeze-out (K^+/π^+=0.31 and K^-/π^-=0.045 for global-time freeze-out), while the experimental values [43] are K^+/π^+=0.192±0.03 and K^-/π^-=0.036±0.008. Thus, as already discussed in Sect. 5.1, the measured integrated ratios indicate a strong approach towards chemical equilibrium even in the strange sector.

We see in Fig. 23 that the p_T-dependence of the K^-/π^- ratio is very well described by our model, while we do not quite reproduce the very strong rise in the K^+/π^+ ratio. This again indicates (as for the kaons at CERN energies) that the measured K^+ spectra are somewhat flatter than predicted. Data for larger m_T and with better statistics would be very valuable for further illuminating this potential problem.

Our method of identification of collective flow relies on the characteristically concave shape of the m_T-spectra which translates into a characteristic flattening of those spectra with increasing hadron rest mass. It remains to be studied, to what extend these features can be obtained by alternative processes: From recent experimental work by the TPC collaboration at SLAC [91] on the pion, kaon and proton p_T-spectra from e^+-e^-- collisions (where p_T is measured relative to the jet axis) it is known that the breaking of color strings (as the basic mechanism behind jet hadronisation) does not produce such a difference in slopes; on the other hand, a slight concave curvature appears to be also inherent in the string fragmentation process.

While the flattening of slopes in central nucleus-nucleus collisions does not appear to be a remnant of the primordial string breaking process (and thus cannot be argued to originate from the very initial pre-equilibrium stages in these collisions), it is, however, also seen in peripheral nucleus-nucleus collisions, in proton-nucleus collisions, and even in very high multiplicity $\bar{p}p$ collisions [92]. In [70] we compared the neutral pion spectra from central and peripheral O+Au collisions at CERN with our model; one observes significant differences in the average slope [75], but both spectra can be explained by our model with the *same* initial density and energy density, taking into account *only* the smaller size of the initial fireball ($A_{\rm FB} = 28$ instead of 68) and the resulting earlier freeze-out in the peripheral case. This would imply that if there is flow in central collisions there has to be also some in peripheral collisions. Csernai and Gutay [93] argue that even in $\bar{p} - p$ collisions there is evidence for collective transverse flow.

Obviously, these latter "conclusions" go far beyond what most of us (including myself) are intuitively inclined to believe, and a very thorough and unprejudiced analysis of the complete body of available data on hadronic and nuclear collisions still has to be performed before one should trust the picture presented here. A comprehensive analysis of all available data on transverse momentum spectra of identified particles from hadron-nucleus and nucleus-nucleus collisions is under way.

Epilogue

We have tried to describe the various dynamical stages of a relativistic heavy-ion collision and their experimental manifestations, under the assumptions of both old-fashioned hadronic dynamics and quark-gluon plasma formation. We found that the initial pre-equilibrium stage is the most difficult one to describe, and we discussed some of the kinetic approaches based on quark-gluon degrees of freedom in Lecture 2.

In Lectures 3 and 4 we discussed the hadronization phase transition and the approach to chemical equilibrium in nuclear collisions both via hadronic final state interactions and via QGP formation. We argued that the study of strange hadrons and antibaryons from nuclear collisions is very important since they tell us about whether or not chemical equilibrium is reached. A large degree of chemical equilibration of strangeness and antimatter is highly unusual in hadronic collisions and may be an indirect signal for "new physics" (e.g. QGP formation). We have quantitatively analyzed K/π ratios and ϕ production from nuclear collisions at CERN and BNL and pointed out that the experiments do indeed point towards a considerable degree of chemical equilibration. We mentioned that both observations are in accord with the hypothesis of QGP formation, but that they can also be explained in a purely hadronic environment. In particular, the observed ϕ enhancement in central collisions can be quantitatively explained in terms of hadronic rescattering processes, provided the latter are also responsible for the J/ψ suppression. However, this explanation requires a very high hadron density in the collision zone, and one may find it unnatural to still talk about hadronic degrees of freedom in such an environment.

We have also discussed predictions for anti-baryon production in the central region, based on a new analysis within chiral models which predict a very strong effective mass effect for baryons in hot and highly excited hadronic matter. We pointed out that these models would receive strong support from any observation of unusually large production rates for antibaryons (or even small anti-nuclei). Again such an observation could also provide indirect evidence for QGP formation, by proving a strong tendency towards chemical equilibration.

In the last Lecture we analyzed the available data on transverse momentum spectra from relativistic nuclear collisions for signs of collective behaviour, in particular of transverse collective flow. We found positive evidence for collective flow, which we interpret as indication that we have already succeeded in generating a very hot and dense quasi-equilibrated initial state ("fireball") in these collisions which, due to its huge internal pressure, begins to expand collectively into the transverse directions. We found that the initial energy densities and baryon densities are high enough to be interesting and to seriously consider the phase transition into a QGP or mixed phase.

All in all, the evidence is mounting that in these collisions a quite new state of sizeable volume (tens of fm^3), very high energy density (GeV/fm^3) and particle density (several $particles/fm^3$), and a large degree of equilibration is created. Such a state has never been seen anywhere else before; whether it also satisfies the criteria for a quark-gluon plasma, further studies will have to show.

Acknowledgements: This work was supported in part by the Deutsche Forschungsgemeinschaft (DFG), grant He1283/3-1, by the Bundesministerium für Technologie (BMFT), grant 06 OR 764, and by the Korean Science and Engineering Foundation (KOSEF), grant 893-0202-002-2. The lecturer (U.H.) would like to thank the organizers of this summer school, in particular Jean Cleymans, for the nice hospitality and financial support.

References

1. W. Greiner (ed.), *The Nuclear Equation of State*, NATO ASI Series B, Plenum Press, New York, 1990
2. H. Stöcker and W. Greiner, Phys. Rep. **137** 277 (1986)
3. S. Daté, M. Gyulassy, and H. Sumiyoshi, Phys. Rev. D **32** 619 (1985)
4. R. Anishetty, P. Köhler, and L. McLerran, Phys. Rev. D **22** 2793 (1980)
5. H. Satz, I. Harrity, and J. Potvin (eds.), *Lattice Gauge Theory '86*, Plenum Press, New York, 1987;
   F. Karsch, Nucl. Phys. B (Proc. Suppl.) **9** 357 (1989)
6. H.-Th. Elze and U. Heinz, Phys. Rep. **183** 81 (1989)
7. J. Schwinger, Phys. Rev. **82** 664 (1951)
8. U. Heinz, Phys. Lett. B **144** 288 (1984)
9. S. K. Wong, Nuovo Cim. **A65** 689 (1970)
10. L. T. Thomas, Phil. Mag. **3** 1 (1927);
    V. Bargmann, L. Michel, and V. L. Telegdi, Phys. Rev. Lett. **2** 435 (1959)
11. U. Heinz, Phys. Rev. Lett. **51** 351 (1983); **56** 93(C) (1986);
    U. Heinz, Ann. Phys. (N.Y.) **161** 48 (1985)
12. E. Wigner, Phys. Rev. **40** 749 (1932)
13. S. R. de Groot, W. A. van Leeuwen, and Ch. G. van Weert, *Relativistic Kinetic Theory*, North-Holland, Amsterdam, 1980
14. H.-Th. Elze, M. Gyulassy, and D. Vasak, Nucl. Phys. B **276** 706 (1986)
15. H.-Th. Elze, M. Gyulassy, and D. Vasak, Phys. Lett. B **177** 402 (1986);
    H.-Th. Elze, Z. Phys. C **38** 211 (1988)
16. U. Heinz, Physica A **158** 111 (1989)

17. D. Vasak, M. Gyulassy, and H.-Th. Elze, Ann. Phys. (N.Y.) **173** 462 (1987)
18. H.-Th. Elze, CERN preprint CERN-TH.5574/89, Oct. 1989
19. H. Weigert and U. Heinz, Regensburg preprint TPR-90-28;
H. Weigert, Diploma thesis, University of Regensburg, 1989
20. C. Cronström, Phys. Lett. B **90** 267 (1980); and references cited therein
21. A. V. Selikhov, I. V. Kurchatov Institute of Atomic Energy preprint IAE-4733/1, submitted to Nucl. Phys. B
22. U. Heinz and J. C. Parikh, in preparation
23. U. Heinz and P. J. Siemens, Phys. Lett. B **158** 11 (1985);
U. Heinz, Ann. Phys. (N.Y.) **168** 148 (1986)
24. S. Mrówczyński, Phys. Rev. D **39** 1940 (1989);and Regensburg preprint TPR-89-9
25. U. Heinz, K. Kajantie, and T. Toimela, Phys. Lett. B **183** 96 (1987); and Ann. Phys. (N.Y.) **176** 218 (1987);
J. Lopez, J. C. Parikh, and P. J. Siemens, Texas A & M preprint, 1985;
T. H. Hansson and I. Zahed, Phys. Rev. Lett. **58** 2397 (1987); and Nucl. Phys. B **292** 725 (1987);
R. Kobes and G. Kunstatter, Phys. Rev. Lett. **61** 392 (1988);
S. Nadkarni, Phys. Rev. Lett. **61** 396 (1988);
R. Kobes, G. Kunstatter, and K. W. Mak, Z. Phys. C **45** 129 (1989);
E. Braaten and R. D. Pisarski, Phys. Rev. Lett. **64** 1343 (1990)
26. U. Heinz, P. R. Subramanian, H. Stöcker, and W. Greiner, J. Phys. G **12** 1237 (1986)
27. R. Hagedorn and J. Rafelski, in: *Statistical Mechanics of Quarks and Hadrons*, p. 237, (H. Satz, ed.), North Holland, Amsterdam, 1981
28. U. Heinz, K. S. Lee, and M. Rhoades-Brown, Mod. Phys. Lett. A **2** 153 (1987)
29. P. R. Subramanian, H. Stöcker, and W. Greiner, Phys. Lett. B **173** 468 (1986)
30. K. S. Lee, M. Rhoades-Brown, and U. Heinz, Phys. Rev. C **37** 1452 (1988)
31. U. Heinz, K. S. Lee, and M. Rhoades-Brown, Phys. Rev. Lett. **58** 2292 (1987)
32. K. Redlich, Z. Phys. C **27** 633 (1985);
L. McLerran, Nucl. Phys. A **461** 245c (1987)
33. H. W. Barz, B. L. Friman, J. Knoll, and H. Schulz, Nucl. Phys. A **484** 661 (1988)
34. P. Koch, B. Müller, and J. Rafelski, Phys. Rep. **142** 167 (1986)
35. J. Ellis, U. Heinz, and H. Kowalski, Phys. Lett. B **233** 223 (1989);
see also T. A. DeGrand, Phys. Rev. D **30** 2001 (1984)
36. R. Matiello, H. Sorge, H. Stöcker, and W. Greiner, Phys. Rev. Lett. **63** 1459 (1989)
37. K. S. Lee, M. Rhoades-Brown, and U. Heinz, Phys. Lett. B **174** 123 (1986)
38. B. Lukács, J. Zimányi, and N. L. Balazs, Phys. Lett. B **183** 27 (1987)
39. C. Greiner, P. Koch, and H. Stöcker, Phys. Rev. Lett. **58** 1825 (1987);
C. Greiner, D.-H. Rischke, H. Stöcker, and P. Koch, Phys. Rev. D **38** 2797 (1988)
40. E. Witten, Phys. Rev. D **30** 272 (1984)
41. J. Rafelski, Phys. Rep. **88** 331 (1982); and Nucl. Phys. A **418** 215c (1984);
N. K. Glendenning and J. Rafelski, Phys. Rev. C **31** 823 (1985)
42. G. Giacomelli and M. Jacob, Phys. Rep. **55** 1 (1979)
43. (a) E802 collaboration, T. Abbott *et al.*, Phys. Rev. Lett. **64** 847 (1990);
(b) E802 collaboration, P. Vincent, M. Sarabura, H. Hamagaki *et al.*, Nucl. Phys. A **498** 67c, 409c, and 415c (1989)
44. E802 collaboration, T. Abbott *et al.*, Phys. Lett. B **197** 285 (1987);
E814 collaboration, J. Barrette *et al.*, Phys. Rev. Lett. **64** 1219 (1990)
45. R. Hagedorn, Nuovo Cim. Suppl. **3** 147 (1965);
R. Hagedorn and J. Ranft, Nuovo Cim. Suppl. **6** 169 (1968)
46. A. Shor, Phys. Rev. Lett. **54** 1122 (1985)

47. NA38 collaboration, A. Baldisseri *et al.*, Annecy preprint LAPP-EXP-89-15, to appear in the Proceedings of the International Europhysics Conference on High Energy Physics, Madrid, Sept. 6-13, 1989;
A. Baldisseri, Thesis, Annecy, 1990; and private communication
48. NA38 collaboration, C. Baglin *et al.*, Phys. Lett. B **220** 471 (1989);
NA38 collaboration, C. Racca *et al.*, in: *Hadronic Matter in Collision 1988*, (P. Carruthers and J. Rafelski, eds.), World Scientific Publ. Co., Singapore, 1989, p. 552
49. T. Matsui and H. Satz, Phys. Lett. B **178** 416 (1986)
50. P. Koch, U. Heinz, and J. Pišút, TPR-90-8, Phys. Lett. B , in press; and TPR-90-7, Z. Phys. C , in press
51. T. H. R. Skyrme, Nucl. Phys. **31** 556 (1962);
E. Witten, Nucl. Phys. B **223** 422, 433 (1983)
52. U. Heinz, P. R. Subramanian, and W. Greiner, Z. Phys. A **318** 247 (1984)
53. J. Ellis, R. A. Flores and S. Ritz, Phys. Lett. B **198** 393 (1987)
54. J. Ashman *et al.*, Phys. Lett. B **206** 364 (1988); and CERN preprint EP/89-73 (1989)
55. G. Adkins, C. Nappi and E. Witten, Nucl. Phys. B **228** 433 (1983)
56. U.-G. Meissner, N. Kaiser and W. Weise, Nucl. Phys. A **466** 685 (1987);
N. Kaiser, U. Vogl, W. Weise and U.-G. Meissner, Nucl. Phys. A **484** 593 (1988)
57. S. J. Brodsky, J. Ellis and M. Karliner, Phys. Lett. B **206** 309 (1988);
J. Ellis and M. Karliner, Phys. Lett. B **213** 73 (1988)
58. T. W. B. Kibble, J. Phys. A **9** 1387 (1976)
59. T. S. Biró and J. Zimányi, Nucl. Phys. A **395** 525 (1983)
60. J. Ellis and H. Kowalski, Phys. Lett. B **214** 161 (1988); and CERN-TH.5316/89 (1989)
61. S. Gavin *et al.*, Phys. Lett. B **234** 175 (1990)
62. J. Gasser and H. Leutwyler, Phys. Lett. B **184** 83 (1987)
63. R. V. Gavai, J. Potvin and S. Sanielevici, Florida State University preprint FSU-SCRI-89-18 (1989)
64. J. Gasser and H. Leutwyler, Phys. Rep. **87** 77 (1982)
65. V. Bernard and U.-G. Meissner, Phys. Lett. B **227** 165 (1989)
66. J. P. Bondorf, S. I. A. Garpmann, and J. Zimányi, Nucl. Phys. A **296** 320 (1978)
67. K. S. Lee, M. Rhoades-Brown, and U. Heinz, Phys. Rev. C **37** 1463 (1988)
68. E. Schnedermann, Diploma Thesis, University of Regensburg, 1989;
U. Heinz, K. S. Lee, and E. Schnedermann, TPR-89-13, in: *Quark-Gluon Plasma*, (R. Hwa, ed.), World Scientific's Advanced Series on Directions in High Energy Physics, Vol. 6, World Scientific, Singapore, 1990
69. K. S. Lee, U. Heinz, and E. Schnedermann, Regensburg preprint TPR-90-18, submitted to Z. Phys. C
70. K. S. Lee and U. Heinz, Z. Phys. C **43** 425 (1989)
71. F. Cooper and G. Frye, Phys. Rev. D **10** 186 (1974)
72. D. Kusnezov and G. Bertsch, Phys. Rev. C **40** 2075 (1989)
73. P. J. Siemens and J. O. Rasmussen, Phys. Rev. Lett. **42** 880 (1979)
74. P. Lévai, B. Lukács, J. Zimányi, and U. Heinz, Z. Phys. A **333** 77 (1989)
75. WA80 collaboration, R. Albrecht *et al.*, GSI preprint 89-03, submitted to Z. Phys. C ;
WA80 collaboration, R. Santo *et al.*, Nucl. Phys. A **498** 391c (1989);
L. Dragon, Thesis, University of Münster, 1989
76. NA35 collaboration, A. Bamberger *et al.*, Phys. Lett. B **203** 320 (1988)
77. P. Lévai, B. Lukács, and J. Zimányi, *Entropy content from strange particle ratios in the E802 experiment*, KFKI preprint KFKI-1989-47/A, 1989;
P. Koch and C. B. Dover, *Production of strange and nonstrange light nuclei in relativistic heavy-ion collisions*, BNL preprint, Dec. 1989
78. C. M. Ko, Phys. Rev. C **23** 2760 (1981)
79. NA35 collaboration, H. Ströbele *et al.*, Z. Phys. C **38** 89 (1988)

80. NA35 collaboration, A. Bamberger *et al.*, Z. Phys. C **43** 25 (1989)
81. NA34 collaboration, T. Åkesson *et al.*, CERN preprint EP/89-111, Z. Phys. C , in press; NA34 collaboration, B. Jacak *et al.*, in Ref. 1
82. J. Sollfrank, P. Koch, and U. Heinz, in preparation
83. NA35 collaboration, J. W. Harris *et al.*, Nucl. Phys. A **498** 133c (1989); NA35 collaboration, R. Renfordt *et al.*, Nucl. Phys. A **498** 385c (1989)
84. B. Alper *et al.*, Nucl. Phys. B **100** 237 (1975); J. W. Cronin *et al.*, Phys. Rev. D **11** 3105 (1975); D. Antreasyan *et al.*, Phys. Rev. D **19** 764 (1979)
85. R. Venugopalan and M. Prakash, Phys. Rev. C **41** 221 (1990)
86. NA35 collaboration, H. Ströbele *et al.*, in Ref. 1
87. E. Fermi, Prog. Theor. Phys. **5** 570 (1950); L. D. Landau, Izv. Akad. Nauk SSSR, Ser. Fiz. **17** 51 (1953); *Collected Papers of L. D. Landau*, p. 569, (D. Ter Haar, ed.), Pergamon, Oxford, 1965
88. J. Stachel and P. Braun-Munzinger, Phys. Lett. B **216** 1 (1989); F. W. Pottag *et al.*, in *Hadronic Matter in Collision 1988*, *loc cit.*, p. 310
89. P. Braun-Munzinger and J. Stachel, Nucl. Phys. A **498** 33c (1989); J. Stachel and P. Braun-Munzinger, Nucl. Phys. A **498** 577c (1989)
90. S. Nagamiya and M. Gyulassy, *Advances in Nuclear Physics*, Vol. 13, p. 201, Plenum, New York, 1984; R. Stock, Phys. Rep. **135** 259 (1986)
91. TPC/Two-Gamma collaboration, H. Aihara *et al.*, Phys. Rev. D **40** 2772 (1989)
92. E-735 collaboration, C. S. Lindsey *et al.*, Nucl. Phys. A **498** 181c (1989); E-735 collaboration, T. Alexopoulos *et al.*, Phys. Rev. Lett. **64** 991 (1990)
93. L. Csernai and L. Gutay, private communication; E. F. Johansen *et al.*, University of Bergen Scientific Report 203/1989

Strings in Ultrarelativistic Collisions

K. Werner

Theory Division, CERN, CH-1211 Geneva 23, Switzerland
On leave of absence from Physics Department, Brookhaven National Laboratory

CERN-TH-5682/90
March 1990

PREFACE

Classical relativistic strings provide a very successful concept, used in models for lepton-induced reactions (e^+e^- annihilation, deep inelastic lepton-nucleon and lepton-nucleus scattering) as well as for hadron or ion induced collisions (hadron-hadron, hadron-nucleus and nucleus-nucleus scattering). The applicability is restricted to ultrarelativistic energies, i.e. the string energies should be in general considerably larger than the mass of the hadrons with the same flavour content as the strings (so for example a uu-d string, having the flavour content of a proton, should have a mass considerably larger than the proton mass).

The diversity of applications for string models is very furtunate: one may build up a hierarchy of models, starting with the easiest case of e^+e^- annihilation, towards the very complex models for nucleus-nucleus scattering. In e^+e^- there is in the simplest case just one string produced, most likely a quark-antiquark (q-$\bar{q}$) string, so the fragmentation of such strings may be studied. In lepton-nucleon scattering in addition to fragmentation we get information about the parton structure of the nucleon. Also we get a new kind of string: diquark-quark (qq-q) strings and even more complicated fellows like quadruquark-antiquark ($qqqq$-$\bar{q}$) strings. Proceeding towards lepton-nucleus scattering we have to deal with two complications: (a) the parton structure of nucleons inside a nucleus may be altered by the nuclear medium and (b) the string fragmentation occurs inside a nucleus, so string fragments may interact with (so far) spectator nucleons. Despite the increasing complexity, for all these lepton-induced reactions at least the basic interaction vertex (involving an exchange-boson and an (anti)quark) can be calculated in perturbation theory. This is not possible in hadron-hadron interactions; so in this case we need a model for string formation before we can use the string fragmentation procedures tested in the models for lepton-induced scattering. The next step to proceed to hadron-nucleus or nucleus-nucleus models is an (almost) straightforward extrapolation of hadron-hadron models; essentially one has to consider in addition just the well known nuclear geometry of nuclei. These simple extrapolations ignore completely string-string interactions or interactions of string fragments with spectator nucleons. Only recently these processes have been included in string models. This very last step is very important, because it provides the link towards collective behaviour (like a Quark Gluon Plasma, an equilibrated system of deconfined quarks and gluons). So string models are nowadays not any more just "background" models, concerning interesting and new physics, they are rather important tools to study the evolution of a nucleus-nucleus collion towards a plasma state.

It is the purpose of this lectures to introduce the above mentioned hierarchy of string models for ultrarelativistic collisions of increasing complexity. We are interested in lepton-induced reactions mainly in order to learn about aspects needed later for the more complicated hadron-induced reactions. Therefore this article should not be considered a complete overview over string fragmentation models for e^+e^-, we are for example not interested in parton showers and multi-jet events, since presently the energies for nucleus-nucleus collisions are not high enough for such details to be relevant.

1 INTRODUCTION

For all theoretical considerations we use so-called "natural units" for physical quantities like time, length, energy, momentum etc. Such units have the advantage of making formulas simpler: factors of c and $\hbar$ disappear. In chapter 1 we introduce these natural units and discuss how they are related to "conventional" ones. In chapter 2 we give a very brief introduction to relativistic mechanics of point particles (zero-dimensional objects). Relativistic string dynamics, to be discussed later, is a generalization which deals with the space-time evolution of one-dimensional objects (=strings). In both cases our objects are elements of the four-dimensional Minkowski space; in the case of point particles we consider trajectories, in the case of strings we consider surfaces in Minkowski space.

1.1 Natural Units

"Conventional" units for length l and energy E are:

$$[l] = \text{fm} \equiv 10^{-10}\text{m} \tag{1.1}$$

and

$$[E] = \text{GeV} = 1.610^{-10}\text{Joule}. \tag{1.2}$$

In order not to introduce new units we multiply time t and momentum p with the velocity of light c, since ct and cp have the dimensions of length and energy, and correspondingly we multiply masses by c^2. So we use ct, cp and c^2m with dimensions

$$\begin{aligned} [ct] &= \text{fm} \\ [cp] &= \text{GeV} \\ [c^2m] &= \text{GeV}. \end{aligned} \tag{1.3}$$

Formulas become considerably simpler by using "natural units", i.e., considering length l and time ct in units of $\hbar c$

$$\begin{aligned} \tilde{l} &\equiv \frac{l}{\hbar c} \\ \tilde{t} &\equiv \frac{ct}{\hbar c} \end{aligned} \tag{1.4}$$

with Planck's constant being

$$\hbar c = 0.197\ \text{GeV}\,\text{fm}. \tag{1.5}$$

So we find the dimensions

$$\begin{aligned} [\tilde{l}] &= [\frac{l}{\hbar c}] = \text{GeV}^{-1} \\ [\tilde{t}] &= [\frac{ct}{\hbar c}] = \text{GeV}^{-1}. \end{aligned} \tag{1.6}$$

Hence by using $\frac{l}{\hbar c}$, $\frac{ct}{\hbar c}$, cp, E and c^2m the only dimension we have to deal with is GeV. In natural units c and $\hbar c$ are obviously unity, i.e.

$$\begin{aligned} c &= 1 \\ \hbar &= 1, \end{aligned} \tag{1.7}$$

which means that these constants do not appear in equations any more. Consider for example the Schrödinger equation

$$i\hbar\frac{\partial\psi}{\partial\tilde{t}} = -\frac{\hbar^2}{2m}\frac{\partial^2}{\partial x^2}\psi. \tag{1.8}$$

By using natural units, i.e., defining $\tilde{t} := \frac{ct}{\hbar c}$, $\tilde{x} := \frac{x}{\hbar c}$, and $\tilde{m} := mc^2$ we get

$$i\frac{\partial\psi}{\partial\tilde{t}} = -\frac{1}{2\tilde{m}}\frac{\partial^2}{\partial\tilde{x}^2}\psi \tag{1.9}$$

where Planck's constant $\hbar$ has disappeared. In the following, unless otherwise noted, we are going to use natural units.

1.2 Relativistic Mechanics

The main postulate of Special Relativity is that for a given space-time point $(t,\vec{x}) = (x_0,\vec{x})$ the quantity

$$s^2 = (x_0)^2 - (\vec{x})^2 \equiv x_0 x_0 - x_i x_i \tag{1.10}$$

is invariant in all inertial frames. By introducing the metric g as the 4×4 diagonal matrix

$$\{g^{\mu\nu}\} \equiv \{g_{\mu\nu}\} \equiv \begin{pmatrix} 1 & & & \\ & -1 & & \\ & & -1 & \\ & & & -1 \end{pmatrix} \tag{1.11}$$

we can write eq. (1.10) as

$$s^2 = x_\mu g^{\mu\nu} x_\nu \tag{1.12}$$

or by using the property of g

$$x^\nu := x_\mu g^{\mu\nu} \tag{1.13}$$

we write

$$s^2 = x^\nu x_\nu . \tag{1.14}$$

So the invariance condition is

$$x'^\mu x'_\mu = x^\mu x_\mu \tag{1.15}$$

or

$$x'^\mu g_{\mu\nu} x'^\nu = x^\mu g_{\mu\nu} x^\nu . \tag{1.16}$$

The linear ansatz

$$x'^\mu = \Lambda^\mu_\nu x^\nu \tag{1.17}$$

leads to

$$\Lambda^\mu_\rho x^\rho g_{\mu\nu} \Lambda^\nu_\sigma x^\sigma = x^\rho g_{\rho\sigma} x^\sigma \tag{1.18}$$

so the Λ has to fulfil

$$\Lambda^\mu_\rho g_{\mu\nu} \Lambda^\nu_\sigma = g_{\rho\sigma} \tag{1.19}$$

which reads in matrix notation

$$\Lambda^T g \Lambda = g. \tag{1.20}$$

There are four types of transformations (fulfilling eq. (1.20)):

(1) Rotations

$$\Lambda = \begin{pmatrix} 1 & \\ & R \end{pmatrix} \tag{1.21}$$

with an orthonormal ($R^T R = 1$) 3×3 matrix R.

(2) Boosts:

$$\Lambda = \begin{pmatrix} \cosh y & & & \sinh y \\ & 1 & & \\ & & 1 & \\ \sinh y & & & \cosh y \end{pmatrix} \tag{1.22}$$

(3) Time inversion

$$\Lambda = \begin{pmatrix} -1 & & & \\ & 1 & & \\ & & 1 & \\ & & & 1 \end{pmatrix} \tag{1.23}$$

(4) Full inversion

$$\Lambda = \begin{pmatrix} -1 & & & \\ & -1 & & \\ & & -1 & \\ & & & -1 \end{pmatrix} \tag{1.24}$$

We can easily understand the meaning of the "angle" y characterizing Lorentz Boosts. Consider the transformation from a comoving system $(t, x, y, z) = (d\tau, 0, 0, 0)$ to some reference frame with $(t', x', y', z') = (dt, dx, dy, dz)$, the differentials indicating that we are in the vicinity of $(0,0,0,0)$, where the two systems coincide. We perform a boost, eq. (1.22), to obtain

$$\begin{aligned} dt &= \cosh y \; d\tau \\ dz &= \sinh y \; d\tau \end{aligned} \tag{1.25}$$

from which we obtain a relation between y and the velocity $v = dz/dt$:

$$v = \tanh y \tag{1.26}$$

which can be inverted as

$$y = \frac{1}{2} \ln \frac{1+v}{1-v} \tag{1.27}$$

which is just the definition of the rapidity y.

In order to express energy and momentum in terms of rapidity y we assume that the energy, being the zero component of an energy momentum four-vector $p = (E, \vec{p})$ is equal to the mass of a system in the comoving frame. So by transforming $(m, 0, 0, 0)$ we obtain the general expression for p. To be a little bit more general, we are not considering the rest frame of a system (particle), but a frame where the longitudinal component p_z vanishes. So we have $p = (m_t, \vec{p_t}, 0)$ with the transverse mass defined as

$$m_t := \sqrt{m^2 + p_t^2}. \tag{1.28}$$

We transform p, using eq. (1.22), to obtain

$$p = (E, p_t, p_z) \tag{1.29}$$

with

$$\begin{aligned} E &= m_t \; \cosh y \\ p_z &= m_t \; \sinh y \end{aligned} \tag{1.30}$$

which, of course, fulfils the mass-shell condition

$$E^2 - p_t^2 - p_z^2 = m^2. \tag{1.31}$$

Using eqs. (1.26,27,30) we can write the rapidity as

$$y = \frac{1}{2}\ln\frac{E+p_z}{E-p_z}. \tag{1.32}$$

One of the most important properties of the rapidity variable is its behaviour under Lorentz boosts. Let y and y' be the rapidities of some particle in the frames F and F', where F moves with the rapidity $\tilde{y}$ relative to F'. We find:

$$\begin{aligned} y' &= \frac{1}{2}\ln\frac{E'+p'_z}{E'-p'_z} \\ &= \frac{1}{2}\ln\frac{E\cosh\tilde{y}+p_z\sinh\tilde{y}+E\sinh\tilde{y}+p_z\cosh\tilde{y}}{E\cosh\tilde{y}+p_z\sinh\tilde{y}-E\sinh\tilde{y}-p_z\cosh\tilde{y}} \\ &= \frac{1}{2}\ln\frac{(E+p_z)(cosh\tilde{y}+\sinh\tilde{y})}{(E-p_z)(\cosh\tilde{y}-\sinh\tilde{y})} \\ &= y+\tilde{y} \end{aligned} \tag{1.33}$$

which means that rapidity (not velocity!) is additional under Lorentz boosts. This makes the rapidity extremely useful for applications.

2 EVOLUTION AND FRAGMENTATION OF STRINGS

In order to keep the article selfcontained we review classical string theory in general, i.e. we discuss how to obtain a string action (2.1), we derive equations of motion for the space-time evolution of strings (2.2), and we discuss the general solution as well as conservation laws (2.3). We then treat the simplest possible string solution, the so called "yo-yo" string (2.4). In section 2.5 we discuss the rules for string breaking, in the general case and in particular for yo-yo strings. Although in classical string theory the time evolution is fixed once a string breakpoint is known, the determination of locations of breakpoints requires further input. For this purpose we introduce in 2.6 a model, where the same symmetry arguments are used which led earlier to the string action (leading to an "area law" for string breaking). We discuss relations of this fragmentation model to others in 2.7. We finally apply our model in 2.8 to compare with e^+e^-, $\bar{\nu}p$, and μp data. e^+e^- annihilation produces essentially quark-antiquark strings, whereas $\bar{\nu}p$ and μp provides a test for diquark-quark fragmentation. The actual weights for the string superpositions are given by the parton-model, a description of which is postponed to chapter 3.

2.1 A gauge invariant string action

In the following chapters we discuss classical string theory (see ref. [2.1]), as the basic tool for later applications. A classical string is a two-dimensional surface in the four-dimensional Minkowski space

$$x = x(\tau,\sigma) \tag{2.1}$$

with a spacelike parameter σ and a timelike one τ. Of course this is only one of infinitely many parametrizations of this surface. A transformation from one parameter space to another

$$\begin{pmatrix} \tau \\ \sigma \end{pmatrix} \longrightarrow \begin{pmatrix} \tilde{\tau}(\tau,\sigma) \\ \tilde{\sigma}(\tau,\sigma) \end{pmatrix} \tag{2.2}$$

is called a gauge transformation, and the group of such transformations is called a gauge group. One assumes that the string action should not depend on the parametrization, so gauge invariance is a necessary requirement. Further restrictions should be locality and covariance. Concerning the question of gauge invariance, it is useful to relate a metric g to a certain string parametrization via (using $\partial_1 \equiv \frac{\partial}{\partial\tau}$ and $\partial_2 \equiv \frac{\partial}{\partial\sigma}$):

$$g_{\alpha\beta} = \partial_\alpha x^\mu \partial_\beta x_\mu \tag{2.3}$$

where α and β assume the values 1 and 2. By using "dot" and "prime" as abbreviations for $\frac{\partial}{\partial\tau}$ and $\frac{\partial}{\partial\sigma}$, the metric can be written as

$$g = \begin{pmatrix} \dot{x}\dot{x} & \dot{x}x' \\ x'\dot{x} & x'x' \end{pmatrix}. \tag{2.4}$$

How does the metric g transform under gauge transformations eq. (2.2)? Defining the two component variable ξ via

$$\xi_1 \equiv \tau; \quad \xi_2 \equiv \sigma \tag{2.5}$$

and using

$$\tilde{\partial}_\alpha \equiv \frac{\partial}{\partial\tilde{\xi}_\alpha} \tag{2.6}$$

we get

$$\begin{aligned} g_{\alpha\beta} &= \partial_\alpha x^\mu \partial_\beta x_\mu \\ &= \tilde{\partial}_i x^\mu \partial_\alpha \tilde{\xi}_i \tilde{\partial}_j x_\mu \partial_\beta \tilde{\xi}_j \\ &= \partial_\alpha \tilde{\xi}_i \tilde{g}_{ij} \partial_\beta \tilde{\xi}_j. \end{aligned} \tag{2.7}$$

Since the components of the Jacobi matrix M of the gauge transformation eq. (2.2) are given as

$$M_{ab} = \partial_b \tilde{\xi}_a \tag{2.8}$$

we can write eq. (2.7) in matrix notation as

$$g = M^T \tilde{g} M \tag{2.9}$$

This leads to the identity

$$\sqrt{|\det g|} = \sqrt{|\det \tilde{g}|}\ |\det M|. \tag{2.10}$$

On the other hand we have

$$d^2\tilde{\xi} = |\det M|\ d^2\xi, \tag{2.11}$$

which together with eq. (2.10) immediately suggests that a ξ integration over $\sqrt{|\det g|}$ is invariant under gauge transformations:

$$\tilde{I} \equiv \int \sqrt{|\det \tilde{g}|}\ d^2\tilde{\xi} = \int \sqrt{|\det g|}\ d^2\xi \equiv I. \tag{2.12}$$

Writing the integral I explicitly as

$$I = \int \sqrt{(x'\dot{x})^2 - x'^2\dot{x}^2}\ d\tau d\sigma \tag{2.13}$$

shows that I is also local and covariant. In fact I is the simplest local, covariant and gauge invariant expression, so a very attractive candidate for a string action. Therefore we define the action of a relativistic string as [2.2]

$$S = \int L \, d\tau d\sigma \tag{2.14}$$

with

$$L = -\kappa \sqrt{-\det g} = -\kappa \sqrt{(x'\dot{x})^2 - x'^2 \dot{x}^2} \tag{2.15}$$

(we used $|\det g| = -\det g$). We will see later that the proportionality constant κ can be identified with the "string tension", the energy per unit length of the string. This action in fact measures the area of the string surface:

$$S = -\kappa \int d^2A \tag{2.16}$$

with d^2A being a string surface element. This becomes obvious when we choose τ to be the time t and σ to be the length of the string: by defining

$$\gamma \equiv (1 - v_\perp^2)^{-1}; \qquad \vec{v}_\perp \equiv \frac{\partial \vec{x}}{\partial t} - \frac{\partial \vec{x}}{\partial l}\left(\frac{\partial \vec{x}}{\partial t}\frac{\partial \vec{x}}{\partial l}\right)$$

and using $\left|\frac{\partial \vec{x}}{\partial l}\right| = 1$ we find

$$S = -\kappa \int \frac{1}{\gamma} \, dl dt \tag{2.17}$$

which is a surface integral as stated in eq. (2.16).

2.2 Equations of motion, conservation laws

We rewrite the action defined in the last chapter more explicitly as

$$S = \int_{\tau_1}^{\tau_2} d\tau \int_0^{\pi} d\sigma \, L \tag{2.18}$$

with

$$L = -\kappa \sqrt{(x'\dot{x})^2 - x'^2 \dot{x}^2} \tag{2.19}$$

We use the convention $\sigma_{\min} = 0$ and $\sigma_{\max} = \pi$, τ_1 and τ_2 are initial and final time. To obtain the equations of motion we require

$$\delta S = 0 \tag{2.20}$$

under infinitesimal variations $\delta x(\sigma, \tau)$ of the string surface. We obtain

$$\delta S = \int_{\tau_1}^{\tau_2} dt \int_0^{\pi} d\sigma \left(\frac{\partial L}{\partial \dot{x}_\mu} \frac{\partial}{\partial \tau} \delta x_\mu + \frac{\partial L}{\partial x'_\mu} \frac{\partial}{\partial \sigma} \delta x_\mu \right). \tag{2.21}$$

Integration by paths yields

$$\delta S = \int_0^{\pi} d\sigma \left[\frac{\partial L}{\partial \dot{x}_\mu} \delta x_\mu \right]_{\tau_1}^{\tau_2} + \int_{\tau_1}^{\tau_2} d\tau \left[\frac{\partial L}{\partial x'_\mu} \delta x_\mu \right]_0^{\pi} - \int_{\tau_1}^{\tau_2} d\tau \int_0^{\pi} d\sigma \left(\frac{\partial}{\partial \tau} \frac{\partial L}{\partial \dot{x}_\mu} + \frac{\partial}{\partial \sigma} \frac{\partial L}{\partial x'_\mu} \right) \delta x_\mu. \tag{2.22}$$

Keeping the initial and final position of the string fixed

$$\delta x_\mu(\tau_1, \sigma) = \delta x_\mu(\tau_2, \sigma) = 0 \tag{2.23}$$

we find the equations of motion

$$\frac{\partial}{\partial\tau}\frac{\partial L}{\partial\dot{x}_\mu}+\frac{\partial}{\partial\sigma}\frac{\partial L}{\partial x'_\mu}=0 \tag{2.24}$$

and the boundary conditions

$$\frac{\partial L}{\partial x'_\mu}=0 \qquad \text{at } \sigma=0,\pi. \tag{2.25}$$

One may obtain conservation laws, being the consequence of the invariance of the Lagrangian density L under the Poincaré group (Lorentz transformations plus translations). In the following we study the consequence of translational invariance. We again consider variations δx, but take into account the equations of motion. We get (see eq. (2.22)):

$$\delta S=\int_{\sigma_1}^{\sigma_2} d\sigma\left[\frac{\partial L}{\partial\dot{x}_\mu}\delta x_\mu\right]_{\tau_1}^{\tau_2}+\int_{\tau_1}^{\tau_2} d\tau\left[\frac{\partial L}{\partial x'_\mu}\delta x_\mu\right]_0^\pi \tag{2.26}$$

where we now consider a small surface $\sigma_1\le\sigma\le\sigma_2$, $\tau_1\le\tau\le\tau_2$. For an infinitesimal translation

$$\delta x_\mu=\epsilon_\mu \tag{2.27}$$

we get

$$\delta S=\left[\int_{\sigma_1}^{\sigma_2} d\sigma\left[\frac{\partial L}{\partial\dot{x}_\mu}\right]_{\tau_1}^{\tau_2}+\int_{\tau_1}^{\tau_2} d\tau\left[\frac{\partial L}{\partial x'_\mu}\right]_{\sigma_1}^{\sigma_2}\right]\epsilon_\mu. \tag{2.28}$$

Because of the invariance of S under arbitrary translations ϵ_μ we find

$$\int_{\sigma_1}^{\sigma_2} d\sigma\left[\frac{\partial L}{\partial\dot{x}_\mu}\right]_{\tau_1}^{\tau_2}+\int_{\tau_1}^{\tau_2} d\tau\left[\frac{\partial L}{\partial x'_\mu}\right]_{\sigma_1}^{\sigma_2}=0. \tag{2.29}$$

With C_0 being the curve surrounding the area $\sigma_1\le\sigma\le\sigma_2$, $\tau_1\le\tau\le\tau_2$, we get

$$\int_{C_0} d\sigma\,\frac{\partial L}{\partial\dot{x}_\mu}+d\tau\,\frac{\partial L}{\partial x'^\mu}=0. \tag{2.30}$$

This is the strings version of conserved currents according to Noether's Theorem. We define the energy momentum currents as:

$$P^\mu_\tau:=-\frac{\partial L}{\partial\dot{x}_\mu} \qquad P^\mu_\sigma:=-\frac{\partial L}{\partial x'_\mu}. \tag{2.31}$$

The energy momentum associated with an arbitrary curve C is defined as

$$P^\mu\equiv P^\mu(C)\equiv\int_C d\sigma\,P^\mu_\tau+d\tau\,P^\mu_\sigma. \tag{2.32}$$

From eq. (2.30) it follows that the momentum of a closed curve is zero

$$P^\mu(C_0)=0, \qquad C_0 \text{ closed}. \tag{2.33}$$

The momentum of the string is defined as

$$P^\mu(\text{string}):=P^\mu(C_{\text{str}}) \tag{2.34}$$

where C_{str} is any curve from one boundary to the other (not necessarily at constant time).

Because of $P^\mu(\text{closed } C) = 0$ and eq. (2.25) $P^\mu_\sigma(\sigma = 0, \pi) = 0$ the definition eq. (2.34) makes sense. This definition in particular implies that the string momentum is time independent, since C_{str} can be chosen at fixed but arbitrary time τ ($C_{str} = C_\tau$):

$$P^\mu(\text{string}) = P^\mu(C_\tau) = \int_{C_\tau} d\sigma \; P^\mu_\tau. \tag{2.35}$$

Using the currents defined in eq. (2.31) we may rewrite the equations of motion eq. (2.24) as

$$\frac{\partial}{\partial \tau} P^\mu_\tau + \frac{\partial}{\partial \sigma} P^\mu_\sigma = 0, \tag{2.36}$$

and the boundary condition eq. (2.25) reads

$$P^\mu_\sigma = 0 \qquad \text{at } \sigma = 0, \pi. \tag{2.37}$$

Our next aim will be to solve these equations of motion.

2.3 Solutions

To solve the equations of motion we choose a gauge which simplifies the equations of motion. The orthonormal gauge

$$x'\dot{x} = 0, \qquad \dot{x}^2 + x'^2 = 0 \tag{2.38}$$

does so. If we completely fix the gauge by setting

$$x_0 = \tau = t, \tag{2.39}$$

eq. (2.38) reads:

$$\vec{x}'\dot{\vec{x}} = 0, \qquad (\dot{\vec{x}})^2 + (\vec{x}')^2 = 1. \tag{2.40}$$

The currents eq. (2.31) are now

$$P_\tau = \kappa \dot{x}, \qquad P_\sigma = -\kappa x' \tag{2.41}$$

and the equations of motion eq. (2.36) are simply wave equations

$$\ddot{x}_\mu - x''_\mu = 0, \tag{2.42}$$

and the boundary conditions eq. (2.37) are

$$x'(t,0) = x'(t,\pi) = 0. \tag{2.43}$$

The solution of eqs. (2.42,43) is

$$\vec{x}(t,\sigma) = \frac{1}{2}[\vec{y}(t+\sigma) + \vec{y}(t-\sigma)] \tag{2.44}$$

where $\vec{y}$ is obviously the trajectory of one endpoint $\vec{y}(t) = \vec{x}(t,0)$, called the directrix. The directrix has to be periodic

$$\vec{y}(t+2\pi) - \vec{y}(t) = \frac{2\vec{P}}{\kappa}. \tag{2.45}$$

Eq. (2.44) means: each point on the string may be obtained by a simple geometrical construction once the directrix $\vec{y}(t)$ is known. From eqs. (2.35,41,44) we also see that the

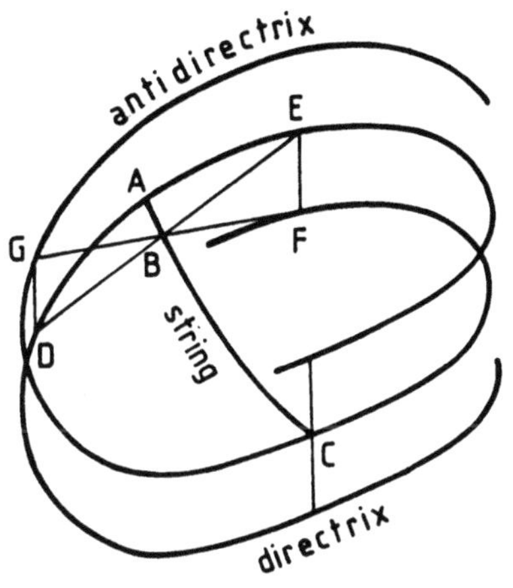

Fig. 2.1. A string with its directrix and antidirectrix. The directrix segment DAE defines the string piece AB and the antidirectrix FCG the string piece CB.

momentum of a piece of string is generated by the momenta of the two corresponding directrix pieces:

$$dP(t,\sigma) = \frac{\kappa}{2}[\dot{y}(t+\sigma)d\sigma + \dot{y}(t-\sigma)d\sigma] \tag{2.46}$$

so the momentum for the string piece corresponding to $[0,\sigma]$ (from A to B in fig. 2.1) is

$$P[0,\sigma] = \int_0^\sigma dP = \frac{\kappa}{2}\int_{t-\sigma}^{t+\sigma} dt'\ \dot{y}(t') = \frac{\kappa}{2}[y(t+\sigma) - y(t-\sigma)] \tag{2.47}$$

being proportional to the distance vector between two directrix points (DE in fig. 2.1). All this shows that the directrix piece from $D = \vec{y}(t-\sigma)$ to $E = \vec{y}(t+\sigma)$ determines the string piece from $A = \vec{x}(t,0)$ to $B = \vec{x}(t,\sigma)$. We also find a section of the directrix related to the other string part from $B = \vec{x}(t,\sigma)$ to $C = \vec{x}(t,\pi)$. However we can equally well relate to this string piece the "antidirectrix" (trajectory of the other end $\vec{x}(t,\pi)$) from F over C to G. It is obvious from fig. 2.1 that putting together the directrix piece D to E and the antidirectrix piece F to G, corresponding to the two string pieces, we recover the full directrix (after a shift of the antidirectrix by $\frac{1}{\kappa}\vec{P}$, which is the constant vector by which directrix and antidirectrix differ).

2.4 The yo-yo string

We are now going to discuss a simple but important example: a one-dimensional directrix, one period of which consists of two linear segments ("yo-yo string"). For a one-dimensional directrix, straight lines with a tilt of 45° against vertical (in space-time) are mandatory, since the string end (represented by $\vec{y}(t)$) moves with the velocity of light (because of eqs. (2.40,43)). From eq. (2.44) it is clear that the corresponding string is a simple straight line ($q\bar{q}$ in fig. 2.2) stretched between directrix and antidirectrix.

It is very instructive to investigate energy and momentum distribution along a one-dimensional yo-yo string. For the following we use $E \equiv p_0$ and $P \equiv p_3$ for energy and longitudinal momentum (no transverse momentum), the space-time coordinates are t and $z \equiv x_3$. From eqs. (2.35, 2.41) we obtain for an arbitary string element

$$dE = \kappa\, d\sigma; \qquad dP = \kappa \dot{z}\, d\sigma. \tag{2.48}$$

The gauge fixing condition $z'\dot{z} = 0$ (from eq. (2.40)) requires either z' or $\dot{z}$ to be zero,

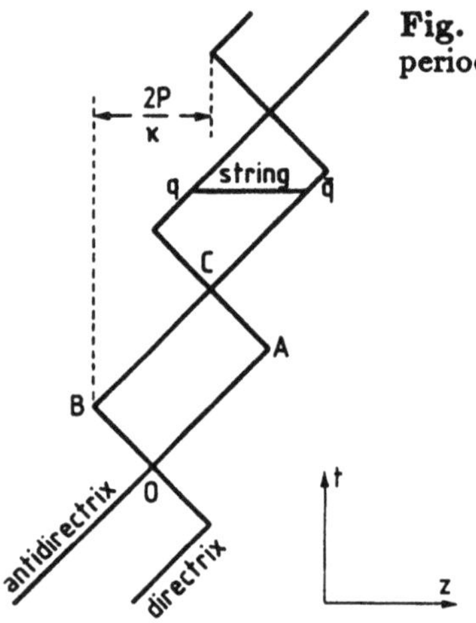

Fig. 2.2. Space-time picture of a "yo-yo" string with period $2P/\kappa$.

an ordinary yo-yo having exactly two points with $z' = 0$: the two endpoints of the string (because of the boundary condition eq. (2.37) every string has to fulfil $\vec{x}' = 0$ at the endpoints). So we obtain from the gauge fixing conditions $z'\dot{z} = 0$ and $\dot{z}^2 + z'^2 = 1$, inside the string

$$\dot{z} \equiv \frac{\partial z}{\partial t} = 0; \qquad |z'| \equiv \left|\frac{\partial z}{\partial \sigma}\right| = 1; \qquad \text{for } \sigma \neq 0, \pi \tag{2.49}$$

and at the endpoints

$$z' = \frac{\partial z}{\partial \sigma} = 0; \qquad |\dot{z}| \equiv \left|\frac{\partial z}{\partial t}\right| = 1; \qquad \text{for } \sigma = 0, \pi \tag{2.50}$$

Since these two domains (characterized by $z' = 0$ and $z' \neq 0$) behave so differently, we are going to discuss their contributions to energy and momentum separately. We use an index g (like glue) for the interior and an index q (like quark) for the endpoints. The energy and momentum of an inner piece of string of length dl are (using eqs. (2.48,59)):

$$dE_g = \kappa\, dl; \qquad dP_g = 0 \tag{2.51}$$

whereas we get from eqs. (2.48,50) for an endpoint, during a time step dt with a corresponding movement dz of the endpoint, the following change of energy and momentum:

$$dE_q = \kappa s\, dt; \qquad dP_q = \kappa s\, dz \tag{2.52}$$

(for $dt > 0$) where $s = +1(-1)$ means the endpoint has absorbed (emitted) a piece of string (or a piece of parameter space, to be precise). Eqs. (2.51,52) demonstrate, among other things, energy conservation: the energy dE_g gained by an endpoint by absorbing a piece of string is equal to the energy loss $-dE_g$ of the string due to its contraction. It is also easy to see from eq. (2.52) that the two endpoints change momentum in an opposite way: $dP_{\bar{q}} = -dP_q$, guaranteeing momentum conservation. We are now going to integrate eq. (2.52). Let us consider one "basic cell" $OACB$ in fig. 2.2. The polygon OBC is half a period of the directrix; instead of the other half we consider the corresponding half period of the antidirectrix OAC (which is equivalent). So $OACB$ defines the string completely. We use for the left end (OBC) the index $\bar{q}$, for the right end the index q. At the turning points the momenta vanish

$$P_q(A) = P_{\bar{q}}(B) = 0 \tag{2.53}$$

and since this implies that at these points the parameter space specifying the endpoints consists of just one point (0 and π respectively), eq. (2.53) also requires the energy to be zero

$$E_q(A) = E_{\bar{q}}(B) = 0. \tag{2.54}$$

Now we can easily integrate eq. (2.52) backwards to point O to obtain the initial energy and momentum (we use $t_O = z_O = 0$):

$$\begin{aligned} E_q(O) &= \kappa\, t_A; \qquad & P_q(O) &= \kappa\, z_A \\ E_{\bar{q}}(O) &= \kappa\, t_B; \qquad & P_{\bar{q}}(O) &= \kappa\, z_B \end{aligned} \tag{2.55}$$

Using light-cone coordinates $x^{\pm} = t \pm z$ and $p^{\pm} = E \pm P$ we get

$$p_q^+(O) = \kappa\, x^+(A); \qquad p_{\bar{q}}^-(O) = \kappa\, x^-(B) \tag{2.56}$$

together with $p_q^-(O) = 0$ and $p_{\bar{q}}^+(O) = 0$. So eqs. (2.56) provide a simple relation between initial momenta and the length of directrix pieces, or in other words, we have a mapping "momentum space" to "real space" via

$$\Delta p = \kappa\, \Delta x. \tag{2.57}$$

2.5 String breaking

We do not know within our classical treatment where a string breaks, but once we know the breakpoint, we know how to proceed. As for the action we assume locality. If a breaking occurs at $x(t,\sigma)$, we have to make sure that for the future as well as the past we have periodic (anti-) directrices, and the directrices for future and past have to match properly in the present. The only way to do so is to periodically continue (independently) the directrix corresponding to one string piece and the antidirectrix corresponding to the other string piece into the future. This fully determines the time evolution of either string piece also for all the future (till the next break at least).

Let us now discuss these "cutting rules" for a yo-yo string (see fig. 2.3). Without interaction the string stretches between directrix $(t, y(t))$ and antidirectrix $(t, \bar{y}(t)) = (t, x(t,\pi))$. Let the point $B = (t, x(t,\sigma))$ be a breakpoint on the string at time t, dividing the string into two segments AB and BC with $A = (t, x(t,0))$ and $C = (t, x(t,\pi))$. The directrix and antidirectrix corresponding to these segments are DEF with

$$\begin{aligned} D &= (t-\sigma, y(t-\sigma)) \\ F &= (t+\sigma, y(t+\sigma)) \end{aligned} \tag{2.58}$$

and GHI with

$$\begin{aligned} G &= (t-(\pi-\sigma), \bar{y}(t-(\pi-\sigma))) \\ I &= (t+(\pi-\sigma), \bar{y}(t+(\pi-\sigma))). \end{aligned} \tag{2.59}$$

Using

$$\bar{y}(t) = y(t-\pi) + \frac{P}{\kappa} \tag{2.60}$$

we verify easily that, after the appropriate shift, the segments DEF and GHI provide a full period of the unperturbed string. As discussed earlier, we obtain the directrices of the two segments after the break by continuation of DEF ($\rightarrow DEFS\cdots$) and of GHI

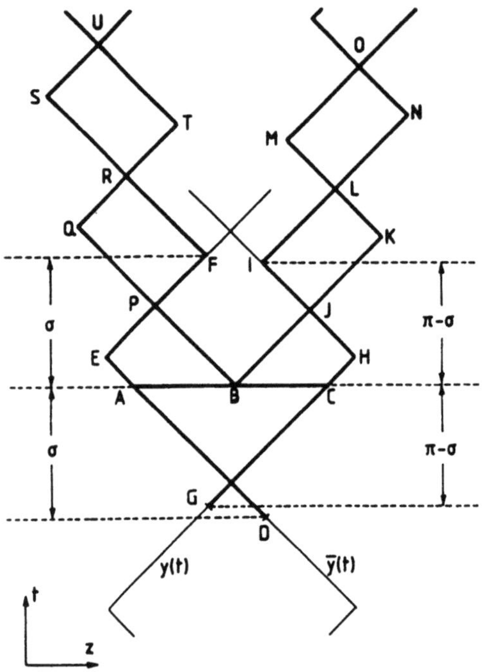

Fig. 2.3. Breaking of a "yo-yo" string. The new directrices are constructed according to the rules of classical string dynamics: we first determine the segments of the (anti)directrix corresponding to the string pieces AB and CB; these segments are then periodically continued into the future.

($\rightarrow GHIN\cdots$). The corresponding antidirectrices can be easily constructed from the relation between directrix y and antidirectrix $\bar{y}$:

$$\bar{y}(t) = \frac{1}{2}(y(t+\pi) + y(t-\pi)), \tag{2.61}$$

so we get $BKM\cdots$ and $BQT\cdots$. We realize the identities

$$\|BJ\| = \|JK\|; \quad \|HJ\| = \|JI\| \tag{2.62}$$

and

$$\|BP\| = \|PQ\|; \quad \|EP\| = \|PF\|, \tag{2.63}$$

which provide a very simple procedure for actually constructing the new directrices in numerical applications.

2.6 A string fragmentation model (AMOR)

We discussed in the last chapters string dynamics, including the case of a breaking string. For a simple one-dimensional yo-yo string this is illustrated in fig. 2.3. We have not yet specified any law determining where the string breaks. This clearly goes beyond any classical treatment. However we can restrict the variety of possible breaking laws by requiring certain properties. In the same way as for the string action S (see section 2.1) it can be shown that the simplest local, covariant and gauge invariant breaking law can be written as

$$dP(\tau,\sigma) \sim \sqrt{-det g}\, d\tau d\sigma \tag{2.64}$$

with dP being the probability for a break at $x(\tau,\sigma)$, and g being the metric (eq. (2.3)). This means that the breaking probability is proportional to the corresponding area on the string surface: $dP \sim d^2A$ or

$$dP = (1-P)\,\alpha\, d^2A \tag{2.65}$$

with the "break probability" α as a parameter. This is the fragmentation law first suggested by Artru and Mennessier [2.3] and later also used by other authors [2.4]. It is so appealing because it is not just a good guess but rather a strict consequence of requiring very plausible properties: locality, covariance and gauge invariance. Another nice feature is that there is only one parameter (α) which should be the same for processes so different as for example diquark fragmentation into baryons or heavy quark fragmentation into heavy mesons.

For the following we restrict ourselves to yo-yo strings, which form a closed group among all possible strings, in the sense that a yo-yo breaks into two yo-yo's again. In fig. 2.4 we show an "elementary cell" of a half period directrix OAC ("quark" q_0) and the corresponding antidirectrix OBC ("antiquark" $\bar{q}_0$). Because of eq. (2.57) the coordinates $x_0^+ = t_A + z_A$ and $x_0^- = t_B - z_B$ are related to the initial momenta of quark (p_0^+) and antiquark (p_0^-) via

$$\|OA\| = x_0^+ = \frac{p_0^+}{\kappa}; \qquad \|OB\| = x_0^- = \frac{p_0^-}{\kappa}. \tag{2.66}$$

Let us consider a break-up at D into a "quark" $(DIJ\cdots)$ and an "antiquark" $(DEF\cdots)$. We may define "break-up momenta" $b_0^\pm$ via

$$\begin{aligned} \|OU\| &= \frac{b_0^+}{\kappa} \\ \|OV\| &= \frac{b_0^-}{\kappa}. \end{aligned} \tag{2.67}$$

With $p_1^\pm$ being the parton momenta of the right substring at E we find

$$\begin{aligned} \|EF\| &= \frac{p_1^+}{\kappa} = \frac{p_0^+ - b_0^+}{\kappa} \\ \|EG\| &= \frac{p_1^-}{\kappa} = \frac{b_0^-}{\kappa}. \end{aligned} \tag{2.68}$$

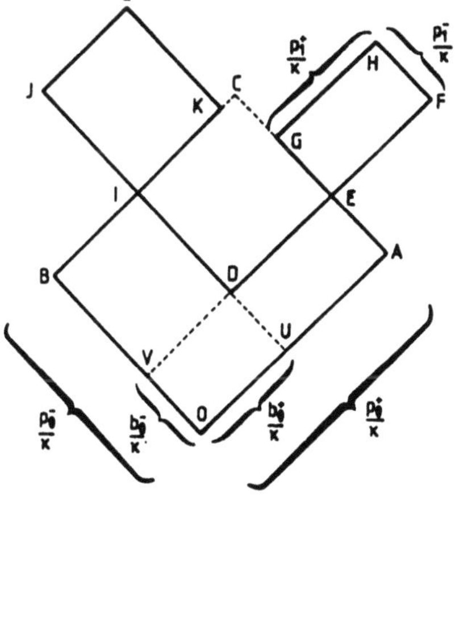

Fig. 2.4. Relation between (anti)directrix segments and parton momenta. In the string model the energy discontinuities (at the endpoints in the case of a yo-yo) can be identified with partons!

A corresponding formula holds for the other substring. The area of absolute past with respect to the break-up point D is given as

$$A \equiv \frac{1}{\kappa^2}\mathcal{A} = \frac{1}{\kappa^2} b_0^+ b_0^- . \tag{2.69}$$

All points D having the same value of A lie on a hyperbola in space-time, given by

$$(t+z)(t-z) = A. \tag{2.70}$$

As the other variable to fix D completely we choose the space-time rapidity

$$\eta \equiv \frac{1}{2}\ln\frac{t+z}{t-z} = \frac{1}{2}\ln\frac{b_0^+}{b_0^-}. \tag{2.71}$$

Using these variables $\mathcal{A}$ and η eq. (2.65) becomes

$$dP = (1-P)\,\frac{\alpha}{\kappa^2}\,d\mathcal{A}\,d\eta \tag{2.72}$$

leading to

$$dP(\mathcal{A}) = \alpha_0\, e^{-\alpha_0 \mathcal{A}}\, d\mathcal{A} \tag{2.73}$$

We are now in a position to define exactly, step by step, how we proceed to fragment a yo-yo string into two substrings. The proton content is completely arbitrary, we treat qq-q strings in the same way as q-$\bar{q}$ strings, or even more complicated structures are possible ($\bar{q}$-$qqqq$, $\bar{q}\bar{q}$-$qqqqq$ etc.). In order to fix the break point D we first determine $\mathcal{A} = \kappa^2 A$ via integrating and inverting eq. (2.73):

$$\mathcal{A} = -\frac{1}{\alpha_0}\ln r \tag{2.74}$$

with $r \in [0,1]$ being a random number. Before fixing D completely by determining η we have to be more specific about the break-up. We create a qq-$\bar{q}\bar{q}$ pair with probability $P_{\rm diq}$ (fit parameter) and a q-$\bar{q}$ with $(1-P_{\rm diq})$. Concerning flavour we create a strange quark with probability $P_{\rm str}$ (fit parameter) and u as well d quarks with $(1-P_{\rm str})/2$. We then look into a resonance table (see VENUS 3 writeup) to determine for each substring the minimum mass $m_{\rm min}$ for the corresponding parton content. So the minimum mass for a $u\bar{d}$ system is the π^+ mass and so on. Suggested by the uncertainty principle we generate transverse momenta $\vec{p}_t$ and $-\vec{p}_t$ for the two partons at D, according to an exponential distribution

$$f(p_t) \sim p_t \exp\left[-\frac{p_t}{2<p_t>}\right] \tag{2.75}$$

with a fit parameter $<p_t>$ to be chosen in the order of the inverse proton size. Taking this value of p_t together with the minimum mass $m_{\rm min}$ we get a minimum transverse mass $\mu_{\rm min}$ for each substring:

$$\mu_{\rm min} = \sqrt{m_{\rm min}^2 + p_t^2}. \tag{2.76}$$

Using for the transverse masses $\mu_\pm$ of the two substrings

$$\begin{aligned} \mu_+^2 &= p_0^+ b_0^- - \mathcal{A} \\ \mu_-^2 &= p_0^- b_0^+ - \mathcal{A} \end{aligned} \tag{2.77}$$

and using

$$\begin{aligned} b_0^+ &= \sqrt{\mathcal{A}}e^{\eta} \\ b_0^- &= \sqrt{\mathcal{A}}e^{-\eta}, \end{aligned} \tag{2.78}$$

we see that the requirement of minimum transverse masses restricts the rapidity η to be

$$\eta_- < \eta < \eta_+ \tag{2.79}$$

with

$$\eta_+ = \ln \frac{\sqrt{\mathcal{A}}\, p_0^+}{\mu_{+\,\min}^2 + \mathcal{A}}; \qquad \eta_- = \ln \frac{\mu_{-\,\min}^2 + \mathcal{A}}{\sqrt{\mathcal{A}}\, p_0^-}. \tag{2.80}$$

For $\eta_+ < \eta_-$ there is no solution, the string cannot be broken. Otherwise we determine the rapidity according to a constant distribution between η_- and η_+:

$$\eta = \eta_- + r(\eta_+ - \eta_-) \tag{2.81}$$

with a random number $r \in [0,1]$. From eq. (2.78) we see that the breakpoint is now fully determined.

This classical picture should be appropriate as long as the two substrings have a large mass. Whenever a small string mass occurs (say below 2 GeV) clearly quantum effects become important, most easily seen by the fact that a string with low mass is a hadron, and hadrons have discrete masses. Since we cannot deal with quantum strings properly, we try to correct for the quantum effect, in our classical model. First of all we introduce a cut-off parameter $m_{\text{clu}} = m_{\text{min}} + m_0$ which prevents strings with mass below m_{clu} from splitting further via the string fragmentation procedure. Such clusters are treated differently as discussed in the following. We are going to treat "exotic" quark configurations (like $\bar{q}qqqq$) later; first we only consider clusters which are — concerning quark and antiquark content — hadrons. We circumvent for all but the lowest hadron states the problem of discrete mass: we allow the resonances to be off-mass-shell. For each quark configuration $(u\bar{d}, u\bar{s}\ldots)$ we have a table of numbers $m_1 < m_2 < \ldots$ where the interval $[m_i, m_{i+1}]$ specifies the mass range for a certain resonance. This feature of resonances with continuous mass simplifies the fragmentation procedure; there is no correction needed to the breaking procedure as described above. However stable hadrons ($\Gamma \ll 1GeV$) have to have discrete masses, so whenever a string breaks into two pieces with at least one of these being a stable hadron (its mass falling into the lowest mass interval) we apply a correction procedure: we construct a new breakpoint D. We recall that D is specified by two parameters: the area $\mathcal{A} = b_0^+ b_0^-$ and the rapidity $\eta = \frac{1}{2}\ln(b_0^+/b_0^-)$. Since we do not want to modify the area law eq. (2.73), we are going to modify η and leave $\mathcal{A}$ fixed, if one stable hadron is involved (two hadrons will be treated later). So when the right substring is a stable hadron, with a required mass m we determine η to be (see eq. (2.80)):

$$\eta = \ln \frac{\sqrt{\mathcal{A}}\, p_0^+}{m^2 + p_t^2 + \mathcal{A}}; \tag{2.82}$$

if the hadron is left, we use

$$\eta = \ln \frac{m^2 + p_t^2 + \mathcal{A}}{\sqrt{\mathcal{A}}\, p_0^-}. \tag{2.83}$$

This guarantees the correct mass. If both substrings are stable hadrons, with masses m_+ and m_-, we have to redetermine both parameters, $\mathcal{A}$ and η, by solving the two equations (using $\mu_\pm = \sqrt{m_\pm^2 + p_t^2}$ and eqs. (2.77,78)):

$$\begin{aligned} \mu_+^2 &= p_0^+ \sqrt{\mathcal{A}} e^{-\eta} - \mathcal{A} \\ \mu_-^2 &= p_0^- \sqrt{\mathcal{A}} e^{\eta} - \mathcal{A} \end{aligned} \tag{2.84}$$

which leads to

$$\mathcal{A} = \frac{1}{2}(p_0^+ p_0^- - \mu_-^2 - \mu_+^2) - \sqrt{\frac{1}{4}(p_0^+ p_0^- - \mu_-^2 - \mu_+^2)^2 - \mu_-^2 \mu_+^2}. \tag{2.85}$$

The other parameter η is determined from eq. (2.84). Because the mass of the hadron before correction is close to the real mass (= mass after correction), the new breakpoint D is in the vicinity of the old breakpoint, so it is really a correction in the sense of a small modification.

We want to stress that the break-up of a string into two substrings occurs completely arbitrarily in the sense that each of the substrings may be a stable hadron, a resonance or a high mass string. This is the major difference from the Lund model, where one fragment has to be a hadron with a discrete mass. So in our model we have a "tree structure": a string decays into two substrings, each substring may then decay into two subsubstrings and so on. The Lund model has a "salami structure": a hadron is chopped off at the end, then another one from the remaining string and so on.

The last step of the fragmentation procedure is resonance decay: all the primary hadrons (from string break-up) decay (if they are unstable) according to standard branching ratios. The off-shellness of resonances poses no difficulties. One has only to consider that some of the partial decays cannot occur because the energy is not available; we simply discard such decay modes. For details of the decay procedure see Ref. [2.5].

Exotic clusters are treated in a procedure called "cluster decay", to be distinguished from ordinary "resonance decay" as well as from "string splitting". Consider a cluster C of n quarks and m antiquarks with $n - m$ being an integer multiple of 3

$$C = (q_1 q_2 \dots q_n \bar{q}_1 \bar{q}_2 \dots \bar{q}_m) \tag{2.86}$$

where q_i, $\bar{q}_i$ are quark flavours $(u, d, s \dots)$. The probabilities to choose randomly a baryon (B), an antibaryon $(\bar{B})$, or a meson (M) are

$$P(B) = \frac{1}{N}\binom{n}{3}; \qquad P(\bar{B}) = \frac{1}{N}\binom{m}{3}; \qquad P(M) = \frac{1}{N} nm \tag{2.87}$$

with

$$N = \binom{n}{3} + \binom{m}{3} + nm. \tag{2.88}$$

According to these probabilities we are randomly selecting a hadron H, and then performing the decay

$$C \rightarrow H + C'. \tag{2.89}$$

With m_C being the cluster mass and m_H and $m_{C'}$ being the masses of the two decay products, we find the available centre-of-mass momentum to be

$$p_{cm} = \frac{1}{2m_C}\sqrt{(m_C^2 - m_H^2 - m_{C'}^2)^2 - (2\, m_H\, m_{C'})^2}. \tag{2.90}$$

The momenta of the two decay products in the cm system of the cluster C are then $\pm p_{cm}\vec{u}$, with a random unit vector $\vec{u} \in S_2$. The mass m_H is a discrete hadron mass, $m_{C'}$ is either a hadron mass or — if C' is still an exotic state — a mass chosen randomly in the possible mass range. In the latter case, the cluster decay procedure is repeated for C':

$$C' \to H' + C'' \tag{2.91}$$

and so on, till we are left just with ordinary hadrons.

Before coming to applications, we want to state some general remarks about our fragmentation model. A major motivation for keeping resonances off-shell is the fact that this fragmentation model was mainly constructed to be used in a model for hadronic interactions (VENUS 3). In particular for such reactions, interactions of produced resonances of the type

$$R_1 + R_2 \to R \to R_3 + R_4 + \cdots \tag{2.92}$$

may occur, i.e. the two resonances fuse into a highly excited resonance R before decaying again. Even if R_1 and R_2 were on-mass-shell, the fused object R will in general not be, so one has to deal with off-shell resonances anyhow. If only discrete masses are considered, one has to correct again and again, which makes the whole approach questionable. Interactions of the type (2.92) are included in the model, but are, however, completely negligible for all examples to be discussed in section 2.9, therefore we do not discuss rescattering here.

Also with view to applications for hadronic collisions we included "exotic clusters". Such objects are already needed for deep inelastic lepton-nucleon scattering: in the simplest case the vector boson couples to one of the valence quarks of the nucleon to produce a diquark-quark string (see fig. 2.7). However it also happens that the boson kicks off a sea quark, leaving back in the remaining nucleon the corresponding antiquark, and thus we get a q-$qqq\bar{q}$ string. After fragmentation we might be left with an exotic $\bar{q}qqqq$ system. Considering rescattering of resonances, we have to deal with exotics as well: by fusing, for example, a uuu and a $u\bar{s}$ we get the five quark state $uuuu\bar{s}$. We refer to the fragmentation model introduced in this section as AMOR (Artru-Mennessier Off-shell-Resonance model)

2.7 Other fragmentation models

In this chapter we would like to comment on the relation of AMOR to other fragmentation models. AMOR, as well as some other models [2.4] based on the Artru-Mennessier model [2.3], takes the string picture seriously and provides a covariant, gauge invariant (= reparametrization invariant) energy and momentum conserving string breaking procedure. On the other extreme there is the Field-Feynman model [2.7]. Instead of strings here one considers two independent partons moving in opposite directions, inspired by the experimentally observed jet-structure of produced particles. The model is not covariant and of course not gauge invariant. And since the two jets are independent, energy and momentum are not conserved. The Lund model [2.8] is somewhat in between. It is similar to the Field-Feynman model in the sense that again two partons are considered, and one fragmentation step amounts to forming a hadron which contains one of the partons. However the two partons are linked by a colour field, which makes it possible to achieve energy and momentum conservation as well as covariance.

Whereas AMOR allows a string to split into two substrings with arbitrary masses (string → string + string), the Lund model requires one of the substrings to be an on-shell hadron (string → string + hadron). In both approaches the iterative procedure terminates whenever the string masses are below some cut-off. Those strings are identified with stable hadrons or known resonances. In our model we do not force a resonance to have its mean mass. This concept has two major advantages: in the first place correction procedures usually necessary to force particles on their mass shell are reduced to a minimum, and secondly final-state interactions (e.g. resonance + resonance → resonance) in the case of multi-string fragmentation (heavy ion collisions) can be implemented quite naturally.

Unlike some other fragmentation models [2.4, 2.8, 2.9, 2.10, 2.11, 2.12], our model does not include perturbative parton showers, since we are not interested in the fragmentation involving very high mass partons created in hard scattering processes. We are mainly interested in strings formed in soft hadronic collisions which might be considered to be unexcited yo-yo strings, without any gluon kinks.

2.8 Comparison with data

We applied AMOR (also referred to as VENUS 3, since it is part of the VENUS 3 model for nuclear collisions) to calculate particle production in e^+e^-, $\bar{\nu}p$ and μp reactions (figs. 2.5-7). An e^+e^- event $\{e^+e^-\}$ is related to quark-antiquark string fragmentation as a simple superposition (see fig. 2.5):

$$\{e^+e^-\} = \frac{1}{\sum_i e_i^2} \sum_i e_i^2 \{q_i - \bar{q}_i\} \tag{2.93}$$

where the weights are determined from the quark charges. The e^+e^- energy determines which flavours have to be considered. In the following figures we use the convention: data are dots, model results are histograms. In fig. 2.8 we consider inclusive spectra for e^+e^- at 14 GeV: data and model agree almost perfectly for the distributions of transverse momentum p_t of charged particles, rapidity y of charged particles and the energy fraction x of photons. The transverse momentum shows essentially the exponential behaviour of the input distribution eq. (2.75), however the correct concave shape of the photon x-distribution is quite remarkable. Photon spectra are strongly affected by η resonance decay, so we may consider this agreement as some support of our low mass cluster treatment (concerning the production of flavour neutral resonances, see [2.5]). The deviation from an ideal plateau in the rapidity distribution in fig. 2.8, namely the shallow peak at large $|y|$, is due to the decay of heavy mesons: there are plenty of c-$\bar{c}$ strings, forming $\bar{q}c$ and $\bar{c}q$ mesons on the outside of the string, therefore we get a contribution at large rapidities. In figs. 2.9,10 we show the same distributions, however for higher energies (22 and 34 GeV). The photon energy fraction distributions are energy independent (scaling), and data and model agree nicely. The transverse momentum distribution in the model is almost energy independent, whereas the data show an increase of the p_t tails with energy. This effect is well known to be due to a perturbative parton cascade, not included in our model. By looking at the rapidity distributions we see that these missing high p_t particles are located at central rapidities. The tails of the rapidity distributions are always reproduced properly.

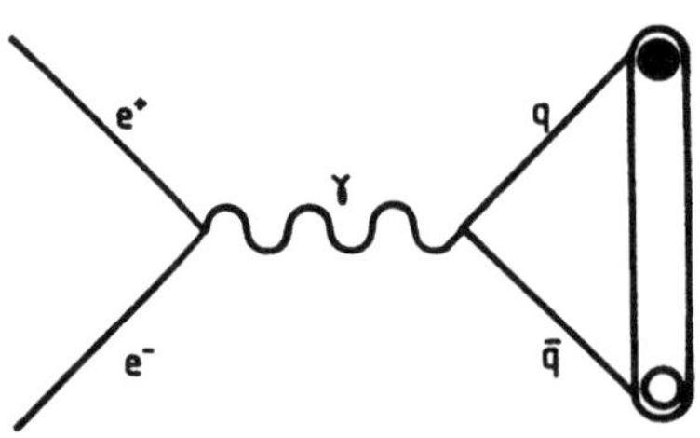

Fig. 2.5. String formation in e^+e^- annihilation.

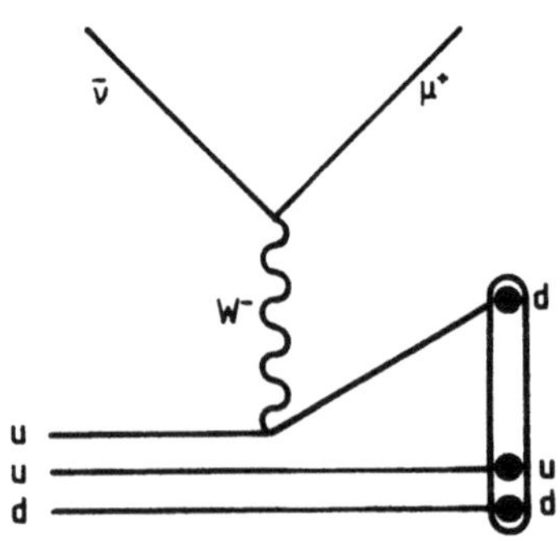

Fig. 2.6. String formation in $\bar{\nu}p$ scattering.

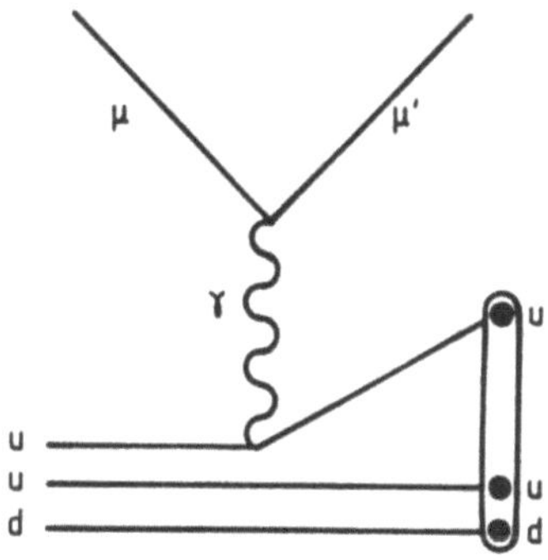

Fig. 2.7. String formation in μp scattering.

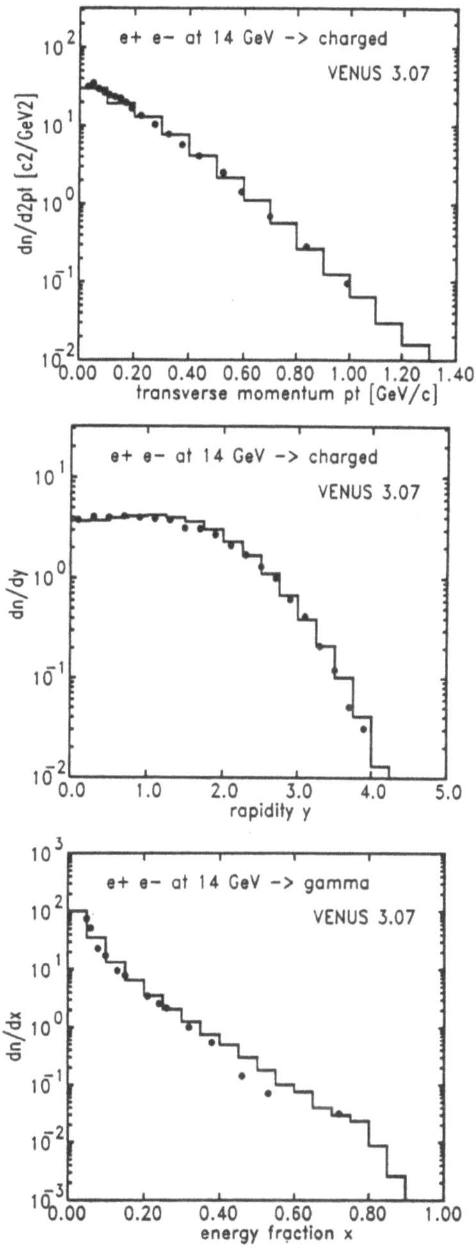

Fig. 2.8. Inclusive spectra for e^+e^- annihilation at 14 GeV: transverse momentum (upper plot) and rapidity distributions (middle plot) of charged particles, and energy fraction distributions of photons (lower plot). The data (dots) are from ref. [2.15].

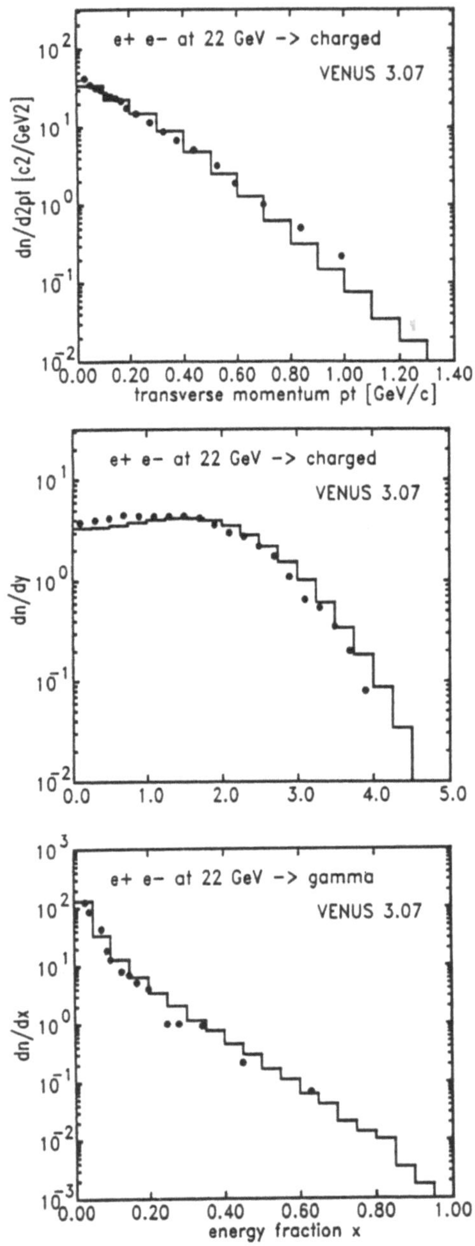

Fig. 2.9. Same as fig. 2.8, but 22 GeV.

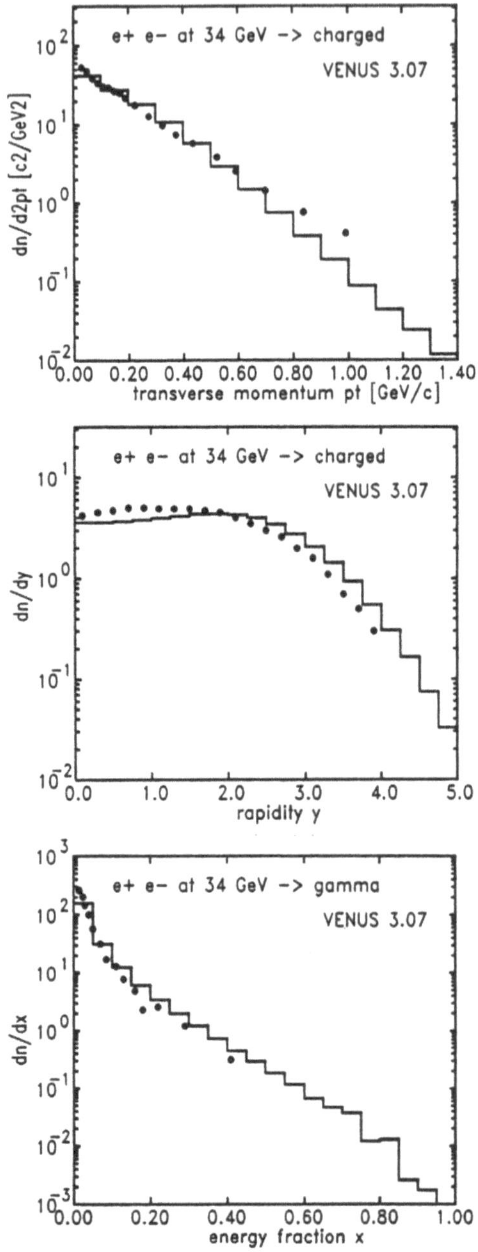

Fig. 2.10. Same as fig. 2.8, but 34 GeV.

We are now turning to antineutrino proton scattering: $\bar{\nu} + p \rightarrow \mu^+ +$ **"string"** , see fig. 2.6. Considering only the contributions with $\Delta s = 0$, we have two possibilities: The W^- boson couples to a u quark which transforms into a d quark, forming a d-ud string together with the remaining diquark $p - u = ud$ (see fig. 2.6); the other case occurs when the W^- couples to a $\bar{d}$ quark from the sea, to be transformed into a $\bar{u}$ quark, forming correspondingly a $\bar{u}$-$uudd$ string. For the experiments we are studying, the kinematic cuts

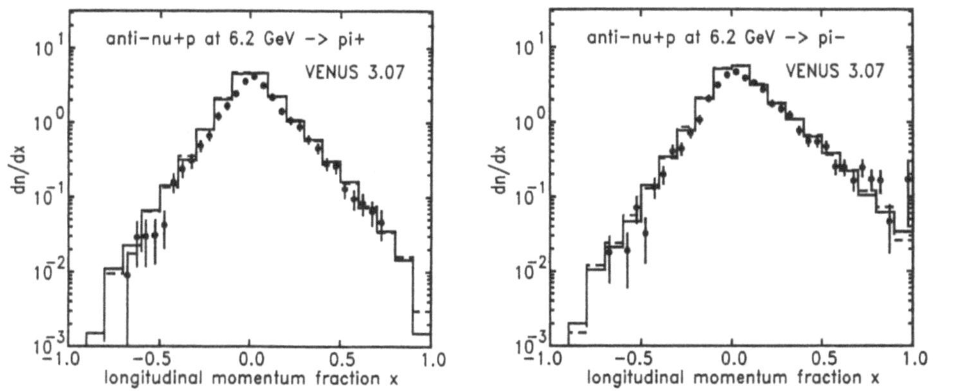

Fig. 2.11. Longitudinal momentum distribution for π^+ (left plot) and π^- (right plot) for $\bar{\nu}p$ scattering at 6.2 GeV (string mass) The dashed curve is a calculation including string-string interactions (no effect). The data (dots) are from ref. [2.16].

are such that the latter contribution can be neglected, so $\bar{\nu}p$ scattering is simply d-ud string fragmentation:

$$\{\bar{\nu}p\} = \{d\text{–}ud\} \tag{2.94}$$

In fig. 2.11 we show distributions of the longitudinal momentum fraction x for $\bar{\nu}p$ scattering at 6.2 GeV (d-ud string fragmentation with a string mass of 6.2 GeV). Momentum fraction refers to the maximum possible momentum of a produced particle in the string cm system, along the longitudinal axis (experimentally the string momentum can be reconstructed since in the reaction $\bar{\nu} + p \to \mu^+ +$ "string" the momenta of $\bar{\nu}$ and μ^+ are known). The data can be at least as equally well reproduced as with the former VENUS fragmentation routine (Field-Feynman procedure, see [2.13]), although now much fewer parameters enter. The distributions in fig. 2.11 are only affected by one parameter: the string breaking probability α_0 in eq. (2.73). Since on the diquark side (negative x) the first particle has to contain the diquark, being a baryon usually, the pions are at least second in the chain, and therefore most likely slower than the baryon. On the other side, where the quark fragments (positive x), a pion is usually first and fastest. For these reasons we get the forward-backward asymmetry for the π^-, seen in fig. 2.11. The dashed line contains final state interactions, not showing any effect.

We finally turn to muon proton scattering ($\mu + p \to \mu' +$ "string" , see fig. 2.7). The intermediate boson couples to either a quark or an antiquark, forming a q-$(p-q)$ or $\bar{q}$-$(p-\bar{q})$ string. The probability of a certain quark (antiquark) flavour i being involved is [2.14]:

$$P_i = N \int dx_B\, dy\, \frac{2\pi\alpha^2}{(-q^2)^2}\, s\, x_B(1+(1-y)^2)\, e_i^2\, q_i(x_B) \tag{2.95}$$

where the usual variables are used:

$$x_B \equiv \frac{-q^2}{2pq}; \qquad y \equiv \frac{pq}{pk} \tag{2.96}$$

with k,q and p being the four-momenta of the incoming muon, the photon and the proton; $q_i(x_B)$ is the parton momentum distribution function, e_i is the parton charge, and N is

Table 2.1: The weights of individual string contributions for the μp reactions discussed in this section.

string	$P\,[\%]$
$u{-}ud$	67.66
$d{-}uu$	8.46
$u{-}\bar{u}uud$	8.80
$d{-}\bar{d}uud$	2.20
$s{-}\bar{s}uud$	0.94
$\bar{u}{-}uuud$	8.80
$\bar{d}{-}duud$	2.20
$\bar{s}{-}suud$	0.94

the normalization. The integration of eq. (2.95) has to be performed over the appropriate acceptance area in the x_B-y plane, in order to compare with a specific experiment. Since the quark flavour i in eq. (2.95) may refer to valence quarks, sea quarks or sea antiquarks, we get contributions from q-qq, q-$qqq\bar{q}$ and $\bar{q}$-$qqqq$ strings. The weights P used for the following figures, are given in table 2.1. For qualitative arguments it is important to notice that quark-diquark (q-qq) strings are dominant, and among those u-ud strings are most likely. We first consider μp scattering at 11.4 GeV. In fig. 2.12 we show longitudinal momentum distributions of pions and kaons. Again we see the strong forward-backward asymmetry for π^+ and K^+, since for $u-ud$ strings only at the forward end of the string, π^+ and K^+ can be produced at the end. The pion distributions are steeper than the kaon distributions which is obvious after writing eq. (2.72) for leading particles as

$$dP(x,m^2) = \frac{\alpha_0}{\Delta\eta}\exp\left[-\alpha_0\, m^2\frac{1-x}{x}\right]\frac{dx}{x}dm^2 \tag{2.97}$$

where x is the momentum fraction and m the mass of the particle. We see from eq. (2.97) that heavier particles have flatter distributions. In fig. 2.13 momentum fraction distributions of baryons are shown: protons, antiprotons, lambdas and antilambdas. We again observe forward-backward asymmetries, which may be summarized as follows: the more the quark content of a produced hadron and a string end differ, the faster the distribution approaches zero for $|x| \to 1$. The reason is that for large differences in the quark content, many other particles in the chain of produced hadrons are closer to the string end. Since the diquark on the minus side differs by just one quark from a proton or a lambda, these distributions are rather flat for $x < 0$. The quark on the other side differs by two quarks from p and Λ, so the distributions fall faster towards $x \to 1$. The difference between the forward quark and $\bar{p}$ or $\bar{\Lambda}$ is 4 quarks, so the distributions fall faster than for p and Λ for $x \to 1$. The largest difference in quark content is, however, observed for the diquark side ($x < 0$) and for $\bar{p}, \bar{\Lambda}$ production: the difference of 5 quarks results in a very fast drop for $x \to -1$. In the former VENUS fragmentation (Field-Feynman) these quark counting arguments were used as an input, by using different splitting functions for different particle species, depending on the difference in quark content. Now we get this behaviour for free, just from the relation between the quark content and the number of steps from an end-point to produce a particle. To avoid confusion: the fragmentation procedure in VENUS

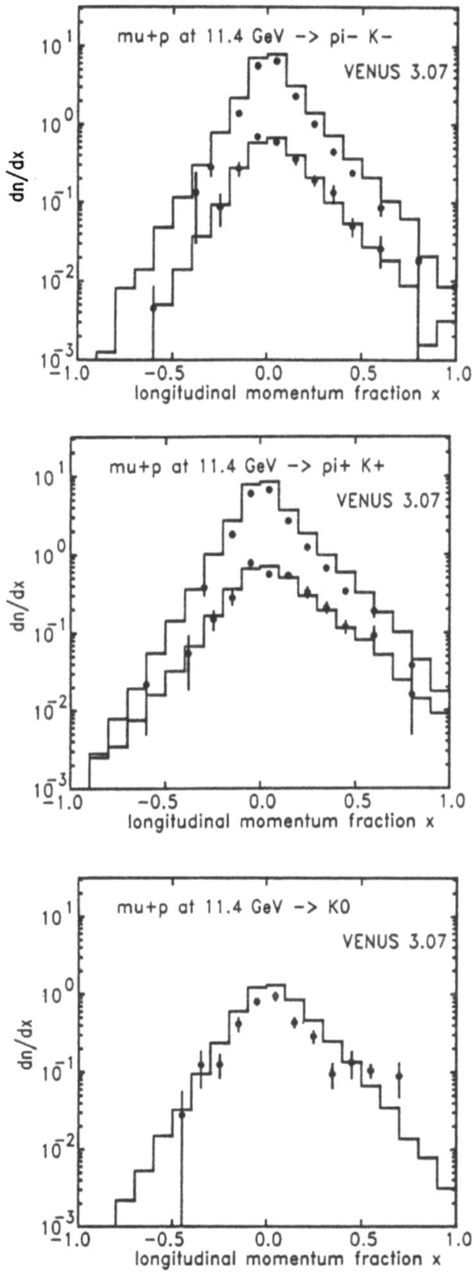

Fig. 2.12.

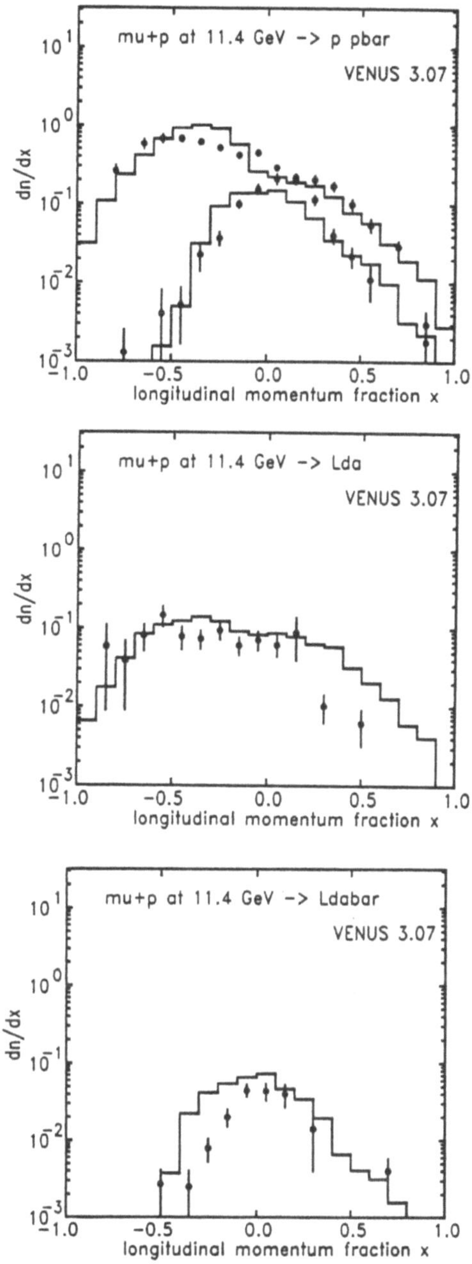

Fig. 2.13.

Fig. 2.12. Longitudinal momentum distributions of pions and kaons for μp scattering at 11.4 GeV (string energy). The data (dots) are from ref. [2.17].

Fig. 2.13. Longitudinal momentum distributions of protons, antiprotons, lambdas and antilambdas for μp scattering at 11.4 GeV (string energy). The data (dots) are from ref. [2.17].

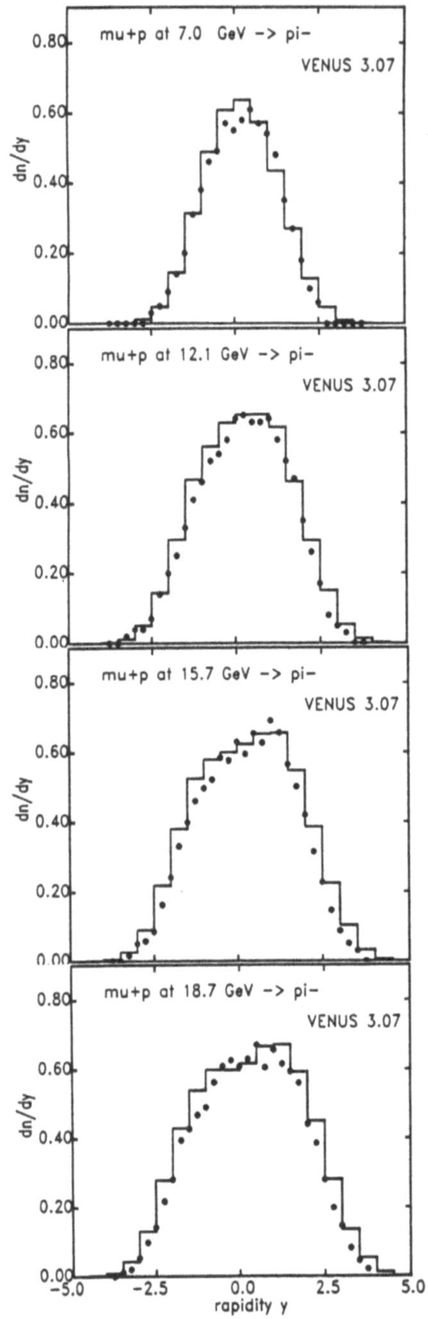

Fig. 2.14

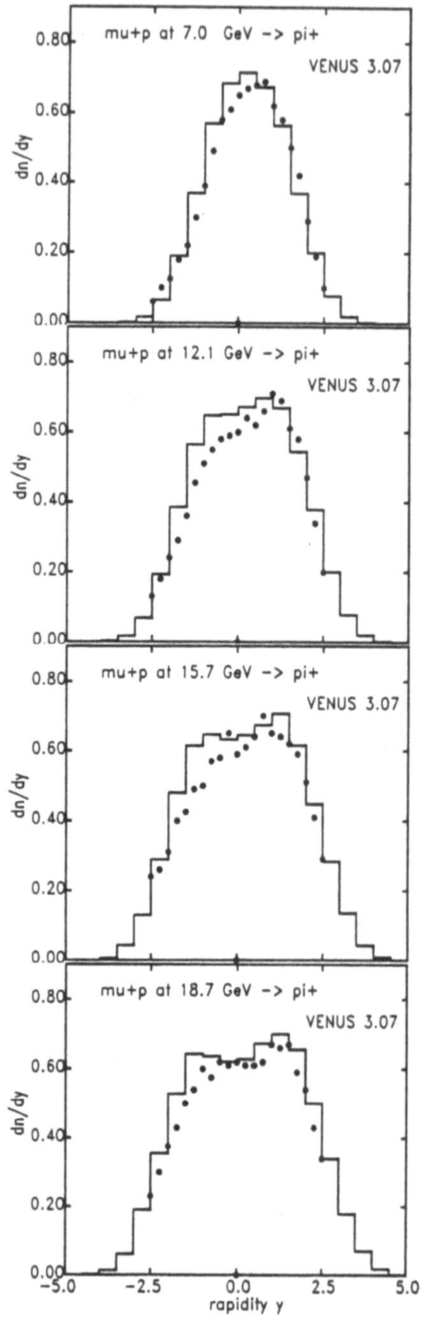

Fig. 2.15

Fig. 2.14. Rapidity distributions of π^- for μp scattering at 7.0 GeV, 12.1 GeV, 15.7 GeV and 18.7 GeV (string energy). The data (dots) are from ref. [2.18].

Fig. 2.15. Rapidity distributions of π^+ for μp scattering at 7.0 GeV, 12.1 GeV, 15.7 GeV and 18.7 GeV (string energy). The data (dots) are from ref. [2.18].

3 breaks the string at arbitrary points, not successively from an endpoint. Nevertheless we can analyse results by considering a chain of produced particles from one string end to the other.

An alternative variable, streching the central region and compressing the fragmentation region, is the rapidity

$$y = \frac{1}{2} \ln \frac{E + p_l}{E - p_l} \tag{2.98}$$

where E is the energy and p_l the longitudinal momentum of a produced particle. In figs. 2.14,15 we display rapidity distributions of π^+ and π^- in μp reactions at string energies of 7 GeV, 12.1 GeV, 15.7 GeV and 18.7 GeV. Whereas the height of the distributions ($y(0)$) remains almost invariant with energy, the width increases clearly. So we observe a rapidity plateau, expected from Lorentz invariance. There is more π^+ than π^- production, since there are more u than d quarks in the string.

2.9 References

[2.1] J. Scherk, Rev. Mod. Phys. 47, 123 (1975)

[2.2] Y. Nambu, Proc. Intl. Conf. on Symmetries and Quark Models, Wayne State Univ., 1969

[2.3] X. Artru and G. Mennessier, Nucl. Phys. B70, 93 (1974)

[2.4] T.D. Gottschalk and D.A. Morris, Nucl. Phys. B288, 729 (1987)

[2.5] P. Koch and K.Werner, in preparation

[2.6] K. Werner and P. Koch, in preparation

[2.7] R.D. Field and R.P. Feynman, Nucl. Phys. B136, 1 (1978)

[2.8] B. Andersson, G. Gustafson, G. Ingelman and T. Sjöstrand, Phys. Rep. 97, 31 (1983)

[2.9] B.R. Webber, Nucl. Phys. B238, 492 (1984)

[2.10] R. Odorico, Nucl. Phys. B228, 381 (1983)

[2.11] R.D. Field and S. Wolfram, Nucl. Phys. B213, 65 (1983)

[2.12] G.C. Fox and S. Wolfram, Nucl. Phys. B168, 285 (1980)

[2.13] K. Werner, Phys. Rev. D39, 780 (1989)

[2.14] K. Werner, Acta Physica Polonica B19, 481 (1988)

[2.15] H. J. Behrend et al, Zeit. Phys. C20, 207 (1983); S.L. Wu, Phys. Rep. 107, 59 (1984)

[2.16] P. Allen et al, Nucl. Phys. B214, 369 (1983)

[2.17] EMC, M. Arneodo et al, Phys. Lett. 150B, 458 (1985)

[2.18] EMC, M. Arneodo et al, Zeit. Phys. C31,1 (1986)

3 STRING FRAGMENTATION IN A MEDIUM

In this chapter we introduce a model for a string fragmenting in a "medium". The medium consists of hadronic objects, like fragments from other strings or simply nucleons. We describe the model quite generally, although in this chapter we only apply it for a nuclear medium, to investigate lepton-nucleus scattering. The model is an ingredient of VENUS 3 — a string model for hadronic collisions — and is therefore sometimes referred to as VENUS 3 as well.

Let us briefly discuss the main ideas behind the model. At the formation point a string is very short, most of the string energy being concentrated at the endpoints. Given this initial condition the string will — according to the classical evolution equations — become eventually very long. However this does not really happen, since the string is going to break after a typical time of 1 fm/c. This means that the length of a string segment will in general not exceed 1 fm, which is of the order of a nucleon size. Therefore concerning reinteraction of string segments among each other and with spectator nucleons, we consider all these objects as pointlike. Thus we have just trajectories of string segments (rather than surfaces) and define to have an interaction whenever two trajectories come closer than a certain distance r_0. How do we realize an interaction? Whereas "String Flip" is the dominant mechanis at high energies, low energy scattering occurs predominantly via fusion. We therefore assume that reinteraction of segments occurs exclusively via fusion, the fused object decaying again after the appropriate time (if it does not interact before).

As mentioned already the model accounts for interactions among string segments and also for interactions of such segments with spectator nucleons. Whereas in hadron-nucleus and nucleus-nucleus collisions both aspects of the model are equally important, in lepton-nucleus scattering we essentially only have to deal with segment-spectator interactions (see discussion below). Therefore we restrict ourselves in this chapter to lepton-nucleus scattering as a first test of the medium fragmentation model, before proceeding towards hadronic interactions.

According to the parton model, a high energy lepton interacts with a nucleon as follows: via boson exchange a considerable amount of momentum is transferred to one of the quarks of the nucleon. The latter one is henceforth transformed into an object consisting of a quark and a diquark with a large relative momentum: a string (see fig. 3.1). Therefore lepton-nucleon scattering allows us to study the fragmentation of diquark-quark (qq-q) strings in vacuum. For lepton-nucleus scattering the initial part of the interaction is as discussed above: due to the very small cross section the lepton will interact with just one nucleon, by transferring momentum to a quark, resulting in a quark-diquark string (see fig. 3.2). But since the string has been formed somewhere inside the nucleus, the space-time evolution of the string will be different from the vacuum case. The whole string or later string fragments may interact with spectator nucleons. So lepton-nucleus scattering provides the unique opportunity to study space-time evolution of strings inside a nuclear medium.

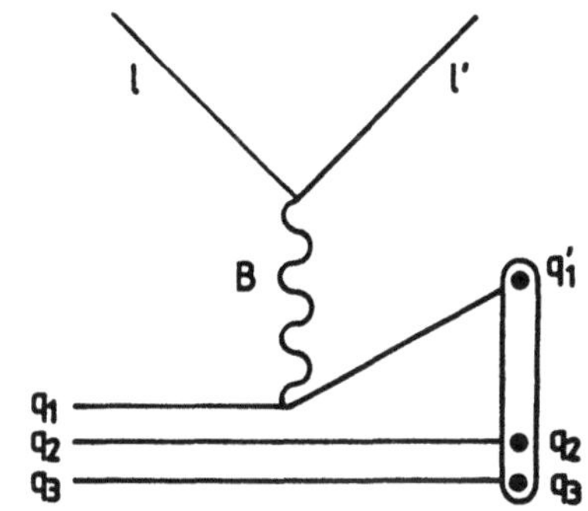

Fig. 3.1

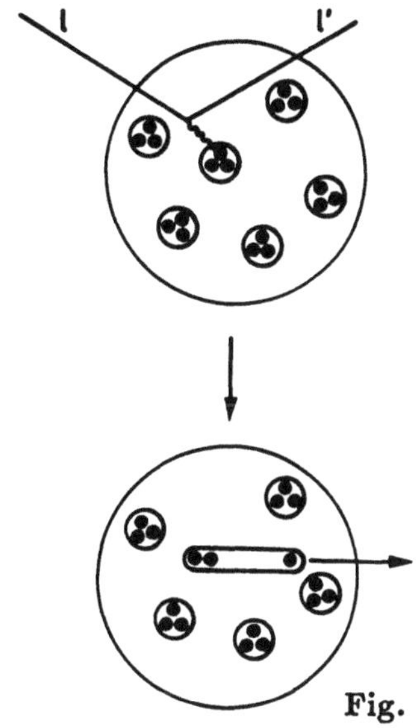

Fig. 3.2

Fig. 3.1. Lepton-nucleon scattering in the Parton Model: the intermediate boson B transfers momentum to a quark, thus forming a quark-diquark string.

Fig. 3.2. In lepton-nucleus scattering a string is formed inside the nucleus. The kicked-off quark moves forward, the diquark is close (in momentum) to the spectator nucleons.

In section 3.1 we introduce the model for the fragmentation of strings in a medium (interacting strings), in section 3.2 we discuss the parton model of neutrino-nucleus scattering, and in chapter 3.3 we apply the model to calculate particle production in neutrino-neon scattering, and compare with data.

3.1 Model for string fragmentation in a medium

In this section we introduce a model for the fragmentation of a string in a medium of pointlike hadronic objects. These objects may be nucleons in a nucleus or string segments from the fragmentation of other strings. We first fragment a string S_0 according to the vacuum fragmentation procedure described earlier, see fig. 3.3. The polygons $O_0R_0U_0$ and $O_0L_0U_0$ are a half period of directrix and antidirectrix ("elementary cell"). The breakpoint B_0 defines the elementary cells $O_1R_0U_1B_0$ and $O_2B_0U_2L_0$ for the two substrings S_1 and S_2. Any further breaks have to occur inside these cells since the absolute past of B_0 (inside of $O_0O_1B_0O_2$) is excluded from further breaks, otherwise B_0 would not be possible! The absolute future (inside of $B_0U_1U_0U_2$) is of course also excluded. In the next step, the substrings S_1 and S_2 take the role of the original one. The breakpoints B_1 and B_2 (see fig. 3.3(b)) determine the string decays $S_1 \rightarrow S_{11} + S_{12}$ and $S_2 \rightarrow S_{21} + S_{22}$. The procedure is continued till all substrings S_α have a mass smaller than some cut-off.

From the example in fig. 3.3 it becomes clear that the sequence of breakpoints is in general not time-ordered in a certain Lorentz frame: the second-generation breakpoint B_1 happens to be earlier than the first-generation break at B_0. This is a consequence of

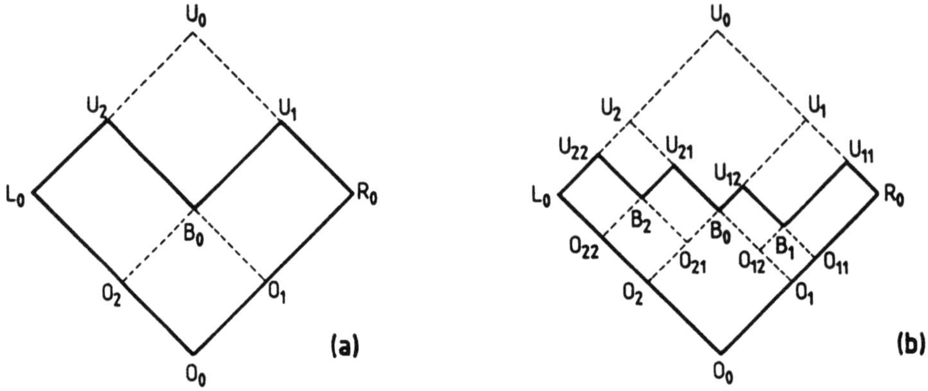

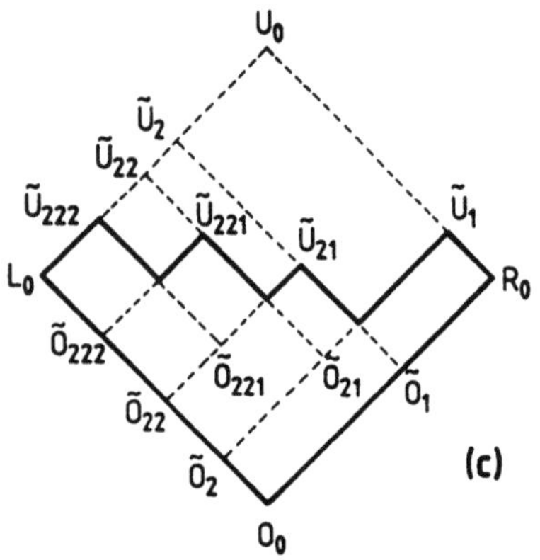

Fig. 3.3. Consecutive string breaking: the first break occurs at B_0 (a), the substrings then break at B_1 and B_2 (b). In this example the second break at B_1 occurs in our reference system earlier than the first break at B_0. In (c) we reconstruct time ordered decays.

our covariant fragmentation scheme. At this point we want to give up covariance and choose an appropriate Lorentz frame. In this frame we determine the real time sequence of string breakups:

$$(B_{\alpha_1})_0 < (B_{\alpha_2})_0 < (B_{\alpha_3})_0 < \cdots, \tag{3.1}$$

where the index zero $()_0$ indicates the time component. In fig. 3.3 we have for example $(B_1)_0 < (B_0)_0 < (B_2)_0$. Keeping the sequence eq. (3.1), we check at these times $(B_{\alpha_i})_0$ whether a break between two consecutive string segments occurred later; if so, the break is considered (at this time) non-existent and the two string pieces are joined again. In this way we obtain a time-ordered decay, the first break occurring at $t = (B_{\alpha_1})_0$:

$$S_0 \to \tilde{S}_1 + \tilde{S}_2 \tag{3.2}$$

the next one at $t = (B_{\alpha_2})_0$, depending whether $\tilde{S}_1$ or $\tilde{S}_2$ decays next as

$$\tilde{S}_1 \to \tilde{S}_{11} + \tilde{S}_{12} \qquad \text{or} \qquad \tilde{S}_2 \to \tilde{S}_{21} + \tilde{S}_{22} \tag{3.3}$$

and so on. In the example of fig. 3.3 we have at $t = (B_1)_0$

$$S_0 \to \tilde{S}_1 + \tilde{S}_2 \equiv S_{11} + \{S_{12} + S_{21} + S_{22}\} \tag{3.4}$$

then at $t = (B_0)_0$

$$\tilde{S}_2 \equiv \{S_{12} + S_{21} + S_{22}\} \to \tilde{S}_{21} + \tilde{S}_{22} \equiv S_{12} + \{S_{21} + S_{22}\} \tag{3.5}$$

and finally at $t = (B_2)_0$

$$\tilde{S}_{22} \equiv \{S_{21} + S_{22}\} \rightarrow \tilde{S}_{221} + \tilde{S}_{222} \equiv S_{21} + S_{22}. \tag{3.6}$$

We construct geometrically, as shown in fig. 3.3(c), the "origins" $\tilde{O}_\alpha$ for the string segments $\tilde{S}_\alpha$. Since we keep track of all parton momenta, we can easily determine the momenta $\tilde{p}_\alpha$ of the string segments. We define a track of a segment to be a straight line, going through the origin, with the direction given by the velocity of the segment:

$$(x_\alpha(t))_i = (\tilde{O}_\alpha)_i + \frac{(\tilde{p}_\alpha)_i}{(\tilde{p}_\alpha)_0}[t - (\tilde{O}_\alpha)_0] \tag{3.7}$$

with $i = 1,2,3$ being the space components, $i = 0$ being the time component. This track corresponds approximately to the diagonal $\tilde{O}_\alpha \tilde{U}_\alpha$ of the elementary cell, only approximately because of the transverse momenta involved, which are however small. We have to consider of course that the string segment $\tilde{S}_\alpha$ lives on this line only for a time interval $t_\alpha^1 < t < t_\alpha^2$, limited by the breaktime t_α^1 where this segment occurred first, and the breaktime t_α^2 where $\tilde{S}_\alpha$ further decays. The actual track

$$\{(x_\alpha(t))_i \,;\, t_\alpha^1 < t < t_\alpha^2\} \tag{3.8}$$

in general does not contain the origin $\tilde{O}_\alpha$.

So we have the following situation: between the time $(O_0)_0$ where the string was formed and the time $(B_{\alpha_1})_0$ of the of the first break, we have just one track, the track $\vec{x}_0(t)$ of the whole string. After the first break, till the next one at $(B_{\alpha_2})_0$, we have two tracks corresponding to the two substrings $\tilde{S}_1$ and $\tilde{S}_2$, and so on. The final string segments have no upper limits (so far, till we consider resonance decay); they follow their track till $t = \infty$.

As the next step we have to consider resonance decay (a detailed description is given in [3.1], see also section 2.6), which also limits the lifetime of final string segments. Knowing the mass and quark content of a segment, we identify it with a resonance according to our resonance table. Also from the table we obtain the lifetime τ_R of the resonance R, which therefore lives on the straight line trajectory between the formation time t_R and $t_R + \gamma\tau_R$, with the Lorentz dilation factor γ. At the latter time we decay the resonance according to a standard decay table (see [3.1]), thus at that point the resonance track splits into two (or more) daughter tracks and so on.

We have thus a complete space-time evolution of the string and its decay products from the string formation time till infinity. This is an improvement on earlier space-time studies [3.2], where we only considered resonances, but not, however, their "parent" string segments. So in a way particles were created at some stage from "nothing", whereas now we keep track of energy and momentum from the beginning till the end.

All this exercise of constructing time-ordered string decays is only necessary if we consider interactions at an early stage, before the last breakup. For interactions of segments it is obviously important which event happens earlier than some other event (we mean events of two segments coming close to each other and interacting). We are now considering how string fragmentation as described above is modified in a medium of pointlike hadronic objects H_β, the latter ones moving on tracks $\vec{x}_\beta(t)$. Whenever at some time t such a track comes close to the track $\vec{x}_\alpha(t)$ of a string segment $\tilde{S}_\alpha$, i.e.

$$|\vec{x}_\alpha(t) - \vec{x}_\beta(t)| \leq r_0, \tag{3.9}$$

then these two objects interact. The two are most likely close in momentum space (for the cases where we want to apply this model: lepton-nucleu or nucleus-nucleus scattering), so for simplicity we assume all interactions to result in fusion of the two objects:

$$\tilde{S}_\alpha + H_\beta \to R \tag{3.10}$$

where the quark content and momenta of R are just the sums of these quantities of the two ingredients. According to its mass and quark content this object R is identified with the help of a resonance table, its lifetime τ_R is determined, and after the time interval $\gamma\tau_R$ this object R decays again into resonances or hadrons:

$$R \to R_1 + R_2 + \cdots. \tag{3.11}$$

In fact we perform this procedure of checking the distance and fusing (if necessary) for all string segments, decay products, "medium" particles, results of fusions (eq. (3.10)), or decay products of such fused objects (eq. (3.11)), so we include all particles present at a certain time. Of course if a segment $\tilde{S}_\alpha$ is involved in an interaction eq. (3.10) at a time t, all the decay products of $\tilde{S}_\alpha$, which would otherwise occur later, are non-existent! Technically speaking, all the breaks of the vacuum string are first done, even if some of them later turn out to be irrelevant, due to a fusion process eq. (3.10) of some parent string segment.

Theoretically there is the possibility that string segments from one single string reinteract and fuse; however, this happens so rarely that this process is irrelevant. The dominant reinteraction process for string fragmentation in a nucleus (lepton-nucleus scattering) is the following: low mass hadrons from string decay, which are close in momentum space to the spectator nucleons, interact with these nucleons, fuse into moderately excited resonances, and decay again into two or three hadrons, the knocked-off baryon being slow in the nuclear rest frame. In this way the multiplicity is not drastically increased, apart from the additional baryons, which without interactions would remain spectators.

3.2 Parton model of neutrino-nucleus scattering

This section provides the link between string fragmentation in a nucleus — for which we constructed a model — and lepton-nucleus scattering — where data exist. Let us first consider lepton-nucleon scattering (see fig. 3.1). We use the following variables: $k \equiv (k_0, \vec{k})$ and $k' \equiv (k_0', \vec{k'})$ are the momenta of the incoming and the outgoing leptons, $q \equiv k - k'$ is the momentum of the exchanged boson, we use as usual $Q^2 \equiv -q^2$ for the virtual mass of the boson, θ and ϕ are the polar and azimuthal angles of the scattered lepton relative to the beam axis in the nucleon rest frame and W^2 is the square of the "string mass", i.e. the mass of "boson + nucleon". For the experiment which we are going to consider [3.3] we know average values for k_0, Q^2, and W^2. From momentum conservation at the boson-nucleon vertex we have (with m_N being the nucleon mass)

$$-Q^2 + m_N^2 + 2m_N(k_0 - k_0') = W^2 \tag{3.12}$$

from which we obtain the energy k_0' of the scattered lepton. In the nucleon rest frame we have

$$q = \begin{pmatrix} k_0 - k_0' \\ -k_0' \sin\theta \sin\phi \\ -k_0' \sin\theta \cos\phi \\ k_0 - k_0' \cos\theta \end{pmatrix} \tag{3.13}$$

with θ to be obtained from

$$Q^2 = 2\,k_0\,k_0'\,(1-\cos\theta). \tag{3.14}$$

The string momentum p is the sum of the boson and nucleon momentum, so in the nucleon rest system we have

$$p = q + (m_N, 0, 0, 0). \tag{3.15}$$

We use the string cm system as the reference system where the fragmentation procedure as described in the last chapter is being performed. We boost the target nucleus into this system via

$$\begin{aligned} x_0' &= \beta_0 x_0 + \sum_{i-1}^{3} \beta_i x_i \\ x_j' &= \beta_j x_0 + x_j + \frac{\beta_j}{1+\beta_0} \sum_{i=1}^{3} \beta_i x_i \end{aligned} \tag{3.16}$$

with

$$\beta_\mu = \frac{p_\mu}{\sqrt{p^2}}, \qquad \mu = 0,1,2,3 \tag{3.17}$$

being the string velocity.

To provide some numbers: for the neutrino-neon experiment with which we want to compare [3.3] , we have an incident neutrino energy of 43 GeV, a momentum transfer Q^2 of 5.4 GeV^2 and a string mass squared W^2 of 27 GeV^2. From eq. (3.12) we obtain an energy transfer $q_0 = k_0 - k_0'$ of 16.76 GeV and the scattering angle from eq. (3.14) turns out to be 3.96°. So the W-boson moves with almost half of the incident neutrino energy almost in beam (=forward) direction. Since this momentum is transferred to one quark of the hit nucleon (see fig. 3.1), this quark consequently moves with a large momentum forward (see fig. 3.2). After the Lorentz boost eq. (3.16), in the string frame, the nucleons move almost along the string axis in backward direction (in the diquark direction). We expect therefore rescattering effects in the backward region (diquark fragmentation region). This will be investigated in the next chapter.

Finally we have to specify the flavour content of strings in charged current neutrino-nucleus scattering. Using the variables

$$x_B \equiv \frac{Q^2}{2\,m_N(k_0 - k_0')}, \qquad y \equiv \frac{k_0 - k_0'}{k_0}, \tag{3.18}$$

the probability $P(q_\alpha \to q_\beta)$ for a certain current $\langle q_\beta | J | q_\alpha \rangle$ is, if q_α and q_β are quarks, given as

$$P(q_\alpha \to q_\beta) = N\,c(q_\alpha \to q_\beta) \int dx_B\,dy \frac{G^2}{\pi} s\,x_B\,q_\alpha(x_B), \tag{3.19}$$

for antiquarks q_α and q_β as

$$P(q_\alpha \to q_\beta) = N\,c(q_\alpha \to q_\beta) \int dx_B\,dy \frac{G^2}{\pi} s\,x_B\,(1-y)^2\,q_\alpha(x_B), \tag{3.20}$$

where N is a normalization constant, $q_\alpha(x)$ are the (anti)quark momentum distribution functions, and $c(q_\alpha \to q_\beta)$ represents Cabibbo mixing, the non-zero coefficients being

$$c(q_\alpha \to q_\beta) = \begin{cases} \cos^2\theta_C & \text{for } d \to u,\ \bar{u} \to \bar{d} \\ \sin^2\theta_C & \text{for } d \to c,\ \bar{c} \to \bar{d} \\ \sin^2\theta_C & \text{for } s \to u,\ \bar{u} \to \bar{s} \\ \cos^2\theta_C & \text{for } s \to c,\ \bar{c} \to \bar{s} \end{cases} \tag{3.21}$$

with the Cabibbo angle θ_C. To each current ($q_\alpha \to q_\beta$) there corresponds a string with q_β and $(N_0 - q_\alpha)$ at the ends, where N_0 is the nucleon involved. So for example a current $d \to u$ in neutrino-proton scattering provides a u-uu string. Since there are no $\bar{c}$ quark in a nucleon, only six of the eight possibilities of eq. (3.21) are realized. The numerical values for the corresponding weights, calculated according to eqs. (3.19,20,21), taking equal probability for neutron and proton targets (symmetric nuclei), are given in table 3.1. For qualitative arguments later we note that we have more than 50 % ud-u strings, and more than 25 % uu-u strings; everything else has small weight.

3.3 Comparison with data

In this section we apply our model for medium-string-fragmentation to calculate inclusive particle spectra for neutrino-neon scattering, and compare with data. Since our fragmentation model is part of the more general string model VENUS 3 for hadronic scattering, we refer to all calculations as VENUS 3. We generate strings with mass 4.5 GeV according to table 3.1, to compare with the data of [3.3]. Rescattering is crucially affected by the interaction radius r_0 in eq. (3.9). In the following we are comparing two scenarios (using $r_i = 0.7\,\text{fm}$ for baryons and $r_i = 0.35\,\text{fm}$ for mesons, i=1,2):

$$r_0 = 0 \qquad \longleftrightarrow \qquad \text{rescattering turned off}$$

$$r_0 = r_1 + r_2 \qquad \longleftrightarrow \qquad \text{rescattering}$$

where our default values may not be the final answer; it is really a best guess rather than an optimized parameter. In the following figures we shall use dashed lines if using $r_0 = 0$ and solid lines if using $r_0 = r_1 + r_2$, points represent data.

We first consider distributions of the longitudinal momentum fraction $x = p_l/p_l^{\max}$, where p_l is the longitudinal momentum of a produced particle in the string frame and $p_l^{\max}$ is the maximum possible momentum. Positive x is defined to be the direction of the (anti)quark kicked off by the intermediate boson, negative x is the direction of the remainder, which is most likely a diquark (see table 3.1). As discussed in the last chapter, the target spectator nucleons reside (in momentum) close to the "diquark", therefore we expect effects at $x < 0$.

In fig. 3.4 we display x-distributions of pions with (solid) and without (dashed) final state interactions. Let us first discuss the π^- distribution. For $x > 0$ we see almost no difference between the solid and dashed histogram, there is no rescattering

Table 3.1: The weights of individual string contributions for the ν–Ne reactions discussed in this section. $p(n)$-string refers to a proton(neutron) being the target.

current	p-string	$P\,[\%]$	n-string	$P\,[\%]$
$d \to u$	$u-uu$	27.37	$u-ud$	58.21
$d \to c$	$c-uu$	1.44	$c-ud$	3.06
$s \to u$	$u-\bar{s}uud$	0.13	$u-\bar{s}udd$	0.15
$s \to c$	$c-\bar{s}uud$	2.43	$c-\bar{s}udd$	2.88
$\bar{u} \to \bar{d}$	$\bar{d}-uuud$	1.88	$\bar{d}-uudd$	2.23
$\bar{u} \to \bar{s}$	$\bar{s}-uuud$	0.10	$\bar{s}-uudd$	0.12

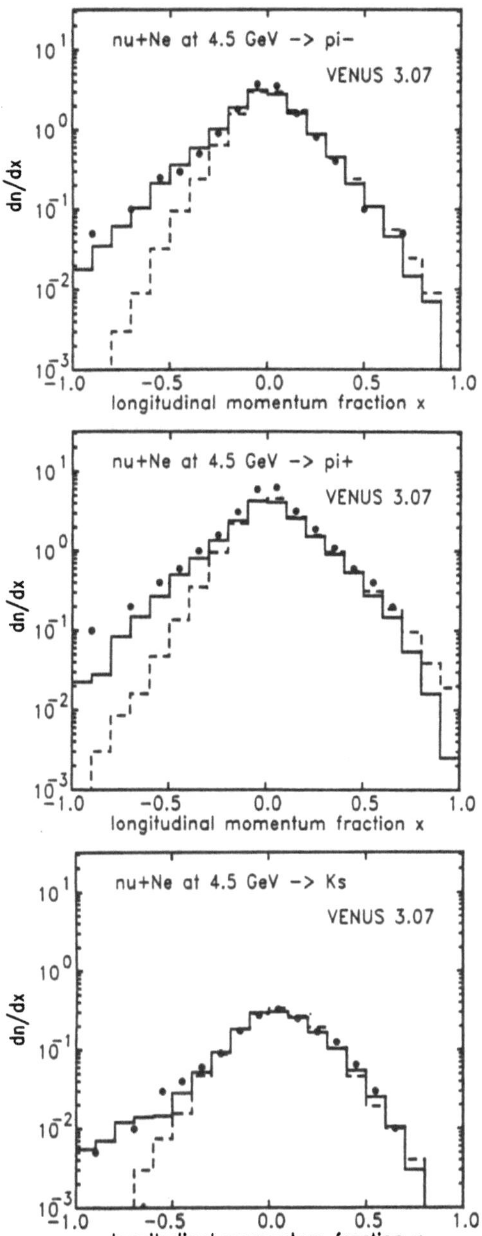

Fig. 3.4

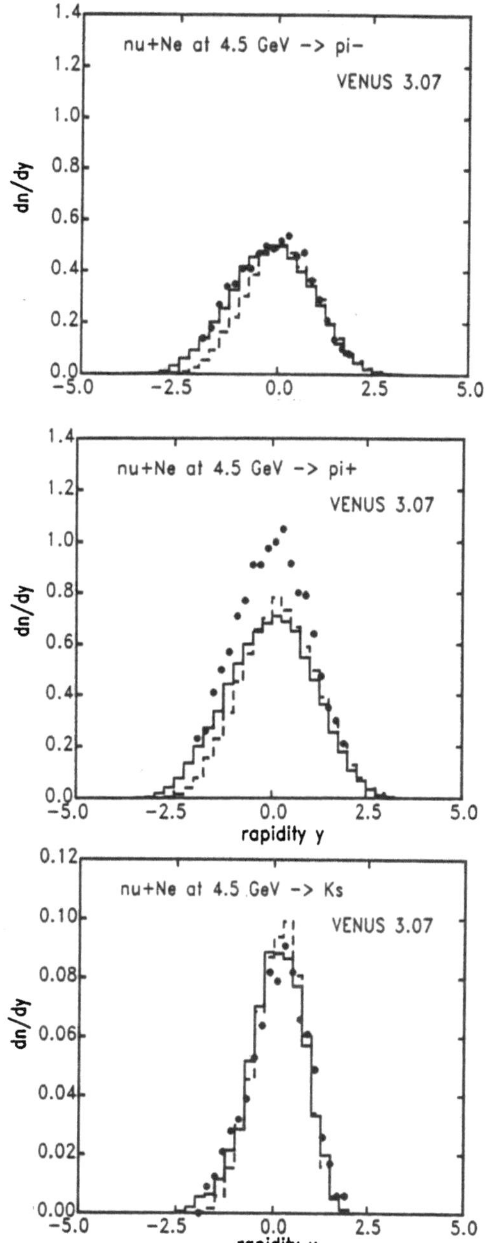

Fig. 3.5

Fig. 3.4. Longitudinal momentum distributions of π^- (upper plot) π^+ (middle) and K_s (lower) in ν-Ne scattering. The points are data, the solid histograms are VENUS results with final state interactions, for the dashed ones interactions are turned off.

Fig. 3.5. Rapidity distributions of π^- (upper plot) π^+ (middle) and K_s (lower) in ν-Ne scattering. The points are data, the solid histograms are VENUS results with final state interactions, for the dashed ones interactions are turned off.

effect — as we expected — since $x > 0$ corresponds to the forward side of the string, where the fragmentation occurs outside the nucleus. However, since the diquark end of the string fragments well inside the nucleus, we see a huge effect for $x < 0$: in the tails the two histograms differ by one order of magnitude. So there are considerably more π^- produced in the diquark fragmentation region, if rescattering is considered. The overall multiplicity however changes only slightly, since this quantity is dominated by the production in the central region, around $x = 0$ (note the logarithmic scale in fig. 3.4). The VENUS results with interaction reproduce the π^- data over the full x range. Concerning the π^+ distributions the situation is quite similar: taking into account final state interactions considerably improves the agreement between calculation and data. However there still seem to be considerably more particles in the data. Since this excess is also observed for $x \geq 0$, it cannot be due to a wrong interaction model. Because we can reproduce lepton-nucleon data from other experiments which seems to support our vacuum fragmentation model, we are tempted to blame particle misidentification for the difference.

In the lower plot of fig. 3.4 we investigate kaon production. Also for kaons (K_s) we observe something similar as for pions: rescattering has no effect in forward direction, but increases particle production in the backward tails. However the latter effect is smaller than for pions.

In fig. 3.5 we show rapidity distributions corresponding to the x-distributions discussed above. We observe the same features as discussed before: a satisfactory improvement in the backward hemisphere by considering final state interactions, a smaller effect for kaons than for pions, an excess of π^+ in the data.

3.4 References

[3.1] P. Koch and K. Werner, in preparation

[3.2] K. Werner, Phys. Lett. B219, 111 (1989)

[3.3] N. J. Baker et al, Phys. Rev. D34, 1251 (1986)

4 SOFT HADRON-HADRON INTERACTIONS

Soft interaction refers to little tranfer of transverse momentum (few hundred MeV). Whereas hard processes (large momentum tranfer) can be treated in perturbation theory, there is presently no method to calculate soft processes, which requires models. We consider in this chapter models which assume that a hadron-hadron collision at ultra-relativistic energies results in two (or more) strings which are (approximately) aligned along the beam axis. The string evolution and fragmentation is then performed as described in the last chapters. We introduce and compare in section 4.1 three different methods to form strings in hadron-hadron collisions. In section 4.2 we discuss the string model VENUS which employs one of these methods ("Colour Exchange"), where strings are formed as a result of rearrangement of colour singlet structures due to the interaction. The VENUS model has been tested against many experimental data concerning very different observables. Due to time and space limitations we only consider one application here: we explain, why a string model like VENUS obeys "KNO scaling" which is an experimentally verified scaling law for multiplicity distributions. We show that KNO scaling occurs due to a scaling law for string mass distributions, the latter scaling law being a consequence of the parton strucure of the nucleon.

4.1 String formation procedures

It is generally assumed that a high energy interaction of two hadrons results in (at least) two strings which are oriented along the beam axis. Only a small transverse momentum component is allowed, of order of r_N^{-1} with r_N being the size of the nucleon. We restrict ourselves first to the case of exactly two strings. To be more precise: by string formation we mean the production of strings with large mass (being much larger than the mass of the original hadrons). In this sense we would also talk about string formation in a model where the incident hadrons are already considered as strings — low mass strings however.

In this section we want to introduce and compare three different models for string formation: (a) "String Rearrangement" [2.3, 4.1, 4.2] which seems to be the natural method to form massive strings in a model where all objects — in particular the incident hadrons — are considered as ideal classical strings; (b) "String Flip" or "Colour Exchange" which is the method used in the VENUS model [4.3], to be described in detail later; (c) "Longitudinal Excitation" which is the method adopted by the Lund group [4.4]. These models differ considerably concerning the ideas behind them and in particular concerning the "language" usually used to introduce them. Therefore we try in this chapter to find a common language — the string language introduced in chapter 2 — to learn about real differences between the models.

We were discussing classical string dynamics in chapter 2 in great detail. A disadvantage of such a complete description is that important statements may be overseen by the reader, so we are going to repeat (without reference) the crucial results, to be relevant for the forthcoming discussion.

(1) A string is completely characterized by the trajectory of one endpoint (directrix). This trajectory is periodic, so even one period of the directrix defines the string for ever. Correspondingly the string is completely given by less than a period of the directrix, however supplemented by the complementary part of the antidirectrix (trajectory of the other endpoint). The simplest possible (1-dimensional) so called yo-yo string is correspondingly completely defined by an "elementary rectangle" $OACB$ (in space-time), see fig. 4.1, with OAC and OBC representing a half period of directrix and antidirectrix respectively.

(2) The trajectories in fig. 4.1 are parallel to the light-cone directions, so it is convenient to use light-cone coordinates $x^\pm = t \pm x_3$ and momenta $p^\pm = E \pm p_3$. The endpoints moving along OA and OB can be identified with partons (massless relativistic point particles), starting at O with momenta p^+ and p^- respectively. These momenta are related to the width and length of the rectangle in fig. 4.1 via

$$\|OA\| = x^+ = \frac{p^+}{\kappa}\ ; \qquad \|OB\| = x^- = \frac{p^-}{\kappa}\ . \tag{4.1}$$

So we have a mapping from momentum to real space and vice versa via $\Delta p = \kappa \Delta x$, and therefore the elementary rectangle $OACB$ in fig 4.1 has a meaning

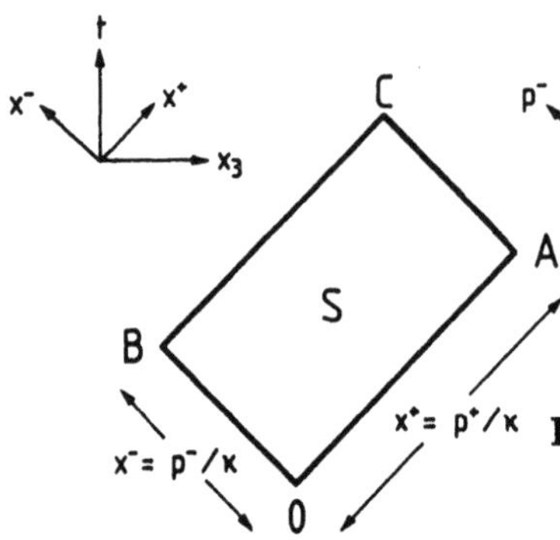

Fig. 4.1

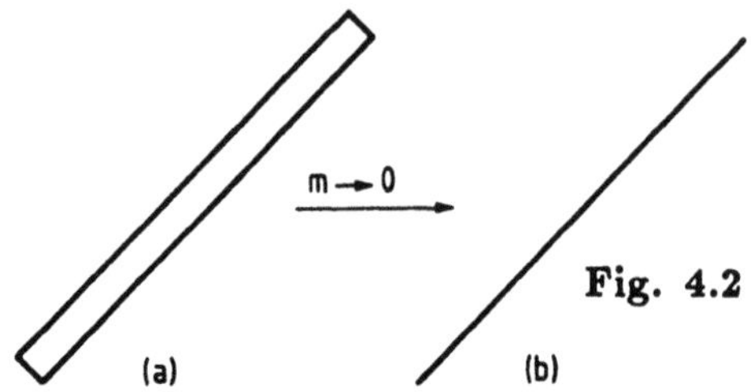

Fig. 4.2

Fig. 4.1. An "elementary rectangle" characterizing a yo-yo string. OAC and OBC are the trajectories of the endpoints, and the string stretches in between. These trajectories may be periodically continued into future or past.

Fig. 4.2. A yo-yo string with small mass (a) and zero mass (b).

both in space-time and in momentum space (often we do not have to specify what space we are working in).

(3) The area S of the rectangle is proportional to the string mass:

$$S = \frac{p^+ p^-}{\kappa^2} = \frac{E^2 - p_3^2}{\kappa^2} = \frac{m^2}{\kappa^2} \tag{4.2}$$

This relation also implies that for a "fast" string with $p^+ >> m$ we obtain a very small p^- i.e. $p^- << m$, so the rectangle looks almost linear, as demonstrated in fig. 4.2a (we consider a fast forward-moving string here). In the limit of $p^+ \to \infty$ (or equivalently $m \to 0$), the rectangle degenerates into a line, being thus identical to the trajectory of a point particle (see fig. 4.2b).

(4) A break of a yo-yo string into two substrings is completely determined by specifying a breakpoint D inside the rectangle $OACB$ (see fig. 4.3). The two "subrectangles" for the two substrings are constructed as shown in fig. 4.3. So the original rectangle S is divided into four rectangles: S_1 and S_2 representing the two substrings, the absolute past P and the absolute future F. The relative size of these areas measures the relation of potential (=mass) and kinetic energy production. If for example a break occurs early, close to O, the areas S_1 and S_2 are small, F however is large. This means that the two strings were formed with a small mass but with a large relative kinetic energy.

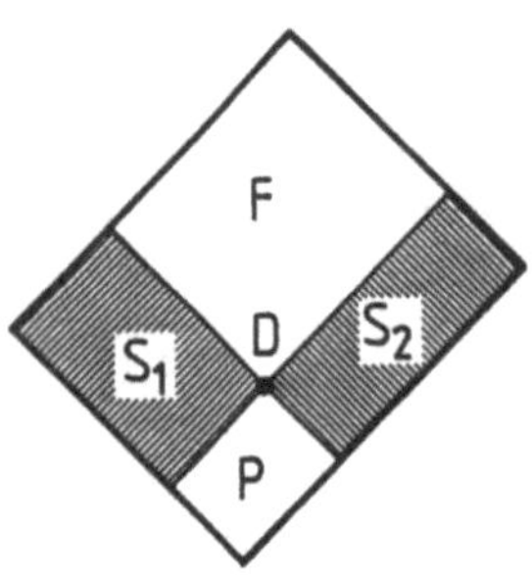

Fig. 4.3. String breaking at D. The hatched areas S_1 and S_2 represent the two substrings, the areas F and P are absolute future and absolute past of the breakpoint D.

We are now in a position to introduce the different string formation procedures and discuss differences and similarities between them.

"String Rearrangement"

We already discussed string breaking, i.e. the division of a string S into two substrings S_1 and S_2:

$$S \rightarrow S_1 + S_2 \tag{4.1}$$

which provides the mechanism of particle production from a massive string. The inverse process, namely fusion of two strings S_1 and S_2 into one

$$S_1 + S_2 \rightarrow S \tag{4.2}$$

provides a possible interaction mechanism of two strings. Fusion requires however that the two strings meet exactly at their endpoints which from a purely geometrical point of view is not very likely. More frequent is the case that the two strings meet at interior points, as indicated in fig 4.4. A possible local interaction would be a combination of breaking and fusion: each of the two strings S_1 and S_2 breaks at the interaction point into two substrings, i.e. S_{11}, S_{12} and S_{21}, S_{22} (see fig 4.4). Then two substrings from different original strings fuse together, say S_{11} and S_{22} and correspondingly S_{12} and S_{21}. The whole process (breaking and subsequent fusion)

$$S_1 + S_2 \rightarrow S_{11} + S_{12} + S_{21} + S_{22} \rightarrow S_1' + S_2' \tag{4.3}$$

with

$$S_1' = S_{11} + S_{22}\,; \qquad S_2' = S_{12} + S_{21} \tag{4.4}$$

is called "Rearrangement" (see [2.3, 4.1, 4.2]) and sketched in fig 4.4.

From now on we restrict ourselves to yo-yo strings, which are fully defined by one rectangle in t-x_3 space (t: time, x_3: longitudinal space coordinate). We assume the two incident strings to be aligned, otherwise we would no longer have yo-yo's after the "Rearrangement" process. We furthermore assume the two strings to have small masses and large momenta (in opposite directions in the string-string centre-of-mass system. The situation is then as shown in fig. 4.5a: the forward string is represented by a rectangle S_2 elongated along the x_+ axis and the rectangle S_1 representing the backward

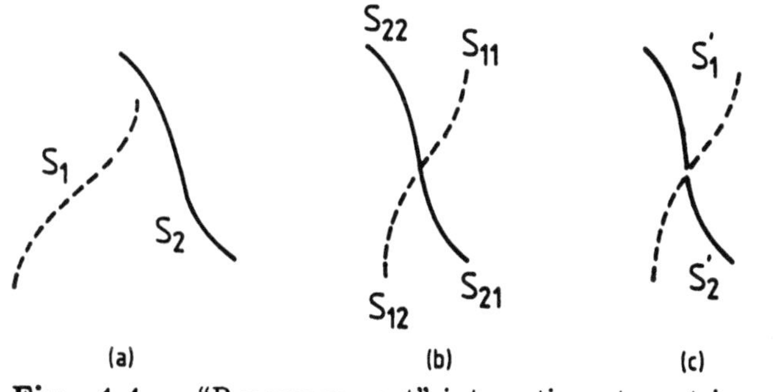

Fig. 4.4. "Rearrangement" interaction: two strings approach each other (a), touch at an internal point with the effect of cutting each string into two substrings (b), and fusing the pieces from different original strings together (c).

Fig. 4.5. "Rearrangement" interaction for yo-yo strings: breaking (cutting) procedure (a) and fusion (b).

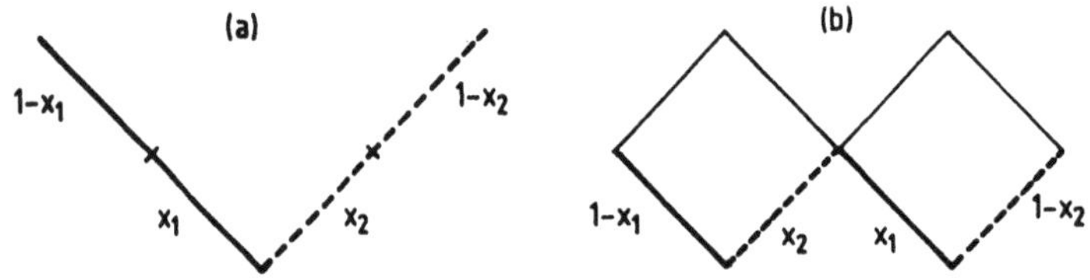

Fig. 4.6. "Rearrangement" interaction for massless yo-yo strings: breaking (cutting) procedure (a) and fusion (b). Compare with fig. 4.5 for the case of finite masses.

moving string is elongated along the x_- axis. The first stage of the interaction — string breakup — is indicated by the hatched areas S_{11} and S_{12} representing the breakup of S_1 and correspondingly S_{21} and S_{22} representing the breakup of S_2. The space-time diagram of fusion is just the same as for breaking, so we construct from the rectangles S_{11} and S_{22} a greater one S_1' as shown in fig. 4.5b. In the same way we construct S_2' from S_{12} and S_{21}. Since we fuse the strings at their turning points, the fused strings are simple yo-yo's again, which is not guaranteed for fusion of two yo-yo strings, even if they are aligned.

An important feature of the above "Rearrangement" process can be seen from fig. 4.5: although we started with two strings (S_1 and S_2) with small masses, the resulting strings (S_1' and S_2') may have a very large mass. The areas representing S_1' and S_2' are in particular large when the original strings break up somewhere in the middle (concerning the elongated axis); this is realized in fig. 4.5.

Let us consider the limit of massless incident strings. As seen from fig. 4.6 the rectangles degenerate to lines, and the string breaking procedure amounts to dividing each line into two segments. We may now characterize the substrings (segments) simply by their lengths relative to the length of the corresponding original string, so we introduce fractions x_1 and $1 - x_1$ for S_{11} and S_{12} and fractions x_2 and $1 - x_2$ for S_{21} and S_{22}. For simplicity of notation we work in the S_1-S_2 centre-of-mass system, so the lengths corresponding to these two strings are the same, and they may be rescaled to 1, so we only have to care about fractions for the following discussion. The second stage of the interaction — fusion — amounts now simply to constructing a rectangle of lengths $1 - x_1$ and x_2 for S_1' and correspondingly a rectangle of lengths $1 - x_2$ and x_1 for S_2', as shown in fig. 4.6. In this figure we see that even having incident strings with zero mass produces in general strings with large mass (provided of course the momenta of the incident strings were large). So "Rearrangement" provides the possibility to transform relative kinetic energy into string mass ("potential energy" in the sense that this energy might be used for particle production).

"String Flip" or "Colour Exchange"

The interaction principle to be discussed in the following was introduced in connection with the Dual Parton Model [4.5]. For pedagogical reasons we use a somewhat different approach, motivated by studies of the Strong Coupling QCD expansion [4.6]. Let us for simplicity first consider interactions between two mesons with large relative momentum ($>> m_\pi$). The situation before the collision is shown in fig. 4.7a (lhs): the projectile quark q_p and antiquark $\bar{q}_p$ are linked together by a string; a corresponding picture applies for the target meson. We are not considering gluons explicitly; they are replaced by strings, since we think that soft processes are represented better by an effective theory of (anti)quarks and strings rather than (anti)quarks and gluons. On a lattice, a link (or string) U between two neighbouring lattice sites is related to the gluon field A via [4.6]

$$U = \exp\{iag\frac{\lambda^\alpha}{2}A^\alpha\}, \tag{4.5}$$

and the theory is then formulated in terms of U rather than A, so the lattice world is another example where (anti)quarks and strings rather than (anti)quarks and gluons are the basic constituents.

Let us specify the meson-meson interaction: we assume that "some effective string interaction" acts in such a way that as the final result the two strings are "flipped", i.e. after the interaction two strings link the (former) projectile quark q_p and the (former) target antiquark $\bar{q}_t$ and correspondingly $\bar{q}_p$ and q_t, see fig. 4.7a (rhs). On the lattice the plaquette operator tr$UUUU$ provides such a transition [4.6]. This "String Flip" interaction may be formally represented as

$$(\bar{q}_p q_p) + (\bar{q}_t q_t) \rightarrow (\bar{q}_p q_t) + (\bar{q}_t q_p) \tag{4.6}$$

where a bracket () indicates that the objects inside are linked by a string, i.e. they form a colour singlet. Since the rhs and lhs of eq. (4.6) differ by the exchange

$$q_p \leftrightarrow q_t \tag{4.7}$$

the interaction eq. (4.6) is also referred to as "Colour Exchange" (between the quarks q_p and q_t). It is not a particle or momentum exchange, just colour exchange: since eq. (4.6) represents the colour singlet structure, all the momenta of the partons are completely unchanged. We know that hadron-hadron interactions are mostly "soft",

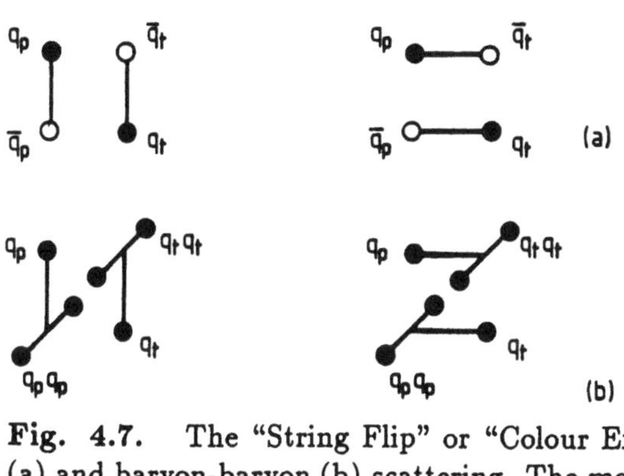

Fig. 4.7. The "String Flip" or "Colour Exchange" principle for meson-meson (a) and baryon-baryon (b) scattering. The momenta of all partons are unchanged, just the strings linking them together are modified ("flipped").

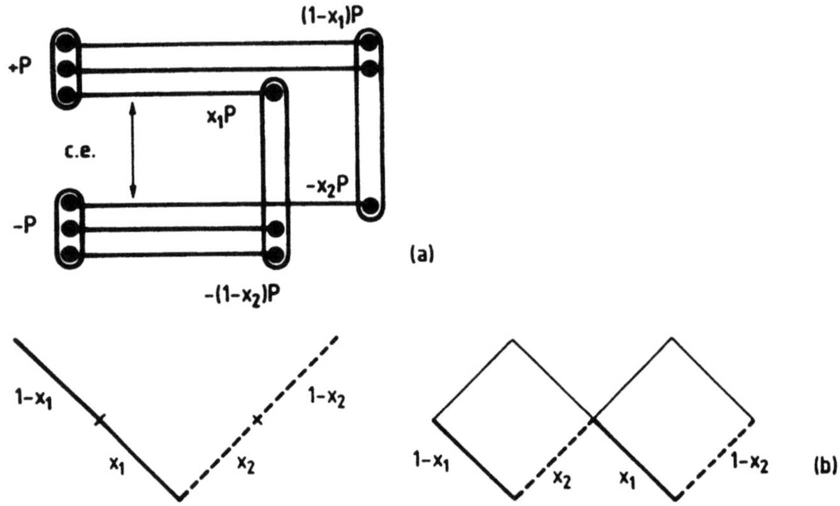

Fig. 4.8. (a) Quark line diagram of "String Flip" or "Colour Exchange" (c.e.) interaction for baryon-baryon scattering. Due to the interaction, parton from different incident baryons are grouped together to singlets (closed lines). Also indicated is how the partons (quarks and diquarks) share the initial momenta $\pm P$. (b) Representation in momentum space — renormalized to $P = 1$ — of the situation in (a): the lhs represents the incident baryons and how their momenta are shared by the quark and diquark; the rhs represents the putting together of partons from different original baryons.

little transverse momentum is transferred (few hundred MeV); "String Flip" or "Colour Exchange" provides an interaction mechanism which is also "soft" longitudinally, since no longitudinal momentum is transferred at all (rearrangement of strings does the job), so the picture seems to be consistent.

The "Colour Exchange" ("String Flip") principle can equally well be applied to baryon-baryon scattering, just the antiquark in the meson has to be replaced by a diquark, see fig 4.7b. Here the string which keeps the diquark together is not involved in the interaction. Formally the baryon-baryon interaction may be written as

$$(q_p q_p q_p) + (q_t q_t q_t) \rightarrow (q_p q_p q_t) + (q_t q_t q_p) \tag{4.8}$$

Here the last of the projectile and target quarks have been exchanged.

Another useful representation of the "Colour Exchange" process is the quarkline diagram, shown in fig. 4.8a: quarks (dots) surrounded by closed lines are meant to be coupled to colour singlets. Due to a colour exchange (arrow), singlet structures are changed, leading to singlets consisting of a quark and a diquark from different original baryons. We also indicate the momentum balance in fig. 4.8a: in their centre-of-mass system the baryons initially have momenta $+P$ and $-P$. These momenta are shared between a quark and a diquark, so the quarks have momenta $x_1 P$ and $-x_2 P$ and the diquarks $(1-x_1)P$ and $-(1-x_2)P$, with x_1, x_2 being numbers between 0 and 1. The new singlet structures (qq-q strings) are now composed each of a diquark and quark moving in opposite directions — having momenta $(1-x_1)P$ and $-x_2 P$ for one string and $-(1-x_2)P$ and $x_1 P$ for the other.

We are now going to plot the situation from fig. 4.8a in a momentum space diagram. Since for the moment we ignore transverse momenta, we consider the p_0-p_3 plane (p_0: energy, p_3: longitudinal momentum) or equivalently the p^+-p^- plane ($p^\pm = p_0 \pm p_3$). Neglecting baryon masses we get for the forward-moving baryon $p^+ = 2P$, $p^- = 0$ — represented by the dashed line parallel to the p^+ axis in fig 4.8b. The backward-moving baryon fulfils $p^+ = 0$, $p^- = 2P$ — represented by the solid line parallel to the p^- axis in fig. 4.8b. The momentum sharing between quark and diquark as discussed in the last paragraph is indicated by splitting the two lines into two segments each, with fractions x_1, $(1-x_1)$ and x_2, $(1-x_2)$. Putting together diquarks and quarks with momentum fractions $(1-x_1)$, x_2 and $(1-x_2)$, x_1 is indicated on the rhs of fig. 4.8b. Comparing with fig. 4.6 we realize that the two diagrams are completely identical, the only difference being that fig. 4.6 represents space-time (x_0-x_3) whereas fig. 4.8b represents momentum space (p_0-p_3). However, we mentioned earlier that because of the mapping $\Delta p = \kappa\, \Delta x$ it does not matter which space we are considering, in particular since we are using renormalized lengths and momenta. This means that "Colour Exchange" is identical to "Rearrangement", however giving structure to the endpoints: we have diquarks and quarks at the endpoints.

We have not yet specified in either case the momentum (or length) sharing distribution $f(x_1, x_2)$. In the VENUS model — which uses the "Colour Exchange" mechanism — we use

$$f(x_1, x_2) = f(x_1) f(x_2); \qquad f(x) = \frac{g(x)}{x^\alpha} \tag{4.9}$$

with $\alpha > 0$ and $0 < g(0) < \infty$. Different options for f are possible which may lead to quite different predictions concerning distributions of measurable quantities.

"Longitudinal Excitation"

"Longitudinal Excitation" [4.4] means the following: the interaction between two hadrons with large relative momentum results in a transfer of longitudinal momentum between the hadrons such that the result is two longitudinally excited objects (strings), see fig. 4.9a,b. Each string contains exactly the quarks of one of the incident hadrons, so the singlet structure of the hadrons is not changed like for "Colour Exchange" interaction, see fig. 4.9c,d.

For the following we restrict ourselves to yo-yo strings (in this section we are in general not so much interested in giving a complete description of the model, but rather want to understand the basic ideas). The interaction is defined in momentum space. Let us again consider two incident baryons in their centre-of-mass system, having momenta $\pm P$, and correspondingly the light-cone momenta $p^+ = 2P$, $p^- = 0$ (dashed line in fig. 4.10a) and $p^+ = 0$, $p^- = 2P$ (solid line in fig. 4.10a). The prescription for "Longitudinal Excitation" can be formulated as follows (see fig. 4.10): divide the lines into two segments each with fractions x_1, $1 - x_1$ and x_2, $1 - x_2$ according to some distribution function $f(x_1, x_2)$. Construct rectangles (representing the new strings) with lengths $1 - x_1, x_2$ and $1 - x_2$, x_1 (see fig. 4.10b). This looks like the "Rearrangement" (fig. 4.6) or "Colour Exchange" (fig. 4.8b) diagram, but there is a substantial difference (see fig. 4.10b): the right string consists of only dashed parton lines, the left one only of solid ones. This indicates: all the partons (quarks and diquarks) of each string are coming from one original baryon. This implies that the energy in the (dashed) segment of length x_2 in fig. 4.10a is used to bend one end of the solid line representing the other baryon

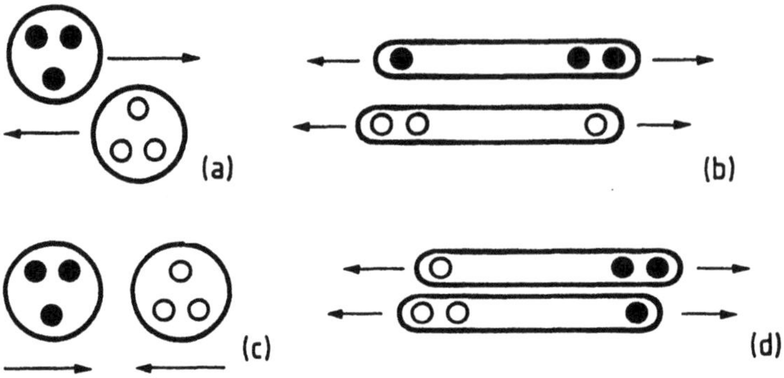

Fig. 4.9. "Longitudinal Excitation" (a,b) requires all partons in a string to originate from one baryon, whereas "Colour Exchange" (c,d) provides strings with partons from different baryons.

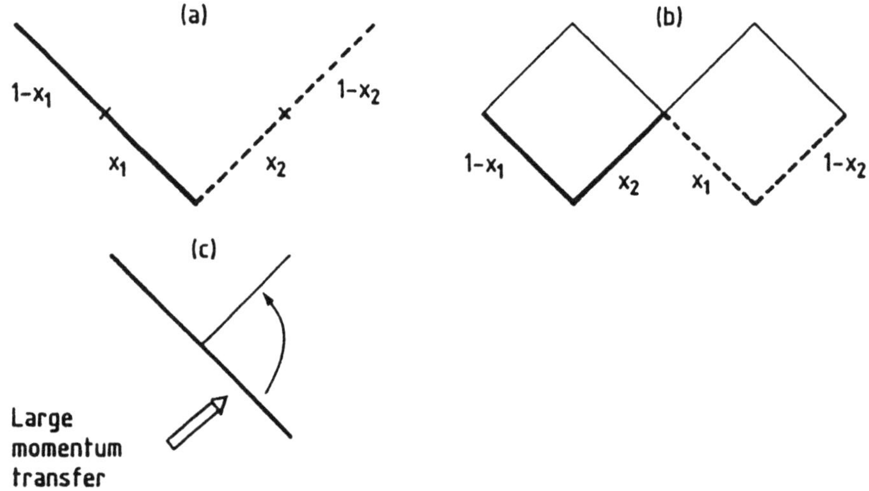

Fig. 4.10. Momentum space picture of a "Longitudinal Excitation" (again original momenta are renormalized to one): the incident momenta are split into two parts (a); one momentum share is then transferred to the other string and vice versa (b), so the straight line representing an incident baryon is bent by a massive momentum transfer acting on one end of the string (c).

(fig. 4.10c). This is obviously a very violent process in momentum space, contrary to "Colour Exchange" where no longitudinal momentum is transferred. Although the physical picture of the two approaches is very different, they are quite similar since the momentum space diagrams (figs. 4.8a,4.10ab) are so similar; a measurable difference is provided only by the different flavour content of the strings.

A substantial difference occurs of course through the fixing of the splitting distributions $f(x_1, x_2)$, which in [4.4] is given as

$$f(x_1, x_2) = f(x_1)f(x_2); \qquad f(x) = \frac{1}{x} \tag{4.10}$$

On has to keep in mind that not only the choice of f is important but also the cut-off procedure to avoid small x values.

4.2 The string model VENUS

The string model VENUS employs the "Colour Exchange" principle to form strings in energetic hadron-hadron collisions. We are only going to sketch the model in the following; we do not intend to give a complete description, the interested reader is referred to [4.3]. The dominant process for baryon-baryon collisions in VENUS is the "Colour Exchange" ("String Flip") process eq. (4.8)

$$(q_p q_p q_p) + (q_t q_t q_t) \to (q_p q_p q_t) + (q_t q_t q_p) \tag{4.11}$$

where a string flip between the projectile proton $(q_p q_p q_p)$ and the target proton $(q_t q_t q_t)$ (left-hand side of eq. (4.11)) results in two diquark quark strings (right-hand side of eq. (4.11)) with a large relative momentum between the quark q and the diquark qq. Since a nucleon may have a more complicated structure, like $(qqq)(q\bar{q})$, in VENUS also "diffractive scattering"

$$(q_p q_p q_p)(q_p \bar{q}_p) + (q_t q_t q_t) \to (q_p q_p q_p) + (q_t \bar{q}_p) + (q_t q_t q_p) \tag{4.12}$$

contributes, where the "surviving" nucleon $(q_p q_p q_p)$ keeps a large fraction of the initial momentum, thus giving rise to a diffractive peak at large momenta.

Altogether we have (per def.) four "One–Colour Exchange" contributions (involving colour exchange between quarks):

$$\begin{array}{lll} (a) & (q_p q_p q_p) + (q_t q_t q_t) & \to (q_p q_p q_t) + (q_t q_t q_p) \\ (b) & (q_p q_p q_p)(q_p \bar{q}_p) + (q_t q_t q_t) & \to (q_p q_p q_p) + (q_t \bar{q}_p) + (q_t q_t q_p) \\ (c) & (q_p q_p q_p) + (q_t q_t q_t)(q_t \bar{q}_t) & \to (q_p q_p q_t) + (q_p \bar{q}_t) + (q_t q_t q_t) \\ (d) & (q_p q_p q_p)(q_p \bar{q}_p) + (q_t q_t q_t)(q_t \bar{q}_t) & \to (q_p q_p q_p) + (q_t \bar{q}_p) + (q_t q_t q_t) + (q_p \bar{q}_t)\,. \end{array} \tag{4.13}$$

These basic VENUS "One–Colour Exchange" contributions are referred to as (a) non-diffractive (b,c) single diffractive and (d) double Pomeron exchange (contribution (a) having by far the largest weight). There are in addition four corresponding contributions for "One–Colour Exchange" between antiquarks ([4.7]):

$$\begin{array}{lll} (\hat{a}) & (q_p q_p q_p q_p \bar{q}_p) + (q_t q_t q_t q_t \bar{q}_t) & \to (q_p q_p q_p q_p \bar{q}_t) + (q_t q_t q_t q_t \bar{q}_p) \\ (\hat{b}) & (q_p q_p q_p)(q_p \bar{q}_p) + (q_t q_t q_t q_t \bar{q}_t) & \to (q_p q_p q_p) + (q_p \bar{q}_t) + (q_t q_t q_t q_t \bar{q}_p) \\ (\hat{c}) & (q_p q_p q_p q_p \bar{q}_p) + (q_t q_t q_t)(q_t \bar{q}_t) & \to (q_p q_p q_p q_p \bar{q}_t) + (q_t q_t q_t) + (q_t \bar{q}_p) \\ (\hat{d}) & (q_p q_p q_p)(q_p \bar{q}_p) + (q_t q_t q_t)(q_t \bar{q}_t) & \to (q_p q_p q_p) + (q_p \bar{q}_t) + (q_t q_t q_t) + (q_t \bar{q}_p)\,. \end{array} \tag{4.14}$$

Once considering a complex nucleon structure, consistency requires us to take into account also "Multiple–Colour Exchange". In VENUS, for example, the following "Double–Colour Exchange" occurs:

$$(q_p q_p q_p)(q_p \bar{q}_p) + (q_t q_t q_t)(q_t \bar{q}_t) \to (q_p q_p q_t) + (q_t \bar{q}_p) + (q_p \bar{q}_t) + (q_t q_t q_p), \tag{4.15}$$

leading to four strings rather than the two strings as in the lowest order contribution eq. (4.11). As shown in [4.3] one can set up a consistent model by expanding the whole contribution σ as

$$\sigma = \sum_n w_n \sigma_n \tag{4.16}$$

where σ_n are contributions with n colour exchanges, w_n being the corresponding weight. These weights w_n are assumed to be of the form

$$w_n \sim c^n \tag{4.17}$$

with some constant c.

4.3 KNO scaling

In a string model like VENUS the multiplicity distribution $P(n)$ for pp scattering is given as a superposition $\int dm\, W(m)\, P(m,n)$ of contributions for a given string mass m. We demonstrate that KNO scaling, an experimentally well established scaling law for multiplicity distributions at different energies, follows as a consequence of a scaling property for the string mass distribution $W(m)$, thus relating the KNO scaling function ψ to the quark structure functions, which determine $W(m)$.

As predicted by Koba, Nielsen and Olesen [4.8], multiplicity distributions $P(n)$ in high energy hadron-hadron collisions obey a scaling law (KNO scaling)

$$< n > P(n) = \psi(\frac{n}{< n >}) \tag{4.18}$$

with an energy independent function ψ, which implies energy independent moments

$$\frac{< n^k >}{< n >^k} = \psi_k = \int dz\, z^k \psi(z)\ . \tag{4.19}$$

Although recent collider experiments [4.9] indicate violations of KNO scaling, it is nevertheless very important to understand why this scaling law is approximately valid up to ISR energies [4.10]. KNO scaling can be derived as an asymptotic form of negative binomial (NB) distributions

$$P(n,\bar{n},k) = \frac{k(k+1)\ldots(k+n-1)}{n!} \frac{\bar{n}^n k^k}{(\bar{n}+k)^{n+k}}\ , \tag{4.20}$$

which were found to fit experimental multiplicity distributions for various pseudorapidity intervals from ISR energies up to collider energies [4.11]. Negative binomial distributions have been proposed from stochastic cell models [4.12, 4.13] or as a consequence of the recurrence relation [4.14]

$$(n+1)\,P(n+1) = g(n)\,P(n) \tag{4.21}$$

with a linear function $g(n)$

$$g(n) = \frac{\bar{n}}{\bar{n}+k}(k+n) \tag{4.22}$$

which might be realized by a cascade process, for example [4.14]. Also parton branching models [4.15] lead to NB distributions and KNO scaling in a certain limit.

We follow a different approach. KNO scaling is typical for hadronic collisions, whereas e^+e^- produces simple Poissonian multiplicity distributions [4.16]. Contrary to e^+e^- annihilation, hadronic collisions provide a superposition of strings with different masses, due to the parton structure of the nucleon. Thus, our aim will be to trace KNO scaling back to properties of string mass distributions, which amounts to relating the scaling function ψ to the quark structure functions.

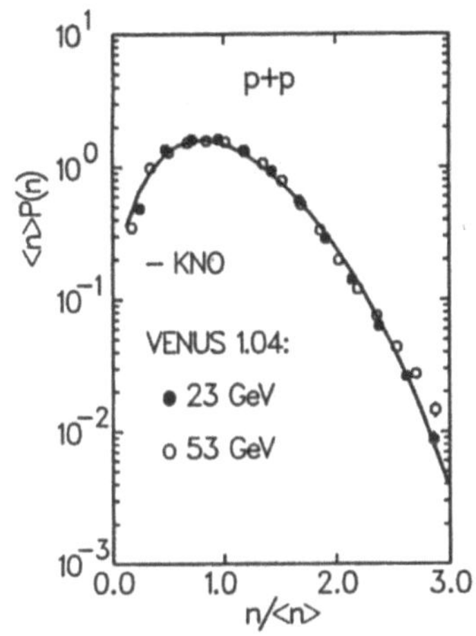

Fig. 4.11. Multiplicity distributions versus the scaling variable $n/ < n >$: VENUS results for 23 and 53 GeV are nearly identical and also agree with the empirical KNO function of [4.22]. Thus, the multistring model VENUS approximately shows KNO scaling!

As demonstrated in fig. 4.11, the VENUS model shows approximate KNO scaling, the reason for which we are going to explain in the following. According to eq. (4.16) we can write the multiplicity distribution in VENUS as

$$P(n) = \sum w_i \, P_{2i}(n) \ , \tag{4.23}$$

$P_j(n)$ being the multiplicity distribution of events with j strings. Whereas the energies of the j strings are correlated via

$$\sum_k E_k^j = E_0 \ , \tag{4.24}$$

the string masses are only weakly correlated, so P_j is approximately a folding of 1-string multiplicity distributions P_1:

$$P_j = P_1 \otimes \ldots \otimes P_1 \ . \tag{4.25}$$

The P_1 in eq. (4.25) may be different from each other, since different kinds of string are involved (diquark-quark or quark-antiquark strings). Scaling of individual distributions P_1 leads to scaling for P_j if the ratios of individual mean values are energy independent. Since eqs. (4.23,25) indicate — provided the w_i are energy independent — that scaling for $P(n)$ can be traced back to scaling for $P_1(n)$, we restrict ourselves in the following to string multiplicities $P_1(n)$.

As promised above, we want to draw a connection between string mass scaling and KNO scaling. So, what is the mass distribution $W(m)$ of strings in our model? Consider a string stretching between partons with momenta xP and $-x'P$ respectively, P being the proton momentum in the pp cm system. Neglecting parton masses, the string mass is given as

$$m^2 = 4P^2 x x' = s x x' \ , \tag{4.26}$$

x and x' being momentum fractions according to quark structure functions. Eq. (4.26) implies an energy independent distribution of $m/\sqrt{s}$, i.e.,

$$W(m) = \frac{1}{\sqrt{s}} \phi(\frac{m}{\sqrt{s}}) \ , \tag{4.27}$$

with an energy independent, universal function ϕ, the form of ϕ depending on the quark

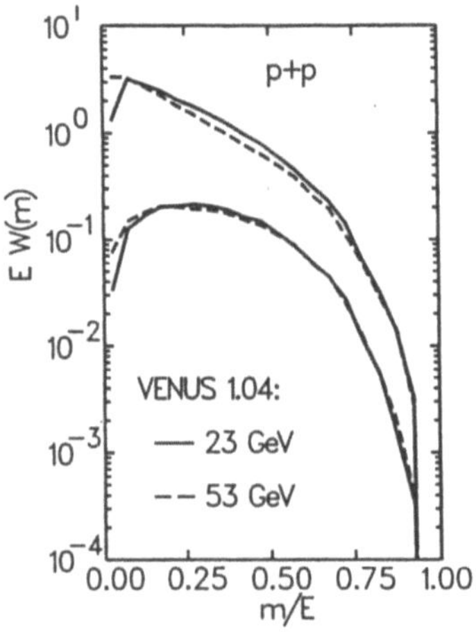

Fig. 4.12. Mass distributions $W(m)$ versus the scaling variable m/E ($E = \sqrt{s}$ is the *pp* cm energy). The upper curves represent all quark-diquark strings, whereas the lower curves consider only valence quark-diquark strings. The VENUS results for 23 and 53 GeV agree more or less (string mass scaling).

structure functions. In fig. 4.12 we show the mass distribution of diquark-quark strings as obtained by the multistring model VENUS. Apart from low mass cut-off effects, the distributions are indeed energy independent, as suggested by eq. (4.27).

In the following we show that string mass scaling (eq. (4.27)) leads to KNO scaling (eq. (4.19)). In order to connect string-masses and multiplicity, we decompose the string multiplicity distribution $P_1(n)$ as

$$P_1(n) = \int dm\ W(m)\, P(m,n)\ , \tag{4.28}$$

$P(m,n)$ being the multiplicity distribution of a string with mass m. With eq. (4.28) we obtain

$$\begin{aligned} \frac{<n^k>}{<n>^k} &= \frac{\int dn\, n^k\, P_1(n)}{[\int dn\, n\, P_1(n)]^k} \\ &= \frac{\int dm\ W(m)\ \int dn\, n^k\ P(m,n)}{[\int dm\ W(m)\ \int dn\, n\ P(m,n)]^k} \\ &= \frac{\int dm\ W(m) <n^k>_m}{[\int dm\ W(m) <n>_m]^k} \end{aligned} \tag{4.29}$$

with

$$<n^k>_m = \int dn\, n^k\ P(m,n)\ .$$

For Poisson distributions $P(m,n)$ we have

$$<n(n-1)\ldots(n-k+1)>_m = (<n>_m)^k\ ,$$

which leads for large $<n>_m$ to

$$<n^k>_m = (<n>_m)^k\ . \tag{4.30}$$

As shown in [4.16], $<n>_m$ can be parametrized as

$$<n>_m = \alpha m^\beta\ , \tag{4.31}$$

which is less popular than a $\log(m)$ fit, but nevertheless valid. Using eq. (4.30) and eq. (4.31), we obtain from eq. (4.29):

$$\frac{<n^k>}{<n>^k} = \frac{\int dm\ W(m)(m^\beta)^k}{[\int dm\ W(m) m^\beta]^k}\ . \tag{4.32}$$

Using the mass scaling law eq. (4.27), we obtain

$$\frac{<n^k>}{<n>^k} = \frac{\int \frac{dm}{\sqrt{s}}\phi(\frac{m}{\sqrt{s}})(m^\beta)^k}{[\int \frac{dm}{\sqrt{s}}\phi(\frac{m}{\sqrt{s}})m^\beta]^k}$$

or

$$\frac{<n^k>}{<n>^k} = \frac{\int dz\phi(z)z^{\beta k}}{[\int dz\phi(z)z^\beta]^k}, \tag{4.33}$$

the latter one being energy independent since ϕ is energy independent. According to eq. (4.19) this means KNO scaling!

We have demonstrated that scaling of string masses as $m/\sqrt{s}$ leads to KNO scaling. String mass scaling should be obtained in all "Dual Parton–like" models, [4.17, 4.18, 4.19, 4.20, 4.21], not only in VENUS, as demonstrated.

Originally [4.8] KNO scaling was proposed as a consequence of Feynman scaling of inclusive cross sections. In the above considerations, Feynman scaling manifests itself in the fact that in the quark structure functions $q(x)$ only momentum fractions x occur. This leads to energy independent x and x' distributions in eq. (4.26 and thus to string mass scaling.

4.4 References

[4.1] K. Sailer, B. Müller, W. Greiner, preprint UFTP 230/1989, 1989

[4.2] E. A. Remler, in Proc. "Gross Properties of Nuclei and Nuclear Excitations", Hirschegg, Austria, January 1987, p. 24

[4.3] K. Werner, Z. Phys. C42, 85 (1989)

[4.4] B. Andersen, G. Gustafson and B. Nielsson-Almqvist, Nucl. Phys. B281, 289 (1987)

[4.5] A. Capella, U. Sukhatme and J. Tran Thanh Van, Z. Phys. C3, 329 (1980)

[4.6] K. Werner, preprint CERN-TH-5538/89, to appear in Phys. Lett. B

[4.7] K. Werner, Phys. Rev. Lett. 62, 2460 (1989)

[4.8] Z. Koba, B. Nielsen, P. Olesen, Nucl. Phys. B40, 317 (1972)

[4.9] G. J. Alner et al (UA5 collaboration), Phys. Lett. 138B, 304 (1984)

[4.10] W. Thome et al, Nucl. Phys. B129, 365 (1977)

[4.11] G. J. Alner et al (UA5 collaboration), Phys. Lett. 160B, 199 (1985)

[4.12] A. Giovannini, Nuovo Cimento 15A, 543 (173)

[4.13] P. Carruthers, C. C. Shih, Phys. Lett. 127B, 242 (1983)

[4.14] A. Giovannini, L. Van Hove, Z. Phys. C30, 391 (1986)

[4.15] I. Sarcevic, Phys. Rev. Lett. 59, 403 (1987)

[4.16] M. Derrick et al, Phys. Rev. D34, 3304 (1986)

[4.17] A. Capella and J. Tran Thanh Van, Z. Phys. C10, 249 (1981)

[4.18] A. Capella, T. A. Casado, C. Pajares, A. V. Ramallo and J. Tran Thanh Van, Z. Phys. C33, 541 (1987)

[4.19] J. Ranft and S. Ritter, Z. Phys. C27, 413 (1985)

[4.20] P. Aurenche, F. W. Bopp and J. Ranft, Z. Phys. C23, 67 (1984)

[4.21] T. P. Pansart, in proc. of the Fifth International Conference on Ultrarelativistic Nucleus–Nucleus Collisions, Asilomar; eds. L. Schroeder and M. Gyulassy, Nucl. Phys. A461, (1987)

[4.22] P. Slattery, Phys. Rev. Lett. 29, 1624 (1972)

5 NUCLEUS-NUCLEUS SCATTERING

We demonstrate in section 5.1 that the VENUS model for hadron-hadron scattering can almost straightforwardly be extrapolated towards nucleus-nucleus collisions. The basic assumption is that a projectile nucleon moves through the target (and vice versa) on a straight line. A "Colour Exchange" occurs whenever a projectile and target nucleon come close to each other (we call these interactions "primary"). This implies that after the first interaction, a nucleon is not a nucleon any more, rather a "leading diquark" or some other coloured object (in general: nucleon minus the parton involved in the colour exchange). Technically all primary interactions are performed before fragmentation sets in. This is very reasonable at high energies ($>$ 100 GeV incident energy), however, at considerably lower energies this concept becomes questionable. The fragmentation of strings is done according to the procedure introduced in chapter 3. That implies a full space-time treatment and taking into account "secondary interactions" between string fragments among each other and with spectators. The latter type of secondary scattering ("cascading") is studied in section 5.2.

5.1 VENUS model for nucleus-nucleus scattering

The VENUS model for *NN* scattering as described in the last chapter serves as a good basis to make a nearly straightforward extrapolation towards nucleus-nucleus collisions. The *NN* model is more than just a fit to *NN* data — few basic assumptions and few parameters are sufficient to reproduce a large number of experimental data. So we want to use the same concept, namely "Colour Exchange" as the source of string production, also in nucleus-nucleus collisions.

We assume the nucleons to move through the other nucleus on a straight line. Although the nature of a nucleon may change on its way through the nucleus, we assume the geometrical size to be unchanged. That means that we use a constant nucleon-nucleon cross section σ_{NN} to determine whether a projectile nucleon interacts with a target nucleon: they have to come closer than

$$r_{\text{i.a.}} = \sqrt{\sigma_{NN}/\pi}. \tag{5.1}$$

Each interaction means "Colour Exchange" and string formation as in NN collisions. The "nucleon" (whatever its nature is after the first collision) is assumed to make its way through the nucleus and be outside already before it starts to hadronize into observable particles. This is reasonable as long as the hadronization time

$$\tau_h = \tau_0 \ \cosh y = \tau_0 \gamma \tag{5.2}$$

is larger than the reaction time

$$\tau_r = 2R \ \coth y \tag{5.3}$$

R being the nuclear radius and y being the rapidity of a projectile nucleon. A condition for the applicability of our model is therefore

$$\tau_h > \tau_r \ . \tag{5.4}$$

For 200 GeV incident energy per projectile nucleon we obtain $\tau_h \approx 200\tau_0$ and $\tau_r \approx 2R$, so for $\tau_0 \approx 1$ fm the condition $\tau_h > \tau_r$ is fulfilled for all nuclei. At 14.5 GeV we find $\tau_h \approx 15\tau_0$, which means the condition eq. (5.4) may be "just" fulfilled, so this energy is close to the borderline of applicability of the model.

Let us consider a projectile nucleus (mass number A) with high enough energy to fulfil the condition $\tau_h > \tau_r$ hitting a target nucleus (mass number B). Our reference system is the centre-of-mass system of one projectile and one target nucleon, further on referred to as the NN cm system, having a rapidity

$$y_{NN} = \frac{1}{2} \cdot \frac{1}{2} \log \frac{E+P}{E-P} = \frac{1}{2} \log \frac{\sqrt{m^2+P^2}+P}{m} \approx \frac{1}{2} \log \frac{2P}{m} \ , \tag{5.5}$$

where E, P and m are the projectile nucleons energy, momentum and mass. The NN energy in this system is

$$\sqrt{\sigma_{NN}} = \sqrt{(\sqrt{m^2+P^2}+m)^2 - P^2} \approx \sqrt{2Pm} \ , \tag{5.6}$$

the last equalities being true in the ultrarelativistic limit $P \gg m$. In the numerical calculations we always use the exact formulas! For 200 AGeV incident energy, for example, we obtain from eq. (5.5) and eq. (5.6) $y_{NN} = 3.0$ and $\sqrt{\sigma_{NN}} = 19.4$ GeV; for 14.5 AGeV incident energy we get $y_{NN} = 1.7$ and $\sqrt{\sigma_{NN}} = 5.4$ GeV. From eq. (5.6) we obtain for the initial energies and momenta of the projctile nucleons in the NN cm system $(1 \leq i \leq A)$

$$\begin{aligned} E^i_{\text{proj}} &= \frac{1}{2}\sqrt{\sigma_{NN}} \approx \sqrt{\frac{Pm}{2}} \\ P^i_{\text{proj}} &= \sqrt{\frac{\sigma_{NN}}{4} - m^2} \approx \sqrt{\frac{Pm}{2}} \end{aligned} \tag{5.7}$$

and of the target nucleons $(1 \leq j \leq B)$

$$E^j_{\text{targ}} = \frac{1}{2}\sqrt{\sigma_{NN}} \approx \sqrt{\frac{Pm}{2}}$$
$$P^j_{\text{targ}} = -\sqrt{\frac{\sigma_{NN}}{4} - m^2} \approx -\sqrt{\frac{Pm}{2}} \quad . \tag{5.8}$$

Correspondingly, the coordinates of the nucleons are labelled x^i_{proj}, y^i_{proj}, z^i_{proj} with $1 \leq i \leq A$ and x^j_{targ}, y^j_{targ}, z^j_{targ} with $1 \leq j \leq B$. The nucleons are distributed isomorphically according to a Woods-Saxon density distribution

$$\rho(r) = \frac{\rho_0}{1 + \exp\left[\frac{r - r_0}{a}\right]} \ , \tag{5.9}$$

with the parameters

$$a = 0.54$$
$$r_0 = 1.19\ A^{1/3} - 1.61\ A^{-1/3} \quad . \tag{5.10}$$

Although not very important, we take into account the nucleon's hard core

$$(x^k - x^l)^2 + (y^k - y^l)^2 + (z^k - z^l)^2 \geq (2r_c)^2 \ , \tag{5.11}$$

with a core radius $r_c = 0.4$ fm. A nucleon is assumed to move on a straight line through the other nucleus, making an interaction whenever it comes close enough to a nucleon:

$$(x^i_{\text{proj}} + b_x - x^j_{\text{targ}})^2 + (y^i_{\text{proj}} + b_y - y^j_{\text{targ}})^2 \leq \frac{\sigma_{NN}}{\pi} \ , \tag{5.12}$$

σ_{NN} being the inelastic NN cross section (we use $\sigma_{NN} = 3.1\ \text{fm}^2$ for 200 GeV) and $(b_x,\ b_y)$ being the impact parameter. As mentioned earlier, we use a constant value of σ_{NN}, regardless of how many collisions precede the actual collision, even though a nucleon after a collision is usually not a nucleon any more.

Two quantities characterizing the geometrical aspect of a nucleus-nucleus collision are the number ν of nucleon-nucleon interactions and the number ν_{p} of participation projectile nucleons. The distributions of these quantities show a qualitative change by going from a symmetric collision ($A \approx B$) to a very asymmetric one ($A \ll B$ or $B \ll A$). In fig. 5.1 we show ν_{p} distributions for oxygen-induced reactions on carbon, copper and gold targets. For the nearly symmetric system, O + C, the distribution peaks at $\nu_{\text{p}} = 1$, dropping continuously for increasing ν_{p}. One participant is only likely, when all projectile nucleons are peripheral with respect to the target — so a maximum at $\nu_{\text{p}} = 1$ means a wide region of peripheral collisions as a consequence of surface diffuseness. As the target size increases, a second maximum develops at $\nu_{\text{p}} = 16$, which means all projectile nucleons participate. This maximum is due to the fact that for a target much larger than the projectile there is a wide range of impact parameter values which lead to a "central collision" where all projectile nucleons participate. Related to this qualitative change of ν_{p} distribution is a qualitative change of the distribution of the collision number ν as a function of the asymmetry. As shown on the left part of fig. 5.1 (consider only the solid lines here) with increasing target size a shoulder appears corresponding to the maximum of the ν_{p} distribution as discussed above. Regardless of the target size, all ν distributions peak at small ν, meaning that peripheral collisions dominate. Therefore, it is rather useless to consider mean values of observables, since always a large portion of uninteresting peripheral collisions contribute — unless triggers are used which require a certain amount of interaction.

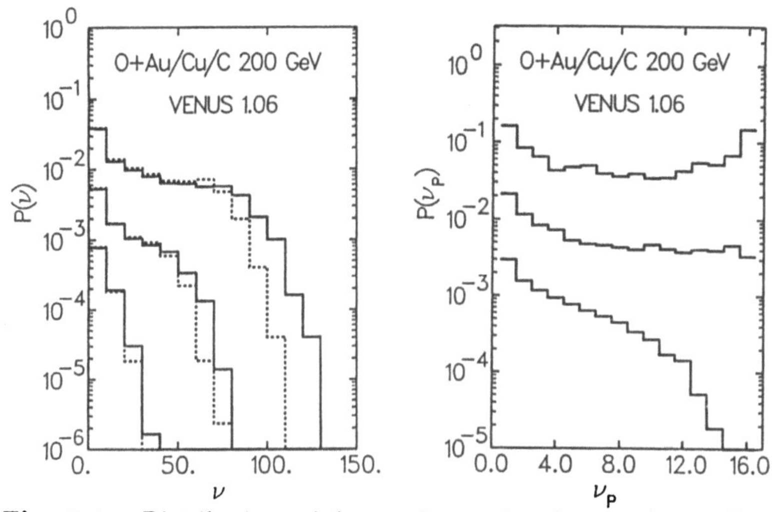

Fig. 5.1. Distributions of the number ν of nucleon-nucleon collisions (left plot, solid lines) and distribution of the number ν_P of participating projectile nucleons (right plot) for O + Au, O + Cu, O + C (from top to bottom) at 200 AGeV. The curves are displaced by a factor of 10.

So far we have talked about collisions between nucleons in nucleus-nucleus scattering without specifying what the interaction really looks like and what the nature of the nucleon is after one or more collisions. We are going to discuss that in the following. Let us recall the definition of an "ordinary" nucleon-nucleon interaction as given in chapter 4. A collision is realized by one or more colour exchanges between a projectile and a target (anti)quark. As a result, a certain number of q-$\bar{q}$ strings emerge and, most important, one "projectile-like" and one"target-like" string, the latter ones (let's call them leading strings) carrying the baryon number and most of the energy of the reaction. A "projectile-like" string, for example, consists of a fast forward-moving nucleon remnant, tied per colour force to slowly backward-moving partons from the other nucleon. By nucleon remnant we mean a nucleon, reduced by some quarks and/or antiquarks, so in general a coloured object. Therefore, in a nucleus-nucleus collision we have to consider interactions between nucleon remnants rather than between nucleons, yet the interaction can be defined in exactly the same way: an interaction between nucleon remnants is realized by "Colour Exchange" involving quarks or antiquarks of the remnants. This definition implies that we only consider those (anti)quarks which have not been struck in an earlier collision to participate in the "Colour Exchange". As for NN collisions, we have four basic "One–Colour Exchange" contributions σ_a, σ_b, σ_c and σ_d involving quarks (eq. (4.13)), and four basic "One–Colour Exchange" contributions $\sigma_{\hat{a}}$, $\sigma_{\hat{b}}$, $\sigma_{\hat{c}}$ and $\sigma_{\hat{d}}$ involving antiquarks (eq. (4.14)) leading to the following "Multiple–Colour Exchange" interaction between projectile-nucleon remnant i and target-nucleon remnant j:

$$\sigma(i,j) = \sum_{k=1}^{\infty} w_k \left[\sum_{m \in M} \alpha_m \, \sigma_m(i,j) \right]^k \tag{5.13}$$

with w_k being the probability for k colour exchanges ($\sum w_k = 1$), α_m being the probability for the basic contribution σ_m ($\sum \alpha_m = 1$), and M being the set $\{a, b, c, d, \hat{a}, \hat{b}, \hat{c}, \hat{d}\}$.

(1d2c3a)

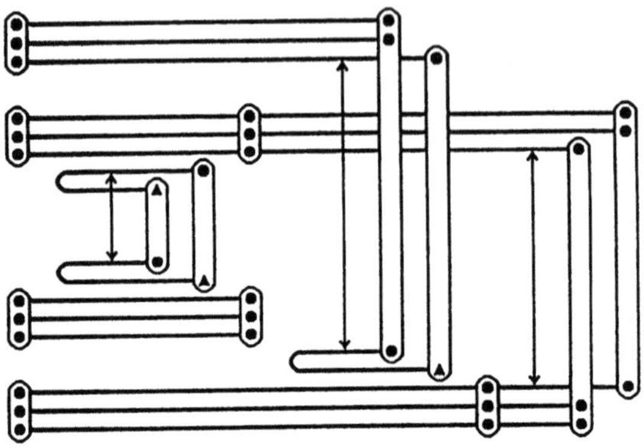

Fig. 5.2. An example for a "Colour Exchange" diagram in nucleus-nucleus collisions with two projectile and two target nucleons participating.

If i_μ and j_μ are the projectile and target nucleon involved in the μth collision of an event, the whole contribution is (using eq. (5.13)):

$$\prod_\mu \sigma(i_\mu, j_\mu) = \prod_\mu \sum_{k=1}^{\infty} w_k \left[\sum_{m \in M} \alpha_m \, \sigma_m(i_\mu, j_\mu) \right]^k . \tag{5.14}$$

In fig. 5.2 we show as an example for three collisions ($\nu = 3$) with two projectile and two target nucleons involved, the contribution $\sigma_d(2,1)\ \sigma_c(1,2)\ \sigma_a(2,2)$. The entire nucleus-nucleus collision can be written as (see eq. (5.14))

$$\sigma_{AB} = \sum_{\{i,j\}} P(\{i,j\}) \prod_\mu \sum_{k=1}^{\infty} w_k \left[\sum_{m \in M} \alpha_m \, \sigma_m(i_\mu, j_\mu) \right]^k \tag{5.15}$$

where $P(\{i,j\})$ is the probability to have a sequence $\{i,j\} = i_1, j_1;\ i_2, j_2\ \ldots$ of nucleon-nucleon collisions. Equation 5.15 can also be written as

$$\sigma_{AB} = \sum_{\{i,j\}} \sum_{k_1 k_2 \ldots} \sum_{m_1^1 \ldots m_{k_1}^1} \sum_{m_1^2 \ldots m_{k_2}^2} \cdots \left(P(\{i,j\}) \prod_\mu w_{k_\mu} \prod_{k=1}^{k_\mu} \alpha_{\{m_k^\mu\}} \right) \times \left(\prod_\mu \prod_{k=1}^{k_\mu} \sigma_{\{m_k^\mu\}}(i_\mu, j_\mu) \right) . \tag{5.16}$$

String properties (momenta and flavour content) for the contributions

$$\prod_\mu \prod_{k=1}^{k_\mu} \sigma_{\{m_k^\mu\}}(i_\mu, j_\mu)$$

of eq. (5.16) are calculated according to exactly the same prescription as corresponding diagrams in NN collisions, described in chapter 4. However, now, instead of the sharing functions $f(x)$ in eq. (4.9) we have to use a multi-parton-distribution function $f(\vec{x})$ which we assume to be of the form

$$f(\ldots x_n \ldots \bar{x}_m \ldots) = \prod_n f_n(x_n) \prod_m \bar{f}_m(\bar{x}_m)\ \theta\left(\sum_n x_n + \sum_m \bar{x}_m\right) \tag{5.17}$$

where $f_n(x)$ and $\bar{f}_m(\bar{x})$ are the sharing functions (or single parton distribution functions) f defined earlier for the NN model in eq. (4.9).

5.2 Cascading in nuclear collisions

It is a commonly accepted picture for the initial stage of an ultrarelativistic heavy ion collision to cut the nuclei geometrically into spectator and participant regions. A small nucleus colliding centrally with a big one would drill a cylindrical hole through the latter, so that all the nucleons in the cylinder and all projectile nucleons participate in the interaction, whereas all target nucleons outside the cylinder just spectate.

In a string model (like VENUS), the participant region is the zone where strings are formed: so we have a central cylinder with longitudinal oriented strings, surrounded by the spectator nucleons outside the cylinder. The strings expand along their longitudinal axes, and break for the first time after typically 1 fm/c. Whenever a string breaks, the resulting two substrings acquire transverse momentum. The transverse momentum component allows the string segments to move outside the participant cylinder and interact with spectator nucleons. Such secondary interactions with spectators are referred to as "cascading".

The string model VENUS used the fragmentation procedure described in chapter 3, i.e. secondary interactions among produced particles as well as interactions of produced particles with spectators are included in the model. Presently there exists only one other model [5.1] which fully includes both these aspects. Other models dealing with secondary interactions are described in refs. [5.2, 5.3]. Although interactions of produced particles among each other in particular concerning the formation of a quark gluon plasma are important, we concentrate here on the other aspect — cascading. It is the purpose of this section to apply VENUS to calculate observables which are sensitive to cascading, and compare the results with data.

The model of interacting string fragmentation has (in chapter 3) already been applied to neutrino-neon scattering which is essentially the fragmentation of a diquark-quark string inside a nucleus. So this model has been tested already, the parameters have been fixed, and it can therefore without further freedom be applied to the more complicated proton-nucleus and nucleus-nucleus collisions.

Rescattering is crucially affected by the interaction radius r_0 in eq. (3.9). In the following we are comparing two scenarios (using $r_i = 0.7\,\text{fm}$ for baryons and $r_i = 0.35\,\text{fm}$ for mesons, i=1,2):

$$r_0 = 0 \qquad \longleftrightarrow \qquad \text{rescattering turned off}$$

$$r_0 = r_1 + r_2 \qquad \longleftrightarrow \qquad \text{rescattering}$$

where our default values may not be the final answer; it is really a best guess rather than an optimized parameter. In the following figures we shall use dashed lines if using $r_0 = 0$ and solid lines if using $r_0 = r_1 + r_2$, points represent data.

In figs. 5.3,4 we show rapidity distributions of negative pions and protons in p-Ag reactions at 100 GeV (data are from [5.4]). We observe that the total number of pions remains almost unchanged by turning on interactions. However, we observe a shift from central to backward rapidity. We see a much greater effect for protons: the number of observed protons is considerably increased, and we even observe protons at negative rapidity, which is not possible in p-p collisions. Without rescattering (dashed histogram) the protons have the following origin in the model: for each collision (on the

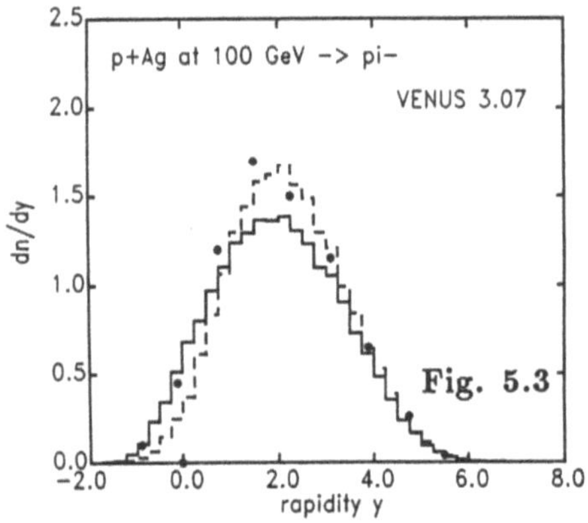

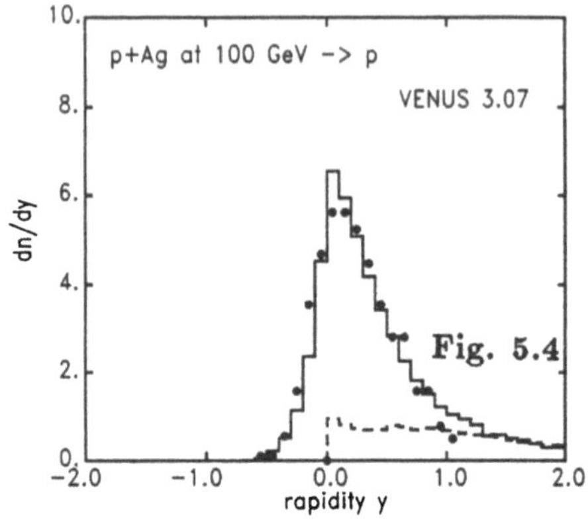

Fig. 5.3. Rapidity distributions of pions for p-Ag scattering at 100 GeV. Points are data, the dashed histogram represents VENUS without rescattering, the solid histogram shows VENUS results with rescattering.

Fig. 5.4. Rapidity distributions of protons for p-Ag scattering at 100 GeV. Same conventions as in fig. 5.3.

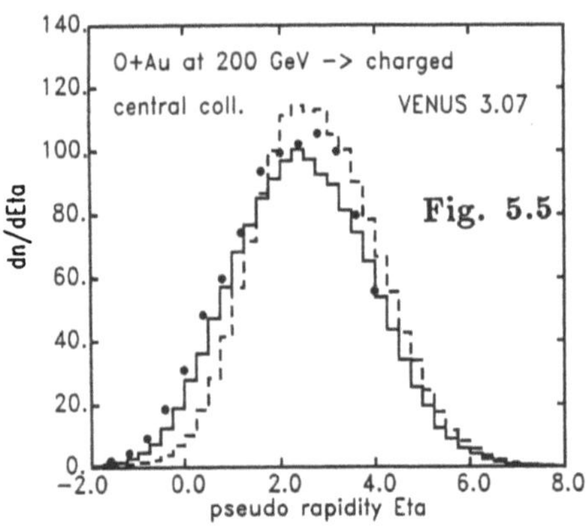

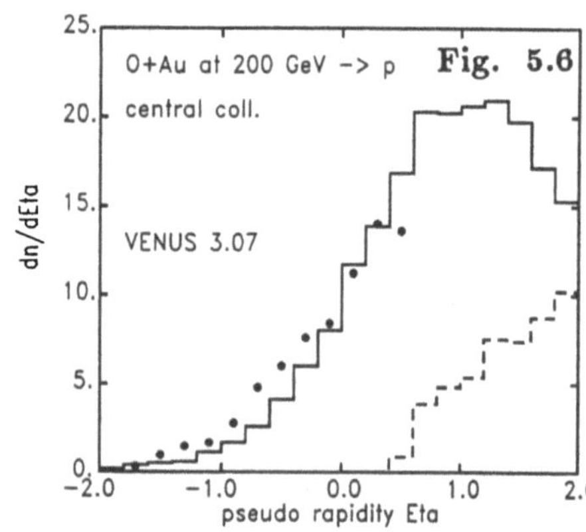

Fig. 5.5. Pseudo-rapidity distributions of charged particles for O-Au scattering at 200 GeV. Points are data, the dashed histogram represents VENUS without rescattering, the solid histogram shows VENUS results with rescattering.

Fig. 5.6. Pseudo-rapidity distributions of protons for O-Au scattering at 200 GeV. Same conventions as in fig. 5.5.

average ≈ 3) we have a diquark-quark string, the diquark facing backwards. Each of these diquarks leads after fragmentation to a baryon, so with roughly half of these being protons we expect approximately 1.5 protons in the fragmentation region ($y < 3$), which is more or less the integral of the dashed histogram (fig. 5.4). Concerning rescattering the most important process is the following: a produced particle (baryon or meson) hits a target spectator, the two hadrons fuse and decay again. This leads to the large increase of baryon number in fig. 5.4 and the shift in the pion distribution in fig. 5.3. We observe qualitatively the same behaviour for the reaction O+Au, see figs. 5.5,6 (data are from [5.5, 5.6]).

Another variable widely used in experiments is the transverse energy E_t which sums up weighted energies of all particles in a given rapidity interval. This quantity is

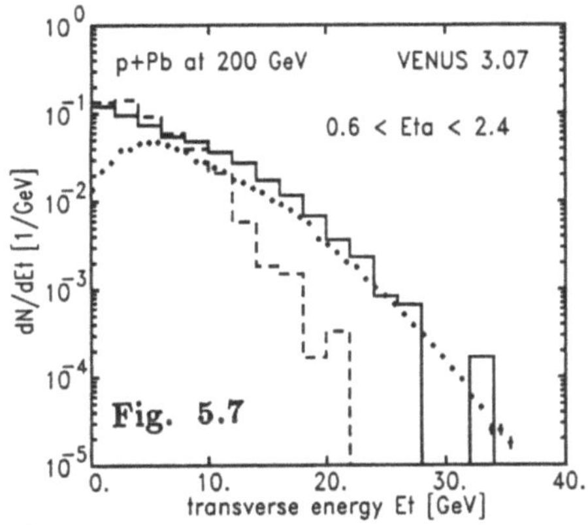

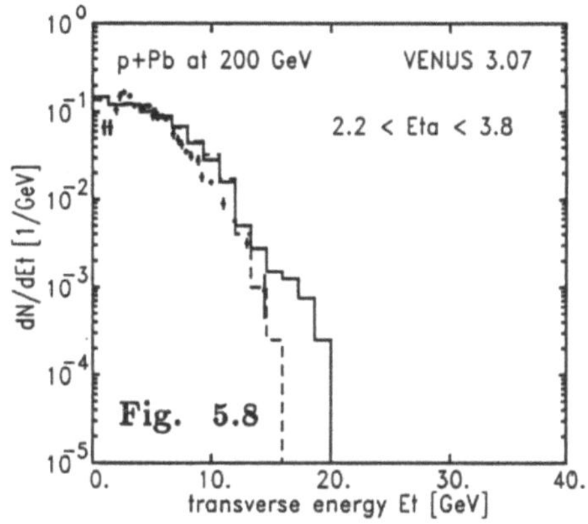

Fig. 5.7. Transverse energy distributions for p-Pb collisions at 200 GeV for a backward rapidity interval. Points are data, the histograms represent VENUS with (solid) and without (dashed) rescattering.

Fig. 5.8. Transverse energy distributions for p-Pb collisions at 200 GeV for a central rapidity interval. Same conventions as in fig. 5.7. The data actually represent p-Au scattering.

essentially a convolution of multiplicity and single particle transverse momentum distribution. By comparing with the rapidity distributions in figs. 5.3-6 we try to understand qualitatively the effect of rescattering on E_t-distributions. Fig. 5.7 shows E_t-spectra for a pseudo-rapidity interval $0.6 < \eta < 2.4$ in p-Pb collisions at 200 GeV (data are from [5.7]). From our multiplicity studies we conclude that the pion multiplicity should not change much in this interval (compare fig. 5.3); however, one effect of the rescattering process eqs. (3.10,11) is the increase of proton number, and another one the transformation of longitudinal momentum into transverse one [5.8]. Both effects lead to a broader E_t-distribution — as observed in fig. 5.7. Fig. 5.8 demonstrates that for a central rapidity interval $2.2 < \eta < 3.8$ there is almost no effect. In this interval however the multiplicity in a rescattering scenario is decreased. This is obviously just compensated by an increase of the average p_t [5.8]. Both histograms (VENUS with and without rescattering) fit the data [5.9] equally well, although for other observables there is quite a difference between the two scenarios. This shows the limited value of the E_t variable to be used to discriminate between models.

We now turn to E_t-distributions for nucleus-nucleus collisions. In fig. 5.9 we show E_t-spectra for a backward rapidity range ($-0.1 < \eta < 2.9$) for O-W scattering at 200 GeV (the data are from [5.10]). The solid histogram represents as usual full VENUS (with rescattering), we also show VENUS results without rescattering (dashed). We see the same effect as for p-A at backward rapidities: rescattering leads to wider distributions, due to more $< p_t >$ per particle and more protons. For central-forward rapidities we see again — as for p-A at central rapidities in fig. 5.8 — almost no effect due to rescattering, see fig. 5.10 (the data are from [5.11]).

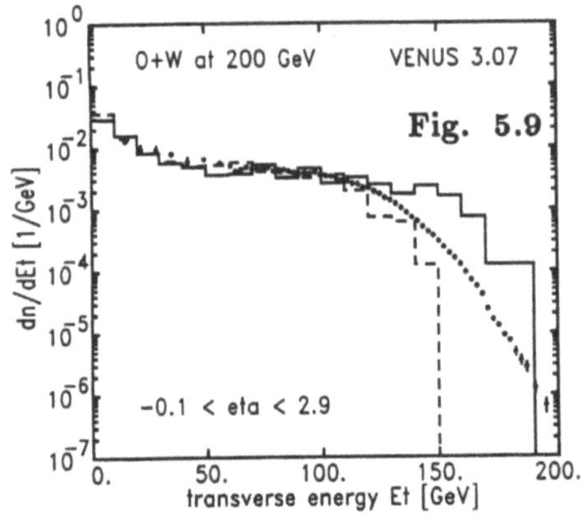

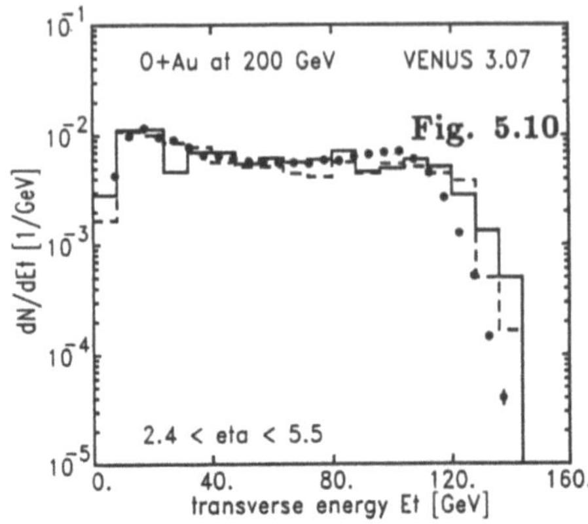

Fig. 5.9. Transverse energy distributions for O-W collisions at 200 GeV for a backward rapidity interval. Points are data, the histograms represent VENUS with (solid) and without (dashed) rescattering.

Fig. 5.10. Transverse energy distributions for O-Au collisions at 200 GeV for a central-forward rapidity interval. Same conventions as in fig. 5.9.

5.3 References

[5.1] H. Sorge, H. Stöcker and W. Greiner, Nucl. Phys. A498, 567c (1989)

[5.2] J. Ranft, Phys. Rev. D37, 1842 (1988)

[5.3] A. Shor and R. Longacre, Phys. Lett. B218, 100 (1989)

[5.4] W. S. Toothacker et al, Phys. Lett. B197, 295 (1987)
J. Whitmore et al, Proc. of "Hadronic matter in collision 1988", P. Carruthers and J. Rafelski, eds., World Scientific Publ. Co. 1989, p. 130

[5.5] R. Albrecht et al, Z. Phys. C38 (1988) 51

[5.6] R. Albrecht et al, preprint GSI-89-59 (1989)

[5.7] T. Akesson et al, Z. Phys. C38 (1988) 397

[5.8] K. Werner and P. Koch, in preparation

[5.9] A. Bamberger et al, Phys. Lett. B184, 271 (1987)

[5.10] T. Akesson et al, Z. Phys. C38 (1988) 383

[5.11] R. Albrecht et al, Phys. Lett. B199 (1987) 297

Part II

Topical Lectures

The Fractal Structure of Multihadron Production at High Energies

H. Satz

Theory Division, CERN, CH-1211 Geneva 23, Switzerland, and
Fakultät für Physik, Universität Bielefeld, D-4800 Bielefeld 1, Fed. Rep. of Germany

We first give a short general introduction to intermittency and fractal structure, and discuss how self-similar patterns can arise in particle production. Then we show that at high energies, self-similar jet production leads to a universal power law behaviour for the multiplicity moments as a function of the relative rapidity $Y/\delta y$, once the rapidity intervals δy are large enough to be outside the usual resonance region ($\delta y \geq 1-2$). Data from the UA5 experiment at CERN (for $\sqrt{s} = 200, 546$ and 900 GeV) are found to agree very well with the predicted universal behaviour.

1. Introduction

The study of multihadron production up to ISR energies ($\sqrt{s} \leq 60$ GeV) has revealed a number of general features, such as Feynman scaling or KNO scaling. With the advent of very high energy accelerators, these phenomenological laws have reached their limits of validity, and new general features of very high energy multihadron production have to be established. At sufficiently high energies one would moreover hope to reach a range in which such features can be related to predictable consequences of QCD.

In this talk, I want to discuss a new pattern of multiparticle production which sets in clearly only for $\sqrt{s} > 50$ GeV; the results I will present are based on joint work with I. Sarcevic[1]. We find at these very high energies a universal power law behaviour of the multiplicity moments as a function of the rapidity interval. Such a behaviour follows from a self-similar, or fractal, pattern of multihadron production.

At very high energies, the multiplicity distribution dN/dy can be measured over a large range $0 \leq y \leq Y$. We can therefore subdivide this range into different size intervals δy which are all larger than that characteristic for the usual resonance correlations. If the invariant mass $M_{\pi\pi}$ of a two-pion system is of order $M_{\pi\pi} \simeq 0.5 - 1.0$ GeV, the separation in rapidity of the two pions is about $\delta y_0 \simeq 1 - 2$. The intervals we want to consider are to be larger than this value δy_0.

Thus our considerations address a range of rapidity intervals complementary to that studied in recent "intermittency" investigations.[2] There one looks for a power-law dependence of multiplicity moments in rapidity intervals $\delta y \to 0$, i.e., for $\delta y < \delta y_0$. It has recently been asserted that the data in that δy range can be accounted for in terms of conventional short-range correlations[3,4], without any necessity for intermittent behaviour. We shall not deal with this question here, however, and instead concentrate

Phase Structure of Strongly Interacting Matter
Editor: J. Cleymans

on the possibility of self-similar behaviour for large rapidity separation, i.e., in the jet region.

The idea of self-similar multihadron production is not new. Hagedorn[5] proposed more than 20 years ago that hadronic collisions would produce "fireballs", consisting of fireballs, which in turn consist of fireballs, and so on, with always the same structure down to the observed hadrons as the end of the cascade. Once we are away from this final step, which has an intrinsic scale fixed by the hadron masses, all steps are self-similar in their decay pattern. Such a fireball picture did not account for the jet structure observed in multihadron production and therefore is cited here only as an illustration. Within the framework of QCD, however, we expect that a jet will decay into jets, which in turn lead to jets, and so on, down to a final hadronisation stage – giving again rise to a self-similar pattern.

Since the study of self-similar structures in particle physics was started quite recently[2], we will spend the next two sections with a simple introduction to intermittency and fractal structure. Following that, we shall show how a decay cascade can be related to these concepts. Finally, in section 5, we shall apply our formalism to high energy multihadron production and show that CERN collider data indeed show the behaviour expected in self-similar jet production.

2. Intermittency

We first want to illustrate how moments provide a measure of fluctuations, and then define what is meant by intermittency. Consider N objects (balls, particles, events, etc.), to be distributed over a d-dimensional unit volume $V \equiv R^d$ according to some given pattern. Subdivide this volume into n equal cells (bins, blocks) of size L^d, so that $n = (R/L)^d \equiv r^d$; r is our basic linear scale measure. In the m-th cell, there are k_m objects, and $\sum_{m=1}^{n} k_m = N$. We now define the normalized multiplicity moments of order l,

$$f_l(r) \equiv [\frac{1}{n} \sum_{m=1}^{n} k_m^l] \,/\, [\frac{1}{n} \sum_{m=1}^{n} k_m]^l. \tag{1}$$

The probability to find an object in the m-th cell is $P_m(r) = k_m/N$, so that we can rewrite eq.(1) as

$$f_l(r) = r^{d(l-1)} \sum_{m=1}^{n} P_m^l. \tag{2}$$

Since $\sum P_m^l \leq \sum P_m = 1$, we have

$$1 \leq f_l(r) \leq r^{d(l-1)}; \tag{3}$$

in particular, $f_l(r=1) = 1$. A suitable quantity for the measure of fluctuations is, as we shall see shortly, the ratio

$$\lambda_l(r) \equiv [\ln f_l(r)/(l-1)\ln r], \tag{4}$$

which, using eq.(3), fulfills

$$0 \leq \lambda_l(r) \leq d. \tag{5}$$

Let us look at $\lambda_l(r)$ for two extreme distributions.

If the N objects are equidistributed over all the n cells, then $P_m = 1/n$, giving us $f_l(r) = 1 \ \forall \ l$. Hence we have

$$\lambda_l(r) \ = \ 0 \quad \forall \ l \tag{6}$$

for an equidistribution. On the other hand, if all objects are placed in one single cell, then $\sum P_m^l = 1$ and $\lambda_l(r)$ attains its upper limit,

$$\lambda_l(r) \ = \ d \quad \forall \ l. \tag{7}$$

Thus the larger $\lambda_l(r)$ is, the more does the distribution fluctuate. If $\lambda_l(r)$ remains non-zero even in the limit of arbitrarily large $r = R/L$, (i.e., either small cell size or large volume)

$$\lambda_l \ \equiv \ \lim_{r\to\infty} \lambda_l(r) \ > \ 0 \ \forall \ l, \tag{8}$$

then we speak of intermittency[2] and call λ_l the (normalized) intermittency index. Intermittency thus implies a power-law dependence of the moments f_l on the scale measure r,

$$f_l \sim r^{\lambda_l}, \tag{9}$$

with λ_l as the power.

Let us note that these considerations remain non-trivial only if $N >> n$. If we keep N fixed as $n \to \infty$, then eventually all cells except some fixed number (independent of n) are empty, giving us $\lambda_l = d \ \forall \ l$. This is a trivial form of intermittency: for fixed N and small enough cell size, all distributions fluctuate. Our extreme case of all objects in one cell is also of this type. What we are really looking for are situations in which the probability moments

$$g_l(r) \ \equiv \ \sum_{m=1}^{n} P_m^l \tag{10}$$

do not simply become constant for large r, but instead retain an r-dependence. In terms of $g_l(r)$, the intermittency indices (8) become

$$\lambda_l \ = \ d \ - \ \lim_{r\to\infty} [\ln \ g_l(r)/(l-1) \ln \ r], \tag{11}$$

so that for non-trivial intermittency, the second term on the r.h.s. of eq.(12) should not vanish. In the following section, we shall see that this implies distributions which have a fractal structure.

3. Fractal Structure

We begin again with a distribution function of some objects in a given d-dimensional space $V = R^d$, such as the distribution of probabilities in the previous section. For simplicity, let us look at the one-dimensional case. We subdivide the line R into $n = r = R/L$ intervals $m = 1, 2, ..., r$; as before, r fixes our linear scale. We now want to count the number $n_F(r)$ of cells for which $P_m(r) \neq 0$. For a smooth distribution, $n_F(r) \sim r$. If this remains true even for $r \to \infty$, then the set of r segments of the line R becomes the one-dimensional continuum. On the other hand, if the objects are found in a finite number N of cells, no matter how large r is, or if we only have a fixed

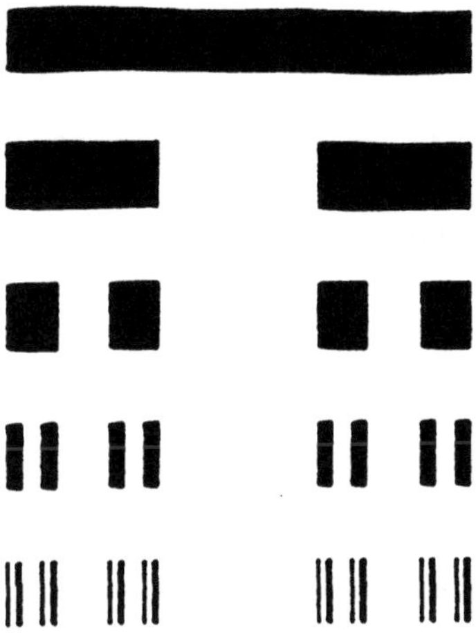

Fig. 1. The Cantor Set.

number N of objects, then the set of segment for which $P_m(r) \neq 0$ becomes for $r \to \infty$ a set of measure or dimension zero. The extreme cases thus give $d = 1$ and $d = 0$ for the dimension of the set of segments for which $P_m \neq 0$ in the limit of large r . There are, however, distributions which fall between these extremes[6]. To see this, we define as generalized dimension d_F the quantity

$$d_F \equiv \lim_{r\to\infty} [ln\ n_F(r)/ln\ r]. \tag{12}$$

For the dimension of any smooth distribution in a one-dimensional space, we therefore recover $d_F = 1$, and for that of a distribution built up of spikes associated to a finite number of space points, we have again $d_F = 0$. The situation is changed, however, for ***self-similar*** distributions such as the Cantor set (fig. 1): the irregularities of this distribution are the same for all scales. After r steps, the Cantor set needs a scale of size $L = (1/3)^r$ to resolve its structure, and we need $n_F = 2^r$ such scale elements. As a result, we now find

$$d_F = [\ln 2/\ln 3] < 1 \tag{13}$$

for the generalized dimension of the Cantor set. The dimension (13) thus "geometrizes" self-similar fluctuating behaviour by associating it to a uniform distribution in a space of ***fractal dimension*** d_F. We shall therefore call ***fractal***[7] those systems, for which $d_F \neq 1$.

These considerations are readily generalized from one to d dimensions. In terms of $r = R/L = n^{1/d}$, we define

$$d_F \equiv \lim_{r\to\infty} [\ln n_F(r)/\ln r] = d \lim_{n\to\infty} [\ln n_F(n)/\ln n], \tag{14}$$

so that now a smooth distribution in a d-dimensional space gives us $d_F = d$, and fractal behaviour implies $d_F < d$. Note that this does not require d_F to be fractional, i.e., non-integer.

The probability $P_m(r)$ to find an object in the m-th of n cells in a topological space V, which we had considered in the previous section, is now replaced by the probability

$P_m(n_F)$ to be in the m-th of n_F elements of a fractal space. Since the distribution in this space is uniform, we simply have

$$P_m(n_F) = 1/n_F. \tag{15}$$

In view of eq.(14), the number n_F is related to (sufficiently large) r or n by

$$n_F(n) \ = \ r^{d_F} \ = \ n^{d_F/d}. \tag{16}$$

While in the previous section we had an irregular distribution of objects in n cells of the d-dimensional space V, we now have a uniform distribution over $n_F \leq n$ elements in a d_F-dimensional fractal space. The multiplicity moments (10) thus become

$$g_l(r) \ = \ g_l(n_F) \ = \ \sum_{m=1}^{n_F} P_m^l \ = \ n_F^{(l-1)}, \tag{17}$$

and hence, using eqs.(16),

$$g_l(r) \ = \ r^{(l-1)d_F}. \tag{18}$$

Inserting this into eq.(11), we obtain[8]

$$d_F \ = \ d \ - \lambda_l \tag{19}$$

for the fractal dimension of any system showing intermittent behaviour.

From eq.(19), we see immediately that in the absence of intermittency ($\lambda_l = 0$), $d_F = d$; at a sufficiently small scale, the original distribution in V is in this case smooth. Further, we note that the "trivial" intermittency obtained by distributing a fixed number of objects into an ever increasing number of cells, or all objects into one, leads with $\lambda_l = d \ \forall \ l$ to $d_F = 0$: it records the fluctuations due to a set of measure or "dimension" zero.

Eq.(19) also shows that the fractal structure for non-trivial intermittency is of the same type as that of the Cantor set, with $d_F < d$. Let us elaborate this point a little. When we go from a uniform distribution to one which has all objects in one cell even for $n \to \infty$, i.e., from the least to the most fluctuating distribution, the intermittency indices $0 \leq \lambda_l \leq d$ vary from their smallest to their largest value. To obtain a system with a fractal dimension $d_F < d$, we need objects in infinitely many cells, but such that the number of cells with $P_m \neq 0$ does not fill the entire d-dimensional volume. The "fractal dust" created in the Cantor set after r steps, with $r \to \infty$, is just of this type.

Let us now see how multiparticle production, in particular cascade production, fits into this picture.

4. A Simple Cascade Pattern*

Fig. 2 gives a specific decay pattern for a heavy "particle" (excited state, fireball or jet) of "mass" m_0. After r decay steps, we have 2^r particles. The two particles, into

---

* Similar arguments concerning cascade decay are given in ref. 9.

which m_0 decays, each have a mass $m_1 = (1/\alpha)m_0$, with $\alpha \geq 2$. Energy conservation gives us

$$m_0^2 = 4[m_1^2 + \mathbf{p_1}^2] \tag{20}$$

where $\mathbf{p_1}$ is the momentum of each of the two particles of mass m_1 in the c. m. of their "parent" m_0. Thus $\alpha = 2$ corresponds to break-up without kinetic energy; if the two particles have non-zero momenta in the c.m. of their parent, then $\alpha > 2$. Continuing the decay cascade, we get by self-similarity $m_r = (1/\alpha)m_{r-1} = (1/\alpha)^r m_0$ as the mass of each of $n(r) = 2^r$ particles after r steps.

To see how this is related to fractal structure, imagine a line of length m_0. For $\alpha = 2$, the $n(r)$ masses m_r cover this line completely: they add up to m_0. If $\alpha > 2$, however,

$$n(r)m_r = (2/\alpha)^r m_0 < m_0; \tag{21}$$

in fact, for $r \to \infty$, $n(r)m_r \to 0$, even though the number $n(r)$ of particles becomes infinite. So the larger r, the less does $n(r)m_r$ cover the line of length m_0. The dimension of the line $n(r)m_r$ for $r \to \infty$ is therefore less than one; $d = 1$ would require that $n(r)m_r$ covers a finite segment of m_0 even for infinite r. It is also larger than $d = 0$, since we have an infinite number of masses. Following the theory of fractal structures[7] as sketched above, we therefore define as generalized dimension[6]

$$d_F \equiv \lim_{r\to\infty}\{\frac{\ln n(r)}{\ln(m_0/m_r)}\} = \frac{\ln 2}{\ln \alpha}. \tag{22}$$

For $\alpha = 2$, we recover with $d_F = 1$ the one-dimensional continuum corresponding to $n(r)m_r = m_0$; for $\alpha > 2$, however, $0 < d_F < 1$. Let us note at this point that the structure of the process remains the same if we keep m_r fixed at the same value $m_r = \mu$, but instead let $m_0 = \alpha^r\mu$ go to infinity as $r \to \infty$. In both cases we arrive at an infinite self-similar cascade: Either we have a fixed mass m_0 decaying into smaller and smaller masses $m_r = (1/\alpha)^r m_0$, or we have asymptotic decay products of mass μ originating from a heavier and heavier starting mass $m_0 = \alpha^r\mu$. The latter case will turn out to be the relevant one for the high energy limit of multihadron production.

5. High Energy Multihadron Production

We now have to connect the cascade decay pattern to multiplicity moments. Consider the multiplicity distribution dN/dy in the rapidity range $0 \leq y \leq Y$; here $Y = Y(s)$ denotes the maximum (laboratory) rapidity possible at a given c.m.s. collision energy $\sqrt{s}$. Now divide this range into $\bar{n}$ intervals (bins) of size δy. The normalized standard moment of order l is then defined as

$$C_l(\delta y) = \langle \frac{1}{\bar{n}}\sum_{m=1}^{\bar{n}} k_m^l / [\frac{1}{\bar{n}}\sum_{m=1}^{\bar{n}} k_m]^l \rangle, \tag{23}$$

where

$$k_m \equiv \int_{y_m}^{y_{m+1}} dy(dN/dy) \tag{24}$$

is the number of particles in the m-th rapidity bin of a given event. Since

$$\sum_{m=1}^{\bar{n}} k_m = \int_0^Y dy(dN/dy) = N \tag{25}$$

is the total number of particles in the event under consideration, we can rewrite Eq. (23) in the form

$$C_l(\delta y) = (Y/\delta y)^{l-1} \langle \sum_{m=1}^{\bar{n}} P_m^l \rangle. \tag{26}$$

Here $P_m \equiv k_m/N$ is the probability to find a particle in the m-th bin, and we have used $\bar{n} = Y/\delta y$. In both Eqs. (23) and (26), the average $\langle \ldots \rangle$ is taken over all available events.

In a typical multihadron production event, the probability P_m will fluctuate considerably from bin to bin and thus is difficult to relate to any theoretical scheme. For the case of a self-similar cascade, however, we can map the P_m from the usual one-dimensional rapidity space, where they fluctuate strongly, onto a fractal space, in which the fluctuations are "smoothed out" at the expense of reducing the dimension from $d = 1$ to $d_F \leq 1$. In the fractal space, we have $n \leq \bar{n}$ intervals, with $P_m = 1/n$. For a jet, the overall multiplicity grows as $\ln s \sim Y$, so that we identify $m_o = Y$ and $m_r = \delta y$. From Eq. (22), we then get $P_m = (\delta y/Y)^{d_F}$, and hence

$$\sum_{m=1}^{n} P_m^l = (\delta y/Y)^{(l-1)d_F}. \tag{27}$$

Inserting this is Eq. (26), we have

$$C_l(\delta y) = (Y/\delta y)^{(l-1)(1-d_F)} \tag{28}$$

Thus we obtain the result that multihadron production through a self-similar cascade will lead to multiplicity moments having a power-law dependence on the ratio $(Y/\delta y)$. In other words, we expect the ratio of logarithms,

$$\ln C_l(\delta y)/\ln(Y/\delta y) = (l-1)(1-d_F), \tag{29}$$

to become a constant when $Y/\delta y \to \infty$. The simple fractal model we had considered with the cascade of Fig. 2 leads to the l-dependence of Eq. (29). The concept of fractal dimension has been extended, however, to also include situations in which d_F depends on l (multifractals).[6] The essential point of Eq. (29) therefore is that for large $(Y/\delta y)$ (i.e., away from the end of the cascade), the ratio of the logarithms in Eq. (10) becomes a constant; this constant could be different for different l.

In the discussion of the cascade of Fig. 2, we had seen that there are two ways to make m_0/m_r large. Similarly, we can increase $Y/\delta y$ by letting $\delta y \to 0$ at fixed Y or by letting $Y \to \infty$ for $\delta y \geq \delta y_0$, where δy_0 denotes the lower bound needed to exclude the "resonance" regime. In a number of intermittency studies the first alternative was investigated.[2] As already mentioned, there are now claims [3,4] that this will only bring us into the region of the usual short-range two-particle correlations, which signal the end of the cascade. We want to investigate here if multihadron production takes place

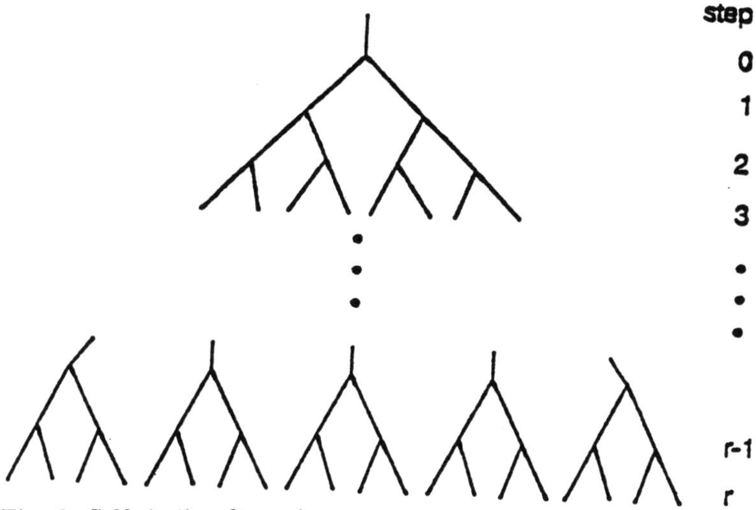

Fig. 2. Self-similar Cascade.

in form of self-similar cascade as $\sqrt{s} \to \infty$. If we keep $\delta y > \delta y_0$, the moments of the multiplicity distribution must then vary with $Y/\delta y$ according to Eq. (29).

Let us now estimate the lower bound δy_0 for the rapidity bins to be used. For a two-pion system of invariant mass $M_{\pi\pi} = 0.5$ GeV, the rapidity separation between the two pions is $\delta y \sim 1$; we have here assumed each pion to have transverse mass $m_T = (p_T^2 + m_\pi^2)^{1/2} = 0.5$ GeV. For $M_{\pi\pi} = 1$ GeV, we get $\delta y \sim 2.3$, so that rapidity windows $\delta y \geq \delta y_0 \sim 1 - 2$ ought to put us above the end region of the cascade.

In Fig. 3, we now illustrate schematically the prediction resulting from our consideration. To have a well defined case, we look at $C_l(Y/\delta y)$ for $\delta y_0 \sim 1.8$ (i.e. $M_{\pi\pi} \sim 0.7$ GeV) and some fixed l. As $\sqrt{s}$ varies from 10 GeV to 1800 GeV, the end of the power-law behaviour predicted by Eq. (10) shifts from $\ln(Y/\delta y) = 0.5$ to 1.5. Beyond this end point, the logarithmic moments will flatten out; how much, depends on the details of the short range correlations. If we increase our $M_{\pi\pi}$ cut-off to 1 GeV, i.e., take $\delta y_0 = 2.3$, then for $\sqrt{s} = 10$ GeV, we have power-law behaviour up to $\ln(Y/\delta y) \sim 0.25$ only. To have a reasonable range within which to test the power-law (Eq. (29)), we therefore need $\sqrt{s} \sim 50$ GeV or higher. The dashed line in Fig. 3 shows the predicted form as $\sqrt{s} \to \infty$.

Before we can apply our considerations to actual data, we have to address two somewhat technical points. Since even at the highest available energies, the multiplicities per bin become quite small when $\delta y \to 1 - 2$, we should, before we study the effect of self-similar cascading, try to eliminate purely statistical fluctuations (which lead to a behaviour similar to the "trivial" intermittency of section 2.). In Ref. 2, it is proposed to do this by considering the factorial moments F_l, rather than the standard moments C_l. For $l = 2 - 5$, the two forms are related by

$$F_2 = C_2[1 - (1/\bar{N}C_2)]$$
$$F_3 = C_3[1 - (3C_2/\bar{N}C_3) + (2/\bar{N}^2C_3)]$$

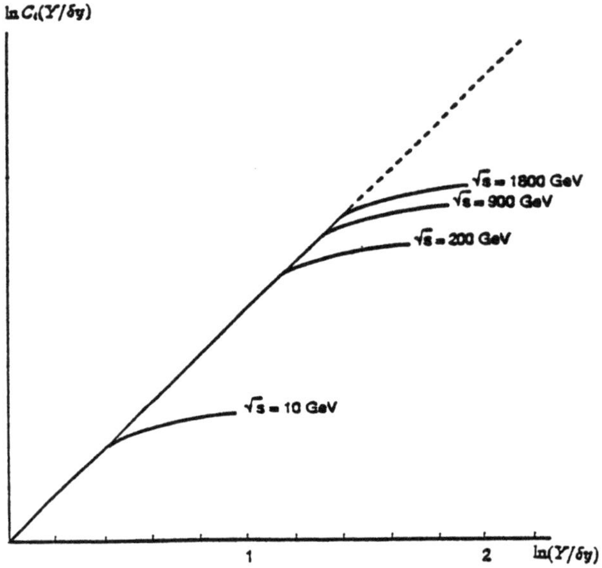

Fig. 3. Schematic illustration of the multiplicity moments $C_l(Y/\delta y)$ at fixed l and $\delta y_0 = 1.8$ for different incident energies $\sqrt{s}$. The dashed line shows the continuation for $\sqrt{s} \to \infty$.

$$F_4 = C_4[1 - (6C_3/\bar{N}C_4) + (11C_2/\bar{N}^2C_4) - (6/\bar{N}^3C_4)]$$
$$F_5 = C_5[1 - (10C_4/\bar{N}C_5) + (35C_3/\bar{N}^2C_5) - (50C_2/\bar{N}^3C_5) + (24/\bar{N}^4C_5)] \quad (30)$$

Here $\bar{N}$ is the average multiplicity. We see that for $\sqrt{s} \to \infty$, and hence $\bar{N} \to \infty$, the two forms become equivalent, since C_l/C_{l+1} also vanishes in that limit. Following Ref. 2, we shall consider the data for the factorial moments; it turns out, however, that up to $\delta y \sim \delta y_0$, the two forms do not differ very much. Below δy_0, the standard moments are strongly enhanced by statistical fluctuations.

The second problem arises because the available data have not been analysed such as to give the moments C_l as we have defined them in Eq. (23), or the equivalent F_l. Instead, experimental data are presented in terms of the moments

$$\tilde{C}_l(\delta y) = \langle \frac{1}{\bar{n}} \sum_{m=1}^{\bar{n}} k_m^l \rangle / \langle \frac{1}{\bar{n}} \sum_{m=1}^{\bar{n}} k_m \rangle^l \quad (31)$$

and the corresponding $\tilde{F}_l$: numerators and denominators are separately averaged over all events. As a consequence, they do not become unity for $\delta y = Y$, and different incident energies lead to different starting points.

We strongly suggest a reanalysis of the data, normalizing event by event and then averaging. To use the existing data, we are forced to shift the curves by their value at $\delta y = Y$, so that we will now look at the behaviour of

$$\ln \tilde{F}_l(\delta y) - \ln \tilde{F}_l(\delta y = Y) \equiv \ln f_l(\delta y); \quad (32)$$

by definition, ln $f_l(\delta y)$ will then vanish for $\delta y = Y$ at all energies and for all l.

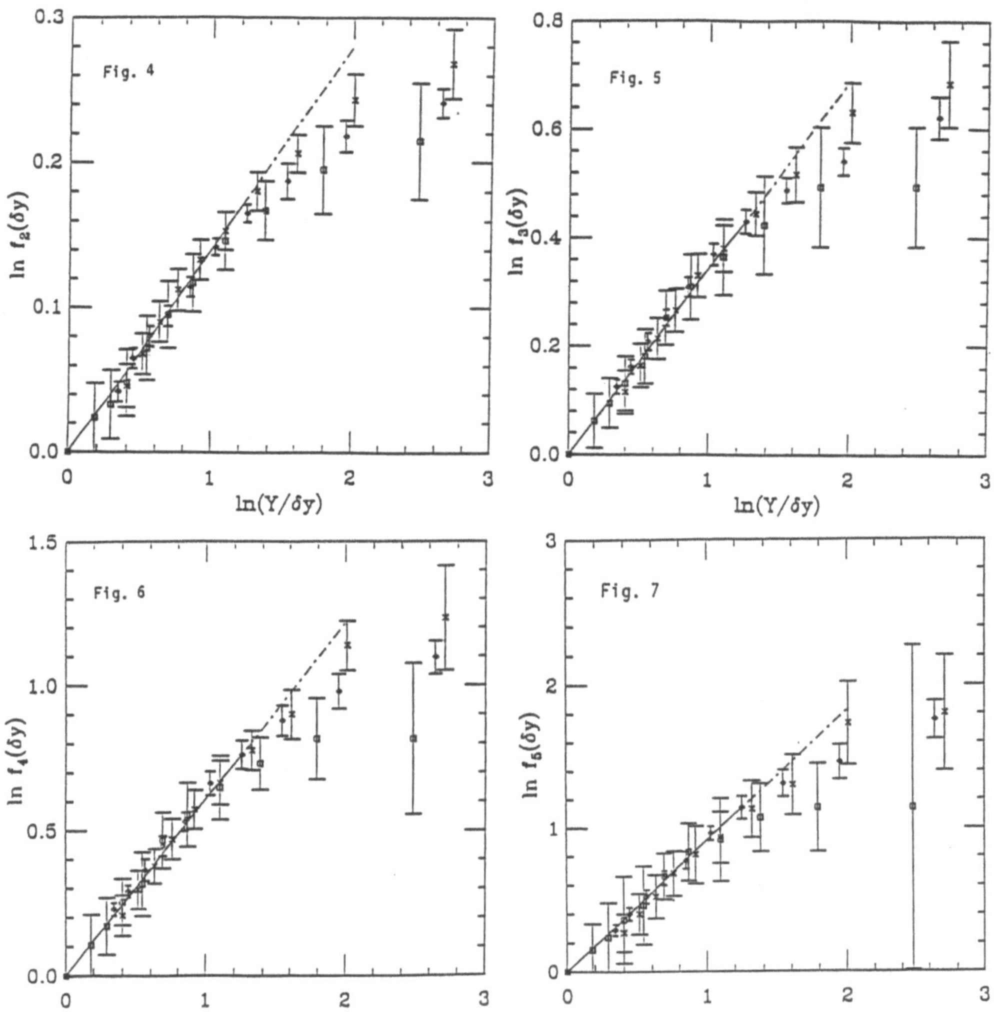

Figs. 4-7. Experimental data on multiplicity moments $\ln f_l(\delta y)$ ($l = 2,3,4,5$) as a function of $\ln(Y/\delta y)$ for the energies $\sqrt{s} = 200$ GeV (squares), $\sqrt{s} = 546$ GeV (diamonds) and $\sqrt{s} = 900$ GeV (crosses) and the straight line fit to the data at large δy.

A complete experimental determination of moments for $l = 2-5$, with the definition (32), is given by the UA5 Collaboration, for $\sqrt{s} = 546$ GeV[10] and for $\sqrt{s} = 200$ and 900 GeV[11]. According to our considerations, the results for ln $f_l(\delta y)$ at different energies but for fixed l should fall on one universal curve, up to $\ln(Y/\delta y) \sim 1.2$(200 GeV), 1.3 (546 GeV) and 1.4 (900 GeV), if we use $\delta y_0 = 1.8$ as smallest bin size. In Figs. 4-7, we see that the data agree very well with this prediction; there is no noticeable systematic deviation from straight-line behaviour, nor any indication for different slopes at different energies. We should emphasize here that the production pattern which we have observed is an intrinsic irregularity of the data – it does not involve any theoretical input. Such a behaviour does arise naturally, however, in a fractal production scheme.

Let us now define a slope measure by

$$\lambda_l \equiv \frac{1}{(l-1)}[\ln f_l(\delta y)/\ln(Y/\delta y)]. \tag{33}$$

A fractal cascade then gives us, from Eq. (29), the l-independent result

$$\lambda_l = (1 - d_F); \tag{34}$$

however, a multifractal picture still allows λ_l to depend on l. Straight-line fits to the data (Figs. 4-7) give the following values for the slope measures

$$\lambda_2 = 0.140 \pm 0.008 \tag{35a}$$

$$\lambda_3 = 0.172 \pm 0.009 \tag{35b}$$

$$\lambda_4 = 0.203 \pm 0.013 \tag{35c}$$

$$\lambda_5 = 0.228 \pm 0.016 \tag{35d}$$

Before drawing any conclusions about this noticeable but not very strong l-dependence, we would prefer to see if it persists when the experimental moments are re-determined according to Eq. (23) or the corresponding $F_l(\delta y)$, rather than the presently used averaging according to Eq. (31).

We note that our considerations predict forthcoming data from the Tevatron ($\sqrt{s} =$ 1800 GeV) to fall on the curves in Figs. 4-7. The region of power-law behaviour in this case is expected to persist to a somewhat higher value of $\ln(Y/\delta y)$. It will be very interesting to see whether this holds true also for multihadron production at LEP, or whether $q\bar{q}$ initial state leads to a different hadronization pattern.

On a theoretical level, we now obviously face the challenging problem of deriving self-similar multihadron production in general, and the observed intermittency measures λ_l in particular, from the multi-jet structure of QCD. For first steps in this direction, see ref.12.

Acknowledgements

Most of this paper is based on joint work and many discussions with I. Sarcevic. In addition, I would like to thank A. Białas, S. Gupta, A. Krzywicki, R. Peschanski, J. Seixas and T. Sjöstrand for discussions and helpful remarks. The support of a NATO Collaborative Research Grant is gratefully acknowledged.

References

1. I. Sarcevic and H. Satz, Phys. Lett. 233B (1989) 251.
2. A. Białas and R. Peschanski, Nucl. Phys. B 273 (1986) 703 and B308 (1988) 803; for a recent survey, see W. Kittel and R. Peschanski, *Review on Intermittency in Particle Multiproduction*, Preprint Nijmegen HEN-325 / Saclay SPhT/89-143 (1989).
3. P. Carruthers and I. Sarcevic, Phys. Rev. Lett. 63 (1989) 1562.
4. A. Capella, K. Fiałkowski and A. Krzywicki, Phys. Lett. 230B (1989) 149
5. R. Hagedorn, Nuovo Cim. Suppl. 3 (1965) 147.
6. F. Hausdorff, Math. Ann. 79 (1919) 157.
7. B. Mandelbrot, *The Fractal Geometry of Nature*, Freeman & Co. Publ., New York (1982).
8. P. Lipa and B. Buschbeck, Phys. Lett. 223B (1989) 465.
9. W. Ochs and J. Wosiek, Phys. Lett. 214B (1988) 617.
10. G. J. Alner *et al.*, Phys. Rep. 154 (1987) 247.
11. R. E. Ansorge *et al.*, Z. Phys. C 43 (1989) 357.
12. R. C. Hwa, *Intermittency in the ϕ^3 Branching Model*, Oregon Preprint OITS-404 (1989);
C. B. Chiu and R. C. Hwa, *Intermittency in Branching Models*, Oregon Preprint OITS-424 (1989).

Q-Stars

B.W. Lynn

Department of Physics, Stanford University, Stanford, CA 94305, USA, and Theory Division, CERN, CH-1211 Geneva 23, Switzerland

Hadronic effective field theories describing ordinary nuclei also contain Q-Ball solutions which can describe a new state of matter, "baryon matter". Baryon matter is stable in very small chunks as well as in stellar-sized objects, since it is held together by the strong force instead of just gravity. Larger chunks, "Q-Stars", in which gravity is important, model neutron stars. A wide variety of Q-Star models, all consistent with known nuclear physics, allow compact objects to have masses much larger, or rotation periods much shorter, than is conventionally believed possible. Smaller chunks of nuclear density baryon matter could also be astrophysically important components of the universe, and at late times would have many properties similar to those of strange matter chunks.

1. Introduction

There are many surprising possibilities lurking in the non- perturbative sector of field theories. Here we consider the possibility of a new state of matter which arises from the discovery of solutions to effective field theories describing, among other things, ordinary nuclei. Effective field theories of interacting baryons and mesons have successfully reproduced measured properties of nuclei as well as results of scattering experiments.[1-3]) We have found[4-6]) non-topological classical solutions to such theories: fermion Q-Balls, or Q-Stars in the case where their self-gravity is important and gravitational effects are included. Q here stands for the conserved charge (baryon number) which stabilizes the matter against decay. The properties of such a state of baryon matter can be different from and essentially independent of the characteristics of ordinary nuclei studied in the laboratory. In particular, neutron stars may have a large binding energy per nucleon, hundreds of MeV, due only to nuclear forces, and may be more massive or able to rotate faster than is suggested by currently accepted limits.[7,8]) Chunks of baryon matter, varying in size from 10^{-12}cm to several kilometers, may also exist. Further, the calculated

Phase Structure of Strongly Interacting Matter
Editor: J. Cleymans

characteristics of Q-Stars and macroscopic chunks are fairly insensitive to the effective field theory used; classes of theories give the same state of matter.

2. Constructing Solutions

There is a simple graphical method for finding these large baryon-number solutions. Consider the following Lagrangian for a baryon ψ interacting with a scalar and vector field, with some potential for the scalar:

$$\mathcal{L} \;=\; \bar{\psi}\left[i\partial\!\!\!/ - m(\sigma) - g_{\rm v}V\!\!\!/\right]\psi + \tfrac{1}{2}(\partial_\mu\sigma)^2 - \; U(\sigma) \;+\; \tfrac{1}{2}m_{\rm v}^2 V_\mu V^\mu \;. \tag{2.1}$$

In this model it is a good approximation to neglect the dynamics of the vector field[1]), and for many fermions we can make the Thomas-Fermi approximation ($dm/dr \ll m^2$). Then we have a Fermi sea of baryons described by a Fermi momentum k_F slowly varying in space, and a chemical potential $\varepsilon_F = (k_F^2 + m^2)^{1/2} + (g_{\rm v}/m_{\rm v})^2 k_F^3/3\pi^2$. Using the identity $\langle\bar{\psi}\psi\rangle = -(\partial P_\psi/\partial m)$,[6]) the classical equation of motion for static σ becomes

$$\nabla^2\sigma \;=\; -\frac{\partial}{\partial\sigma}(P_\psi - U) \tag{2.2}$$

where P_ψ is the pressure of the baryons. This is equivalent to Newton's $F = ma$ for a mechanical 'particle' at 'position' σ at 'time' r moving in a potential $V_{eff} = P_\psi - U$, with 'friction' from the $(2/r)\partial\sigma/\partial r$ term negligible for large r.

The Q-Ball solution occurs when σ rolls between degenerate maxima of the potential V_{eff}, where one of the maxima is the vacuum. The σ field starts off infinitesimally close to the top of the first hill, which is some value σ_{inside}, stays at that value out to a large radius (since the top of the hill is flat), and then quickly rolls to the vacuum. The large baryon number Q-Ball thus has a flat interior, a radius which is a free parameter, and a thin surface (of order the scalar Compton wavelength). The density is determined by ε_F and σ_{inside}, which are fixed (with algebraic equations) by requiring degenerate maxima: $V_{eff} = 0$ and $\partial V_{eff}/\partial\sigma = 0$. A Q-Ball must still have total energy $E < Qm_N$, where m_N is the free nucleon mass, in order to be bound. We can graph $P_\psi(\sigma)$ and $U(\sigma)$ for any Lagrangian and examine the picture to see if a solution exists. As ε_F is increased, the P_ψ

curve centered at $m(\sigma) = 0$ expands until it touches U at some σ_{inside}. If it is tangent to U at a σ_{inside} different from the vacuum, and if $P_\psi - U$ is less than zero between these values, σ can roll to the vacuum and the theory has a large baryon number Q-Ball. Further details can be found in Refs. 6 and 9.

One example of nuclear effective theories is Walecka's Quantum Hadro-Dynamics.[1]) The proton and neutron are fermion fields ψ_i whose effective mass is $m(\sigma) = g_s\sigma$ with $m_N = g_s\sigma_0$; the scalar field has a quadratic potential $U = \frac{1}{2}m_s^2(\sigma - \sigma_0)^2$, and the vector field is the ω particle. The large baryon number Q-Ball in the theory occurs for approximately equal numbers of neutrons and protons, and is called infinite symmetric nuclear matter when $N = Z$. The small baryon number solutions of the theory (nuclei) depend on the friction term in eq.(2.2): the rolling starts above $P_\psi - U = 0$ and friction brings σ to rest at the vacuum value where $m = m_N$. The parameters are fit to ^{40}Ca and the bulk properties of nuclear matter, giving $(m_N g_v/m_v)^2 = 195.9$, $(m_N g_s/m_s)^2 = 267.1$, $m_s = 0.518 m_N$, and insensitivity to m_v. The chemical potentials are $\varepsilon_F = 923MeV$ for infinite symmetric nuclear matter and slightly larger for ^{40}Ca. We translate these results into our framework in Fig. 1, which displays the ^{40}Ca and infinite symmetric nuclear matter solutions.

Infinite symmetric nuclear matter is not realizable in nature because of Coulomb forces. Realistic solutions have electrons balancing the charge of the protons so that Coulomb forces are negligible. This requires neutrons, protons, and electrons in a configuration satisfying local charge neutrality and β-decay equilibrium: $k_{F,p} = k_{F,e}$ and $\varepsilon_{F,n} = \varepsilon_{F,p} + \varepsilon_{F,e}$. For this case (which gives a baryon pressure P_ψ^{neutral}), a large baryon number Q-Ball does not exist in Walecka's theory, since P_ψ^{neutral} can not be made tangent to U in such a way that $P_\psi^{\text{neutral}} - U < 0$ between σ_{inside} and σ_0. Therefore, in Walecka's theory, the large baryon number solutions, *i.e.*, neutron stars, are bound only by gravity. For other potentials the situation can be very different.

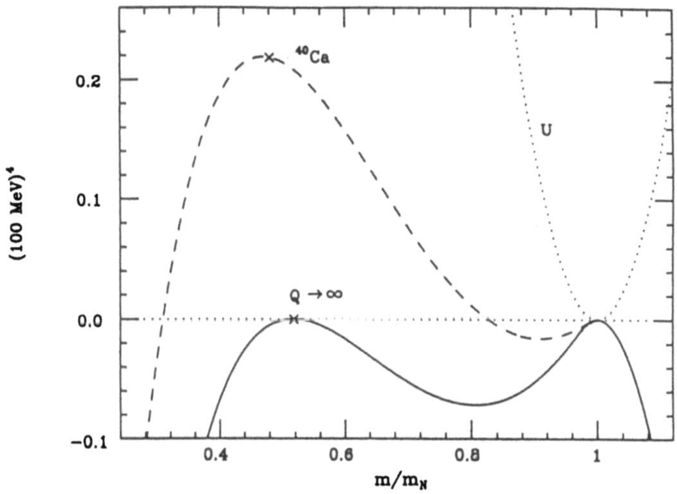

Figure 1. Graphical representation of Q-Balls in Walecka's nuclear theory[1]): $U(\sigma)$ (dots) is the potential for the scalar field, P_ψ is the baryon pressure, and $P_\psi - U$ (dashes for ^{40}Ca, solid for the large baryon number Q-Ball) is the effective potential that σ 'rolls' in, starting at $r = 0$ (marked by crosses).

The models that we have found in the nuclear physics literature consider only renormalizable potentials, $U(\sigma)$, and a Yukawa coupling $m(\sigma) = g_s\sigma$. An effective field theory, for which the fields are not fundamental quarks and gluons, but rather composites such as pions, nucleons and vector mesons, can only be used at energies low enough for the composite particles to appear point-like,[3]) and thus only below about seven times nuclear density. The potential and the coupling $m(\sigma)$ can have all forms not ruled out by a symmetry of the underlying theory (QCD in this case), and must be determined either from experiment or from the underlying theory. If effective m and U are determined in this way, quantum corrections are included in their definitions.

Boguta and Strocker[2]) have examined phenomenological potentials with quadratic, cubic, and quartic terms and found a large class of models which reproduce known bulk properties of nuclear matter, including saturation density, binding energy, bulk compressibility modulus, and optical potential[2]). In Fig. 2 we compare their potentials with Walecka's potential. The region 'a-b-c-d' is the only part of the potential relevant for nuclear data. U is not constrained outside $0.5 \lesssim m(\sigma)/m_N \leq 1$ and, since this is an ef-

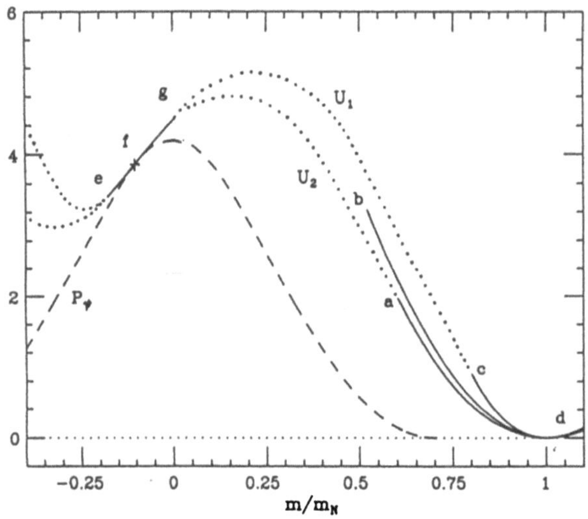

Figure 2. Potentials U (solid lines) given in Refs. 1 and 2.

fective theory, U might have any shape. We display two of the infinite number of possible potentials (dotted lines) consistent with nuclear data that do admit large baryon number Q-Balls. They touch P_ψ^{neutral} at the same point ('f') and since the Q-Ball properties are determined by $m(\sigma_{inside})$ and ε_F, both of these potentials describe Q-Balls with the same volume properties (their thin surfaces differ). Their properties can not be determined from experiments performed on nuclei since σ_{inside} is outside the region relevant to nuclei.

This scenario predicts the existence of stable macroscopic chunks of high density baryon matter. They can be any size above a minimum Q_{bulk} below which the gradient energy of the surface becomes important. For $Q \lesssim Q_{\text{bulk}}$ the friction term in eq. (2) can not be neglected and the differential equation must be solved to determine the Q-ball properties. As Q decreases, both the surface energy and the Coulomb energy due to deviations from charge neutrality tend to decrease the binding energy. At some model-dependent baryon number, Q_{min}, the Q-ball may become energetically unbound. If there is no suppressed phase transition for production of nuclear-sized chunks of non-strange baryon matter, then Q_{min} must be larger than the baryon number of the lowest trans-uranic elements in order that ordinary nuclei not decay into baryon matter.[10]) In this case, baryon matter might be observed as very neutron-rich isotopes of ordinary atoms:

when Coulomb effects are included, a chunk of baryon matter will develop some net positive charge. If there is a suppressed phase transition, then $Q_{\min}$ need not be large, and baryon matter might be observed as atoms whose nuclear binding energy is greater than in ordinary atoms. The calculation of $Q_{\min}$ in any given theory is therefore very interesting. Experiments on small baryon matter chunks, if they could be observed, could lead to definite predictions of the properties of neutron stars.

3. Astrophysical Considerations

The Q-Ball state is a perfect fluid (in the local rest frame the non-diagonal, non-isotropic parts of $T^{\mu\nu}$ are all proportional to derivatives of the field, which vanish in the interior in the large-Q limit) with an equation of state $\mathcal{E}(P)$ determined by $\partial V_{eff}/\partial\sigma = 0$. We may therefore include the effects of gravity by integrating the Oppenheimer-Volkoff equations with the equation of state $\mathcal{E}(P)$ to form Q-Stars. Before we calculate Q-Stars in a specific hadronic theory, we can estimate their size. Gravity will become important when $GM/R \sim 1$. If σ_0 is the vacuum value of σ, the Q-Ball has energy density $\sim \sigma_0^4$ so the radius $R \sim m_{pl}\sigma_0^{-2} \sim 100$ km, and the mass $M \sim m_{pl}^3\sigma_0^{-2} \sim 10\ M_\odot$ for $\sigma_0 = 100 MeV$.

Consider for simplicity the special 'chiral' case in which the Q-Ball has massless nucleons in the interior: $m(\sigma_{inside}) = 0$. When the baryon number is large enough ($\approx 10^{57}$) gravity will be important and a neutral 'chiral Q-Star' will form; This Q-Star has the convenient feature that a constant $\sigma = \sigma_{inside}$ is a solution to the equation of motion so that the equation of state for a chiral Q-Star depends only on the vector repulsion strength and $U_0 \equiv U(\sigma_{inside})$:

$$\mathcal{E} - 3P - 4U_0 + \alpha_v\left(\mathcal{E} - P - 2U_0\right)^{3/2} = 0\ . \tag{3.1}$$

This equation of state holds for a star with neutrons, protons, and electrons satisfying beta-equilibrium and local charge neutrality (neglecting the electron mass and the neutron-proton mass difference) with $\alpha_v \equiv (g_v/m_v)^2 3^{1/2}/\pi$. It follows from $m = 0$ (the solution to $\partial V_{eff}/\partial\sigma = 0$) and the expressions for the total energy and pressure[6])

$$\begin{aligned} \mathcal{E} &= \frac{k_{F,n}^4 + k_{F,p}^4 + k_{F,e}^4}{4\pi^2} + \frac{1}{2}\frac{g_v^2}{m_v^2}\left(\frac{k_{F,n}^3 + k_{F,p}^3}{3\pi^2}\right)^2 + U_0 \\ P &= \frac{k_{F,n}^4 + k_{F,p}^4 + k_{F,e}^4}{12\pi^2} + \frac{1}{2}\frac{g_v^2}{m_v^2}\left(\frac{k_{F,n}^3 + k_{F,p}^3}{3\pi^2}\right)^2 - U_0\ . \end{aligned} \tag{3.2}$$

Only the nucleons couple to the σ and vector fields, so in the surface charge separation results and equations more complicated than (2.2) determine the surface structure. We neglect the details of this region since it is about 10^{-20} times smaller than the stellar radius.

Integrating the Oppenheimer-Volkoff equations for (3.1) gives the solid lines in the mass *vs.* stellar radius plot, Fig. 3, for two values of U_0 with $\alpha_v = 1.23 \times 10^{-4} MeV^{-2}$ (Walecka's value for this model). We also show for comparison (dashed lines) two neutron star models, pion condensation and Walecka's, which are among the least stiff and the

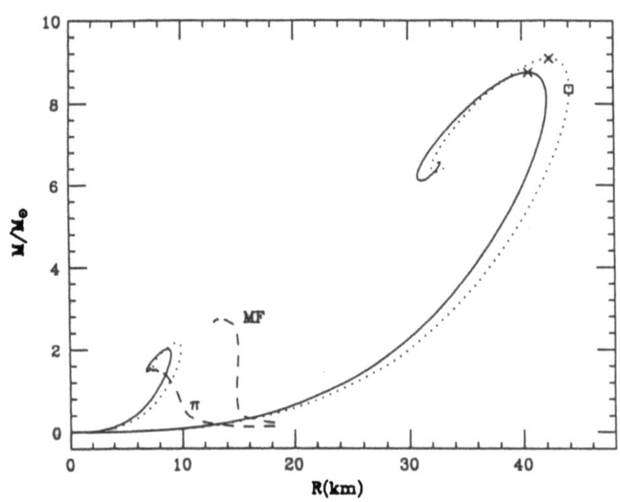

Figure 3. Mass *vs.* stellar radius for chiral (solid lines) and representative non-chiral (dotted lines) Q-Stars. Large mass Q-Stars are shown for theories with $U_0 = (85MeV)^4$; small mass Q-Stars are shown for theories with $U_0 = (200MeV)^4$. Central energy density increases along the curliques. The crosses give the maximum stable masses (for clarity, only the large mass case is labeled). Conventional neutron star models of varying stiffness (dashed lines) are shown for comparison.

most stiff, respectively, of conventional neutron star models. All conventional models begin with high density bulk neutral matter unbound in the absence of gravity: as the mass and baryon number of such a neutron star increase the stellar radius decreases. In contrast, fermion Q-Balls are bound even in the absence of gravity, and since their density is fixed by the theory, Q-Balls satisfy both M and $Q \propto R^3$. This causes Q-Stars to increase in size as they become more massive until eventually gravity becomes strong enough to overcome the inherent stiffness. The 'curlique' mass-radius plot is thus generic for Q-Stars.

Both the stellar radius and mass increase as U_0 decreases. For a general non-renormalizable chiral potential, U_0 is unknown, and cannot be determined from nuclear data when the solutions for nuclei in that theory lie away from σ_{inside}. This uncertainty means that the upper limit for the neutron star mass might be large. $M_{\max}$ can not be arbitrarily large because (3.1) is only valid at densities high enough for the nuclear interactions to be important (which holds in the interior of a Q-Star). A lower limit on the density at which (3.1) is applicable places an upper limit on $M_{\max}$. If we assume the Harrison-Wheeler equation of state below, for example, white dwarf densities ($< 10^{10}\mathrm{gm/cm^3}$) this gives $M_{\max} \lesssim 10^3 M_\odot$. We know of no experimental constraints on the equation of state of a large number of baryons above white dwarf densities; it follows that Q-Balls and Q-Stars may have densities lower than nuclear and masses up to $\sim 10^3\ M_\odot$.

Using $\alpha_v U_0^{1/2} = 1.23$ (Walecka's value when $U_0 = (100MeV)^4$), the upper mass limit on a neutron star with equation of state (3.1) is

$$M_{\max} \; = \; 6.68\ U_{100}^{-1/2}\ M_\odot\ , \tag{3.3}$$

where $U_{100} \equiv U_0/(100MeV)^4$. The star also reaches its maximum baryon number at this mass: $Q_{max} = 22.5\ U_{100}^{-3/4} \times 10^{57}$, giving a binding energy of $608\ U_{100}^{1/4}\ MeV$ per nucleon. Note that the scaling with U_0, including $R \sim U_0^{-1/2}$, holds only if α_v is simultaneously scaled as $U_0^{-1/2}$.

Q-Stars can have masses much larger than the upper mass limits to neutron stars estimated by current theories ($M_{\max} < 2.3M_\odot$[11]) and larger than the theoretical Rhoades-Ruffini limit of $3.2M_\odot$. The latter limit was calculated variationally, assuming the equation of state was known below some density $\mathcal{E}_0$ taken to be about $4.6 \times 10^{14}\mathrm{gm/cm}^3 \approx 1.6\mathcal{E}_{\mathrm{nuc}}$. We have shown that a new equation of state, such as (3.1), can hold below nuclear density and still be consistent with nuclear physics. Scaling $\mathcal{E}_0$ down to the minimum density allowed by the white dwarf argument gives $M_{\max} \sim 10^3 M_\odot$. Thus heavy compact objects such as Cygnus X-1 ($> 3.4\ M_\odot$, probably $9 - 15\ M_\odot$[12]) and LMC X-3 ($> 6 - 14\ M_\odot$[13]) are Q-Star candidates.

Even small Q-Balls are stable against dispersal into free particles, because of their positive binding energy, and against adiabatic radial pulsations, since their adiabatic index $\Gamma \sim U_0(P_\psi + P_e - U)^{-1} \gg 4/3$. The turning point for the stability of this mode occurs at the point where $dM/d\mathcal{E}_c = 0$, indicated by crosses in Fig. 3, so Q-Stars before the maximum mass in Fig. 3 are stable. For a rotating Q-Star the relativistic analogue of the Maclaurin spheroid analysis as applied to other neutron star models[8]) shows (Fig. 4) that certain Q-Stars can rotate at extremely high rates, easily above the 0.5 ms limits on other equations of state. The inclusion of vector repulsion can increase the maximum rotation rate. Because the Q-Star is at about six times nuclear density for $P_{rot} = 0.5\mathrm{ms}$[14]), (although the associated Q-Ball can be less dense; see Fig. 4) an effective field theory of composite baryons and mesons might not be applicable. At such high densities, either quark Q-Star models or an interior of (asymptotically) free quarks supported by pressure from a baryon Q-Star shell might apply.

The chiral baryon Q-Ball equation of state (neglecting vector repulsion) has a similar form to the MIT bag model's strange stars and nuggets[15]) (neglecting gluon exchange). The potential $U(\sigma)$ dynamically generates the analogue of bag pressure, but its magnitude is of course unrelated.

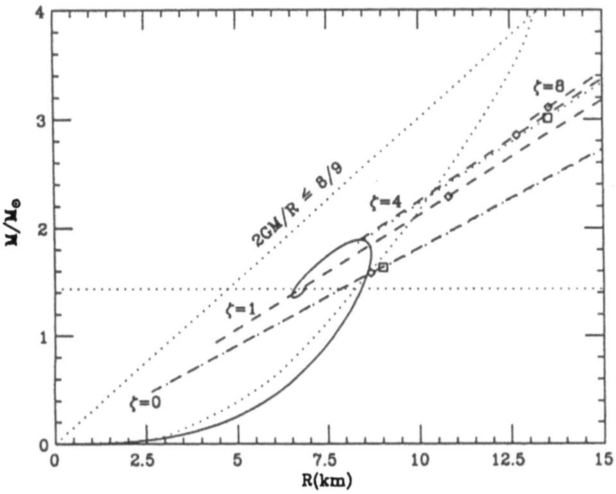

Figure 4. The curved dotted line is a lower limit for stable rotation at 0.5ms. The horizontal dotted line gives the mass of the binary pulsar PSR 1913+16. The four straight dashed lines are the maximum masses for nonrotating chiral Q-Stars in different field theories, for fixed $\zeta \equiv 3^{-1/2}\pi\alpha_v U_0^{1/2}$, and the superimposed dotted lines are non-chiral Q-Stars for $\zeta = 0$ and 4. The continuous parameter along the lines is U_0 (the diamonds and boxes give the points for $U_0 = (160\text{MeV})^4$ on the chiral and non-chiral lines respectively; both M and R scale as $U_0^{-1/2}$ for fixed ζ). The lower limits shown for each line are the points where the binding energy of the non-rotating Q-Ball vanishes. Q-Stars inside the region between the dotted lines can rotate at 0.5ms and are massive enough to be candidates for the binary pulsar. The solid curlique shows chiral Q-Stars of various sizes in a particular theory, showing that the Q-Ball is less dense and can rotate less rapidly than the Q-Star.

We examined first the special case of chiral Q-Balls and Q-Stars, for which the nucleons are massless inside, because this case simplifies the mathematics and illustrates the general features of these models. We know of no reason why the potential should necessarily be so accommodating, although a wide class of theories does contain chiral Q-Balls. (Lee and Wick's abnormal nuclear matter[16]) is a chiral Q-Ball in the subclass of theories with quartic potentials and $m = g\sigma$. This specific subclass is thought to be incompatible with nuclear data,[1,17]) but we have shown that when U is not restricted to being quartic, abnormal nuclear matter can exist in a theory which also describes ordinary nuclei.) Non-chiral Q-Stars can also be solved, and are displayed in Fig. 3.

4. Conclusions

The Q-Star model differs in important respects from conventional neutron star models, raising interesting astrophysical questions. The electromagnetic properties of Q-Stars might be very different from those of neutron stars in conventional models. The neutrino cooling rate should also be much higher and a comparison with observed thermal X-ray fluxes from supernova remnants would be useful. Q-Stars might have masses much greater than observed neutron star masses; the clustering of these masses at $\approx 1.4 M_{\odot}$ reflects the Chandrasekhar mass for pre-collapse stellar cores, not the maximum stable neutron star mass. Only a restricted class of effective field theories in which U wiggles such that there are two separate Q-Star phases (with different chemical potentials) can describe both neutron stars with large masses ($\gg 3M_{\odot}$) and neutron stars with short rotation periods ($P_{rot} < 0.5$ms). Further, a conventional neutron star phase (neutrons, protons, and electrons bound only by gravity) is always also present in any field theory which contains Q-Stars. To which phase a supernova remnant would collapse, Q-Star or conventional, will depend on the theory.

Acknowledgements: I thank the organizers for an enjoyable and stimulating conference, and the CERN Theory Group for its hospitality this year. This work was done in collaboration with Stephen Selipsky and Safi Bahcall, and was supported in part by the U.S. National Science Foundation, contract NSF-PHY-86-12280.

REFERENCES

1. B.D. Serot and J.D. Walecka in: Advances in Nuclear Physics, vol. 16, ed. J.W. Negele and E. Vogt (Plenum, New York, 1985), p. 1;
   B.D. Serot and J.D. Walecka, Phys. Lett. B 87 (1979) 172;
   J.D. Walecka, Annals Phys. 83 (1974) 491;
   F.E. Serr and J.D. Walecka, Phys. Lett. B 79 (1978) 10.
2. J. Boguta, Phys. Lett. B 106 (1981) 241;
   J. Boguta and S.A. Moszkowski, Nucl. Phys. A 403 (1983) 445;
   J. Boguta and H. Strocker, Phys. Lett. B 120 (1983) 289.

3. H. Georgi, Weak Interactions and Modern Particle Theory (Benjamin-Cummings, Menlo Park, CA, 1984).
4. For the theory of boson Q-Balls see
R. Friedberg, T.D. Lee and A. Sirlin, Phys. Rev. D 13 (1976) 2739;
S. Coleman, Nucl. Phys. B 262 (1985) 263.
5. The original Q-Star ansatz is given in B.W. Lynn, Nucl. Phys. B 321 (1989) 465; see also S.B. Selipsky, D.C. Kennedy, and B.W. Lynn, Nucl. Phys. B 321 (1989) 430.
6. For the theory of fermion Q-Balls and Q-Stars see S. Bahcall, B.W. Lynn, and S.B. Selipsky, Nucl. Phys. B 325 (1989) 606.
7. C.E. Rhoades and R. Ruffini, Phys. Rev. Lett. 32 (1974) 324;
J.B. Hartle, Phys. Reports 46, No. 6 (1978) 201.
8. S. Shapiro, S.A. Teukolsky and I. Wasserman, Ap. J. 272 (1983) 702; Cornell preprint CRSR-919, May 1989;
J.L. Friedman, J.R. Ipser and L. Parker, Nature 312 (1984) 255.
9. S. Bahcall, B.W. Lynn, and S.B. Selipsky, Nucl. Phys. B 331 (1990) 67;
S. Bahcall, B.W. Lynn, and S.B. Selipsky, Astrophys. J., in press.
10. B.W. Lynn, A.E. Nelson, and N. Tetradis, "Strange Baryon Matter", Stanford preprint SU–ITP–860, July 1989.
11. M. Prakash, T.L. Ainsworth and J.M. Lattimer, Phys. Rev. Lett. 61 (1988) 2518.
12. J.N. Bahcall, in: Physics and Astrophysics of Neutron Stars and Black Holes, eds. R. Giacconi and R. Ruffini (North Holland, Amsterdam, 1978).
13. B. Paczynski, Ap. J. Lett. 273 (1983) L81.
14. R.V. Wagoner and C. Perez, private communication.
15. E. Witten, Phys. Rev. D 30 (1984) 272;
E. Farhi and R. L. Jaffe, Phys. Rev. D 30 (1984) 2379;
C. Alcock, E. Farhi and A. Olinto, Ap. J. 310 (1986) 261.
16. T.D. Lee and G.C. Wick, Phys. Rev. D 9 (1974) 2291;
T.D. Lee, Rev. Mod. Phys. 47 (1975) 267.
17. G. Baym and C. Pethick, Ann. Rev. Astron. Astrophys. 17 (1979) 415;
V.R. Pandharipande and R.A. Smith, Phys. Lett. B 59 (1975) 15;
J. Kunz, D. Masak, U. Post and J. Boguta, Phys. Lett. B 169 (1986) 133.

Specific Heat of Strongly Interacting Matter

N.J. Davidson[1], *H.G. Miller*[1], *R.M. Quick*[1], *B.J. Cole*[2], *R.H. Lemmer*[2], *and R. Tegen*[2]

[1]Department of Physics, University of Pretoria, Pretoria 0002, South Africa
[2]Department of Physics, University of the Witwatersrand, P O WITS 2050, South Africa

I Introduction

Strong analogies exist between the spectra of baryons and nuclei. Aside from the difference in scale, in both cases the low--lying spectrum is collective and therefore sparse and increases exponentially with increasing exicitation energy. In the case of deformed nuclei this change in level density gives rise to a peak in the specific heat which has been associated with the shape transitions in the nucleus found in mean field calculations [1]. These "phase" transitions are associated with the change from a low--lying collective spectrum to a denser more random spectrum due to the independent particle degrees of freedom at higher excitation energies. One observes a similar change in level density in the hadron spectrum [2] -- namely low--lying sparse baryon resonances which lie on (almost) linear trajectories [3,4] corresponding to the collective rotational bands in the nuclear case which are built on a few low--lying band heads. The origin of the change in the level density in the hadron case requires a reliable QCD description of the hadrons, which is not unfortunately not available at present.

In the following lectures the effect of these changes in the level densities will be discussed in some detail, in the nuclear case for deformed nuclei and in

Phase Structure of Strongly Interacting Matter
Editor: J. Cleymans

hadronic matter. In particular, in both cases the specific heat has been calculated and peaked structures have been identified which arise from the changes in the experimental level densities.

II Nuclei

One of the interesting questions which arises in the finite--temperature description of nuclei is whether or not shape transitions really do occur. Finite--temperature mean--field calculations have addressed this question but have not been able to provide a definitive answer because of the large fluctuations inherent in such calculations. The vanishing of an order parameter such as the gap parameter in superconducting nuclei [5] or the quadrupole moment in deformed nuclei [6] has been offered as evidence of existence of a phase transition in such systems. When the fluctuations, which are partly thermal and partly due to the nuclear structure approximation, are calculated they are generally large enough to obscure the presence of a such a phase transition [7--9].

Perhaps a more meaningful quantity to study in this respect is the specific heat [10]. Model studies in an SU(2) x SU(2) system show that, in the thermodynamic limit, this system exhibits a singularity in the specific heat characteristic of a true phase transition [11,12]. For this reason shape transitions are sometimes referred to as phase transitions. Furthermore the remnant of this singularity remains in the form of a peak in finite systems of this type. The presence of this peak in the specific heat has been used to map out the phase structure in such a model [13].

In the FTHF approximation the thermodynamic potential

$$\Omega = <H>_T - T\,S - \mu\,N \qquad (1)$$

is minimized with respect to the Hartree–Fock orbitals and the single–particle thermal occupation probabilities, f_ν, subjected to the constraint

$$\Sigma\, f_\nu = N. \qquad (2)$$

Here $< >_T$ denotes the ensemble average at temperature T; the chemical potential is given by μ, the number of particles by N and the entropy by S, where

$$S = -\Sigma\,[f_\nu \ln f_\nu + (1-f_\nu)\ln(1-f_\nu)]. \qquad (3)$$

FTHF calculations have been performed for the two nuclei ^{20}Ne and ^{24}Mg in both the 1s–0d valence space and in a larger model space which includes the 0s, 0p and 1s–0d shells with no core. In addition we have performed the calculation for ^{24}Mg in a still larger model space, namely the 0s, 0p, 1s–0d and 1p–0f shells with no core. Only the thermal response of the states of total isospin I = 0 was considered in the present FTHF calculations. For both nuclei, numerical calculations performed in the FTHF approximation have been compared with exact canonical–ensemble results obtained from shell–model eigenfunctions in the sd–shell valence space. Of particular interest is the comparison of results for the specific heat [10,13–15] which, as we have recently pointed out [10,12], provides a means of detecting the presence of a shape transition. The number of states in the systems studied is not large, but recent model studies of quantum spin chains have demonstrated that quantum systems with few degrees of freedom display quantum statistical behaviour. Numerical studies of such systems [16] have shown that they could be adequately described by the canonical ensemble even though only 2^7 states were present and the density of states was too irregular to be described by a Boltzman distribution.

In the FTHF approximation the specific heat [12,14] is given by

$$C = \frac{\partial < \hat{H} >_T}{\partial T}. \tag{4}$$

For the mean--field calculations in ^{20}Ne and those for ^{24}Mg in the sd shell only, the derivative has been determined by fitting a piecewise smooth curve to the values of $< \hat{H} >$ and differentiating. This leads to slight uncertainties close to the transition temperature depending on the precise fit used and one observes minor changes in the height of the peak for different fits. For the no--core mean--field calculations in ^{24}Mg the specific heat has been calculated analytically directly from the FTHF equations [17].

In the canonical ensemble, the partition function is given by

$$Z(\beta) = \sum_{J,I,\nu(J)} (2J+1)\, e^{-\beta E_{J,\nu(J)}} \tag{5}$$

where $\beta = 1/T$ and $\nu(J)$ labels the states in each irreducible representation with angular momentum J. Here only the ground state is populated at zero temperature and not the entire yrast band. The latter case is more appropriate for nuclei formed in heavy ion collisions since the kinetic energy can be converted into collective energy before thermalization. The ensemble average of the energy is formally defined by $<E> = -\partial(\ln Z)/\partial\beta$ and the specific heat is calculated as the square of the fluctuation in the energy multiplied by β^2; alternatively

$$C_N = \frac{\partial}{\partial T} < E > \qquad (6)$$

where the subscript N indicates that the specicific heat is evaluated in the canonical ensemble, with the number of particles N in the system fixed.

The exact diagonalisation of the effective Hamiltonian in the sd shell was performed by means of the Lanczos algorithm incorporated in the Glasgow shell–model code [38]. Because of the presence of the Boltzmann factor in eq. (5), the lowest–lying eigenstates are most heavily weighted in the exact canonical ensemble calculations. Previous calculations [19] have demonstrated that using roughly 10% or less of the eigenspectrum produces results at sufficiently low temperatures which are almost indistinguishable from those obtained with the complete eigenspectrum. In the present work we have used only those eigenstates of ^{24}Mg which lie in the first 10–15 MeV of excitation energy for each of the even angular momenta; all the eigenstates of ^{20}Ne in the sd shell were included.

The effective interactions of the Iowa State group [20–23] have been used in all the calculations; these effective interactions were constructed from a two–particle G matrix with the Reid soft–core interaction used as the bare two–particle interaction. For the sd–shell calculations we used the Vary–Yang interaction [22], with additional third–order corrections to the G matrix to provide a more complete accounting of the core polarization effects [23], together with the following single–particle energies:

$$\epsilon_{d_{5/2}} = -5.00 \text{ MeV} \qquad \epsilon_{d_{3/2}} = 0.08 \text{ MeV} \qquad \epsilon_{s_{1/2}} = -4.13 \text{ MeV}.$$

In the larger model spaces consisting of either the 0s,0p and 1s--0d or the 0s, 0p, 1s--0d and 1p--0f shells, the Hartree--Fock orbitals were expanded in a harmonic oscillator basis with oscillator frequency $\hbar\omega = 14$ MeV. A realistic effective Hamiltonian, including folded--diagram corrections to the two--particle G matrix, was used. Since the matrix elements were originally calculated for ^{16}O with A = 16 particles, they were scaled here to include an appropriate A dependence [20,21]. The matrix elements of the scaled Hamiltonian are given by

$$<\hat{H}> = \frac{\omega'}{\omega}\left[\frac{A}{A'} <\hat{T}_{rel}> + <\hat{V}_{eff}>\right], \qquad (7)$$

where A' = 20 and $\hbar\omega' = 12.06$ MeV for ^{20}Ne and A' = 24 and $\hbar\omega' = 13.05$ MeV for ^{24}Mg; in addition, $\hat{T}_{rel}$ denotes the relative kinetic energy and $\hat{V}_{eff}$ the effective two--body interaction. However, in order to reproduce the experimental binding energy and r.m.s radius at zero temperature, it proves necessary to adjust $\hat{H}$ slightly; the matrix elements of $\hat{T}_{rel}$ are multiplied by 0.98 for ^{20}Ne, by 0.99 for ^{24}Mg in the 0s, 0p, 1s--0d shells and by 0.98 for ^{24}Mg in the 0s, 0p, 1s--0d, 1p--0f shells, and those of $\hat{V}_{eff}$ by 1.07 for ^{20}Ne and by 1.04 and 1.051 for ^{24}Mg in the respective model spaces. These correction factors are remarkably close to unity, indicating that the scaling procedure is indeed reasonable for the calculation of bulk properties. In addition, it will be shown that the results for the larger model spaces show the same behaviour as those for the 1s--0d shell. In this model space we have an unscaled interaction and performing the full diagonalization yields an eigenspectrum that is in reasonably good agreement with the experimental one [23].

At T = 0 in all the model spaces the ground--state solutions of the HF equations in ^{20}Ne are axially symmetric, while for ^{24}Mg the ground--state solutions have ellipsoidal symmetry. We also found in all model spaces an axially--symmetric solution for ^{24}Mg which is nearly degenerate with the elipsoidally--symmetric ground state. All of these solutions have in addition time reversal symmetry. Although it is not clear what shape ^{24}Mg should have [24--28], the presence in the experimental data [26,28] of two nearly degenerate low--lying bands with K = 0 and 2 seems to suggest that a ground--state solution without axial symmetry is not unreasonable [24].

The ensemble averages of the energy for ^{20}Ne and ^{24}Mg in all model spaces are given in figures 1 and 2, respectively. For comparison, the ensemble average of the energy in the canonical ensemble calculated in the sd shell is also given. Although one expects the results from the different calculations to be displaced in energy, simply because of the discrepancy at zero temperature, as previously discussed [19], there is a marked difference in shape between the mean--field results and those obtained in the canonical ensemble, even at low temperatures. For T $\lesssim$ 0.5 MeV the FTHF results are much less sensitive to variations in the temperature than those obtained from the canonical ensemble. This is probably due to the fact that the gap between those lowest--lying single--particle orbitals that have the highest occupation probabilites and the rest of the single--particle orbitals is too

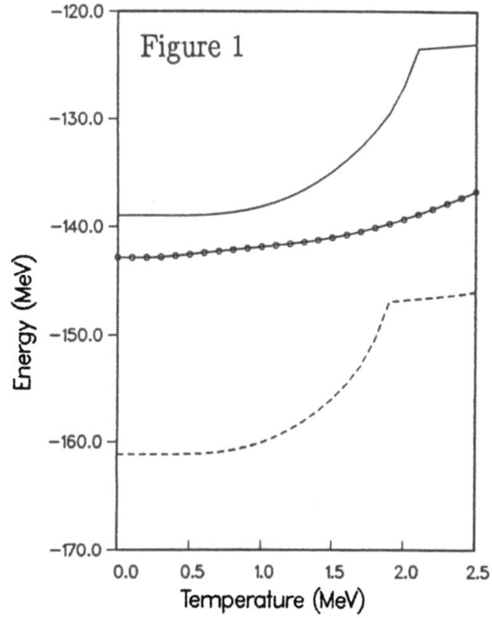

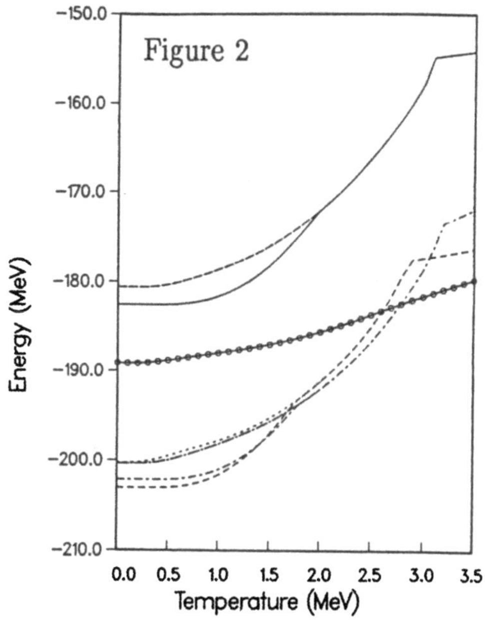

Figure 1. The ensemble average of the energy as a function of temperature for ^{20}Ne. The circles denote the results from the canonical ensemble calculation within the 1s--0d shell and the solid and dashed lines denote the results from the FTHF calculations within the 1s--0d and the 0s, 0p and 1s--0d shells respectively.

Figure 2. The ensemble average of the energy as a function of temperature for ^{24}Mg. The circles denote the results from the canonical ensemble calculation within the 1s--0d shell and the solid, dashed and chain--dashed lines denote the results from the FTHF calculations within the 1s--0d shell, the 0s, 0p and 1s--0d shells and the 0s, 0p, 1s--0d and 1p--0f shells respectively. The long dashed, dotted and chain--dotted lines indicate the results for the higher lying axial solution below the transition from ellipsoidal to axial symmetry.

large and does not decrease sufficiently rapidly with increasing temperature. In the canonical ensemble, however, the low--temperature response is strongly influenced by the energy of the first 2^+ state of the ground--state rotational band [14]. At higher temperatures the FTHF results rise quite steeply; with increasing temperature the single--particle spectra become more degenerate in energy and this higher level density leads to a more pronounced thermal response of the system. On the other hand, the lower--lying eigenspectra of both ^{20}Ne [19] and ^{24}Mg [23] are still too sparse to yield a strong thermal response in the canonical ensemble at these temperatures, whereas the FTHF approximation appears to overestimate the level density of the exact eigenspectrum at the excitation energies which are important at these higher temperatures. These discrepancies can be attributed to the number fluctuations in the mean field calculation [19].

In order to further investigate the model--space independence in the FTHF approximation we have calculated the specific heat [9,17]; the results are shown in figures 3 and 4. In the FTHF appproximation prominent peaks are seen in the specific heat at the same temperatures at which changes in slope are observed in the ensemble average of the energy (see figures 1 and 2). In ^{24}Mg the low--temperature peaks in the specific heat correspond to an average change in shape of the system from ellipsoidal to axially symmetric [24]. The remaining peaks in the specific heat of ^{20}Ne and ^{24}Mg correspond to an average change of shape in both systems from axially symmetric to spherically symmetric. The position of these peaks in the specific heat occurs at slightly higher temperatures in the sd shell than in the no--core calculation. The peaks in the calculation including the pf shell lie slightly higher.

In the canonical--ensemble average one can also see a broad peak in the specific heat (see figures 3 and 4) in both nuclei at the temperature roughly corresponding to the critical temperature at which a deformed--to--spherical

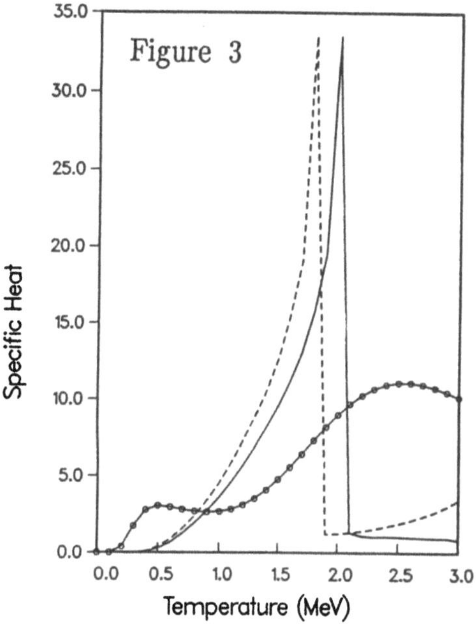

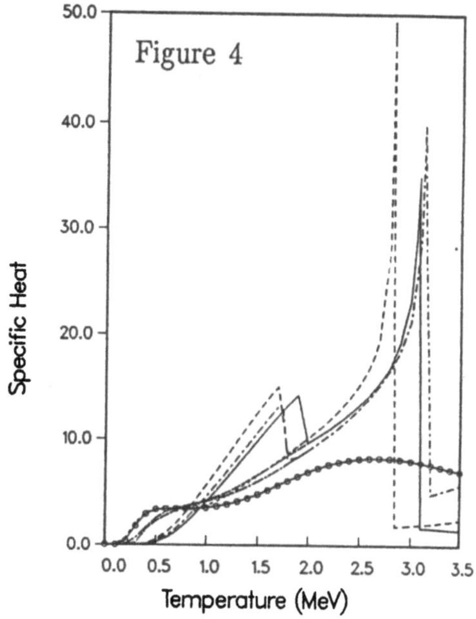

Figure 3. The ensemble average of the specific heat as a function of temperature for ^{20}Ne. The circles denote the results from the canonical ensemble calculation within the 1s--0d shell (multiplied by a factor of 2) and the solid and dashed lines denote the results from the FTHF calculations within the 1s--0d and the 0s, 0p and 1s--0d shells respectively.

Figure 4. The specific heat as a function of temperature for ^{24}Mg. The circles denote the results from the canonical ensemble calculation within the 1s--0d shell (multiplied by a factor of 2) and the solid, dashed and chain--dashed lines denote the results from the FTHF calculations within the 1s--0d shell, the 0s, 0p and 1s--0d shells and the 0s, 0p, 1s--0d and 1p--0f shells respectively. The long dashed, dotted and chain--dotted lines indicate the results for the higher lying axial solution below the transition from ellipsoidal to axial symmetry.

shape transition is predicted in the FTHF approximation in both model spaces. There is also a shoulder in the specific heat of both nuclei in the canonical ensemble at $T \simeq 0.5$ MeV which is probably due to the presence of the ground--state rotational band [9,14]. If one calculates the specific heat for the ground state rotational band only it exhibits a small peak for both nuclei at $T \simeq 0.5$ MeV and appears to go asymptotically with increasing temperature to roughly unity as expected [14] In the full canonical--ensemble results this asymptotic behaviour is obscured by the presence of the larger peak.

Canonical ensemble calculations have been performed with other realistic effective interactions. In figure 5 the low--lying part of the positive--parity eigenspectra of ^{20}Ne for the various sd--shell effective interactions [16,18,29,30] are compared with the experimental spectrum. More energy levels occur in the experimental spectrum as no attempt has been made to isolate the states arising primarily from sd--shell configurations. All of the interactions provide a reasonable description of the low--lying spectrum.

The specific heats as a function of temperature calculated in the canonical ensemble from the eigenstates in ^{20}Ne of the various effective interactions are given in figure 6. In all cases the specific heats calculated from the complete eigenspectrum are almost identical and exhibit the same structure: a small peak at $T \simeq 0.5$ MeV and a much larger peak at $T \simeq 2.4$ MeV. The temperature at which the maximum of the larger peak occurs differs by less than a few hundred keV for the different effective interactions.

If only the states in the ground--state rotational band are used in the calculation of the specific heat this is sufficient to reproduce the smaller peak at the lower temperature and to yield roughly unity at higher temperatures. This behaviour is identical for all of the effective interactions and is shown in figure 6 for the PW interaction. The asymptotic behaviour thus gives the

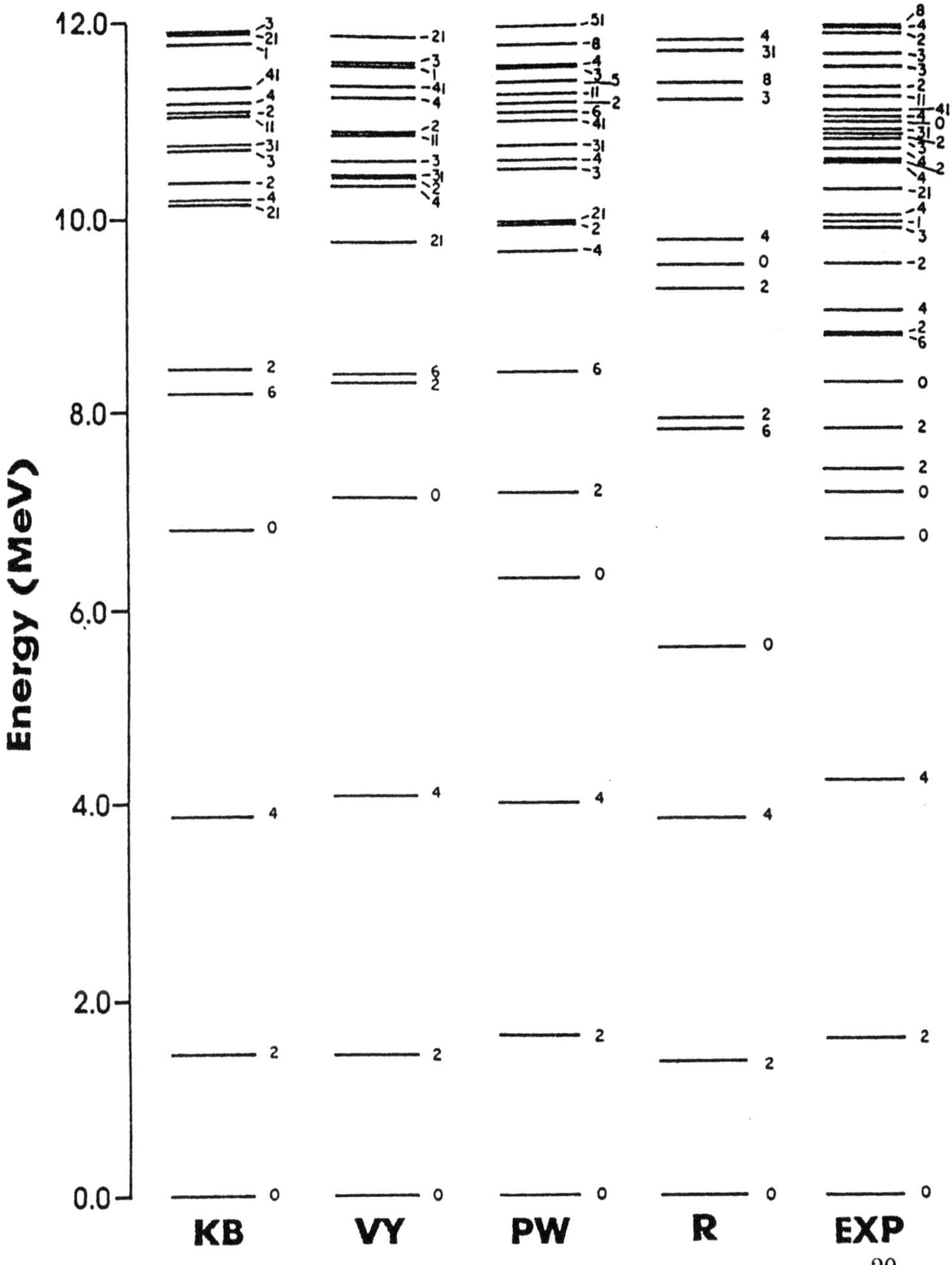

Figure 5. Shell--model spectra for the positive--parity states in ^{20}Ne calculated with the Kuo Brown (KB) interaction [16], the Vary--Yang (VY) interaction [29], the Preedom--Wildenthal (PW) interaction [18] and the Rosenfeld (R) interaction [18]. The experimental spectrum [32] (EXP) includes only the positive--parity states.

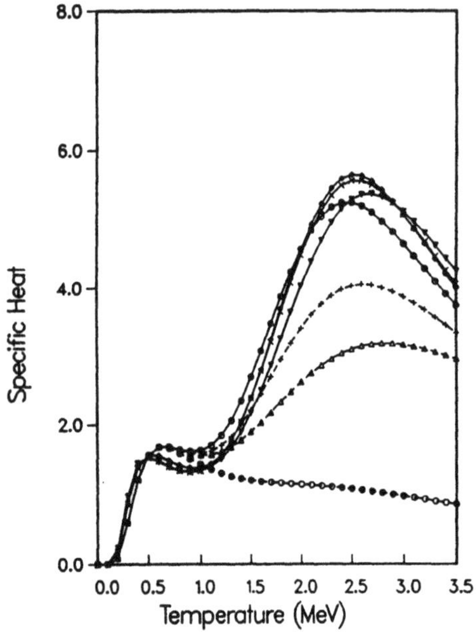

Figure 6. The specific heat as a function of temperature in the canonical ensemble calculated, solid lines, from the complete eigenspectrum of the KB [16] (◇), VY [29] (×), PW [18] (○) and R [30] (▽) interactions and, dashed lines, using the eigenstates of the PW [18] interaction in the ground--state rotational band (○), and the following sets of eigenstates of J and I: even J I = 0 (△) and all J I = 0 (+).

correct value for the number of relevant degrees of freedom, namely two for a rotor -- each degree of freedom contributes one--half in units of the Boltzmann constant to the asymptotic value of the specific heat. Note also that if only the states in the ground--state rotational band are used in the calculation of the specific heat the larger peak at T $\simeq$ 2.4 MeV is no longer present. As soon as the states with higher excitation energies are used in the calculation of the specific heat a prominent peak appears at T $\simeq$ 2.5 -- 3.0 MeV. Regardless of which subset of eigenstates of J and I are included in the ensemble the specific heat shows a large peak which obscures the high--temperature contribution from the states in the ground--state rotational band. Furthermore as more states are included in the ensemble this peak becomes more pronounced. This behaviour may be understood in the following manner: as the system is heated up a transition occurs in the states populated in the canonical ensemble from those associated with the purely collective ground--state rotational band to those of a more random nature. This peak in the specific heat arises from the change in the level density associated with the thermal excitation (see figure 5). It signals a change in the relevant degrees of freedom of the system. This phase transition must occur in all deformed nuclei and is more evident when more states are included in the ensemble. In the FTHF approximation the transition manifests itself as a deformed--to--spherical shape transition, the spherical shape arising from the increased degeneracy of the single particle orbitals . In this manner the thermal mean field calculations give rise to a change in the single particle level density which in turn leads to a change in the many particle level density [31].

The specific heat has also been calculated using the experimental energy spectrum of ^{20}Ne [32] (see figure 7). The specific heat was computed using just the positive--parity states and states of both parities to see the effect of the negative--parity states which are not included in the shell model calculations. In both cases the specific heat curves obtained resemble in shape those obtained using the shell model eigenstates. The differences in the magnitude of the peaks are probably due to the fact that at higher excitation energies the experimental level spectrum is denser than those obtained theoretically (see figure 5). For the positive--parity states the position of the maximum is in good agreement with the shell--model results;

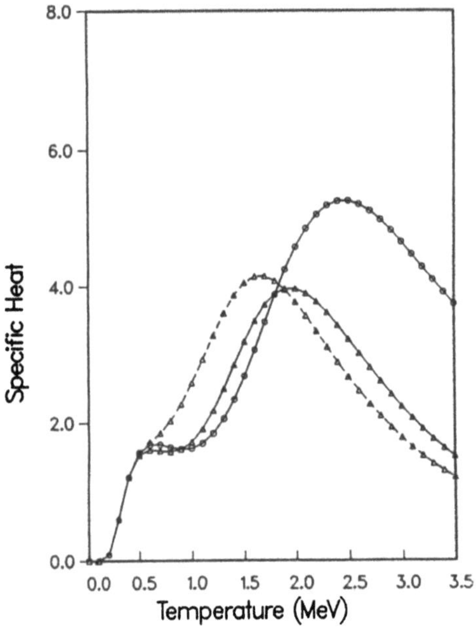

Figure 7. The specific heat as a function of temperature calculated using the complete eigenspectrum of the PW interaction (—ө—), all of the experimental energy levels (- ▲ -) with excitation energy below 14.5 MeV [32] and only those with positive--parity (—▲—)

including the negative--parity states shifts the position of the maximum to a slightly lower temperature.

In calculations with the experimental spectrum only those energy levels whose excitation energy is less than 14.5 MeV have been considered since above this energy the assignment of spins and parities of the various levels is less certain. As the excitation energy, E_X, increases it becomes more difficult to enumerate the states of the A--nucleon system, since the density of states grows rapidly with increasing excitation energy. It is therefore convenient to replace eq.(1) by

$$Z(\beta, A) = \sum_{\nu<N} g_\nu \exp[-\beta E_\nu] + \int_{E_N}^{\infty} dE_X \, \rho(E_X, A) \exp[-\beta E_X] \quad (8)$$

where $\rho(E_x,A)$ is the density of states at excitation energy E_x. In principle it should not matter where the cut--off N is taken, provided $\rho(E_x,A)$ adequately represents the density of states of the A--nucleon system above the threshold energy E_N.

The construction of the density of states function is still subject to many uncertainties and, despite several recent investigations [33--35], a formula for $\rho(E_x,A)$ valid for large E_x remains elusive. We have employed a modification of the usual Fermi--gas model prediction [36]

$$\rho(E_x, A) = g_A \frac{\exp\left[\,2\sqrt{E_x a_A}\,\right]}{E_x^{5/4}}. \tag{9}$$

Although this formula is not expected to give an accurate description of the density of states of the A--nucleon system for all relevant excitation energies, recent work [37] indicates that such formulae are adequate for light nuclei up to about 8 MeV in temperature. The level density parameter a_A and normalization factor g_A for a particular nucleus can be estimated from the known experimental level density for that nucleus. We have adjusted their values to describe the measured density of levels in the region of the particle--emission thresholds; lack of reliable spin/parity assignments precludes an accurate determination of these parameters. Nonetheless, the qualitative results presented here should not be influenced by the precise form chosen for $\rho(E_x,A)$, provided it has both the correct magnitude near the threshold E_N and an adequate energy dependence.

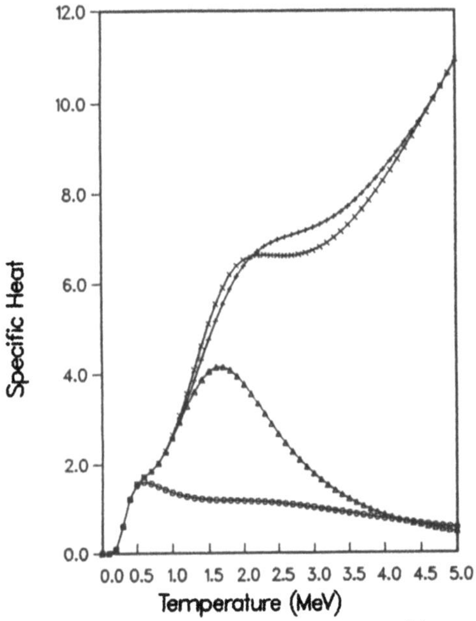

Figure 8. The specific heat for ^{20}Ne as a function of temperature calculated with the ground--state rotational band only (o), with the experimental point spectrum up to 14.33 MeV (Δ), and the experimental point spectrum plus continuum starting at 10 MeV (×) and 14.33 MeV (+). Parameters for the continuum calculation are $g_A = 1.10$, $a_A = 1.33$.

We present results for the specific heat of ^{20}Ne in Figure 8. The solid curve (o) was produced using eq.(5) with the experimentally determined energies of the ground--state band up to spin J = 8; the small peak evident at 0.5 MeV is a feature of all rotational spectra [14] and its position depends only on the energy of the first excited state. Truncation of the spectrum causes the specific heat eventually to fall below the expected asymptotic value of unity, which corresponds to two degrees of freedom for a rotor. The curve labelled (Δ) results from the entire experimental discrete spectrum of ^{20}Ne up to 14.33 MeV excitation energy. The previously--mentioned peak at 0.5 MeV now appears as a shoulder on a larger peak at 1.7 MeV, which was associated with a shape transition in [3]. Again, the asymptotic behaviour of the specific heat is masked by the effect of truncation of the spectrum, although the asymptotic value appears to be unity rather than 3A/2.

This same basic structure also occurs in the specific heat calculated with the discrete spectrum of ^{20}Ne plus a continuum beginning at 14.34 MeV (see the curve labeled '+' in Figure 8), although the the peak at T = 1.7 MeV is now replaced by a plateau. The continuum itself produces a contribution to the specific heat which rises monotonically (and rapidly) as temperature is increased; all nuclear structure effects are superimposed on this. Asymptotically, the curves which include the effects of the continuum should approach 3A/2, where A is the number of nucleons, but this cannot be investigated here due to the deficiencies of eq.(9) at large excitation energy.

The curve in Figure 8 labeled (×) illustrates the effect of varying the continuum threshold E_N; it correspond to the use of the continuum plus complete experimental spectrum with E_N equal to 10 MeV. Whatever the value of E_N, a plateau region is seen at a temperature of about 2.0--3.5 MeV. Other choices of the many--body density of states, eq.(9), lead to a definite peak in the specific heat rather than a plateau, although increasing the value of a_A too drastically can wash out the structure by allowing thermal excitation of the continuum to occur too rapidly, or even introduce new structure due to a mismatch in the level density at excitation energy E_N. We wish to emphasize that the plateau evident in Figure 8 is really the same structure as the peak discussed in [3] and illustrated by the curve (Δ) in Figure 8; the structure is in both cases produced by a rather abrupt increase in the level density of the discrete spectrum, below the threshold E_N.

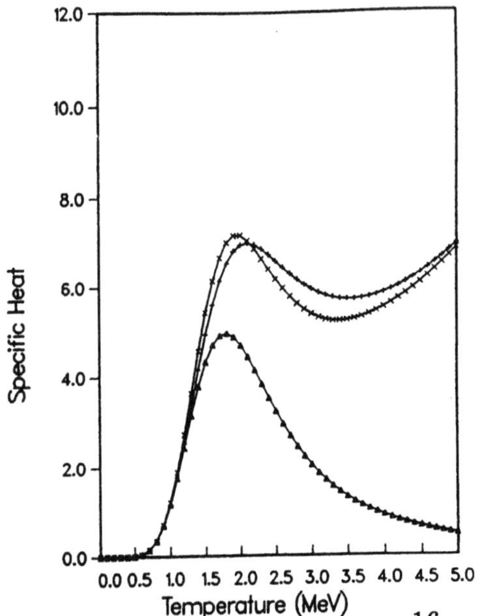

Figure 9. The specific heat for ^{16}O as a function of temperature calculated with the experimental point spectrum up to 13.87 MeV (Δ) and the experimental point spectrum plus continuum starting at 10 MeV (×) and 13.87 MeV (+). Parameters for the continuum calculation are $g_A = 0.90$, $a_A = 0.94$.

The specific heat of ^{16}O is shown in Figure 9; compared to the corresponding curves for ^{20}Ne, the major difference is the absence of the small shoulder at 0.5 MeV, since ^{16}O is not a rotational nucleus. The structure in the specific heat, which is now a broad peak rather than a plateau, occurs at a temperature of about 2 MeV, as in the case of ^{20}Ne; we again associate this structure with a change in the level density in the spectrum of ^{16}O. Again, the addition of a continuum to the experimental discrete spectrum does not alter the original conclusions, indicating the global nature of this phenomenon.

The prominent structure seen in the specific heat of ^{20}Ne was originally associated with a shape transition [3] since similar calculations with shell--model eigenenergies, and also FTHF calculations, indicated clearly that a deformed--to--spherical shape transition had taken place. The nature of the phase transition in ^{16}O is, however, clearly different, since this nucleus is not deformed at zero temperature. Nevertheless, in each case we can associate the structure in the specific heat with an increase in the nuclear level density, which occurs at an excitation energy of approximately 10 MeV in most light nuclei with even N and Z, and at considerably lower energies in other nuclei. The nature of any low--temperature phase transition is therefore related to the nuclear structure effects which produce the corresponding change in level density. We would like to point out that a phase transition of the type described here will occur for any nucleus which has a low--lying collective spectum, regardless of the nature of the collectivity. Eventually at some excitation energy the energy level spectrum will be become denser and more random and a peak will occur in the specific heat which indicates that the relevant degrees of freedom in the system are no longer purely collective.

III Hadronic Matter

As was pointed out in the introduction, there are strong analogies between the spectra of hadrons and nuclei. In the case of deformed nuclei this change in level density gives rise to a peak in the specific heat which has been associated with shape transitions in the nucleus (see above). The similarity between the hadronic and nuclear spectra leads us to the question of whether the specific heat of a hadron gas will also show structure.

Firstly, however, let us assume that structure is present in the specific heat of a hadron gas, and ask what the significance of such a structure might be. In contrast to the nuclear calculations, we are working in the grand canonical ensemble with a gas of particles, each with a single particle spectrum, rather with the spectrum of the many body states in the canonical ensemble as was the case in the nuclear calculations. There are thus two possible sources of structure:

1. The form of the gas itself, and
2. the spectra of the particles that make up the gas.

To investigate whether the first possible source does, in fact, lead to structures in the specific heat, we consider the case of a free gas of massless hadrons with hard core baryon repulsion (see Cleymans et al. [39]).

For such a gas, the energy and baryon number densities are given by

$$E = E'/(1 + n_b V_0) \tag{10}$$

where

$$E' = g_b \left(\frac{7\pi^2}{120} T^4 + \frac{\mu^2 T^2}{4} + \frac{\mu^4}{8\pi^2}\right) + g_m \frac{\pi^2}{30} T^4 \tag{11}$$

$$n_b = g_b \left(\frac{\mu T^2}{6} + \frac{\mu^3}{6\pi^2}\right) \tag{12}$$

and V_0 is the baryon hard core volume. In the above, g_b and g_m are the baryon and meson degeneracies respectively, μ is the chemical potential required to conserve baryon number density, and T is the temperature.

By simply looking at the temperature dependence of the individual terms in (11), it is tempting to conclude that there will be structures in the specific

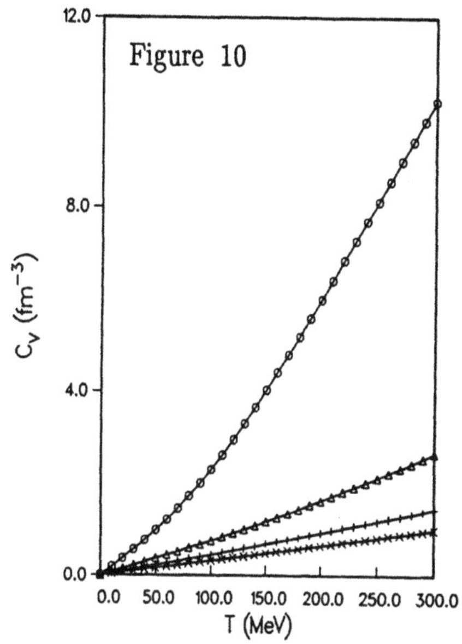

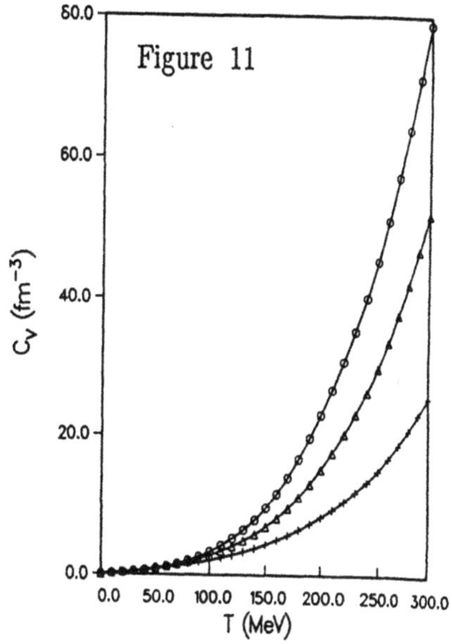

Figure 10 Specific heat of a massless hadron gas as a function of temperature for fixed chemical potential μ. The individual graphs correspond to the following values of μ:

$\mu = 500$ MeV (o)
$\mu = 1500$ MeV (Δ)
$\mu = 2500$ MeV (+)
$\mu = 3500$ MeV (×)

Figure 11 Specific heat of a massless hadron gas as a function of temperature for fixed baryon number density n_b. The individual graphs correspond to the following values of n_b :

$n_b = 0.2\ fm^{-3}$ (o)

$n_b = 0.3\ fm^{-3}$ (Δ)

$n_b = 0.4\ fm^{-3}$ (+)

heat for the above model of the gas, and that structures in the specific heat of a hadron gas are thus due to form of the gas rather than the spectra of the constituents. However, calculations reveal that no structure appears in the specific heat for either constant chemical potential μ or for constant baryon number density n_b (see Figures 10 and 11). The reason for this is that the coefficients multiplying the terms with different temperature dependences are not of the same order of magnitude. There is thus no interplay of terms which would give rise to structures in the specific heat. In fact, for reasonable values of μ, the major contributions come from the meson term. Even neglecting the mesons altogether, and only using the baryon terms in (11) does not result in structures. This lack of structure in the specific heat obtained from calculations with a massless hadron gas therefore seems to indicate that any structures which do occur using the experimental hadron spectrum must be due to this spectrum of the constituents of the gas, rather than the form of the gas itself.

To investigate the role of the spectrum in the phase structure of a hadron gas, we have to include a reasonable portion of the experimental spectrum. In the calculations which follow, we include only the reliable (3-- and 4--star) baryon resonances and the meson resonances marked with a dot by the Particle Data Group [4]; this leads to a natural cut--off at $\Lambda = 2$ GeV, beyond which the resonances are generally less well established. For masses above 2 GeV, we find that the (model dependent) continuum contributions are fairly small, and we ignore them here in order to simplify the presentation. The problems of including continua will discussed in more detail below.

Using the experimental hadron spectrum up to 2 GeV as input, the specific heat per unit volume for a relativistic hadron gas, in which the baryon number density and strangeness are conserved, is calculated as a function of temperature in the grand canonical ensemble. For large values of the baryon number density (5--10 times normal nuclear density), we find that this quantity exhibits structure around $T = 140$ MeV due to a peak in the specific heat per unit volume in the baryon sector at this temperature. This peak appears to be due to a combination of the constraint on the baryon number density and the change in the level density of the baryon spectrum. It should be noted that a reasonable portion of the baryon spectrum must be included in the calculation to obtain this structure. If the low--lying experimental mass spectrum is not properly taken into account no structure in the specific heat occurs.

The above interpretation of the structure is analogous to similar calculations in the nuclear realm. Structures in the specific heat obtained from canonical ensemble calculations which are due to changes in the level density of the nucleus correspond to shape transitions in deformed nuclei seen in finite temperature mean field calculations [3]. In the hadron case, the structure is again the result of a change in the level density. However, the reason for this change is not apparent. In the nuclear case, the change is due to a shift from collective to single particle degrees of freedom within the nucleus. To explain the origin of the change in level density in the hadron case requires a reliable QCD picture of hadrons with masses below 2 GeV. Unfortunately, QCD cannot yet explain this portion of the hadronic mass spectrum.

To include the interactions between the relativistic hadrons in a thermodynamic system an S--matrix formulation, as in ref. [40], would be desirable. Here we include these interactions by means of an effective repulsive potential, which in the mean field approximation becomes proportional to the particle density of the hadron type (mesons, baryons or antibaryons) [41]. We assume that the strength of this interaction for the baryons and antibaryons is the same. The constant of proportionality for the baryon sector, K_b, is taken to be either 0.68 or 1.7 GeV fm^3; the lower value is obtained from the Reid potential [42], while the latter value has been obtained from the strength of the isoscalar vector field which is required to stabilize relativistic nuclear matter [43]. The disparity between these numbers can be seen as reflecting the fact that nuclear medium effects, which are absent in the phenomenological nucleon--nucleon potentials, are accounted for approximately in the nuclear matter calculation by demanding saturation at the proper density.

To find the interaction strength for the mesons, we make use of the Weinberg effective Lagrangian for the π--π interaction, which yields a value for the proportionality constant K_m of approximately 0.6 GeV fm^3. We have not attempted to include interactions between hadrons of different types (i.e. meson--baryon, meson--antibaryon and baryon--antibaryon) in these calculations, since the form of these interactions is not well known. We thus prefer to leave these interactions out entirely; this may not be too harsh an approximation, since the coupling strengths are expected to be small, and should therefore not alter the qualitative behaviour seen in the present calculations.

The single particle energy of a hadron of mass m_i and momentum $\vec{p}$ is then

$$\epsilon_i(\vec{p},m_i,n) = (\vec{p}^2 + m_i^2)^{\frac{1}{2}} + Kn \tag{13}$$

where n is the particle density of the hadron type (either meson, baryon or antibaryon) and K the corresponding coupling constant. The second term gives the contribution of the repulsive interaction to the self energy of the particle in the Hartree approximation [41]. Since we are working at temperatures which are low in comparison with the hadron masses, we use Maxwell--Boltzmann (MB) statistics for all the hadronic states rather than Fermi--Dirac (FD) or Bose--Einstein (BE) statistics. (This statement is not strictly true for the lighter mesons, in particular the pion. However, a simple calculation shows that for reasonable values of the meson--meson potential, the errors associated with the use of MB rather than BE statistics for the pion are at the 1--5% level.)

The thermodynamic potential per unit volume for an interacting relativistic hadron gas in a Hartree approximation is

$$\Omega = -Tn - \frac{1}{2}\sum_j K_j n_j^2, \qquad (14)$$

where the sum runs over hadron types. Here T is the temperature, n is the total hadron density given by

$$n = \frac{1}{2\pi^2}\sum_i g_i \int_0^\infty dp\, p^2 \exp[-(\epsilon_i - \mu_b b_i - \mu_s s_i)/T] \qquad (15)$$

where the sum extends over the experimental hadron states up to $\Lambda = 2$ GeV, g_i is the degeneracy, b_i is 0 for mesons, 1 for baryons and --1 for antibaryons, and s_i is the strangeness of the ith state. The chemical potential μ_b ensures conservation of the baryon number, while μ_s ensures that strangeness is conserved. (Although we will work with strangeness zero, the asymmetry induced in the baryon sector by the strange baryons at non--zero baryon number densities requires the inclusion of a non--zero μ_s to ensure that the total strangeness does, in fact, remain zero.). From equation (15), it is obvious that the total hadron density n can be divided into contributions from the meson and baryon sectors, $n = n_{ms} + n_{bs}$, where

$$n_{ms} = \frac{1}{2\pi^2} \sum_i^m g_i \int_0^\infty dp\, p^2 \exp[-(\epsilon_i - \mu_s s_i)/T] \tag{16a}$$

$$n_{bs} = \frac{1}{2\pi^2} \sum_i^b g_i \int_0^\infty dp\, p^2 \exp[-(\epsilon_i - \mu_b b_i - \mu_s s_i)/T] \tag{16b}$$

where the sums extend over all meson states in (16a), and over all baryon and antibaryon states in (16b). It should be noted at this stage that n_{bs} is the sum of the baryon and antibaryon densities n_b and $n_{\bar{b}}$, and is not the conserved quantity $n_B = n_b - n_{\bar{b}}$, the baryon number density. The determining equations for n_{bs} and n_{ms} are coupled through the chemical potential μ_s. We must thus solve for n_{ms}, μ_b and μ_s simultaneously at each temperature. Since the defining equations are non--linear, the system has been solved by an iterative procedure.

Given Ω, eqns. (14--16), we can calculate all thermodynamic quantities of interest. In particular, the energy density is given by the fundamental relation

$$E = \left(\frac{\partial}{\partial\beta}[\beta\Omega]\right)_z , \tag{17}$$

where $\beta = \frac{1}{T}$ and z is the fugacity. This yields (see eq. (14))

$$E = \frac{1}{2\pi^2} \sum_{ij} g_i \int_0^\infty dp\, p^2 (\epsilon_i - \tfrac{1}{2} K_j n_j) \exp[-\beta(\epsilon_i - \mu_b b_i - \mu_s s_i)] \tag{18}$$

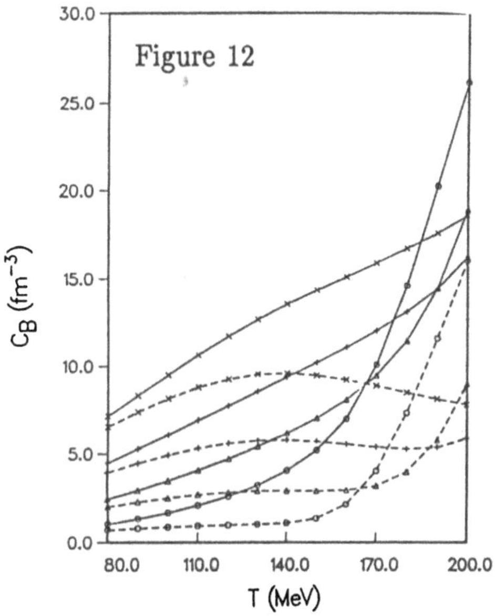

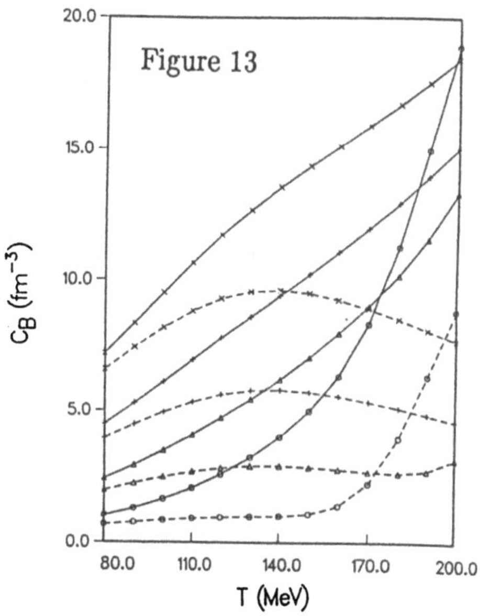

Figure 12 Specific heat per unit volume C_B versus temperature T for a hadron gas for a baryon potential strength of $K_b = 0.680$ GeV fm^3 and a meson potential strength of $K_m = 0.600$ GeV fm^3. The individual graphs correspond to the following values of the baryon density n_b in multiples of normal nuclear density $n_0 = 0.17$ fm^{-3}:

$n_B = n_0$ (○)
$n_B = 3n_0$ (Δ)
$n_B = 6n_0$ (+)
$n_B = 10n_0$ (×)

For each value of n_B, the solid curve corresponds to the hadron specific heat per unit volume, while the dashed curve is the specific heat per unit volume due to the baryon sector only.

Figure 13 As Figure 1, but for the case of $K_b = 1.700$ GeV fm^3.

The specific heat per unit volume at fixed baryon number density, C_B, is found from

$$C_B = \left(\frac{\partial E}{\partial T}\right)_{n_B}. \tag{19}$$

In Figures 12 and 13 C_B is plotted as a function of temperature T for two values of the baryon potential strength, $K_b = 0.680$ GeV fm^3 and

$K_b = 1.7$ GeV fm^3 respectively, for a selection of baryon number densities for the temperature range 80--200 MeV. For both values of K_b, we have set $K_m = 0.6$ GeV fm^3. Structure appears in the specific heat curve for the baryon sector at a temperature of approximately 140 MeV for both values of K_b. At low values of the baryon number density, this structure is a flattening of the C_B versus T curve around T = 140 MeV, which develops into a peak for larger values of the baryon number density. For higher values of the baryon number density, structure appears in the total specific heat per unit volume (i.e. in the hadronic specific heat, rather than just the baryon specific heat) in the form of a change in slope or slight shoulder. The structure at T ≈ 140 MeV becomes more pronounced as the baryon density increases (in the range 1 -- 10 times normal nuclear density [44]). The curve for $K_b = 0$ GeV fm^3 (not shown) also shows a flattening in the baryon sector around T ≈ 140 MeV. The position of the structure is remarkably insensitive to the choice of the baryon number density or the values of K_b and K_m.

The results do not show any truncation effects resulting from the neglect of hadron states above 2 GeV. (This can be clearly seen in Figs. 12 and 13 for the lower baryon number densities; for the higher values of n_B, continuing the calculations to higher temperatures shows that the specific heat per unit volume for the baryon sector starts to rise again above 250 MeV.) This is encouraging, as it avoids the problem of including a continuum which would introduce more parameters. In addition, the continuum is derived for hadron densities (baryons and mesons), and cannot be easily separated into baryon and meson contributions. It is thus not straightforward to consistently constrain the baryon density within these continuum models. We find, however, that the continuum above 2 GeV contributes very little, firstly because the experimental resonance spectrum is already quite dense in the region of 2 GeV, and follows an exponential form closely, and secondly because the relevant information seems to be contained in the medium--to--low--lying part of the resonance spectrum, which is model independent.

We have also performed calculations in which the total hadron density is constrained, and others with only mesons present with no constraint on the density. Although these situations are not physical, these calculations are useful in that they indicate that the structure found in C_B is related to the constraint on the number density, whether baryon or hadron. We thus find that, in the physical situation of constrained baryon number density and free meson density, the structure in the specific heat appears only in the baryon sector. The meson sector contributes a smooth background onto which the baryon--related structure is superimposed.

At large values of the baryon number density, as would be encountered in relativistic heavy ion collisions, the structure in the hadron (i.e. baryon plus meson) specific heat is indicative of a change in character of the hadron gas from baryon--dominated to meson--dominated (see Figs. 12 and 13). The rising tail of the hadron specific heat is predominantly due to the contributions from the meson sector, while the low temperature behaviour is determined almost entirely by the baryon sector. It should be remembered, however, that the structure in the hadron specific heat is due to the peak in the baryon specific heat. At low baryon number densities, C_B displays a plateau in the baryon sector. The apparent lack of structure in the total C_B indicates that this structure is washed out by the contribution from the meson sector. Thus the structure in the baryon specific heat becomes submerged in the smooth mesonic contribution, so that although the gas has changed character, there is no direct signal for this in the hadron specific heat per unit volume.

In summary, the specific heat per unit volume for a relativistic hadron gas has been calculated in the grand canonical ensemble using the experimental hadron resonance spectrum up to 2 GeV as input. In the physical situation of conserved baryon number density and zero strangeness density, a broad structure is apparent in the specific heat per unit volume in the baryon sector, the position of which is rather insensitive to either the baryon number density or the strengths of the repulsive hadron--hadron interactions used in the calculations. This structure appears to be related to the change in the level density of the baryon resonance spectrum, in analogy to the situation in deformed nuclei. Since it only appears in the baryon sector, we must also conclude that it is also due to the constraint of conserved baryon number density. The calculations presented can provide only a qualitative picture of the behaviour of the specific heat of a hadron gas; however, we believe that the structure in the specific heat per unit volume of a hadron gas at high baryon number densities is a real effect which may provide a signature for a change in character of the hadron gas, namely from baryon--dominated to meson dominated with increasing temperature. At low baryon number densities, meson domination of the gas occurs at lower temperatures, and the hadron specific heat per unit volume no longer provides an indication of this change in character.

Acknowledgements

We acknowledge the financial support of the Foundation for Research Development, Pretoria.

References

[1] See for example, J Cleymans, R. V. Gavai and E. Suhonen, Phys. Rep. 130, 217 (1986).

[2] See for example V. Bernard, Phys. Rev. D35, 1601 (1986).

[3] H. G. Miller, B. J. Cole and R. M. Quick, Phys. Rev. Lett. 63, 1922 (1989).

[4] Review of Particle Properties, Phys. Lett. B204, 1 (1988).

[5] A.L. Goodman, Nucl. Phys. A 352, 30 (1981).

[6] H.G. Miller, R.M. Quick, G. Bozzolo and J.P. Vary, Phys. Lett. B118, 13 (1986).

[7] J.L. Egido, P. Ring, S. Iwasaki and H. J. Mang, Phys. Lett. B154, 1 (1985).

[8] A.L. Goodman, Phys. Rev. C29, 1887 (1984).

[9] H.G. Miller, R.M. Quick and B.J. Cole, Phys. Rev. C39, 1599 (1989).

[10] K. Tanabe and K. Sugawara–Tanabe, Phys. Lett. B97, 337 (1980).

[11] D.H. Feng, R. Gilmore, and L.M. Narducci, Phys. Rev. C19, 1119 (1979).

[12] R. Rossignoli and A. Plastino, Phys. Rev. C32, 1041 (1985).

[13] E.D. Davis and H.G. Miller, Phys. Lett. B196, 277 (1987).

[14] R.K. Bhaduri and W. Van Dijk, Nucl. Phys. A485, 1 (1988).

[15] R.K. Pathria, in Statistical Mechanics (Pergamon, Oxford, 1982) pp. 76, 159, 195.

[16] T.T.S.Kuo, Nucl. Phys. A105, 7 (1967); T.T.S. Kuo and G.E. Brown, Nucl. Phys. 85, 40 (1966).

[17] R. M. Quick and H. G. Miller, submitted for publication.

[18] B. M. Preedom and B. H. Wildenthal, Phys. Rev. C6, 1633 (1972).

[19] R. M. Quick, H. G. Miller and B. J. Cole, Phys. Rev. C40, 993 (1989).

[20] G. Bozzolo and J.P. Vary, Phys. Rev. Lett. 53, 903 (1985).

[21] G. Bozzolo and J.P. Vary, Phys. Rev. C31, 1909 (1985).

[22] J. P. Vary and S. N. Yang, Phys. Rev. C15, 1545 (1977).

[23] M. S. Sandel, R. J. McCarthy, B. R. Barrett and J. P. Vary, Phys. Rev. C17, 777 (1978).

[24] B. J. Cole, R. M. Quick and H. G. Miller, Phys. Rev. C40, 456 (1989).

[25] P. Bonche, H. Flocard and P. H. Heenen, Nucl. Phys. A467, 115 (1987).

[26] R. K. Sheline, I. Ragnarsson, S. Aberg and A. Watt, J. Phys. G: Nucl. Phys. 14, 1201 (1988).

[27] H. Flocard, P. H. Heenen, S. J. Krieger and M. S. Weiss, Prog. Theor. Phys. 72, 1000 (1984).

[28] P. M. Endt and C. Van der Leun, Nucl. Phys. A310, 1 (1978).

[29] J.P. Vary and S.N. Yang, Phys. Rev. C 15, 1545 (1977); M.S. Sandel, R.J. McCarthy, B.R. Barrett and J.P. Vary, Phys. Rev. C 17, 777 (1978).

[30] L. Rosenfeld, Nuclear Forces (North Holland,Amsterdam,1958).

[31] R. M. Quick, N. J. Davidson, B. J. Cole and H. G. Miller, submitted for publication.

[32] F. Ajzenberg--Selove, Nucl. Phys. A392, 1 (1983).

[33] S. M. Grimes, Phys. Rev. C38, 2362 (1988)

[34] C. Jacquemin and S.K. Kataria, Z. Phys. A324, 261 (1986)

[35] B. Strohmaier, S.M. Grimes and H. Satyanarayana, Phys. Rev. C36, 1604 (1987)

[36] A. Bohr and B.R. Mottelson, Nuclear Structure (Benjamin, NewYork, 1969) Vol 1.

[37] E. Suraud, P. Schuck and R.W. Hasse, Phys. Lett. B164, 212 (1985)

[38] R. R. Whitehead, A. Watt, B. J. Cole and I. Morrison, Adv. Nucl. Phys. 9, 123 (1977).

[39] J. Cleymans, K. Redlich, H. Satz and E. Suhonen, Z. Phys. C33, 151 (1986).

[40] R. Dashen, S. Ma and H. J. Bernstein, Phys. Rev. 187, 345 (1969).

[41] A. L. Fetter and J. D. Walecka, Quantum Theory of Many--Particle Systems. (McGraw--Hill, Inc., 1971).

[42] G. E. Brown and A. D. Jackson, The Nucleon--Nucleon Interaction. (North--Holland, Amsterdam, 1976).

K. A. Olive, Nucl. Phys. B198, 461 (1982).

[43] D. Serot and J. D. Walecka, Adv. Nucl. Phys. 16, 1 (1986).

[44] Such nuclear densities will become accessible with the Relativistic Heavy Ion Collider (RHIC), see:

N. P. Samios and T. W. Ludham, Nucl. Phys. A498, 323C (1989).

S. Nagamiya in Proceedings of the 1989 Intern. Nucl. Phys. Conf., São Paulo (Brazil), Aug. 20--26, 1989.

P. Kienle, ibid.

QCD Transport Coefficients

D.W. von Oertzen

Institute for Theoretical Physics and Astrophysics, Department of Physics, University of Cape Town, 7700 Rondebosch, South Africa

We present the transport coefficients of quark matter, gluon matter and a quark anti-quark mixture, as they arise from relativistic kinetic theory. The results for the simple quark and gluon system are comparable with values obtained from similar calculations in the literature, and we present the interesting kinetic coefficients of a quark anti-quark mixture. The divergent scattering integrals are solved by introducing Coulomb logarithms, their evaluation is discussed.

1 Introduction

Current experimental efforts with ultra relativistic heavy ions offer the possibility to study strongly interacting matter in the laboratory [1]. A possible new phase of this matter is the quark gluon plasma (QGP), a system of deconfined quarks scattering off other quarks and gluons. In order to study such a system and facilitate the interpretation of experimental signatures of the possible deconfinement transition, one needs to understand the QGP evolution, and in particular the properties (like the transport coefficients) of such an exotic phase of matter.

In this article the transport properties of quarks and gluons are discussed. Transport processes arise as a result of collisions between particles, and are responsible for the equilibration of a system. In heavy ion collisions for example, this process is believed to proceed in two steps:

1. The evolution of the system from a completely violent stage where most of the particle creation takes place, to a system where the local density, temperature and hydrodynamic velocity can be specified.

2. A conversion from the local to a globally equilibrated state in which the spatial non uniformities disappear.

The late phases of stage 1 in the evolution of the system is the stage we analyse; after which the use of local hydrodynamics is justified. By using a suitable relativistic kinetic equation which describes the interplay between micro- and macroscopic phenomena, ie. the collisions vs. the hydrodynamical evolution, we find the transport coefficients of the system. The knowledge of these coefficients of the QGP enables one to include the dissipative effects in the plasma hydrodynamics which is needed to predict the rate of expansion of the QGP fluid until hadronization sets in. For example, the kinetic

Phase Structure of Strongly Interacting Matter
Editor: J. Cleymans

or transport coefficients associated with the entropy production are the volume (bulk) and shear viscosities and the heat conductivity, all having a microscopic origin and a macroscopic manifestation. Relativistic kinetic theory, and in particular the relativistic transport theory (see eg. [2] and [3]), provide a natural framework in which to calculate the transport coefficients selfconsistently and study non-equilibrium effects.

Several attempts exist in the literature to compute these coefficients for a relativistic gas using eg. the Grad method [3] and the collision time (relaxation time) approach ([4] and [5]). In this paper we propose to use the **Chapman Enskog (CE)** method, using the formalism as developed in ref. [2], to solve for the transport coefficients of a gas composed of quarks, or gluons or quarks and anti-quarks.

Advantages of the CE method over other methods of solution are that:

- The approximations involving the CE scheme of solution are well understood and no arbitrariness is introduced due to higher orders of approximation or ambiguous truncation procedures.
- The relativistic Onsager relations can be derived within this method (in contrast to the Grad and other methods [3]).
- The difficulty of finding characteristic collision times, especially in mixtures, where this is not a well defined concept anymore, is eliminated (in contrast to the relaxation time approximation [4]) ,
- the method selfconsistently provides one with the linear laws between the thermodynamic forces and the corresponding flows, with the transport coefficients as the constants of proportionality.

Weak points of the CE method include:

- It is less general than eg. the relativistic 14-moment approximation since the form of the local equilibrium distribution has to be specified.
- The weaknesses associated with a Boltzmann like equation, ie.
    1. Its applicability to dilute gases only where two body collisions dominate.
    2. The hypothesis of molecular chaos, ie. no particle correlations.
    3. The distribution function may only vary slowly in space-time. This restriction however defines length- and time-scales of the problem and determines the degree of "maximal non equilibration" which may be studied.

Various accounts of transport theory and the evaluation of transport related phenomena exist in the literature, we refer the reader to an excellent review of quantum transport theory [6], and for recent reviews of the quark gluon transport theory to refs. [7]. The transport coefficients of the quark gluon plasma are also discussed in refs. [8]; references with specific solutions of relativistic transport equations in heavy ion scenarios are given in [9].

In the following section we state the expressions used for the kinetic coefficients of simple systems and mixtures, as derived from a covariant transport equation using the Chapman Enskog method [2]. These are then evaluated in section 3, using the scattering matrix elements and differential cross sections from Appendix A. A discussion of the various divergent integrals and the solution via Coulomb logarithms is presented in Appendix B.

In this paper we use units such that

$$c = \hbar = k_B = 1,$$

the metric $g^{\mu\nu}$ is given by diag $(1, -1, -1, -1)$.

2 Transport Coefficients

We view the QGP as a collision dominated system of classical particles (ie. spin and color only effects the transition rate), interacting through two-body collisions with other particles in the medium. The transport equation to be solved, ie. the relativistic Boltzmann equation, describes the space-time evolution of the phase space density $f = f(x,p)$, is then given by [10]

$$p^{\mu}\partial_{\mu} f = C[f, f], \tag{1}$$

with the collision operator $C[f, f]$:

$$C[f, f] = \frac{1}{2}\int \frac{d^3p_2}{p_2^0}\frac{d^3p_3}{p_3^0}\frac{d^3p_4}{p_4^0}\left[f_3 f_4 - f_1 f_2\right] W(p_3p_4|p_1p_2), \tag{2}$$

where $W(p_3p_4|p_1p_2)$ is the transition rate in the collision process $p_1 + p_2 \rightarrow p_3 + p_4$. As motivated in the Introduction, we use the Chapman Enskog method of solving equation (1) for the transport coefficients of the system, but rather than present the lengthy scheme of solution, we refer the reader to the monograph by *de Groot* and coworkers [2].

For a system where scattering processes between one particle species constitute the dominant collision mode, one has the following transport coefficients:
the **volume viscosity**

$$\eta_v = \frac{T}{\sigma(T)}\frac{(\alpha^2)^2}{A^{22}}, \tag{3}$$

the **heat conductivity**

$$\lambda = \frac{1}{3\sigma(T)}\frac{(\beta^1)^2}{B^{11}}, \tag{4}$$

and the **shear viscosity**

$$\eta = \frac{T}{10\sigma(T)}\frac{(\gamma^0)^2}{C^{00}}, \tag{5}$$

where the factors A^{22}, B^{11} and C^{00} are complicated expressions in terms of generalized collision integrals. We will further investigate these coefficients and their numerical values in section 3.

For a binary mixture of components 1 and 2, one obtains the following kinetic coefficients: the **volume viscosity**

$$\eta_v = \frac{T}{\sigma(T)} x_1 x_2 \frac{\alpha_1^1 \alpha_2^1}{A_{12}^{11}}, \tag{6}$$

the **shear viscosity**

$$\eta = \frac{T}{10\,\sigma(T)\,\Delta_c} [(x_1 \gamma_1^0)^2 C_{22}^{00} - 2x_1 x_2 \gamma_1^0 \gamma_2^0 C_{12}^{00} + (x_2 \gamma_2^0)^2 C_{11}^{00}], \tag{7}$$

the **diffusion coefficient**

$$D_d = \frac{1}{3\,n\,\sigma(T)} \frac{\beta_{11}^0 \beta_{12}^0}{B_{12}^{00}}, \tag{8}$$

the **thermal diffusion** and **Dufour coefficient**

$$\begin{aligned} D_T = D_T' &= \frac{1}{3\,n\,T\,\sigma(T)\,\Delta_B} [\frac{x_1}{x_2} \beta_1^1 \beta_{11}^1 y_2 + \frac{x_2}{x_1} \beta_2^1 \beta_{12}^1 y_3 \\ &+ (\beta_1^1 \beta_{12}^1 + \beta_2^1 \beta_{11}^1) y_4 + \beta_2^1 \beta_{11}^0 y_5 + \frac{x_1}{x_2} \beta_1^1 \beta_{11}^0 y_6] \end{aligned} \tag{9}$$

and the **heat conductivity**

$$\lambda = \frac{1}{3\,\sigma(T)\Delta_B} [(x_1 \beta_1^1)^2 y_2 + 2x_1 x_2 \beta_1^1 \beta_2^1 y_4 + (x_2 \beta_2^1)^2 y_3]. \tag{10}$$

These formulae will be used in the second part of section 3 to calculate some numerical values for a quark anti-quark mixture.

3 Results

In this section we present the expressions for the various transport coefficients, both for simple systems of quarks and gluons, and for a quark anti-quark mixture.

3.1 One Component Systems

3.1.1 The simple quark system

The transition matrix element for a simple quark system, ie. a system where the scattering process

$$q\,q \rightarrow q\,q$$

is the dominant scattering mode, is given in Appendix A, with (31) the final result, the differential cross section in the cm-frame is presented in (36).

Because we are assuming that the quarks are massless, the volume viscosity η_v is zero, as can generally be shown for ultra-relativistic and non-relativistic particles [11]. The transport processes involved are only the heat conduction, with the coefficient of heat conductivity given by (4), and the shear viscous flow, with the coefficient of shear viscosity given by (5). Since (from [2])

$$\beta^1 = \frac{3\gamma}{\gamma - 1} = 12$$

and

$$\gamma^0 = 10\,\frac{h^{(0)}}{T} = 40$$

(massless particles have an enthalpy per particle $h^{(0)} = 4T$ and $\gamma = \frac{c_p}{c_v} = \frac{4}{3}$), and with the choice of a characteristic mean cross section

$$\sigma(T) = \frac{\alpha_s^2}{T^2} \tag{11}$$

one can calculate both B^{11} and C^{00}.

The **heat conductivity** is then given by

$$\lambda = \frac{216}{\pi\,\sigma(T)\,[24\,\log\Lambda_c - \Lambda^3 - 37\,\Lambda]} \tag{12}$$

or, for $\Lambda \to 0$,

$$\lambda = \frac{9}{\pi\,\sigma(T)\,\log\Lambda_c} = \frac{9\,T^2}{\pi\,\alpha_s^2\,\log\Lambda_c}. \tag{13}$$

For the **shear viscosity** one finds

$$\eta = \frac{216\,T}{\pi\,\sigma(T)\,[24\,\log\Lambda_c - \Lambda^3 - 37\,\Lambda]} \tag{14}$$

or, for $\Lambda \to 0$,

$$\eta = \frac{9\,T}{\pi\,\sigma(T)\,\log\Lambda_c} = \frac{9\,T^3}{\pi\,\alpha_s^2\,\log\Lambda_c}. \tag{15}$$

The logarithm appearing in equations (12) to (15) is a Coulomb type logarithm and takes care of the infrared divergences in the scattering matrix element (31). It is discussed and evaluated in Appendix B.

For $T = 200\,\mathrm{MeV}$, $\alpha_s = 0.1$ and $\log\Lambda_c = 1$ (small), one has the numerical values (for (13) and (15)) of

$$\begin{aligned} \lambda &= 11.46\,[GeV^2] = 1.21\;\;10^{18}\,[\frac{kg\;m}{s^3\,K}] \\ \eta &= 2.29\,[GeV^3] = 3.16\;\;10^{13}\,[\frac{kg}{m\;s}], \end{aligned} \tag{16}$$

compared to the terrestrial values of $\lambda \simeq 10^{-2}[\frac{kg\,m}{s^3\,K}]$ and $\eta \simeq 10^{-5}[\frac{kg}{m\,s}]$ [12].

Equations (13) and (15) are similar to the results of a simple system of quark-like particles [13] which has most of the essential features (Coulomb logarithm, high temperature behaviour) but is considerably easier to calculate.

3.1.2 The simple gluon system

The dominant scattering process in a simple gluon system is the process

$$g\,g \rightarrow g\,g$$

for which the cm differential cross section is given in Appendix A, eq. (33).

Using again (4) and (5) for the heat conductivity and shear viscosity respectively, gives

$$\lambda = \frac{10240}{3\,\pi\sigma(T)\,[\,1920\,\log\Lambda_c \;-\; 12\,\Lambda^5 \;-\; 215\,\Lambda^3 \;-\; 2175\,\Lambda\,]} \tag{17}$$

and with $\Lambda \rightarrow 0$

$$\lambda = \frac{16}{9\,\pi\,\sigma(T)\,\log\Lambda_c} = \frac{16\,T^2}{9\,\pi\,\alpha_s^2\,\log\Lambda_c}. \tag{18}$$

For the shear viscosity one obtains

$$\eta = \frac{10240\,T}{3\,\pi\sigma(T)\,[\,1920\,\log\Lambda_c \;-\; 12\,\Lambda^5 \;-\; 215\,\Lambda^3 \;-\; 2175\,\Lambda\,]} \tag{19}$$

and with $\Lambda \rightarrow 0$

$$\eta = \frac{16\,T}{9\,\pi\,\sigma(T)\,\log\Lambda_c} = \frac{16\,T^3}{9\,\pi\,\alpha_s^2\,\log\Lambda_c}. \tag{20}$$

The results eqs.(18) and (20) are not changed by using the factor 3 instead of $\frac{51}{16}$ in the gluon matrix element (see the discussion following (33) in Appendix A), and characteristic values for $T = 200\,\mathrm{MeV}$, $\alpha_s = 0.1$ and $\log\Lambda_c = 1$ are

$$\begin{aligned} \lambda &= 2.26\,[\,GeV^2\,] = 2.41\;\; 10^{17}\,[\frac{kg\;m}{s^3\,K}] \\ \eta &= 0.45\,[\,GeV^3\,] = 6.24\;\; 10^{12}\,[\frac{kg}{m\,s}]. \end{aligned} \tag{21}$$

The *Eucken* relation, ie. the ratio of the heat conductivity to shear viscosity is

$$\frac{\lambda}{\eta} = \frac{1}{T}$$

for both the simple quark and the simple gluon system; this can be compared to the value of $\frac{23}{10T}$ for a simple neutrino system [2].

It should be noted, that the values for the gluon kinetic coefficients are smaller than the corresponding quark coefficients, emphazising the importance to include the effects of quarks in a full quark gluon plasma calculation: this also motivates us to investigate the transport coefficients of a quark anti-quark mixture, as will be done in the next section.

3.2 Two Component Systems

3.2.1 The quark anti-quark mixture

The transition matrix element (32) for the process

$$q\ \bar{q} \rightarrow q\ \bar{q}$$

gives the cm differential cross section (37).

The choice for the mean cross section $\sigma(T)$ is again

$$\sigma(T) = \frac{\alpha_s^2}{T^2}.$$

Without presenting the calculational details we now list the results for the kinetic coefficients for a mixture of massless ($\eta_v = 0$) quarks (fraction x_1) and anti-quarks (fraction x_2): with $\Lambda \rightarrow 0$

$$\lambda = \frac{576\,T^2}{\pi\,\alpha_s^2}\,\frac{x_1x_2\,[5760\,S + 3186] - 2880\,\log\Lambda_{c2} - 1039}{184320\,[x_1x_2\,S^2 + G] + x_1x_2\,[132992\,S + 17169] - 66496\,\log\Lambda_{c1}} \tag{22}$$

$$D_T = 0, \tag{23}$$

$$D_d = \frac{-\,2592\,T^2}{\pi\,\alpha_s^2\,n\,[576\,\log\Lambda_{c2} + 97\,]}, \tag{24}$$

$$\eta = \frac{576\,T^3}{\pi\,\alpha_s^2}\,\frac{x_1x_2\,[5760\,S + 3186] - 2880\,\log\Lambda_{c2} - 1039}{184320\,[x_1x_2\,S^2 + G] + x_1x_2\,[132992\,S + 17169] - 66496\,\log\Lambda_{c1}} \tag{25}$$

The abbreviation

$$S = \log\Lambda_{c1} + \log\Lambda_{c2}\,,\ \ s.t.\ S > 0 \tag{26}$$

and

$$G = -\,\log\Lambda_{c1}\,\log\Lambda_{c2}\,,\ \ s.t.\ G > 0 \tag{27}$$

has been used; the logarithmic factors are discussed in Appendix B.

Again, the *Eucken* relation

$$\frac{\lambda}{\eta} = \frac{1}{T}$$

is recovered, as was already seen in the discussion of one component systems.

It is clear from the results that a multicomponent system is indeed very complex, and it seems practical to postulate a simpler form for the differential cross section, which has most of the quark anti-quark properties (ie. infrared divergences and correct high temperature behaviour) but is less complicated in structure [13].

4 Conclusions and Outlook

We have presented the transport coefficients of simple quark and gluon systems and a quark anti-quark mixture.
The method leading to the above results is laborious, and results from simpler methods, such as e.g. the relaxation time approximation, yield similar results. However, the difference between the Chapman Enskog method as used and others becomes important when studying multicomponent systems: because one does not have to specify a mean collision time for individual particle species and one treats the small angle scattering correctly (i.e. no ad hoc factors such as $1 - \cos\theta$ or similar), we believe that the extra calculational effort is justified. Also, the Chapman Enskog philosophy selfconsistently provides the various conservation equations, and is ideally suited for the description of particle systems in an external field: the treatment can be extended to color fields for the study of a more realistic quark-gluon system.

The results obtained for the simple quark and gluon systems are similar to others in the literature ([4], [14], [15] and [16]), where different methods of solution of the Boltzmann equation have been used.

However, many problems remain, and we close by mentioning a few:

- the rigorous solution of the cut-off problem,
- the evaluation and study of the color conductivity in the presented framework,
- the extension of the calculations to the deconfinement transition.

Acknowledgements: I have benefitted from stimulating discussions with Prof Jean Cleymans, and thank him for his support and for carefully reading the manuscript. Also, I thank Prof Ulrich Heinz for helpful comments and suggestions.

A The Matrix Elements

In this Appendix we present the scattering matrix elements [17], to lowest order in the pertubation expansion, for the processes:

$$q\,q \rightarrow q\,q \tag{28}$$

$$q\,\overline{q} \rightarrow q\,\overline{q} \tag{29}$$

and

$$g\,g \rightarrow g\,g\,, \tag{30}$$

all particles are considered to be massless.

Process (28) can generally be written as

$$q^i_\alpha\, q^k_\beta \rightarrow q^j_\alpha\, q^l_\beta$$

where α, β is the quark flavor index, and i, j, k, l the quark color index: $i, j, k, l = 1, 2, 3$.
The channels available in this process are the t- and the u-channel, for processes where the quark flavor (index α and β) is the same in the in- and outgoing channel, the resultant matrix element squared is

$$|M_{q^i_\alpha\, q^k_\alpha \rightarrow q^j_\alpha\, q^l_\alpha}|^2 = \frac{4}{9} g^4 \left[\frac{s^2+u^2}{t^2} + \frac{s^2+t^2}{u^2} - \frac{2}{3}\frac{s^2}{ut}\right]. \tag{31}$$

The scattering matrix element for process (29), generally written as

$$q^i_\alpha\, \overline{q}^k_\beta \rightarrow q^j_\delta\, \overline{q}^l_\gamma$$

where the indices have the same meaning as in the previous process (channels available are the s- and t-channel) is:

$$|M_{q^i_\alpha\, \overline{q}^k_\alpha \rightarrow q^j_\alpha\, \overline{q}^l_\alpha}|^2 = \frac{4}{9} g^4 \left[\frac{s^2+u^2}{t^2} + \frac{t^2+u^2}{s^2} - \frac{2}{3}\frac{u^2}{st}\right]. \tag{32}$$

The gluon-gluon scattering matrix element, process (30), is the final process under consideration, one has to consider 4 channels , the s-, t-, u-channel and the so called 'seagull-' or '4-point'-channel. One obtains

$$|M_{g\,g \rightarrow g\,g}|^2 = \frac{9}{2} g^4 \left[\frac{51}{16} - \frac{us}{t^2} - \frac{st}{u^2} - \frac{ut}{s^2}\right]. \tag{33}$$

It should be noted that Combridge et al [18] and Shuryak [1] have obtained a different answer than the one presented (their first term in (33) is 3 instead of $\frac{51}{16}$). They chose an explicit polarization vector representation for the gluons and not the Fadeev Popov ghost method used here [17]. However, since a complete analysis of their work could not be found, we will use the result (33), noting, that the lowest order (ie. neglecting terms of order Λ) results are the same (see section 3.1.2).

One still needs to find the differential cross section for the processes (28) - (30), to this end we make use of the connection between the transition rate $W(p_3p_4|p_1p_2)$ as used in the transport equation, the centre of momentum (cm) differential cross section and the transition matrix element squared [19]

$$\begin{aligned} W(p_3p_4|p_1p_2) &= s\,\sigma(s,\theta)\,\delta^{(4)}\,(p_1+p_2-p_3-p_4) \\ &= \frac{1}{8\pi^2}\delta^{(4)}\,(p_1+p_2-p_3-p_4)|M_{p_1+p_2\rightarrow p_3+p_4}|^2. \end{aligned} \tag{34}$$

Furthermore, in the cm-frame, one rewrites the Mandelstam variables [19] for massless particles as

$$\begin{aligned} s &= 4|\vec{p}|^2 \\ t &= -2|\vec{p}|^2(1-x) \\ u &= -2|\vec{p}|^2(1+x), \end{aligned} \tag{35}$$

where $x = cos\theta$ is the cm scattering angle. Using (34) and (35), one finds the cm differential cross section for process (28):

$$\sigma_{cm}(s,x) = \frac{\alpha_s^2}{9\,s}[\frac{4+(1+x)^2}{(1-x)^2} + \frac{4+(1-x)^2}{(1+x)^2} - \frac{2}{3}\frac{4}{1-x^2}]. \tag{36}$$

For process (29)

$$\sigma_{cm}(s,x) = \frac{\alpha_s^2}{9\,s}[\frac{4+(1+x)^2}{(1-x)^2} + \frac{(1-x)^2+(1+x)^2}{4} + \frac{1}{3}\frac{(1+x)^2}{(1-x)}] \tag{37}$$

and from (33) for process (30)

$$\sigma_{cm}(s,x) = \frac{9\,\alpha_s^2}{8\,s}[\frac{51}{16} + 2\frac{(1+x)}{(1-x)^2} + 2\frac{(1-x)}{(1+x)^2} - \frac{1-x^2}{4}]. \tag{38}$$

The cm cross sections (36) to (38) are used in various scattering integrals, in Appendix B we discuss the divergences for $x = +1$ and $x = -1$ and how they are treated in a plasma system.

B The Cut-off Problem

In this Appendix we discuss the infrared divergences appearing in the scattering matrix elements for the processes under consideration, and how we deal with them.

In order to solve the divergent integrals, we introduce a cut-off Λ at the divergent limit, ie.

$$\int_{-1}^{+1} dx \, \frac{x^p}{(1+x)^2} \quad \to \quad \int_{-\Lambda}^{+1} dx \, \frac{x^p}{(1+x)^2}, \tag{39}$$

$$\int_{-1}^{+1} dx \, \frac{x^p}{(1-x)^2} \quad \to \quad \int_{-1}^{+\Lambda} dx \, \frac{x^p}{(1-x)^2}, \tag{40}$$

$$\int_{-1}^{+1} dx \, \frac{x^p}{1+x^2} \quad \to \quad \int_{-\Lambda}^{+\Lambda} dx \, \frac{x^p}{1+x^2}, \tag{41}$$

and relate the cut-off to the physical system.

On solving the integrals (39) - (41), one has two common logarithmic factors appearing:

$$\log(\frac{1+\Lambda}{1-\Lambda}) \to \log \Lambda_{c1} > 0 \tag{42}$$

and

$$\log(\frac{1-\Lambda}{2}) \to \log \Lambda_{c2} < 0. \tag{43}$$

The physical origin of the divergences is clear: In the plasma phase, both quarks and gluons have long range Coulombic forces. But, in the presence of other charged (ie. color or electric) particles, these forces are modified. In the case of electrically charged particles the modification is always a screening with an effective range given by the Debye length λ_D; with the introduction of color charges, anti-screening is another possibility [20]. In order to relate the cut-off to the plasma parameters, one needs to understand characteristic plasma length scales:

- the Debye screening length λ_D and
- the Landau length λ_L.

Traditionally [21] the Coulomb logarithm is then given by

$$\log \Lambda_c = \log(\frac{\lambda_D}{\lambda_L}). \tag{44}$$

Since the Debye length is given by

$$\lambda_D^2 = \frac{T}{(4\pi)^2 \alpha_s n} \tag{45}$$

and the Landau length by

$$\lambda_L = \frac{\alpha_s}{T}, \tag{46}$$

one finds

$$\frac{\lambda_D}{\lambda_L} = \left(\frac{T^3}{(4\pi)^2 \alpha_s^3 n}\right)^{\frac{1}{2}} \simeq \alpha_s^{-\frac{3}{2}}, \tag{47}$$

or

$$\log \Lambda_c = \frac{3}{2} \log(\frac{1}{\alpha_s}) \simeq 3.45. \tag{48}$$

From the integrals (39) - (41) it can be seen that the identification

$$\log \Lambda_c = \log \Lambda_{c1} \tag{49}$$

is justified, giving an estimate for the cut-off Λ:

$$\Lambda \simeq \frac{1 - \alpha_s^{\frac{3}{2}}}{1 + \alpha_s^{\frac{3}{2}}} \simeq 0.94 \tag{50}$$

and therefore

$$\log \Lambda_{c1} \simeq 3.45 \tag{51}$$

$$\log \Lambda_{c2} \simeq -3.50. \tag{52}$$

This intuitive calculation should be analysed more carefully, especially with the underlying non abelian character of a proper quark gluon plasma in mind. It is however beyond the scope of this paper to pursue this matter in greater detail.

References

[1] J.Cleymans, R.V. Gavai and E. Suhonen, Phys.Rep. **130**(1986) 217,
E.V. Shuryak, *"The QCD Vacuum, Hadrons and the Superdense Matter"*, World Scientific 1988

[2] S.R. de Groot, W.A. van Leeuwen and C.G. van Weert, *"Relativistic Kinetic Theory"*, North-Holland 1980

[3] J.M. Stewart, *"Non-Equilibrium Relativistic Kinetic Theory"*, Springer Lecture Notes in Physics **10**(1971)

[4] S. Gavin, Nucl.Phys. **A435**(1985) 826

[5] G. Baym, Phys.Lett. **B138**(1984)18

[6] P. Carruthers and F. Zachariasen, Rev.Mod.Phys. **55**(1983) 245

[7] H.T. Elze and U. Heinz, Phys. Rep. **183** (1989) 82
S. Mrowczynski, preprint Univ. Regensburg TPR-89-9 (March 1989)

[8] W. Czyz and W. Florkowski, Acta Phys.Pol. **B17**(1986) 819
S. Mrowczynski, Acta Phys.Pol. **B19**(1988) 91

[9] A. Bialas and W. Czyz, Phys.Rev. **D30**(1984) 2371
K. Kajantie and T. Matsui, Phys.Lett. **B164**(1985) 373

[10] U. Heinz, Ann. Phys. **161** (1985) 48

[11] L.D. Landau and E.M. Lifshitz, *"Fluid Mechanics"* , Pergamon Press 1963

[12] F. Reif, *"Fundamentals of Statistical and Thermal Physics"* , McGraw-Hill 1965

[13] D.W. von Oertzen and J. Cleymans, S.Afr.Jour.Phys. **13** (1990) 91

[14] G. Baym, H. Monien, C.J. Pethick and D.G. Ravenhall, Phys. Rev. Lett. **64** (1990) 1867

[15] P. Danielewicz and M. Gyulassy, Phys.Rev. **D31**(1985) 53

[16] A. Hosoya and K. Kajantie, Nucl.Phys. **B250**(1985) 666

[17] E. Leader and E. Predazzi, " *An Introduction to Gauge Theories*", Cambridge University Press, 1982

[18] B.L. Combridge, J. Kripfganz and J. Ranft, Phys. Lett. **B 70** (1977) 234

[19] F. Halzen and A. Martin, "*Quarks and Leptons*" John Wiley and Sons, 1984

[20] C.J. Pethick, G. Baym and H. Monien, Nucl. Phys. **A 498** (1989) 313c
C. Gale and J. Kapusta, Phys. Lett. **B 198** (1987) 89

[21] F.F. Chen, "*Introduction to Plasma Physics and Controlled Fusion*"
Second Edition, Plenum Press 1983

Strangeness Production in Heavy-Ion Collisions*

J. Cleymans[1], *H. Satz*[2], *E. Suhonen*[3], *and D.W. von Oertzen*[1]

[1]Department of Physics, University of Cape Town, South Africa
[2]CERN, CH-Geneva 23, Switzerland
[3]Department of Theoretical Physics, University of Oulu, Finland

1 Introduction

Strangeness has been proposed as a signal for quark-gluon plasma formation in relativistic ion collisions [1,2]. It has also been proposed on a variety of occasions that stable multi-strange states could exist in nature, viz. strange matter or strangelets [3,4,5,6]. Subsequent to the original proposal several papers appeared which considerably weakened the early claims made for strangeness production in heavy ion collisions [7,8,9,10]. Representative of many I quote a statement taken from Lee et al. [10]: "... we conclude that there is no natural large difference in flavor composition between the (...) quark-gluon plasma and an *equilibrium* hadron gas.". It thus became clear that if both the hadronic system and the quark-gluon plasma phase are in chemical equilibrium then the amount of strangeness in both phases is similar. However, it was pointed out that, due to non-equilibrium effects one could still expect a large enhancement of strangeness. In the quark-gluon plasma phase the equilibration rate for strange quarks is fast because the strange quark mass is not very large compared to the typical strong interaction scale. In the hadronic phase however the equilibration rate for strange mesons is expected to be slower because the masses involved are higher [11]. Since estimates of non-equilibrium behaviour are much more model dependent it becomes unclear what to expect.

In 1987 the Quark Matter Conference was held at Schloß Neukirchen in West Germany [12]. It was the first conference where experimental data were presented on the production of strange particles in ultra-relativistic ion collisions. The E802 collaboration [13] working at the Brookhaven National Accelerator Center announced the following ratios $K^+/\pi^+ = 19.2 \pm 3\%$ and a $K^-/\pi^- = 3.6 \pm 0.8\%$. These ratios were compared to the corresponding results observed in $p-p$ collisions where especially the K^+/π^+ ratio is much smaller. The question immediately arose: "Does this indicate quark-gluon plasma formation or not?". This talk will be mainly concerned with answering this question. I will first explain on an intuitive level what behaviour should be expected for the K/π ratios and then go into more detail presenting the chemically equilibrated hadronic gas model and a discussion of strangeness conservation in the canonical model.

---

*Talk presented by J. Cleymans at the Summer School in Theoretical Physics "Phase Structure of Strongly Interacting Matter.", Cape Town, January 8-19 ,1990

Phase Structure of Strongly Interacting Matter
Editor: J. Cleymans

2 The K/π Ratios

The basic idea behind the K/π ratios is as follows. In hot hadronic matter we expect strangeness to be suppressed because the strange quark has a heavier mass than the u and d quarks. The number of u and d quarks will be large because we start with dense baryonic matter; at the same time the number of anti-u and anti-d quarks is suppressed by the chemical potential. We thus have very roughly the situation outlined by the equations below

$$\begin{aligned} \frac{K^+}{\pi^+} &\sim \frac{\bar{s}u}{\bar{d}u} \\ &\sim \frac{\bar{s}}{\bar{d}} \\ &\sim \frac{\exp(-m_s/T)}{\exp(-\mu/T)} \end{aligned} \tag{1}$$

while for the corresponding K^-/π^- ratio one has

$$\begin{aligned} \frac{K^-}{\pi^-} &\sim \frac{\bar{u}s}{\bar{u}d} \\ &\sim \frac{s}{d} \\ &\sim \frac{\exp(-m_s/T)}{\exp(\mu/T)} \end{aligned} \tag{2}$$

As one increases the baryon density, the chemical potential μ increases correspondingly; therefore, one expects the K^+/π^+ ratio to increase, and the K^-/π^- ratio to decrease. This is indicated in figure 1.

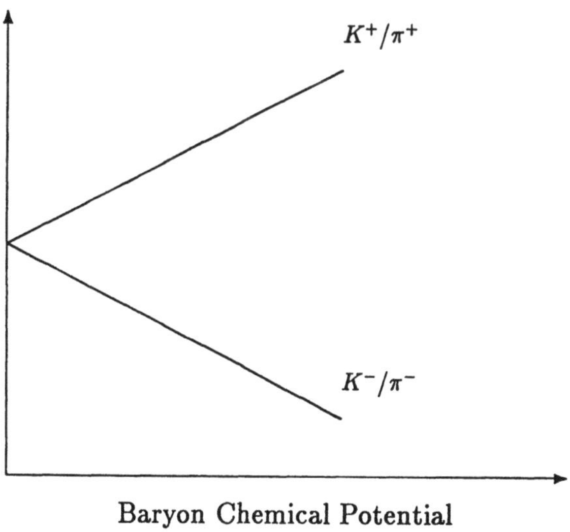

Figure 1: Qualitative behaviour of the K/π ratios as a function of baryon chemical potential (or baryon density).

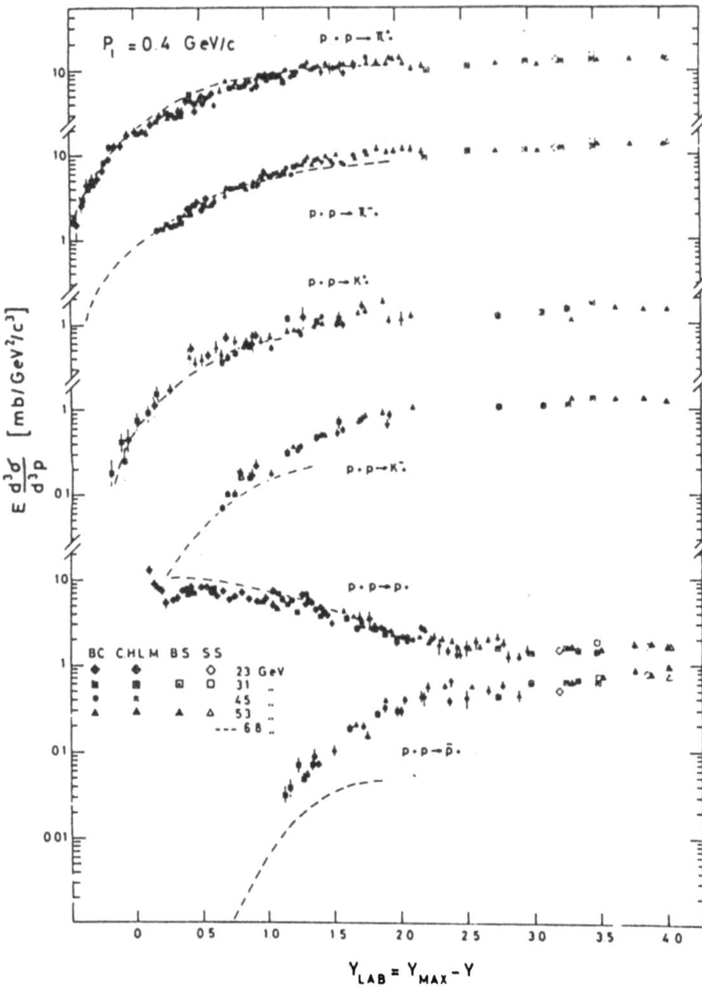

Figure 2: ISR data showing the plateau in the central region of rapidity and the particle yields.

Let us now compare this theoretical expectation with the existing experimental situation. To find data at zero baryon density we go to the central region in rapidity space, because it is known from ISR data obtained in the 1970's that, to a very good approximation equal numbers of protons and antiprotons are produced there. This is shown in figure 2, which is taken from a compilation of ISR data published by Giacomelli and Jacob [18].

We see that, in this baryon-free region, both the K^+/π^+ and the K^-/π^- ratios are equal to each other and approximately equal 0.11. This gives us the reference point at zero chemical potential. We note that data on p-p scattering at lower energies generally leads to a smaller K^+/π^+ ratio, this could be attributed to a smaller interaction volume as has been proposed by Hagedorn [17].

To find data at a larger baryon density we consider the data obtained by the E802 collaboration [13] from Silicon-Gold scattering at the Brookhaven National Laboratory where the following results were obtained : $K^+/\pi^+ = 19.2 \pm 3\%$ and $K^-/\pi^- = 3.6 \pm 0.8\%$ at 14.6 GeV per nucleon. We thus have confirmation of the expected qualitative behavior. The results of the E802 collaboration [13] therefore support the statistical description of matter described above.

We would like to see whether this picture can be used to obtain more quantitative results for the K/π ratios. In particular we want to see whether we can understand these experimental results in a purely hadronic gas description. Previous calculations led to a much higher prediction for the K^+/π^+ ratio [19] than the experiments yielded.

3 Chemically Equilibrated Hadronic Gas

To analyze the K/π ratios we consider a simple thermodynamic model of hadronic matter [21], in which the state is specified by the temperature T and the baryon number density n_B, with zero overall strangeness. Such matter will have a certain composition of particle species, characterized by corresponding "chemical" potentials. In addition to the baryon number chemical potential μ_B, we have to impose overall strangeness conservation. This can be done by introducing a strangeness potential μ_S, [22] since particles of a given strangeness can have different baryon numbers (e.g., $\bar{K}^0$ and Λ). The value of μ_B is fixed by giving the overall baryon number density n_B, and that of μ_S by fixing the overall strangeness, which in our case will be zero. The number densities for all hadron states are then given in terms of μ_B and T .

To keep our model simple, we include only the essential interaction features of a multi-hadron system near freeze-out: resonance production and baryon repulsion. The bulk of meson production is known to take place through intermediate resonance states, and we therefore start with an ideal gas containing all the observed non-strange and strange mesonic and baryonic resonances [20] up to a mass of 2 GeV. To obtain the distributions of the actually observed hadrons, in particular pions and kaons, we then let these resonances decay according to the measured branching ratios. For small or vanishing baryon number, this would suffice to fix our model; at high n_B, however, a strong, short-range repulsion sets in between baryons, even if they have different quantum numbers; this can be described in terms of a hard-sphere picture [23]. As a consequence, the baryon number density n_B becomes

$$n_B = \frac{n_B^0}{[1 + V n_B^0]} \tag{3}$$

where n_B^0 denotes the density calculated for an ideal gas of pointlike baryons. We note that for $n_B^0 \to \infty$, $n_B \to 1/V$, i.e., we have dense packing of hard-sphere baryons with an intrinsic volume V. In the case of several baryon species $\alpha = 1, 2, ..., r$, this becomes

$$n_B = \frac{n_B^0}{[1 + \sum_{\alpha=1}^{r} V_\alpha n_\alpha^0]} \tag{4}$$

using a baryon volume V_α proportional to the mass as suggested, e.g., by the bag

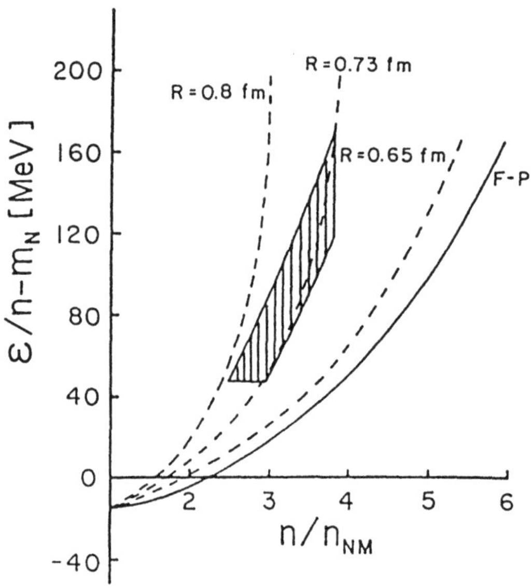

Figure 3: Comparison of different equations of state for dense nuclear matter with experimental estimates .

model. Here we take $V = 4\pi R^3/3$, with $R = 0.5$ and 0.8 as indicative radii for the nucleon. In addition, we shall also consider the case where R=0, to check the role baryon repulsion plays in determining particle number ratios. This completes our picture. The resulting equation of state is compared [24] with experimental estimates in figure 3.

Other proposals have been put forward in particular it has been proposed [1] that the correction be proportional to the energy density instead of the baryon density. It is also known that these proposals are not thermodynamically consistent, in the sense that $-\partial\Omega/\partial\mu = N$ is not satisfied. This question has been addressed recently by Kouno and Takagi [25] and also by Suhonen and Sohlo [26]; a field theoretic treatment based on the Walecka model has also been considered recently [27].

To illustrate the resulting pattern, we note that

$$n^0_{K^\pm} = (2\pi)^{-3} \int d^3p \, \frac{1}{[e^{(E_K \mp \mu_S)/T} - 1]} \tag{5}$$

gives the number density of *thermally produced* kaons, and

$$n^0_\Lambda = (2\pi)^{-3} \int d^3p \, \frac{1}{[e^{(E_\Lambda - (\mu_B - \mu_S))/T} + 1]} \tag{6}$$

that of *thermal, pointlike* Λ's; here $E_K = (\mathrm{p}^2 + m_K^2)^{1/2}$ and $E_\Lambda = (\mathrm{p}^2 + m_\Lambda^2)^{1/2}$. The *actual* production rates are obtained from this by including baryon repulsion according to eq.(2) and by adding to eqs. (3) and (4) the corresponding values from resonance production, such as $K^{0*} \to K^+ + \pi^-$, using the experimentally determined branching ratios.

Requiring the difference between all strange particle number densities to vanish,

$$\sum_i n_i^{(S)} - \sum_i n_i^{(\bar{S})} = 0 \tag{7}$$

fixes μ_S and thus gives us all particle densities in terms of T and μ_B. Summing at fixed μ_B over all baryon contributions,

$$\sum_i n_i^{(B)} - \sum_i n_i^{(\bar{B})} = n_B \tag{8}$$

determines the overall baryon number density n_B. Note that in 8 we sum over the *physical* baryon densities as defined by 4.

Before we turn to the numerical K/π ratios resulting from this model, let us discuss qualitatively the expected behaviour. At $n_B = 0$, we have complete particle-antiparticle symmetry and hence $K^+/K^- = \pi^+/\pi^- = 1$. With increasing n_B, Fermi repulsion forces baryons of a given species to acquire higher momenta and this eventually makes certain processes, e.g. $p \rightarrow \Lambda\ K^+$, energetically favourable. At low temperatures, this results in an increase of K^+/π^+ with n_B. Since processes such as $n \rightarrow \Delta^0$, with the subsequent decay $\Delta^0 \rightarrow p\ \pi^-$, enhance π^- production, but no similar effect exists for antikaons, the K^-/π^- ratio decreases with increasing n_B. For higher temperatures, and hence more available kinetic energy, the role of Fermi repulsion and thus also the variation of K^+/π^+ and K^-/π^- with n_B is reduced.

When the baryon density is increased still further, more higher mass baryon states come into play. These - although some also lead to further kaon production - on the whole enhance pion more than kaon rates. This means that eventually the initial K^+/π^+ gain is overcome, and also this ratio starts falling with n_B. On the other hand, strange baryon resonances now lead to further antikaon production, so that the decrease of the K^-/π^- ratio slows down. At still higher densities one eventually reaches the dense packing limit.

We now discuss our actual calculations. In figure 4 we show the dependence of the K/π ratios on the hard-core radius R as a function of baryon density. As one

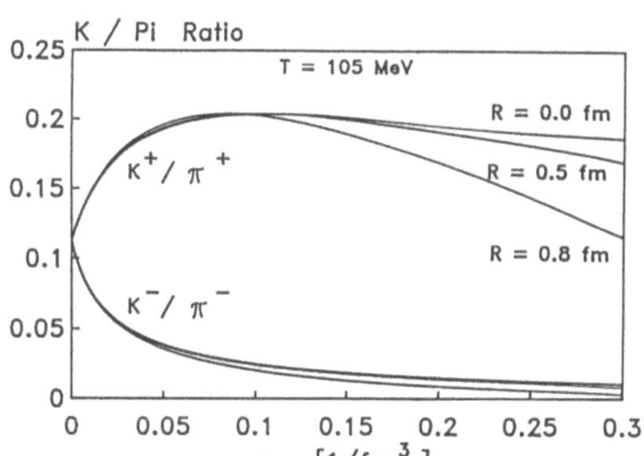

Figure 4: Dependence of the K/π ratios on the hard core radius R as a function of baryon density.

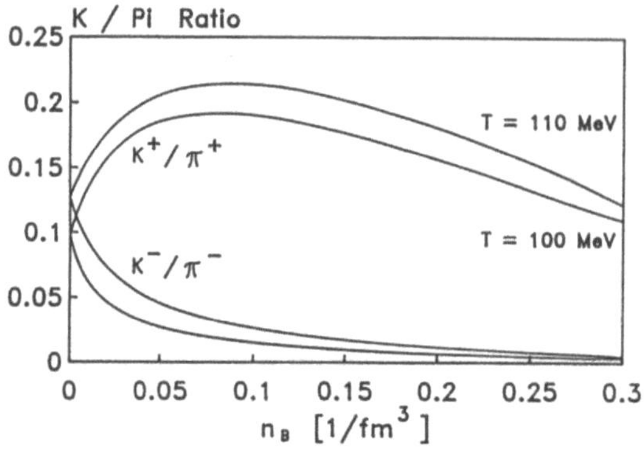

Figure 5: Dependence of the K/π on the baryon density.

can see the K^+/π^+ratio initially rises with increasing baryon density, it then reaches a plateau and starts decreasing. For moderate densities the K^+/π^+ ratio is almost independent of the hard-core radius and after reaching a maximum the ratio becomes very strongly dependent on this radius. The K^-/π^- ratio always decreases for increasing baryon densities.

Figure 5 shows the K/π ratios in their T and n_B dependence for the range of relevance for the BNL data. As seen the data fall into the region of $93 \leq T \leq 112$ MeV and $0.02 \leq n_B \leq 0.12$ fm^{-3}; a fit gives $T \simeq 105$ MeV and $n_B \simeq 0.05$ baryons/fm^3 as freeze-out parameters for a baryon radius of $R = 0.8$ fm.

To test for the dependence of these results on the finite volume corrections we show these results again in figure 6 together with different choices for volume corrections. These include the purely geometric volume correction, a volume correction depending only on the overall baryon density of the gas and one which depends on the energy density [1] of the gas. We also include the case of no volume corrections at all. As can be seen from figure 5, the influence of the different volume corrections is minimal in the low baryon density region (corresponding to the freeze-out value). This is consistent with the assumptions of the model. If the baryon density would have been extremely high the gas would still continue existing and only freeze out much later. These results lead to a temperature which is about ten to twenty percent lower than the temperature value deduced from the transverse momenta spectra

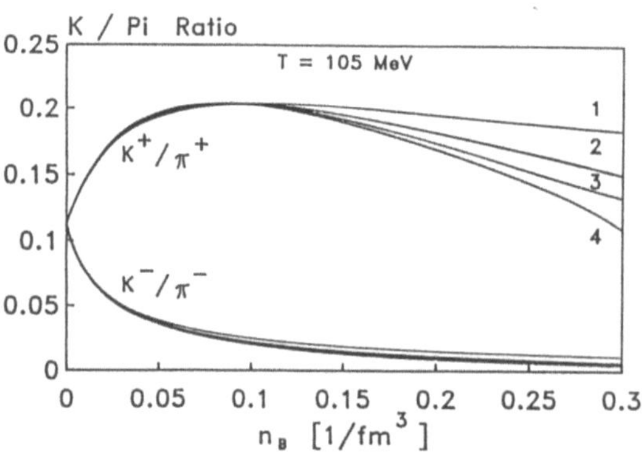

Figure 6: Dependence of the K/π ratios on different choices for the volume corrections.

presented by the E802 collaboration [13]. As our model is crude, this could be due to a variety of factors, such as non-equilibrium, hydrodynamic expansion, interaction between hadrons etc.. In view of all this it is surprising that we come to such a close agreement with the results obtained from the transverse momenta spectra.

The obvious question arises as to how the heavy ion collision results compare to $p-p$ collisions . The K/π ratio is always very sensitive to kinematic factors like transverse momentum, rapidity, total energy. One reliable point seems to be the ISR meassurement in the central region, since this corresponds very closely to zero baryon density and also remains constant over the whole central rapidity region. The ratio measured is 11% for both K/π ratios. This is reproduced by our results. One should remember that finite size effects could be appreciable in $p-p$ collisions [17].

A similar calculation has been performed recently [28] using smaller number of hadrons and without considering their decay products. The following result is obtained :

$$T = 118\text{MeV} \qquad n_B \sim 0.07/fm^3 \tag{9}$$

We view these results as being compatible with ours in view of the fact that they only included a subset of our particle input. We therefore conclude that it is possible to understand the K/π ratios in a statistical picture and that finite volume corrections are not relevant at present. Due to the low value for the temperature, deviations from chemical equilibrium cannot be excluded. The K/π ratios have also been analysed [29] using a covariant microscopic approach.

4 Canonical Ensemble

It was pointed out by Hagedorn [17] some twenty years ago that the production of heavy particles in high energy proton-proton collisions calls for the use of the canonical ensemble. In particular, he showed that the production of anti- He^3 is wrong by seven orders of magnitude when the grand canonical ensemble in its standard form is used instead of the canonical ensemble. The reason for this is that the number of particles as well as the interaction volume are very small and therefore a more correct, albeit much more complicated, statistical treatment is called for. The original presentation of Hagedorn [17] has been considerably developed and expanded by Redlich and Hagedorn [31] where both the baryon number and strangeness were exactly conserved. In the grand canonical ensemble for the density of K-mesons :

$$n_K^+ = \int \frac{d^3p}{(2\pi)^3} \exp\left(-\frac{E_{K+}}{T} + \frac{\mu_s}{T}\right) \tag{10}$$

while in the canonical ensemble for a small system one has:

$$n_K^+ = \left[\int \frac{d^3p}{(2\pi)^3} \exp\left(-\frac{E_{K+}}{T}\right)\right] V \int \frac{d^3p}{(2\pi)^3} \exp\left(-\frac{E_{K-}}{T}\right) \tag{11}$$

i.e. one sees explicitly that strangeness is conserved because each time one asks for a K^+ one must also balance the strangeness by a K^-. Also the density will be suppressed because one has two exponential factors. Strangeness conservation requires that at least two kaons are produced. For a small system this makes a substantial difference. So the question arises how does one incorporate strangeness exactly? We would like to present results in a form appropriate to present heavy ion collisions. It is certainly not the case that everything in such a collision will be either in the form of a quark-gluon plasma or a hadronic gas. We therefore want to treat baryon number in a grand canonical way since some of the protons and neutrons will not take part in the thermalised gas. We thus consider a heat bath of baryons. On the other hand strangeness will be zero everywhere and we want to treat it exactly, i.e. canonically. This type of treatment has been considered first by Rafelski and Danos [30] some ten years ago. Their analysis was limited to particles having strangeness one, zero or minus one. It is also possible to include particles with higher strangeness.

For completeness we will repeat some of the main results of Hagedorn and Redlich and show how they are modified in the case where baryon number is treated grand canonically and strangeness exactly.

5 Exact Strangeness Conservation

Let us consider a gas composed of two kinds of particles namely K^+'s and K^-'s and request that the overall strangeness be zero. The partition function becomes:

$$\begin{aligned} Z = \; & \sum_{n_0^+=0}^{\infty} \frac{1}{n_0^+!} e^{-\frac{\epsilon_0^+ n_0^+}{T}} \cdots \sum_{n_\infty^+=0}^{\infty} \frac{1}{n_\infty^+!} e^{-\frac{\epsilon_\infty^+ n_\infty^+}{T}} \\ & \sum_{n_0^-=0}^{\infty} \frac{1}{n_0^-!} e^{-\frac{\epsilon_0^- n_0^-}{T}} \cdots \sum_{n_\infty^-=0}^{\infty} \frac{1}{n_\infty^-!} e^{-\frac{\epsilon_\infty^- n_\infty^-}{T}} \\ & \delta_{n_0^+ + n_1^+ + \cdots + n_\infty^+ \, , \, n_0^- + n_1^- + \cdots + n_\infty^-} \end{aligned} \tag{12}$$

The Kronecker delta function insures that the overall strangeness is zero. To disentangle the summations one introduces an extra integral:

$$\frac{1}{2\pi} \int_0^{2\pi} d\phi \; e^{i(n-m)\phi} = \delta_{n,m} \tag{13}$$

It is then possible to perform all the summations and one obtains the following result

$$Z = \frac{1}{2\pi} \int_0^{2\pi} d\phi \exp\left\{ V \int \frac{d^3p}{(2\pi)^3} e^{-\frac{E^+}{T} - i\phi} \right\} \exp\left\{ V \int \frac{d^3p}{(2\pi)^3} e^{-\frac{E^-}{T} + i\phi} \right\} \tag{14}$$

or, using a more compact notation

$$Z = \frac{1}{2\pi} \int_0^{2\pi} d\phi \exp(Z^1_{K^+} e^{-i\phi} + Z^1_{K^-} e^{i\phi}) \tag{15}$$

where we have introduced the single-particle partition functions

$$Z^1_{K^\pm} \equiv V \int \frac{d^3p}{(2\pi)^3} e^{-\frac{E^\pm}{T}} \tag{16}$$

Expanding each exponential in a power series it is possible to find the integrals in a straightforward manner. The final result is

$$Z = \sum_{n=0}^{\infty} \frac{1}{(n!)^2} (Z^1_{K^+})^n (Z^1_{K^-})^n \tag{17}$$

where one recognizes the zeroth modified Bessel function:

$$Z = I_0(2\sqrt{Z^1_{K^+} Z^1_{K^-}}) \tag{18}$$

The generalization of the above to a gas containing several types of particles is straightforward. It is by now textbook material [32]. For a gas containing particles with strangeness plus one, zero and minus one:

$$Z = \frac{1}{2\pi} \int_{-\pi}^{\pi} d\phi \exp(S_1 e^{i\phi} + S_{-1} e^{-i\phi} + S_0) \tag{19}$$

here S_1 stands for the sum of all single particle partition function of all particles having strangeness plus one :

$$S_1 = Z^1_{\bar{K}} + Z^1_{\Lambda} + Z^1_{\bar{K}^*} + ... \tag{20}$$

while S_{-1} stands for the sum of all single particle partition functions having strangeness minus one :

$$S_{-1} = Z^1_{K} + Z^1_{\bar{\Lambda}} + Z^1_{K^*} + ... \tag{21}$$

As an illustration we quote the explicit form of Z_{Λ} :

$$Z_{\Lambda} = \frac{V}{(2\pi)^3} \int d^3p \exp((-E_{\Lambda} + \mu_B)/T) \tag{22}$$

with μ_B being the baryon chemical potential. This equation makes it clear that baryon number is being treated grand canonically. This is necessary because the baryon density is varying very strongly as a function of the kinematic variables of the final state products. In the central region of rapidity space for example it is known to be very close to zero. To calculate the above partition function more explicitly we expand each term in a power series :

$$Z = Z_0 \frac{1}{2\pi} \int_{-\pi}^{\pi} d\phi \sum_{m=0}^{\infty} \sum_{n=0}^{\infty} \frac{1}{m!} \frac{1}{n!} S_1^m S_{-1}^n \exp(im\phi) \exp(-in\phi) \tag{23}$$

where Z_0 is the standard partition function for all particles having zero strangeness. After the integral over ϕ we are left with:

$$Z = Z_0 \sum_{n=0}^{\infty} \frac{1}{(n!)^2} (S_1 S_{-1})^n \tag{24}$$

which is the result of reference [30] . In each term of expression 5 one sees explicitly strangeness conservation at work namely each term in the sum is the product of a strangeness plus one particle multiplied by a strangeness minus one particle. The above result can alternatively be written in integral form. This can be obtained in a

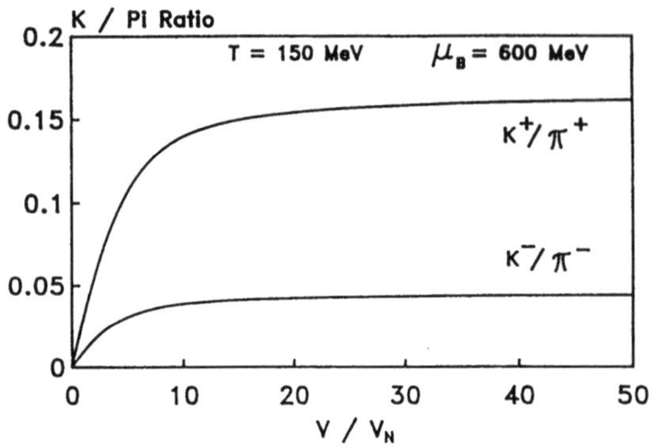

Figure 7: Dependence of the K/π ratios as a function of the interaction volume in the canonical ensemble.

straightforward manner from equation (1):

$$Z = Z_0 \frac{1}{\pi} \int_0^\pi d\phi \ \cos[(S_1 - S_{-1}) \sin \phi] \exp((S_1 + S_{-1}) \cos \phi) \tag{25}$$

this reduces to the expression considered by Hagedorn and Redlich in the case where $S_1 = S_{-1}$. In the present case this reduction does not happen because in the grand canonical ensemble these two expressions are not necessarily equal. The resulting K/π ratios are shown in figure 7 . As one can see for large volumes the results are the same as those obtained in the grand canonical ensemble.

6 Conclusions

We conclude with the following remarks:

1. The K/π ratios fit nicely in a statistical picture of particle production. The BNL data are consistent with the production of a chemically equilibrated hadronic gas.

2. Finite volume corrections are not relevant at present.

3. Kaon production rates are compatible with a chemically equilibrated hadronic gas model. The resulting temperature is lower than the one obtained from transverse momenta spectra by about twenty percent.

4. The canonical ensemble description leads to a strong dependence on the interaction volume for small systems.

Clearly, the observed signals do not yet give evidence for quark-gluon plasma formation. One should remember that the observed hadrons are produced at freeze-out time when the temperature has already dropped considerably and the baryon density is low.

References

[1] J. Rafelski and R. Hagedorn, in: Thermodynamics of Quarks and Hadrons, (ed. H. Satz), North-Holland (1981).

[2] J. Rafelski and B. Müller, *Phys. Rev. Lett.*48 (1982) 1066; **56** (1986) 2334[E].

[3] J.D. Bjorken and L.D. McLerran, *Phys. Rev.* **D20** (1979) 2353.

[4] E. Witten, *Phys. Rev.* **D30** (1984) 272.

[5] E. Farhi and R.L. Jaffe, *Phys. Rev.* **D30** (1984) 2379.

[6] C. Greiner, P. Koch and H. Stöcker, *Phys. Rev. Lett.* **58** (1987) 1825.

[7] K. Redlich, *Z. Phys.* **C27** (1985) 633.

[8] J. Kapusta and A. Mekjian, *Phys. Rev.* **D33** (1986) 1304.

[9] T. Matsui, B. Svetitsky and L. McLerran, *Phys. Rev.* **D34** (1986) 783 and 2047.

[10] K.S. Lee, M. Rhoades-Brown and U. Heinz, *Phys. Rev.* **C37** (1988) 1452.

[11] P. Koch, B. Müller and J. Rafelski, *Phys. Rep.* **142** (1986) 167.

[12] *Quark Matter '87* , Proceedings of the 6th International Conference on Ultra-Relativistic Nucleus-Nucleus Collisions, Eds. H. Satz, H.J. Specht and R. Stock, *Z. f. Physik* **C38** (1988).

[13] T. Abbott et al. *Phys. Rev. Lett.* (to be published, 1990).

[14] J. Harris et al. , NA35 Collaboration, *Nucl. Phys.* **A498** (1989) 133c; M. Gaździcki, *Nucl. Phys.* **A498** (1989) 375c.

[15] E. Quercigh, WA85 Collaboration, *Nucl. Phys.* **A498** (1989) 369c.

[16] A. Baldisseri et al., Annecy preprint LAPP-EXP-89-15.

[17] R. Hagedorn, CERN yellow report 71-12 (1971).

[18] G. Giacomelli and M. Jacob, *Phys. Rep.* **55** (1979) 1.

[19] N.K. Glendenning and J. Rafelski, *Phys. Rev.* **C31** (1985) 823.

[20] Particle Data Group, *Phys. Lett.* **204B** (1988) 1.

[21] J. Cleymans, H. Satz, E. Suhonen and D.W. von Oertzen, CERN-TH 5660/90 and UCT-TP 131/9

[22] P. Koch, J. Rafelski and W. Greiner, *Phys. Lett.* **123B** (1983) 151.

[23] J. Cleymans, K. Redlich, H. Satz and E. Suhonen, *Z. Phys.* **C33** (1986) 2341.

[24] J. Cleymans, E. Suhonen, *Z. f. Phys.* **C37** (1987) 51.

[25] H. Kouno and F. Takagi, *Z. Phys.* **C42** (1989) 209 and preprint TU/89/343, Sendai, June 1989.

[26] E. Suhonen and S. Sohlo, *J. Phys. G: Nucl. Phys.* **13** (1987) 1487.

[27] H.W. Barz, B.L. Friman, J. Knoll and H. Schulz *Phys. Rev.* **D40** (1989) 157.

[28] P. Lévai, B. Lukács and J. Zimányi, preprint KFKI-1989-47/A, Budapest.

[29] R. Mattiello, H. Sorge, H. Stöcker and W. Greiner, *Phys. Rev. Lett.* **63** (1989) 1459.

[30] J. Rafelski and M. Danos, *Phys. Lett.* **97B** (1980) 279.

[31] R. Hagedorn and K. Redlich, *Z. f. Physik* **C27** (1985) 541.

[32] J. Kapusta,*Finite-Temperature Field Theory*, Cambridge University Press, (1989).

[33] I. Lovas, K. Sailer and Z. Trócsányi, *J. Phys. G.: Nucl. Phys.* **15** (1989) 1709.

[34] U. Heinz, K.S. Lee and M. Rhoades-Brown, *Mod. Phys. Lett.* **A2** (1987) 153.

[35] D.B. Kaplan and A.E. Nelson, *Phys. Lett.* **B175** (1986) 57.

Dilepton Production in Nuclear Collisions*

C.A. Dominguez[1] *and M. Loewe*[2]

[1]Institute of Theoretical Physics and Astrophysics, University of Cape Town, Rondebosch 7700, Cape, South Africa
[2]Facultad de Física, Pontifica Universidad, Católica de Chile, Santiago, Chile
*Lecture presented by C.A. Dominguez

Dilepton production rates in nuclear collisions are calculated in order to study their sensitivity to the quark-gluon plasma and to the hadronic phases. This treatment differs from previous work on the subject in two respects: The width of the rho-meson, being exchanged in $\pi\pi$ annihilation, is made temperature dependent, thus taking into account resonance melting as the critical deconfinement temperature T_d is approached. Secondly, we study in addition to the standard scenario where chiral symmetry restoration and deconfinement occur at the same temperature ($T_c = T_d$), an alternative possibility where deconfinement preceeds chiral symmetry restoration ($T_d < T_c$). Results differ substantially from those obtained assuming a temperature independent rho-meson width, and $T_c = T_d$.

1 Introduction

Dilepton pair production in heavy ion collisions has been recognized as an important signal for quark-gluon plasma (QGP) formation [1,2]. Quantitative studies indicate that at high dilepton invariant mass the QGP phase dominates over the hadronic phase, while the latter is the dominant contribution to the production rate at low invariant mass. Depending on the order of the phase transition there is a broad (narrow) region of coexistence between the two phases, this region being broad (narrow) for a first order (second order) transition. Calculations have been performed [1,2] using Bjorken's scaling solution for longitudinal hydrodynamic expansion [3] in order to integrate the space-time dependence of the temperature. For a one-dimensional non-dissipative expansion this solution leads to entropy current conservation

$$s\tau = \text{const} \tag{1}$$

where s is the local entropy density and τ is the proper time. In the case of a first order phase transition, bag model equations of state are usually assumed. At low temperatures and low dilepton invariant masses it is reasonable to expect the hadronic phase to be dominated by $\pi\pi$ annihilation. In this case one may invoke Vector Meson Dominance (VMD) in order to calculate explicitly the contribution of this phase to the dilepton production rate. It has been customary to assume that the rho-meson resonance parameters (mass and width) are temperature independent (see e.g. [2]). In such a model, it has been found that for dilepton invariant masses below 1 GeV the production mechanism is dominated by the hadronic phase, while above 2.5 GeV for a first order phase transition, or 1.6 GeV for second order, the production mechanism

Editor: J. Cleymans

changes with the QGP dominating by up to two orders of magnitude. Hence, if the model is correct dilepton production can provide valuable information on quark-gluon plasma formation, as well as on the nature of the phase transition.

In this paper we reanalyze this important problem. Our approach differs from previous analyses in two respects, to wit. Firstly, we introduce a temperature dependence in the rho-meson width when calculating the hadronic phase contribution. Secondly, we study in addition to the standard case where chiral symmetry restoration and deconfinement occur at the same temperature, the alternative scenario where deconfinement precedes chiral symmetry restoration. The motivation for this is as follows. The rho-meson being a $\pi\pi$ resonance, or alternatively a quark-antiquark bound state, it is expected to melt as the temperature is increased from $T = 0$. At some critical temperature T_d the hadronic width should presumably become infinite, thus signalling quark deconfinement. In some sense the hadronic width can play the role of an order parameter. On the other hand, there is no a-priori reason to expect such behaviour for the resonance mass, which should be treated simply as a parameter. In fact, in chiral perturbation theory, for example, the calculated temperature dependence of the pion mass shows a smooth and negligible dependence on the temperature for $T \lesssim 300$ MeV [4]. The same is true for the temperature dependence of the nucleon and sigma-meson masses in the framework of the linear sigma model [5]. We shall assume this to be the case also for the rho-meson resonance. Coming back to the rho-meson width, a simple model illustrates the temperature behaviour expected on physical grounds. Starting from the kinematical definition of the hadronic width

$$\Gamma_\rho = \frac{g^2_{\rho\pi\pi}}{4\pi}\frac{M_\rho}{12} \tag{2}$$

invoking the KSFR relation [6]

$$g_{\rho\pi\pi} = \frac{M_\rho}{\sqrt{2}f_\pi} \tag{3}$$

and using the chiral perturbation theory expression for $f_\pi(T)$ up to order $0(T^2)$ [4]

$$f_\pi(T) = f_\pi(1 - T^2/8f_\pi^2) \tag{4}$$

one finds [7]

$$\Gamma_\rho(T) = \frac{\Gamma_\rho}{1 - T^2/4f_\pi^2} \tag{5}$$

With $f_\pi \equiv f_\pi(0) = 93$ MeV, the above equation implies a deconfinement temperature $T_d \simeq 2f_\pi \simeq 180$ MeV, quite in line with expectations. Whether this deconfinement temperature coincides with that for chiral symmetry restoration T_c, or instead $T_d < T_c$, is still an open problem. Lattice results [8] seem to favour the former scenario, while a QCD sum-rule estimate of T_d provides some indication that T_d might be smaller than T_c [9]. Hence, we shall study both possibilities here. The proper-time evolution of the system is quite different for the two scenarios. Assuming first order phase transitions, if $T_c = T_d$ the system starts at some high temperature T_o at time τ_o and cools down until it reaches a temperature $T_c = T_d$ at some later time τ_Q. This is the pure quark-gluon phase. From there on the system remains for some time at this fixed temperature, the quark-gluon phase coexisting with the hadronic phase. At some time τ_H the system cools down again until it reaches some final temperature T_f at a time

τ_f. This is the purely hadronic phase. However, if $T_d < T_c$ after cooling from T_o to T_c the system remains for some time in a mixed phase of quarks and gluons and "point" pions, i.e. pions with point interactions not resonating into a rho-meson resonance, with the temperature frozen at T_c. At a time τ_1 the system cools down reaching a temperature $T = T_d$ at some time τ_d, always in a mixed phase. With the temperature fixed at $T = T_d$ the system evolves until at a time τ_H it reaches the purely hadronic phase cooling then down to $T = T_f$ at time τ_f. It should be noticed that if $T_d < T_c$, then at the intermediate temperatures $T_d < T < T_c$ the plasma should still contain some "point" pions because $f_\pi(T)$ does not yet vanish in that region.

The main results of our calculation may be summarized as follows. The introduction of a temperature dependence on the rho-meson hadronic width produces a flattening of the dilepton production rate at low invariant masses. At the same time it enhances this rate (over the Γ_ρ = constant case) for intermediate and high dilepton invariant masses. Physically, this is a consequence of resonance broadening, which when integrated over the proper-time leads to enhancement. Unfortunately, this enhancement makes the hadronic phase contribution comparable to the QGP contribution in the case where $T_c = T_d$. Hence, the mere observation of dilepton production in heavy ion collisions at high dilepton invariant masses would no longer be such a clear signal for quark-gluon plasma formation in such a scenario. Nevertheless, a measurement of the shape of the production rate in the region around the rho-meson mass can provide valuable information on the hadronic phase. This spectrum should also tell us whether the system went through a one step ($T_c = T_d$) or a two step ($T_d < T_c$) mixed phase, as the dilepton production rate turns out to be different in the two different scenarios. Also, when $T_d < T_c$ the enhancement of the dilepton spectrum due to the temperature dependence of the rho-meson width is not as large as in the case $T_c = T_d$, and the QGP contribution at large dilepton invariant masses is an order of magnitude bigger than the hadronic. If this scenario is correct, then experimental data may still offer some insight into the QGP phase. This paper is organized as follows. In Section II we introduce the relevant formulae to be used in calculating the dilepton spectrum. In Section III we present our results and discuss them, paying particular attention to a comparison with previous work.

2 Dilepton Production Rates

The relevant formulae for the dilepton production rates in the different phases have been discussed in detail in [2]. For completeness, we reproduce in the sequel the main results and refer the reader to [2] for more details.
In the QGP phase one has

$$\frac{dN_Q}{d^4x\, d^4p} = \frac{\alpha^2}{4\pi^4}\left(1+2\frac{m^2}{M^2}\right)\left(1-4\frac{m^2}{M^2}\right)^{1/2} e^{-E/T} K_Q(p,T)\sum_i e_i^2 \qquad (6)$$

where N_Q is the number of dileptons, d^4x the infinitesimal space-time volume, p the momentum of the dilepton, M^2 its variant mass, m the lepton mass, α is the electromagnetic coupling, e_i the charge of the quark, and the function $K_Q(p,T)$ at zero chemical potential is given by

$$K_Q(p,T) = \frac{T}{p}\frac{1}{1-e^{-E/T}} \ln\left[\frac{(x_2+e^{-E/T})(x_1+1)}{(x_1+e^{-E/T})(x_2+1)}\right] \tag{7}$$

where

$$x_{1,2} = \exp(-E_\pm/T) \tag{8}$$

$$E_\pm = \frac{1}{2}(E \pm p) \tag{9}$$

Assuming VMD the dilepton production rate in the hadronic phase is

$$\frac{dN_H}{d^4x\,d^4p} = \frac{\alpha^2}{48\pi^4}\left(1+2\frac{\mu_\pi^2}{M^2}\right)\left[1-4\frac{\mu_\pi^2}{M^2}\right]^{3/2} e^{-E/T}|F_\pi(M^2,T)|^2\, K_H(p,T) \tag{10}$$

where μ_π is the pion mass, and

$$K_H(p,T) = \frac{T}{p}\frac{1}{1-e^{-E/T}} \ln\left[\frac{(x_2-e^{-E/T})(x_1-1)}{(x_1-e^{-E/T})(x_2-1)}\right] \tag{11}$$

with $E_\pm$ now given by

$$E_\pm = \frac{1}{2}\left[E\left(1+\frac{\mu_\pi^2}{M^2}\right) \pm p\left(1-\frac{\mu_\pi^2}{M^2}\right)\right] \tag{12}$$

The pion form factor in VMD is

$$|F_\pi(M^2,T)|^2 = \frac{M_\rho^4 + M_\rho^2\Gamma_\rho^2(T)}{(M^2-M_\rho^2)^2 + M_\rho^2\Gamma_\rho^2(T)} \tag{13}$$

For $\Gamma_\rho(T)$ we generalize Eq.5 and write

$$\Gamma_\rho(T) = \frac{\Gamma_\rho}{1-T^2/T_d^2} \tag{14}$$

In order to find the total rates one must integrate over the space-time volume of the system. In the framework of Bjorken's scaling model the relevant integration variables are the proper time τ, the space-time rapidity y and the transverse coordinates $x_\perp$, i.e. $d^4x = \tau\, d\tau\, dy\, d^2x_\perp$, with $E = \sqrt{M^2+p_\perp^2}\cosh y$. One must then specify how does the system cool and expand. This will in turn depend on whether $T_c = T_d$ or $T_d < T_c$. Let us consider first the case where $T_c = T_d$, i.e. there is only one phase transition which we assume in the sequel to be first order. Starting from the QGP phase at temperature T_o for a time τ_o the system cools according to

$$T = T_o\left(\frac{\tau_o}{\tau}\right)^{1/3} \tag{15}$$

The dilepton rate is then given by

$$\frac{dN_Q(\text{QGP})}{dM^2\,d^2p_T\,dY} = \frac{1}{2}\int_{\tau_o}^{\tau_Q} \tau\, d\tau \int_{y_{min}}^{y_{max}} dy\, \frac{dN_Q}{d^4x\,d^4p} \tag{16}$$

where the dilepton rapidity Y is

$$Y = \frac{1}{2} \ln \left(\frac{E + p_z}{E - p_z} \right) \tag{17}$$

and τ_Q is the time at which the system leaves the pure QGP phase to enter the mixed phase, where it will continue to expand but at the fixed temperature $T = T_c = T_d$. The quark-antiquark annihilation contribution to the rate is now given by

$$\frac{dN_Q(\text{mixed})}{dM^2\, d^2p_T\, dY} = \frac{1}{2} \int_{\tau_Q}^{\tau_H} \tau\, d\tau \int_{y_{min}}^{y_{max}} dy\, f(\tau) \frac{dN_Q}{d^4x\, d^4p} \tag{18}$$

with τ_H the time when the system leaves the mixed phase to enter the purely hadronic phase. The function $f(\tau)$ is the fraction of the entropy in the QGP phase

$$s(\tau) = f(\tau)\, s_Q + [1 - f(\tau)]\, s_H \tag{19}$$

Using bag model equations of state one has

$$s_Q = \frac{4}{3} \frac{37\pi^2}{30} T_c^3 \tag{20}$$

$$s_H = \frac{2\pi^2}{15} T_c^3 \tag{21}$$

hence

$$\frac{\tau_H}{\tau_Q} = \frac{s_Q}{s_H} = \frac{37}{3} \tag{22}$$

and τ_Q is obtained from 15 putting $T = T_c$. Using 1 one finds

$$f(\tau) = \frac{s_Q \tau_Q - s_H \tau}{\tau(s_Q - s_H)} \tag{23}$$

In this mixed phase we have in addition to 18 the contribution from the hadronic phase (pion annihilation) given by

$$\frac{dN_H(\text{mixed})}{dM^2\, d^2p_T\, dY} = \frac{1}{2} \int_{\tau_Q}^{\tau_H} \tau\, d\tau \int_{y_{min}}^{y_{max}} dy\, [1 - f(\tau)] \frac{dN_H}{d^4x\, d^4p} \tag{24}$$

Finally, for $\tau > \tau_H$ the system enters the purely hadronic phase obeying the cooling law

$$T = T_c(\tau_H/\tau)^{1/3} \tag{25}$$

The rate then becomes

$$\frac{dN_H(\text{hadron})}{dM^2\, d^2p_T\, dY} = \frac{1}{2} \int_{\tau_H}^{\tau_f} \tau\, d\tau \int_{y_{min}}^{y_{max}} dy\, \frac{dN_H}{d^4x\, d^4p} \tag{26}$$

where τ_f is the final time when the system reaches the freeze-out temperature T_f.

Turning to the alternative scenario where $T_d < T_c$, the system starts as before at a temperature T_o at time τ_o cooling down to $T = T_c$ at $\tau = \tau_Q$. In this pure QGP phase the rate is still given by 16 where

$$\tau_Q = \tau_o (T_o/T_c)^3 \tag{27}$$

but now $T_c \neq T_d$. The system then enters a mixed phase with a QGP contribution and a hadronic contribution from pion annihilation with point interaction, i.e. without ρ-meson resonance formation. The differential rate for the latter is then given by 10, but with $F_\pi(M^2, T) = 1$. This first mixed phase will last from $\tau = \tau_Q$ until $\tau = \tau_1$, where τ_1 is some intermediate time between chiral-symmetry restoration and deconfinement (a free parameter). The quark-antiquark annihilation contribution to the rate is

$$\frac{dN_Q(\text{mixed } 1)}{dM^2\, d^2p_T\, dY} = \frac{1}{2}\int_{\tau_Q}^{\tau_1} \tau\, d\tau \int_{y_{min}}^{y_{max}} dy\, f_1(\tau) \frac{dN_Q}{d^4x\, d^4p} \tag{28}$$

where

$$s(\tau) = f_1(\tau) s_Q + [1 - f_1(\tau)] s_{1QH} \tag{29}$$

and

$$s_{1QH} = s_Q \frac{\tau_Q}{\tau_1} \tag{30}$$

The function $f_1(\tau)$ can then be explicitly written as

$$f_1(\tau) = \frac{\tau_Q}{\tau}\left(\frac{1 - \tau/\tau_1}{1 - \tau_Q/\tau_1}\right) \tag{31}$$

The hadronic contribution becomes

$$\frac{dN_H(\text{mixed } 1)}{dM^2\, d^2p_T\, dY} = \frac{1}{2}\int_{\tau_Q}^{\tau_1} \tau\, d\tau \int_{y_{min}}^{y_{max}} dy [1 - f_1(\tau)] \frac{dN_H}{d^4x\, d^4p} \tag{32}$$

The system then undergoes further cooling to reach a temperature $T = T_d$ at a time $\tau = \tau_d$, with

$$T = T_c (\tau_1/\tau)^{1/3} \tag{33}$$

The rates are obtained from 28 and 32 but without the function $f_1(\tau)$ and the limits of integration in τ changed to $\tau = (\tau_1, \tau_d)$. From $\tau = \tau_d$ until $\tau = \tau_H$ the system expands at the constant deconfinement temperature $T = T_d$ where

$$\tau_d = \tau_1 \left(\frac{T_c}{T_d}\right)^3 \tag{34}$$

The function $f_1(\tau)$ in 28 and 32 is now different

$$f_2(\tau) = \frac{s_Q\, \tau_Q - s_H \tau}{\tau [s_Q \left(\frac{\tau_Q}{\tau_1}\right)\left(\frac{T_d}{T_c}\right)^3 - s_H]} \tag{35}$$

and the rates are given formally by 28 and 32 with $f_1(\tau)$ replaced by $f_2(\tau)$ and the integration region in τ becoming $\tau = (\tau_d, \tau_H)$. The hadronic rate is still calculated with $F_\pi(M^2, T) = 1$. Finally, for $\tau > \tau_H$ the system cools down according to

$$T = T_d (\tau_H/\tau)^{1/3} \tag{36}$$

until it reaches $T = T_f$ at $\tau = \tau_f$ (freeze-out point). Notice that now instead of 22 one has

$$\tau_H = \frac{37}{3}\tau_Q \left(\frac{T_c}{T_d}\right)^3 \tag{37}$$

In this final expansion $F_\pi(M^2, T) \neq 1$.

3 Results and Discussion

We have used the same values of the initial and final parameters chosen in [2] in order to make comparisons easier. We have thus taken $T_o = 284$ MeV, $\tau_o = 1f$, $T_f = 154$ MeV, and specified a beam of ^{16}O with incident energy of 225 GeV per nucleon, on a target of ^{195}Pt. The dileptons were chosen as dimuons. The values of y_{min} and y_{max} are: $y_{min} = -4.32$, and $y_{max} = 4.32$. All rates are to be multiplied by the factor 611 GeV^{-2} which is the effective collision area of ^{16}O [2]. We have studied the sensitivity of the rates to T_o and τ_o by allowing them to change within wide limits. We find that these variations produce a shift of the rate with little or no change in the shape of the spectrum. For the standard scenario ($T_c = T_d$) we choose two typical critical temperatures: $T_c = T_d = 200$ MeV and $T_c = T_d = 246$ MeV. For the second scenario ($T_d < T_c$) we consider $T_c = 200$ MeV, $T_d = 170$ MeV, and $T_c = 246$ MeV, $T_d = 200$ MeV, as a pair of representative choices. As for the time at which chiral-symmetry is restored, τ_1, we have allowed it to vary in the wide range $\tau_Q \lesssim \tau_1 \lesssim 20\tau_Q$, with $\tau_Q \approx 2.86f$ if $T_c = 200$ MeV, and $\tau_Q \approx 1.54f$ if $T_c = 246$ MeV. We find that this variation of τ_1 produces a shift of the dimuon rates with little or no deformation of the shape of the spectrum. Typical results of our calculation are shown in figures 1 and 12.

Figures 1 and 2 show the spectrum for $T_c = T_d = 200$ MeV, and $T_c = T_d = 246$ MeV, respectively, both for $F_\pi = F_\pi(M^2)$, i.e. no temperature dependence in Γ_ρ. Dashed

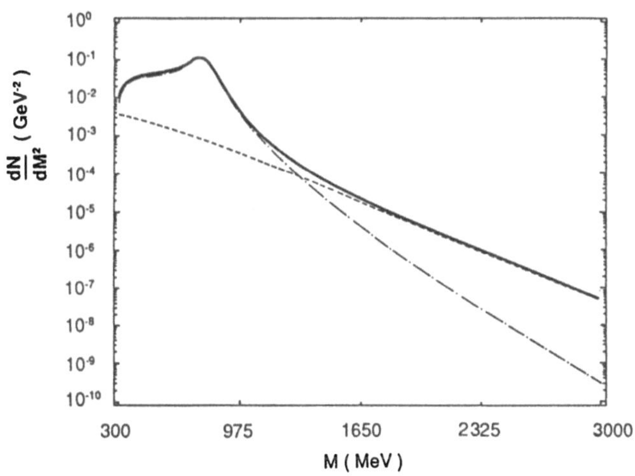

Figure 1: Dimuon spectrum for $T_c = T_d = 200$ MeV, and $F_\pi = F_\pi(M^2)$. Dashed and dash-dotted lines are the QGP and the hadronic contributions, respectively. Solid lines are the total.

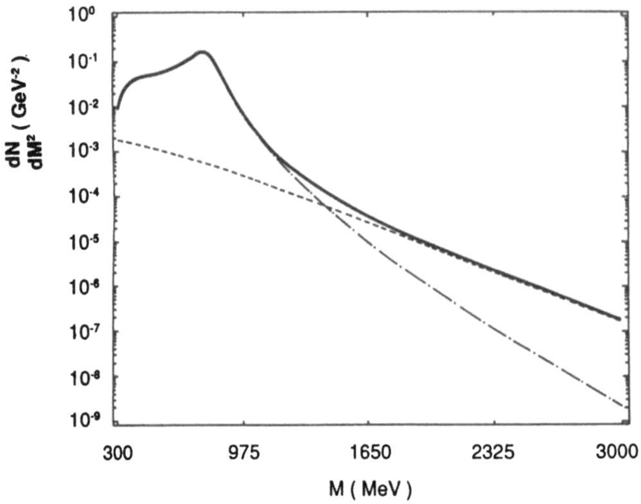

Figure 2: Dimuon spectrum for $T_c = T_d = 246$ MeV, and $F_\pi = F_\pi(M^2)$. Dashed and dash-dotted lines are the QGP and the hadronic contributions, respectively. Solid lines are the total.

curves are the QGP contributions, the dashed-dotted curves the hadronic pieces, and the full curves the total. This is the case considered in [2] and our results are in agreement. Notice that at high invariant mass the QGP contribution is some two orders of magnitude larger than the hadronic. The effect of the temperature dependence of F_π, through $\Gamma_\rho(T)$, is illustrated in figures 3 and 4, where $T_c = T_d = 200$ MeV and $T_c = T_d = 246$ MeV, respectively. At low M there is essentially no trace of the ρ-meson peak, and at high M the QGP contribution is comparable to the hadronic. This is most unfortunate in the sense that experimental data would not be offering much information on the QGP if the scenario $T_c = T_d$ is correct. The enhancement of the hadronic contribution over the case $F_\pi = F_\pi(M^2)$ is due to resonance broadening with increasing temperature. This can easily be understood by simply computing the area under $|F_\pi(M^2, T)|^2$, i.e.

$$\int_{\tau_H}^{\tau_f} |F_\pi(M^2, T)|^2 \, d\tau \tag{38}$$

and comparing it with the area under $|F_\pi(M^2)|^2$. The former is some two orders of magnitude larger than the latter, in the whole range of invariant masses being considered here. Nevertheless, a measurement of the spectrum can offer valuable information on the hadronic phase, in particular on the melting of the ρ-meson resonance. In figures 5 and 6 we show the spectrum in the vicinity of the ρ-meson mass for $T_c = T_d = 200$ MeV and $T_c = T_d = 246$ MeV, respectively. Curve (a) corresponds to $F_\pi = F_\pi(M^2)$ and curve (b) to $F_\pi = F_\pi(M^2, T)$. The effect of resonance broadening is quite a clear signal.

Turning to the alternative scenario where $T_d < T_c$, figure 7 shows the dimuon spectrum for $T_c = 200$ MeV, $T_d = 170$ MeV, and $F_\pi = F_\pi(M^2)$. The dashed curve is the QGP contribution, the dash-dotted line the hadronic contribution and the full line the total. Comparing this spectrum with figure 1 one can see that the height of the ρ-meson

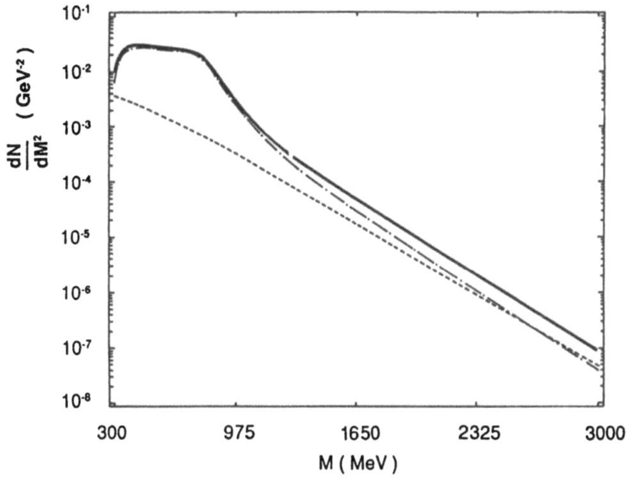

Figure 3: Dimuon spectrum for $T_c = T_d = 200$ MeV, and $F_\pi = F_\pi(M^2, T)$. Dashed and dash-dotted lines are the QGP and the hadronic contributions, respectively. Solid lines are the total.

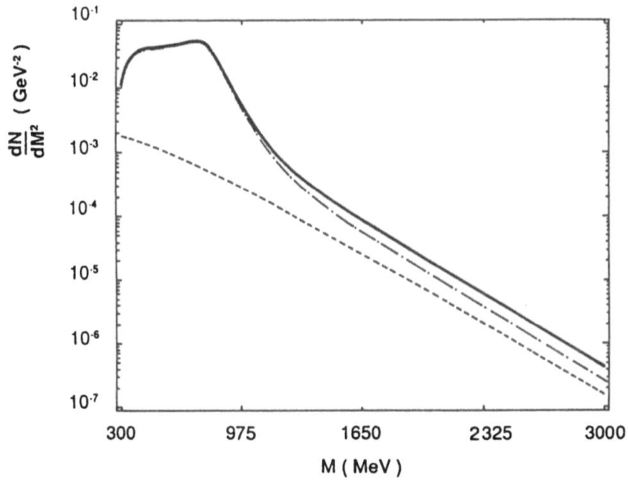

Figure 4: Dimuon spectrum for $T_c = T_d = 246$ MeV, and $F_\pi = F_\pi(M^2, T)$. Dashed and dash-dotted lines are the QGP and the hadronic contributions, respectively. Solid lines are the total.

peak is now reduced. At high M the QGP contribution dominates over the hadronic but only by one order of magnitude, instead of two orders of magnitude as in figure 1. These features prevail if $T_c = 246$ MeV and $T_d = 200$ MeV, always with $F_\pi = F_\pi(M^2)$ as illustrated in figure 8. Comparing with figure 2 we notice that now the total rate at high M is an order of magnitude bigger. Figure 9 shows the results for $T_c = 200$ MeV, $T_d = 170$ MeV, and $F_\pi = F_\pi(M^2, T)$. Comparing with figure 3 we notice that now the QGP contribution is one order of magnitude bigger than the hadronic at intermediate and high values of M. These features prevail if $T_c = 246$ MeV, $T_d = 200$ MeV, and

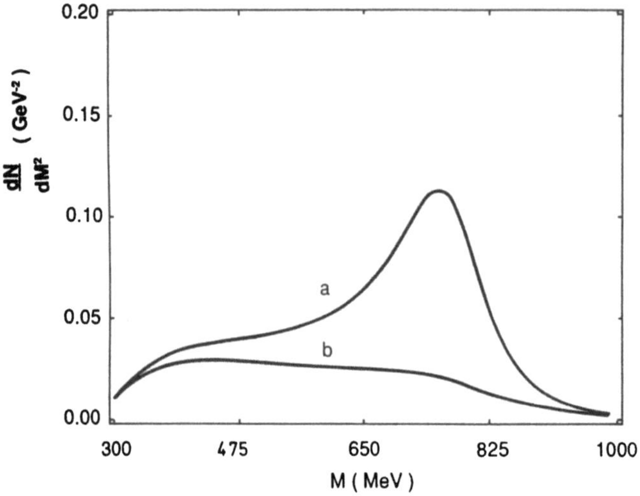

Figure 5: Dimuon spectrum for $T_c = T_d = 200$ MeV, and $F_\pi = F_\pi(M^2)$, (curve (a)), $F_\pi = F_\pi(M^2, T)$, (curve (b)).

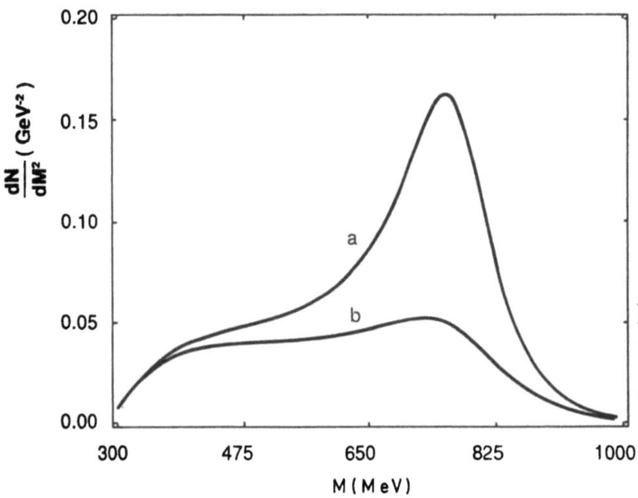

Figure 6: Dimuon spectrum for $T_c = T_d = 246$ MeV, and $F_\pi = F_\pi(M^2)$ (curve (a)), $F_\pi = F_\pi(M^2, T)$ (curve (b)).

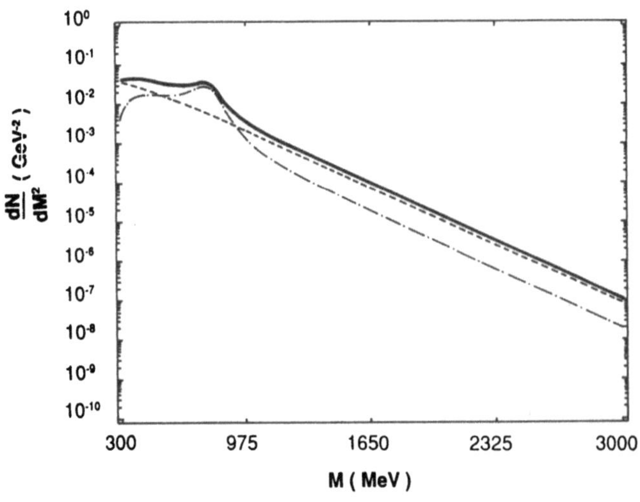

Figure 7: Dimuon spectrum for $T_c = 200$ MeV, $T_d = 170$ MeV, and $F_\pi = F_\pi(M^2)$. Dashed and dash-dotted lines are the QGP and the hadronic contributions, respectively. Solid lines are the total.

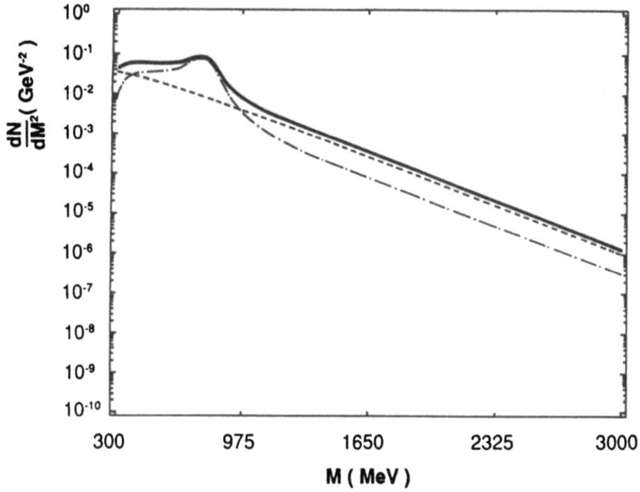

Figure 8: Dimuon spectrum for $T_c = 246$ MeV, $T_d = 200$ MeV, and $F_\pi = F_\pi(M^2)$. Dashed and dash-dotted lines are the QGP and the hadronic contributions, respectively. Solid lines are the total.

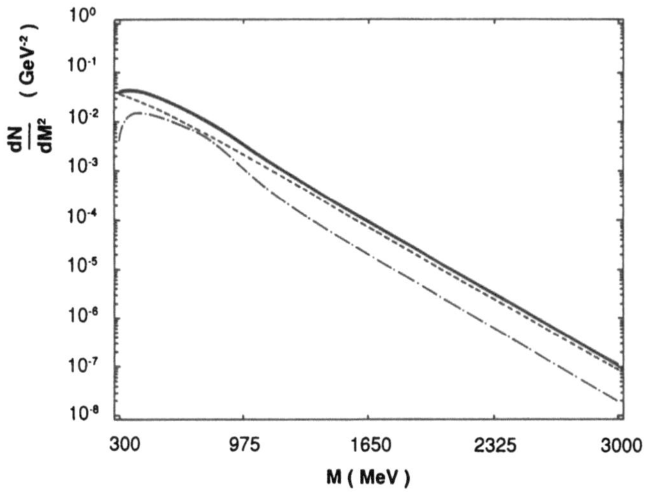

Figure 9: Dimuon spectrum for $T_c = 200$ MeV, $T_d = 170$ MeV, and $F_\pi = F_\pi(M^2, T)$. Dashed and dash-dotted lines are the QGP and the hadronic contributions, respectively. Solid lines are the total.

$F_\pi = F_\pi(M^2, T)$ as shown in figure 10 (which should be compared with figure 4). The differences between the $T_c = T_d$ and the $T_d < T_c$ scenarios in the region of the ρ-meson peak are shown in figures 11 and 12 for $F_\pi = F_\pi(M^2)$ and $F_\pi = F_\pi(M^2, T)$, respectively, both for $T_c = T_d = 246$ MeV, (curves (a)) and $T_c = 246$ MeV, $T_d = 200$ MeV (curves (b)). The differences between the two scenarios show up very clearly, regardless of whether $F_\pi = F_\pi(M^2)$ or $F_\pi = F_\pi(M^2, T)$. The latter two possibilities lead in turn to different shapes of the spectrum.

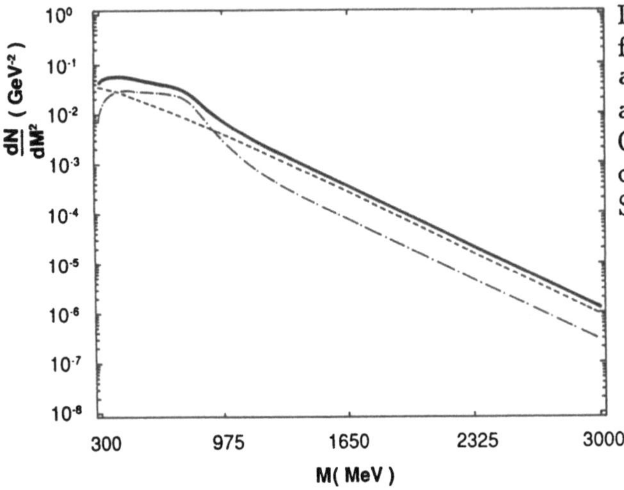

Figure 10: Dimuon spectrum for $T_c = 246$ MeV, $T_d = 200$ MeV, and $F_\pi = F_\pi(M^2, T)$. Dashed and dash-dotted lines are the QGP and the hadronic contributions, respectively. Solid lines are the total.

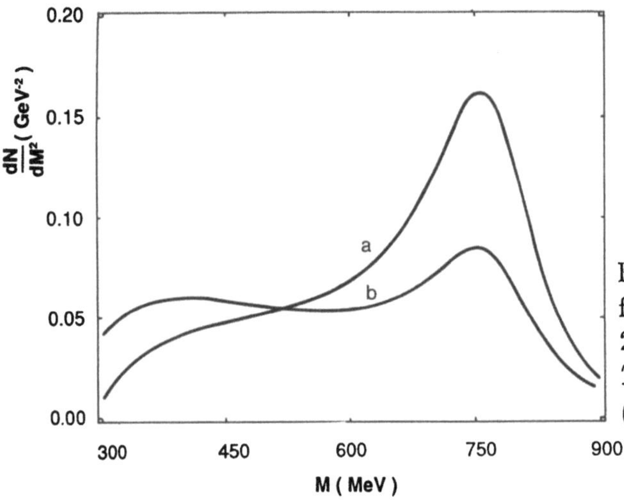

Figure 11: Dimuon spectrum for $F_\pi = F_\pi(M^2)$, $T_c = T_d =$ 246 MeV (curve (a)), and $T_c = 246$ MeV ,$T_d = 200$ MeV (curve (b)).

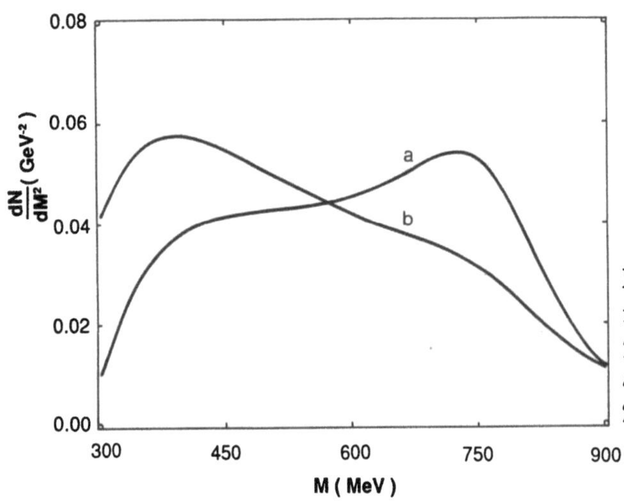

Figure 12: Dimuon spectrum for $F_\pi = F_\pi(M^2, T)$, $T_c =$ $T_d = 246$ MeV (curve (a)), and $T_c = 246$ MeV ,$T_d =$ 200 MeV (curve (b)).

In conclusion, if the $T_c = T_d$ scenario is correct, and $F_\pi = F_\pi(M^2, T)$ as expected on physical grounds, then the dimuon production spectrum would offer very little, if any, information on the QGP phase. However, if $T_d < T_c$ then experimental data at high M, and particularly at low M around the ρ-meson peak, would offer insight into the QGP phase.

Acknowledgements

The authors wish to thank J. Cleymans and K. Redlich for stimulating discussions. The work of (CAD) has been supported in part by FRD, and that of (ML) by FONDECYT 367-88, 367-90.

References

[1] For early work on the subject see e.g. G. Domokos, J.I. Goldman, Phys. Rev. **D23**, 203 [1981]; R.C. Hwa, K. Kajantie, Phys. Rev. **D32**, 1109 [1985]; K. Kajantie, J. Kapusta, L. McLerran, A. Mekjian, Phys. Rev. **D34**, 2746 [1986].

[2] J. Cleymans, J. Fingberg, K. Redlich, Phys. Rev. **D35**, 2153 [1987].

[3] J.D. Bjorken, Phys. Rev. **D27**, 140 [1983].

[4] J. Gaßer and H. Leutwyler, Phys. Lett. **B184**, 83 [1987].

[5] A. Larsen, Z. Phys. **C33**, 291 [1986]; C. Contreras, M. Loewe, Int. J. Mod. Phys. **A** to appear.

[6] K. Kawarabayashi and M. Suzuki, Phys. Rev. Lett. **16** 255, 384[E]; [1966] Riazuddin and Fayyazuddin, Phys. Rev. **147** 1071 [1966].

[7] C.A. Dominguez, in Proceedings of the 6th Adriatic Meeting in Particle Physics, Nucl. Phys. **B** [Proc. Suppl.] to appear; C.A. Dominguez, M. Loewe, in Proceedings of the Madrid Europhysics High Energy Physics Conference, Nucl. Phys. **B** [Proc. Suppl.] to appear.

[8] For a recent review see e.g. F. Karsch, CERN-TH-5498/89 [1989].

[9] C.A. Dominguez, M. Loewe, Phys. Lett. **B233**, 201 [1989].

Quantum Chromodynamics and the Nucleon–Nucleon Interaction*

A. Faessler

Universität Tübingen, Institut für Theoretische Physik,
D-7400 Tübingen, Fed. Rep. of Germany

Quantum Chromodynamics is assumed to be the correct theory for strong interactions. This conviction stems solely from extremely high energy data with momentum transfers of 10 GeV/c and larger where the running coupling constant $\alpha_s(q^2)$ is small and perturbation theory can be used. At low energy hadron and nuclear physics we have a length scale between 0.2 and 20 fm corresponding to momentum transfers of 10 to 1000 MeV/c. This means that we cannot use perturbative methods to solve QCD. There are promises that QCD on the lattice could once yield exact results for these low energy data, but at the moment we have not even reliably results with this approach to describe a single nucleon and therefore we are far away to understand with lattice QCD the nucleon-nucleon interation. We propose here by doing an averaging on the lattice to introduce colour displacement fields for the colour electric and the colour magnetic field strength treating the dielectric constant as a dynamical variable which describes collective quantities of gluons (something similar to collective glueballs). These collective glueballs describe then the long range behaviour of the interaction of gluons with quarks. The short range part which cannot be described adequately by averaging over larger areas of the lattice are further described by plain gluons. The theory has similarities with electrodynamics were one also introduces the electric displacement fields as averaged quantities over a large number of atoms which describe smoothly the electric and the magnetic fields which vary strongly at scales of inner atomic distances. The theory yields absolute colour confinement for quarks and gluons. It contains two parameters, the strong coupling constant α_s and the product of a coupling constant connected with the quark mass term g_m multiplied with the mass of the collective glueballs. This two parameters are adjusted to reproduce the nucleon mass and the radius of 0.6 fm of the quark content of the nucleon. Then we calculate parameter free the nucleon-nucleon ^1S potential. The strength comes out right but the

* Work supported by the Deutsche Forschungsgemeinschaft

Phase Structure of Strongly Interacting Matter
Editor: J. Cleymans

range is slightly too large. This is attributed on one side to the fact that the center of mass corrections are till now only performed for the quarks but not for the collective glueballs and that the model does not yet contain sea quarks. In spite of this minor shortcoming the model gives the most direct bridge between QCD and the nucleon-nucleon interaction which is at the moment available.

1. Introduction

There is a general agreement[1,2] that Quantum Chromodynamics (QCD) is the correct theory describing the strong interaction. This conviction stems from high momentum transfer data with momentum transfers $|q| > 10$ GeV/c. At these and higher momentum transfers the running coupling constant $\alpha_s(q^2)$ is small and perturbation theory can be used. At the moment there are no high momentum transfer data available which cannot be explained by QCD.

Figure 1 shows for example the number of 2-jet events in proton-antiproton collisions in the UA1 colliding beam experiment at CERN with 200 GeV for protons and antiprotons.

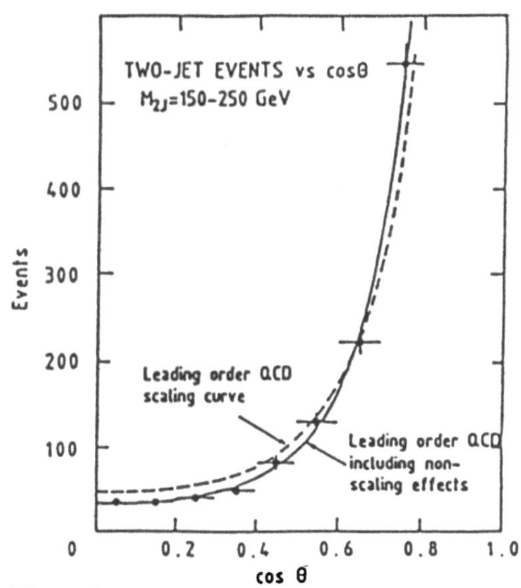

Figure 1
Angular distribution of the 2-jet events measured by UA1[3] at CERN in colliding beam experiments of protons on antiprotons each having 200 GeV. The solid line includes also higher order corrections of QCD and agrees completely with the data.

QCD with higher corrections can explain the data up to the least detail. The situation is quite different if we go to low energy transfer data in hadron and nuclear physics. There a perturbative treatment of QCD is not possible since the running coupling constant $\alpha_s(q^2)$ is of the order of unity. But how could we study the influence of the quark and gluon degrees of freedom on hadrons and nuclei? One approach would be lattice QCD[4]. (See for example the lecture of Redlich and of Satz at this workshop.) Lattice QCD has at present a good chance to describe the gluon degrees of freedom but is not able to describe reliably the quark degrees of freedom. We have for example no trustworthy description of the nucleon by lattice QCD and consequently are still very far away from describing the nucleon-nucleon interaction on the lattice.

Therefore we want to present here an approach which starts from the exact Lagrangian of QCD. Averaging over areas of a scale of 0.1 - 0.2 fm yields effective displacement fields for the colour electric and magnetic field strength and a dielectric constant which we treat as a dynamic field variable. This is very similar to how an applied physisist or engineer would average the microscopic electric and magnetic field strengths which are very rapidly varying fields inside atoms. There only the averaged displacement quantities are used. Due to non-linearity of the non-abelian QCD the gluon fields get highly collective at large distances. These collective gluon excitations (collective glueballs) are described dynamically as a dielectric field. We assume that it describes fully the non-linearity of the gluon fields. Thus we keep only plain gluons (like the photon field in QCD) to describe the short range behaviour. We shall see that this approach yields an absolute colour confinement for the quarks and the gluons and describes the nucleon properties including also Δ resonance quite well by fitting two parameters the strong coupling constant α_s and a constant connected with the product of the mass parameter of the quarks and the mass of the collective glueballs. With the model Lagrangian described in this way we are then able to calculate parameter free the nucleon-nucleon interaction in reasonable agreement with the data.

2. Quantum Chromodynamics

We are starting now with the QCD Lagrangian.

$$\mathcal{L}_{QCD} = -\frac{1}{4}\sum_{c=1}^{8} F^c_{\mu\nu}F^{\mu\nu}_c + \sum_{flavor} \bar{\psi}_f\{\gamma^\mu[\hat{p}_\mu - g_s A_\mu] - m_f\}\psi_f \tag{1}$$

$F_c^{\mu\nu}$ is the anti-symmetric gluon field tensor with the colour octet quantum number c.

$$F_{\mu\nu}^c = \partial_\mu A_\nu^c - \partial_\nu A_\mu^c + g_s f_{ab}^c A_\mu^a A_\nu^b \tag{2}$$

A_μ^c ist the gluon potential with the colour octet quantum number c.

$$A_\mu = \sum_{c=1}^{8} A_\mu^c \lambda_c \tag{3}$$

g_s is the strong coupling strength connected with the coupling constant $\alpha_s = g_s^2/(4\pi)$. The SU(3) structure parameters f_{ab}^c are defined by the commutation relations

$$\lambda_a \lambda_b \;-\; \lambda_b \lambda_a \;=\; i\, f_{ab}^c \lambda_c \tag{4}$$

of the Gell-Mann colour matrices normalized in the way indicated in equation (5).

$$tr\{\lambda_a \lambda_b\} = \frac{1}{2}\delta_{a,b} \tag{5}$$

To bring QCD onto the lattice we use an imaginary time $\tau = it$ and go over to Euclidian space. Since we were starting with a metric $g^{\mu\nu} = (1,-1,-1,-1)$ we are changing now the sign of the kinetic energy term of the gluon field (first term in eq. (1)). The action of the kinetic energy term of the gluon field in a Euclidian metric is now given by the expression:

$$S_g = \frac{1}{2}\int d^4x \; tr\{F_{\mu\nu}F^{\mu\nu}\}$$
$$F_{\mu\nu} = \sum_{c=1}^{8} F_{\mu\nu}^c \lambda_c \tag{6}$$

On the lattice the action of the gluon field (6) goes over to a sum over all plaquettes (s. figure 2):

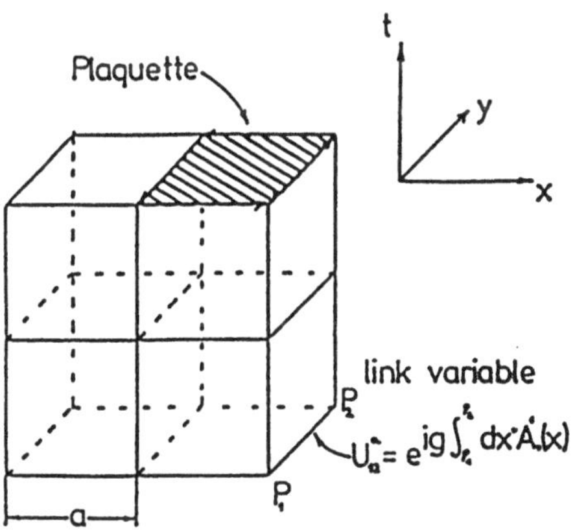

Figure 2

Four dimensional space-time lattice with the spacing "a" indicating a plaquette ijkl. The link operator connecting the different sites in the lattice is given (7) by U_{ji} for transition from i to j.

$$
\begin{aligned}
S_g &= \sum_{\Box} S_{\Box} \\
S_{\Box} &= \frac{2n}{g_s^2}\left[1 - \frac{1}{n} Re\ tr\{U_{il}U_{lk}U_{kj}U_{ji}\}\right] \\
U_{ji} &\equiv\ exp\left\{ig_s \int_{x_i}^{x_j} A_\nu dx_\nu\right\} \\
&\approx 1 + ig_s A_\nu\left(\frac{x_i + x_j}{2}\right)\ a \\
A_\nu &= \sum_{c=1}^{8} A_\nu^c \lambda_c
\end{aligned}
\tag{7}
$$

Using the infinitesimal expression for the link operator U_{ij} one can show the identity of (7) with (6). n is the number of colours ($n = 3$). The symbol Re indicates the real part of the trace.

The leading term in (7) under the trace is the 3×3 unity matrix which yields for colour $SU(n)$ n and therefore cancels which the first term "1". The product of the four link operators which contain only one g_s are automatically zero since the trace of each single colour matrix λ_c is identically zero. The first term being different from zero is obtained according to (5) from the trace of two colour matrices. The denominator $2n$ cancels against the corresponding expression in the numerator and the same also happens to the coupling constant g_s^2. An expansion of the gluon potential around the center of the plaquette or around one of the corners yields then expression (6). An important quantity is the partition function Z which is defind by a functional integral over all possible link variables.

$$
Z = \int \mathcal{D}U\ exp\{-S(U)\} \tag{8}
$$

3. Colour-Dielectric Quarkmodel

The main difficulty in treating a non-abelian gauge theory like Quantum Chromodynamics are apart from the large coupling constant g_s the non-linear terms in the antisymmetric field tensor (2). At large distances where the momentum transfer is small and the running coupling constant g_s is large these terms determine the behaviour. At small distances the linear terms with the four dimensional curl are the leading ones. To separate out the long range behaviour of the gluon field Nielsen and Patkos[5] proposed to average over the lattice. The procedure is shown in figure 3 and has also been used by Pirner[6], by Thomas[7] and by us in Tübingen[8].

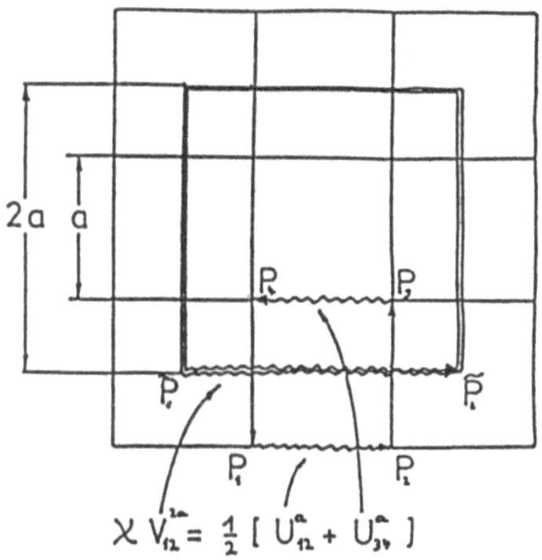

Figure 3
To average on the lattice one goes to larger plaquettes and replaces the link operator $U_{1'2'}$ by averaging over the neighbours. In a two dimensional lattice shown in this figure this is the average value of the link operators $U_{2,1}$ and $U_{3,4}$. In a four dimensional lattice one has eight such neighbours and the averaged link operator $U_{2',1'}$ is given by this averaged value.

The basic idea in treating the highly collective gluonic degrees of freedom at large distances is to average on the lattice several times as indicated once in figure 3. The link operator on the larger lattice (for example "2a" lattice size) is taken as the averaged value of the 8 neighbouring link operators on the four dimensional lattice.

$$U_{2'1'} = \frac{1}{8}\sum_{k=1}^{8} U_k = \chi(x)\, \tilde{U}_{2'1'}$$
$$\tilde{U}^+\tilde{U} = 1 \tag{9}$$
$$0 \leq \chi(x) \leq 1$$

In general if $\tilde{U}_{2'1'}$ is assumed to be unitary $\chi(x)$ is a hermitian 3×3 matrix times a phase. If one averages several times no colour should be prefered and thus $\chi(x)$ gets diagonal with identical elements. We choose the phase to be unity and thus obtain (9). This corresponds to the choice of a special local gauge.

In equation (9) the original averaged link operator $U_{2',1'}$ is not anymore unitary. This can easily be seen from figure 4. Since the averaged link operator is not unitary we introduce as already discussed above a dielectric field $\chi(x)$ which we treat as a dynamical field variable.

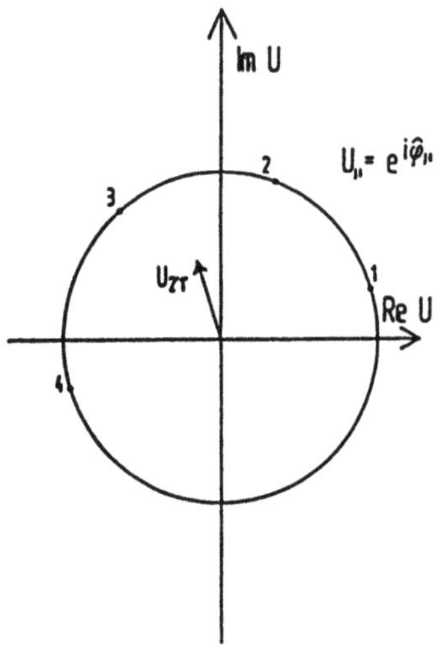

Figure 4
The averaging of different link operators does not lead to a unitary operator for the averaged value. Although the phase of the link operator is here an operator and not an number we can make the analogy to the above four link operators 1, 2, 3 and 4 on an unit circle in the complex U-plain. The averaged value $U_{2',1'}$ does not lie on the unit circle and therefore is not unitary. One renormalizes it so that one introduces a new dielectric field $U = \chi(x)\tilde{U}$. (Exactly χ should be a hermitian 3×3 matrix and a phase. But see discussion in text.)

$$U = \frac{1}{8}\sum_{k=1}^{8} U_k = \chi(x)\,\tilde{U} \tag{10}$$

$\tilde{U}$ is now again a unitary link operator defined so, that it would lie in figure 4 on the unit circle.

The action on the lattice which has to be minimized as a function of the link variables is now given by:

$$S_{eff;g} = \frac{2n}{g_s^2}\sum_{all\,\Box}\left[1 - \frac{1}{n}Re\; tr\{(\chi\tilde{U}_{il})(\chi\tilde{U}_{lk})(\chi\tilde{U}_{kj})(\chi\tilde{U}_{ji})\}\right] = Min. \tag{11}$$

In this expression for the action of the gluons on the lattice n is again the number of colours and g_s the strong coupling constant. The partition function is now given by:

$$Z_{eff} = \int \mathcal{D}\tilde{U}\;\; exp\{-S_{eff,g}(\chi\tilde{U})\} \tag{12}$$

We treat the dielectric field $\chi(x)$ as a dynamical variable. But on the other side we assume that the dielectric field χ handles the degrees of freedom of the gluon field at large distances. Thus we describe the gluons as plain gluons only.

$$\tilde{F}^c_{\mu\nu} = \partial_\mu\tilde{A}_\nu \;-\; \partial_\nu\tilde{A}_\mu \tag{13}$$

To get the full Lagrangian for the gluons we add the kinetic energy $T(\chi)$ and the potential energy $V(\chi)$ of the χ field to the Lagrangian.

$$\begin{aligned}\tilde{\mathcal{L}}_g(x) &= -\frac{1}{4}\sum_{c=1}^{8} \tilde{F}^c_{\mu\nu}\tilde{F}^{\mu\nu}_c \chi^4(x) + T(\chi) - V(\chi) \\ &= \frac{1}{2}\sum_{c=1}^{8}\Big[(\mathbf{E}_c)^2 \; - \; (\mathbf{B}_c)^2\Big]\epsilon(x) + \frac{1}{2}(\partial_\nu\chi)^2 - \frac{1}{2}M_\chi^2\chi^2(x)\end{aligned} \tag{14}$$

Here $\chi(x)$ is the dielectric field introduced by averaging on the lattice. By separating the effective gluon Lagrangian in colour electric and magnetic fields one sees that it is connected with the colour dielectric constant which one also knows in an analogous way from electrodynamics.

$$\mathbf{D}_c = \epsilon\mathbf{E}_c; \quad \mathbf{H}_c = \epsilon\mathbf{B}_c; \quad \epsilon(x) = \chi^4(x) \tag{15}$$

Naturally the quantities shown in eq. (15) are colour fields and have nothing to do with electromagnetic quantities. This is indicated by the subscript c which stands for one of the 8 colour octet indices.

The absolute confinement of the averaged displacement fields of the gluons is seen if (14) is writen in these fields:

$$\tilde{\mathcal{L}}_g(x) = \frac{1}{2\epsilon(x)}\sum_{c=1}^{8}[(\mathbf{D}_c)^2 \; - \; (\mathbf{H}_c)^2] + \frac{1}{2}(\partial_\nu\chi)^2 - \frac{1}{2}M_\chi^2\chi^2(x) \tag{14a}$$

We still have the problem to get an averaged Lagrangian for the description of the quarks and for the quark gluon coupling.

$$\mathcal{L}_{q+qg} = \bar{\psi}\Big[\gamma^\mu\{\hat{p}_\nu \; - \; g_s A_\nu\} \; - \; m_q\Big]\psi \tag{16}$$

We now obtain for the Lagrangian (16) the action by integrating over the four dimensional space. By making the transition to Euclidean four dimensional space and going on the lattice we obtain for the action:

$$S_{q+qg} = a^3 i \sum_{\{i<j\}} \bar{\psi}_i\gamma_\mu\hat{e}_\mu U_{ij}\psi_j \; - \; m_q a^4 \sum_i \bar{\psi}_i\psi_i \tag{17}$$

"a" is again the unit size of the lattice. The sum with the curly brackets runs over the next neighbour sites i, j counting each link only once. The first term contains the kinetic energy of the quarks and a coupling to the gluon field through the link operator U_{ij}.

$$U_{ij} \approx 1 + g_s a\hat{e}_\mu A_\mu\left(\frac{x_i + x_j}{2}\right) \to \chi(x)\tilde{U}_{ij} \approx \chi[1 + ig_s\tilde{a}e_\mu\tilde{A}_\mu] \tag{18}$$

Equation (17) which brings the quarks on the lattice must be naturally subjected to one of the procedures to prevent the doubling of the states as for example discussed in reference 4. Since we immediately will go back from the lattice to the continuum limit using again the same procedure no problem is arising from the doubling of the quark states, if one of the established methods[4] is used.

$$\tilde{S}_{q+gq} = \int d^4x \Big[\sqrt{\chi(x)}\bar{\psi}\gamma_\mu[\hat{p}^\mu - g_s\tilde{A}^\mu]\cdot\sqrt{\chi(x)}\psi - m_q\bar{\psi}\psi\Big] \tag{19}$$

Equation (19) gives now the effective action describing the quarks and the interaction of the quark with the gluons. The distribution of the dielectric collective glueball field $\chi(x)$ is dictated by the requirement of hermiticity. This substitution

$$\tilde{\psi} = \sqrt{\chi(x)}\psi(x) \tag{20}$$

leads now to the model Lagrangian.

$$\begin{aligned} \mathcal{L}_{Model} &= \bar{\tilde{\psi}}\left[\gamma_\mu\{\hat{p}^\mu - g_s\tilde{A}^\mu\} - \frac{g_m}{\chi(x)}\right]\tilde{\psi} \\ &- \frac{1}{4}\sum_{c=1}^{8}\tilde{F}^c_{\mu\nu}\tilde{F}^{\mu\nu}_c\chi^4(x) + \frac{1}{2}(\partial_\mu\chi)^2 - \frac{1}{2}M_\chi^2\chi^2(x) \end{aligned} \tag{21}$$

$\tilde{F}^c_{\mu\nu}$ is the antisymmetric field tensor of the plain gluons (13). g_m is a dimensionless quantity which is up to a scale the current quark mass m_q, which is the only term in the original QCD Lagrangian which violates chiral symmetry. Comparing the original QCD Lagrangian (1) and the model Lagrangian (21) one sees three differences:

(i) The antisymmetric colour field strength tensor (2) is replaced by the field strength of plain gluons (13).

(ii) The mass term $m_q\bar{\psi}\psi$ is replaced by a term which contains in the denominator the dielectric field $\chi(x)$ describing the collective glueballs. This term yields in the model exact colour confinement for the quarks because in all areas where the collective glueball field is zero the effective mass of the quarks goes to infinity.

(iii) To describe the long range behaviour by the collective glueballs $\chi(x)$ we have added a kinetic energy and a potential energy with the mass M_χ for the collective glueball field $\chi(x)$.

It turns out that the kinetic energy of the collective glueball field $\chi(x)$ can be neglected. Making the substitution

$$\chi(x) = g_m\tilde{\chi}(x) \tag{22}$$

one realizes immediately that the model has only two parameters:

$$\alpha_s = \frac{g_s^2}{4\pi} \quad ; \quad g_m \cdot M_\chi \tag{23}$$

4. The Nucleon

To adjust the two parameters (23) we apply the model Lagrangian (21) to the nucleon and fit the two parameters of the model to the nucleon mass (938 MeV) and to the radius of the quark content of the nucleon (0.6 fm). Our model contains now for the nucleon the three valence quarks, the dielectric glueball field which interacts non-linearly with all the three quarks and plain gluons describing the short range interaction between the quarks and also interacting directly with the collective gluons. Opposite to the way in which the model has been used by Nielson and Patkos and by Pirner we are including also plain gluons. Our calculations did show that without the inclusion of plain gluons the mass of the nucleon cannot be reproduced. But it is also quite natural that the dielectric collective glueball field $\chi(x)$ can only describe the long range behaviour of the gluons which is highly non-linear and collective. But the short range perturbative gluon exchange between the quarks cannot be described by the χ field[9].

The model Lagrangian (21) contains as part of the valence quarks for the description of the nucleons also the sea quarks. But in the description we shall give now for the nucleons we will neglect the sea quark admixture. An obvious and relatively simple way of including the sea quarks would be to couple mesons and especially π mesons to the quarks.

For the valence quarks of the nucleon we now make the ansatz

$$\tilde{\psi}_q = exp\Big[-i\{\frac{2}{3}M_N - < \frac{g_m}{\chi(\mathbf{r})} > + < g_s\tilde{A}_0 >\}t\Big] \cdot \begin{pmatrix} \varphi(\mathbf{r},t) \\ \eta(\mathbf{r},t) \end{pmatrix}, \tag{24}$$

which is motivated by removing exactly the center of mass motion for the quarks and by eliminating the small quark amplitudes $\eta(\mathbf{r},t)$ of the Dirac spinor.

$$i\hbar\partial_t\eta(\mathbf{r},t) \approx 0 \tag{25}$$

The expectation values in the exponent of eq. (24) have to be taken with the large upper two spinor components $\varphi(\mathbf{r},t)$ of the solution of the Dirac equation. Thus we end up with a selfconsistency problem. We put now the ansatz (24) into the Dirac equation.

$$[\gamma_\nu\{\hat{p}^\nu - g_s\tilde{A}^\nu\} - \frac{g_m}{\chi(\mathbf{r})}]\tilde{\psi}_q = 0 \tag{26}$$

We eliminate the small amplitude $\eta(\mathbf{r},t)$ under the assumption (25), which is equivalent to a non-relativistic reduction. Not even with the ansatz (24) this is fully justified. But it yields excellent results in the non-relativistic quark model and with the ansatz (24), this approach should even be more justified.

The plain gluon field is now eliminated from the equation in the one-gluon exchange approximation. This yields a quark-quark interaction (the one-gluon exchange potential). Introducing the Hartree-Fock approximation yields for quark 1 the following selfconsistent gluon potential:

$$\begin{aligned} U_g(\mathbf{r}_1) &= 2 \cdot < STC \mid \int d\tau_2 \, \varphi^+(\mathbf{r}_2) \frac{g_s^2}{4\pi} \sum_{c=1}^{8} \lambda_1^c \lambda_2^c \\ & \left[\frac{1}{|\mathbf{r}_1 - \mathbf{r}_2|} - \frac{\pi}{(M_N/3)^2} \delta(\mathbf{r}_1 - \mathbf{r}_2) \{1 + \frac{2}{3} \sigma_1 \cdot \sigma_2\} \right] \varphi(\mathbf{r}_2) \, |STC > \end{aligned} \tag{27}$$

λ_i^c are the Gell-Mann colour matrices for quark $i = 1$ and 2. M_N is the nucleon mass and $\varphi(\mathbf{r}_2)$ are the large amplitudes of the Dirac spinor of the second valence quark. The factor 2 takes into account that valence quark 1 interacts with two other valence quarks 2 and 3. $|STC >$ indicates spin, flavour (isospin) and colour parts of the nucleon wave function of the three valence quarks. Since we want only to consider the ^1S nucleon-nucleon interaction, we have omitted the tensor and the two-body spin-orbit part of the one-gluon exchange potential. The coupled Euler-Lagrange equation which have to be solved selfconsistently for the quarks $\varphi(\mathbf{r})$ and the dielectric collective glueballs $\chi(\mathbf{r})$ can now be derived.

$$\begin{aligned} \left[\hat{\mathbf{p}} \frac{1}{2M(\mathbf{r})} \hat{\mathbf{p}} + U_g(\mathbf{r}) + \frac{g_m}{\chi(\mathbf{r})} - E_q \right] \varphi(\mathbf{r}) = 0 \\ \Delta \chi(\mathbf{r}) - M_\chi^2 \chi(\mathbf{r}) = - \frac{3 g_m}{2\chi(\mathbf{r})} \left\{ \varphi^2 - \left[\frac{1}{2M(\mathbf{r})} \hat{\mathbf{p}} \varphi \right]^2 \right\} \end{aligned} \tag{28}$$

The coupling of the quark wave function φ to the collective glueball field has χ in the denominator of the mass term and therefore guarantees quark confinement. The second equation for the dielectric collective glueball field $\chi(\mathbf{r})$ is coupled to the quark field which is its source. The second term on the right hand side comes from the elimination of the small amplitudes $\eta(\mathbf{r}, t)$ of the Dirac spinor of the quarks. The factor 3 counts the three valence quarks. The total energy of the nucleon at rest includes the energies from the three valence quarks, from which one has to subtract the gluon exchange energy to count this energy only once, and it contains also the kinetic and potential energy of the dielectric collective glueball field χ.

$$E_N = 3 \left[E_q - \frac{1}{2} \int d\tau \varphi^+(\mathbf{r}) U_g \varphi(\mathbf{r}) + \frac{1}{2} \int d\tau \left\{ (\nabla \chi)^2 + M_\chi^2 \chi^2(\mathbf{r}) \right\} \right] \tag{29}$$

$$M(\mathbf{r}) = \frac{1}{3} M_N + \frac{1}{2} \left\{ \frac{g_m}{\chi(\mathbf{r})} - \int d\tau \varphi^+ \frac{g_m}{\chi} \varphi \right\} \tag{30}$$

The mass $M(\mathbf{r})$ in eq. (28) is defined as one third of the nucleon mass times one half of the radial dependent mass term for the quarks minus its averaged value. The mass of the nucleon $M_N = E_N$ given in eq. (29) can now be obtained by a numerical solution of equations (28) or by making an ansatz for the quark wave function φ and the glueball function χ and minimizing the nucleon energy (29). These two procedures yield in a reasonable limit the same results. For minimizing the nucleon energy (29) we make the ansatz:

$$
\begin{aligned}
|N> &= \mathcal{A}\Big\{\prod_{i=1}^{3}\varphi(\mathbf{r}_i)|STC>\Big\} \\
\varphi(\mathbf{r}) &= [\pi b^2]^{-3/4}\, e^{-r^2/(2b^2)} \\
\chi(\mathbf{r}) &\propto \frac{1}{A+Br^4}
\end{aligned}
\tag{31}
$$

The variational parameters are now A and B of the glueball wave function and the oscillator length b for the quark wave function. The results are shown in table 1 without and with the removal of the center of mass motion as mentioned above. The strong coupling constant α_s and the product of the dimensionless parameter g_m and the mass M_χ of the glueball field are adjusted to the nucleon mass (938 MeV) and to the radius of the quark content of the nucleon (0.6 fm).

Table 1
Fit of the nucleon mass (938 MeV) and the radius of the quark content of the nucleon (0.6 fm) by adjusting the strong coupling constant α_s and the product g_m M_χ of the dimensionless current quark mass (measured by a scale which is not known) and the mass of the dielectric collective glueball field χ. The results are shown without and with removal of the center of mass motion.

	without rem. C.M.	with rem. C.M.
$\alpha_s = g_s^2/(4\pi)$	2.7	1.3
$g_m M_\chi$ $[MeV fm^{-2}]$	345	237
$< g_m/\chi >_q$ $[MeV]$	407	221
$< U_g >_N$ $[MeV]$	−1483	−724
E_{3q} $[MeV]$	647	621
E_χ $[MeV]$	291	316
M_N $[MeV]$	(938)	(938)

5. Nucleon-Nucleon Interaction

We calculate now the nucleon-nucleon interaction in the adiabatic (Born-Oppenheimer) approximation. That means we assume that the two nucleons are at a definite distance r. For this configuration we calculate the energy of the two nucleons and subtract from it the energy of the same system when the two nucleons have an infinite distance (figure 5).

$$V_{NN}(r) = E_{NN}(r) - E_{NN}(\infty) \tag{32}$$

The ansatz for the two nucleon wave function assumes that the two nucleons are at a distance r and that the quark wave functions are gaussians with the oscillator length b.

$$|NN>= \mathcal{A}\Big\{\{(\pi b^2)^{-3/4}\}^6 \prod_{k=1}^{3} e^{-(\mathbf{r}_k-\mathbf{r}/2)^2/(2b^2)} \cdot \prod_{k=4}^{6} e^{-(\mathbf{r}_k+\mathbf{r}/2)^2/(2b^2)}|STC>\Big\} \tag{33}$$

All six valence quarks of the two nucleons are antisymmetrized by $\mathcal{A}$. The wave function (33) has still to be projected[10] on to a relative S-wave which we consider here only. (33) contains also the spin, isospin and colour wave functions of the six quarks. Since we want not to include the complications of the tensor force in the one-gluon exchange, we restrict ourself to the ^1S interaction between the two nucleons.

For the radial dependent quark mass and thus also for the dielectric, collective glueball field χ we make the ansatz

$$m_q(\mathbf{r}) = \frac{g_m}{\chi(\mathbf{r})} = A + B\big[\rho^2 + (|z| - \tfrac{r}{2})^2\big]^2 - \frac{r}{2C}\big[4\rho^2 + r^2\big]e^{-C|z|} \tag{34}$$

using cylinder coordinates z and ρ. This ansatz corresponds to the wave function of the glueball field χ given for one nucleon in equation (31). We have made this ansatz on the right and on the left at the positions $r/2$ and $-r/2$. The only difference is that we included an interpolating term (last term in equation (34)) to have a continuous transition from the left to the right hand side. The variational parameters of the six quarks and the glueball wave functions are now A, B, C and the oscillator length b. One calculates now the total energy

$$\begin{aligned} E_{NN}[\mathbf{r};A,B,C,b] = & \\ & \sum_{i=1}^{6} < \psi_{6q};b|\hat{\mathbf{p}}_i \frac{1}{2M(\mathbf{r}_i)}\hat{\mathbf{p}}_i + \frac{g_m}{\chi(\mathbf{r}_i)}|\psi_{6q};b> \\ & + \sum_{i<j=1}^{6} < \psi_{6q};b|V_g(i,j)|\psi_{6q};b> \\ & + \frac{1}{2}\int d\tau\Big[(\nabla\chi)^2 + M_\chi^2\chi^2(\mathbf{r})\Big]_{A,B,C} \end{aligned} \tag{35}$$

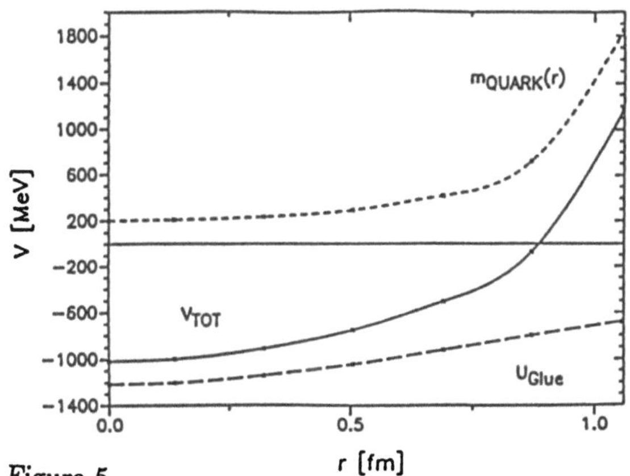

Figure 5
The figure shows the total potential of the quark wave equation (28) V_{TOT} as the sum of the gluon potential U_g and the mass term $g_m/\chi(\mathbf{r})$. In addition the figure shows also the two terms separately. One sees that the radial dependent quark mass is increasing at the surface of the nucleon and goes to infinity outside.

of the two nucleons and minimizes the expression for each distance r as a function of the parameters A, B, C and b. These four parameters are then given as a function of r. The total energy of the two nucleon yields then the nucleon-nucleon potential according to equation (32). The expression (35) for the energy of the two nucleons contains the one-gluon exchange potential

$$V_g(1,2) = \frac{g_s^2}{4\pi}\sum_{c=1}^{8}\lambda_1^c \cdot \lambda_2^c\Big[\frac{1}{|\mathbf{r}_1-\mathbf{r}_2|} - \frac{\pi}{(M_N/3)^2}\delta(\mathbf{r}_1-\mathbf{r}_2)\{1-\frac{2}{3}\sigma_1\cdot\sigma_2\}\Big], \qquad (36)$$

which we have already used in equation (27) to get the self consistent potential of one valence quark due to interaction by one-gluon exchange with the other two quarks.

The result for the ^1S potential calculated according to eq. (32) is shown in figure 6. This figure shows the nucleon-nucleon potential obtained with the help of eq. (35) and (32) calculated using the parameters α_s and g_m M_χ (see table 1) obtained from the numerical solution and from the variational solution of the single nucleon. These results are compared with the ^1S nucleon-nucleon potential of the Paris force[11].

Figure 6 shows that qualitatively the ^1S nucleon-nucleon interaction is reproduced quite well. One obtains only a slightly larger range than in the Paris potential. This can be due to two reasons:

(i) We have not included in this calculation the sea quarks. In a phenomenological description the sea quarks correspond to the meson cloud surrounding the nucleon. We

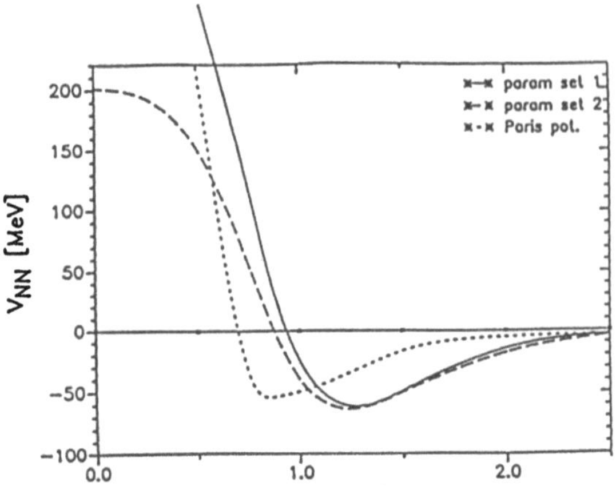

Figure 6

Nucleon-nucleon potential calculated according to eq. (32) and (35) with the parameters α_s and g_m M_χ obtained by fitting the energy of the nucleon and the radius of the quark content of the nucleon by solving numerically equation (28) and by using a variational approach in minimizing the energy of the nucleon given in equation (29). To circumvent the complication of the one-gluon exchange tensor and two-body spin-orbit forces we have calculated only the ^1S interaction. The two results are compared with the corresponding result of the Paris potential[11].

know that such a meson cloud exerts a pressure on the nucleon and is reducing the radius of the quark content of the nucleon. Although we have fitted the radius of the quark content of the nucleon such a pressure due to the sea quarks would modify the parameters adjusted to the nucleon mass and to the radius. We therefore could expect a change in the range of the nucleon-nucleon interaction.

(ii) A second shortcoming of this calculation is that we could not remove the center of mass motion in the nucleon-nucleon interaction connected with the collective glueball field χ. The $\chi(\mathbf{r})$ field (34) is taken relative to the origin and not relative to the center of mass of the six valence quarks. For the six valence quarks we were able to remove exactly the center of mass motion. The fact that this could not be done for the collective glueball field $\chi(\mathbf{r})$ includes zero point oscillations of the origin of the glueball field relative to the center of mass of the six quarks. This increases the range of the nucleon-nucleon interaction.

In general the agreement between the Paris potential and the result obtained here is very satisfactory especially in view of the fact that the model adopted here can be derived in an almost straightforward way from QCD.

6. Summary

The general believe that Quantum Chromodynamics (QCD) is discribing the strong interaction stems from the great success of QCD in the perturbative region, that means at momentum transfers of 10 GeV/c and more.

For hadron and nuclear physics such tests of QCD cannot be performed since the running coupling constant is very large and perturbative QCD cannot be applied. Lattice QCD is at the moment not even able to describe the properties of the nucleon and is therefore far away from describing the fundamental nucleon-nucleon interaction.

We propose here a model inspired by the work of Nielsen and Patkos[5] where we perform an average over the lattice and obtain averaged colour fields in the same way as one obtains in electrodynamics the displacement fields and the dielectric constant by averaging over a large number of atoms.

The averaged link operator on the lattice is not anymore unitary. We ensured unitarity by introducing a dielectric field $\chi(x)$ which describes dielectric, collective glueballs. One expects that this collective colour dielectric field is describing the long range non-linear and non-perturbative behaviour of the gluons. Thus we restrict us to plain gluons in the model hamiltonian (21). In this model hamiltonian the quarks and also the plain gluons are absolutely confined since as soon as the glueball field is zero the mass of the quarks goes to infinity and the energy density of the plain gluons diverges.

By fitting two parameters to the mass and to the radius of the quark content of the nucleon we are able to describe almost quantitatively the ^1S nucleon-nucleon interaction. The range of the interaction comes out slightly too large. We explain this partially by the missing sea quarks and by the fact that for the glueball field we were not able to remove the center of mass motion in the two nucleon problem. But in general the agreement between the Paris potential and the nucleon-nucleon interation calculated here is quite satisfactory, especially in view of the fact that the model Lagrangian (21) can be derived in an almost straightforward way from the exact QCD Lagrangian (1).

I would like to thank Dr. Kurt Bräuer and Dr. A. Drago with whom this work has been performed.

References

1. C. Itzykson, J. B. Zuber, "Quantum Field Theory" New York 1980.
2. P. Becher, M. Böhm, H. Joos, "Eichtheorien der starken und elektroschwachen Wechselwirkung." Teubner, Stuttgart 1981.
3. E. Rademacher, Progr. Part. Nucl. Phys. **14** (1985) 231 and private communications.
4. M. Creutz, "Quarks, Gluons and Lattices": Cambridge University Press, London 1983.
5. H. B. Nielsen, A. Patkos, **B198** (1982) 137.
6. H. J. Pirner, J. Wroldsen, H. Ilgenfritz, Nucl. Phys. **B294** (1987) 905 and J. F. Mathiot, G. Chanfray, H. J. Pirner, to be published in Nucl. Phys.
7. A. G. William, L. R. Dodd, A. W. Thomas, Phys. Lett. 176B (1986) 158 and Phys. Rev. **D25** (1987) 1040.
8. A. Drago, K. Bräuer, A. Faessler, J. Phys. **G15** (1989) L7.
9. K. Bräuer, A. Drago, A. Faessler, Nucl. Phys. A (1990) to be published.
10. K. Bräuer, A. Faessler, F. Fernandez, K. Shimizu, Z. Phys. A.320 (1985) 609 and Zhang Zong-ye, K. Bräuer, A. Faessler, K. Shimizu Nucl. Phys. A443 (1985) 557.
11. M. Lacombe et al. Phys. Rev. **C21** (1980) 861.

Resonant Dimeson Decay Mechanism in K $\rightarrow \pi\pi$ Decays*

R.D. Viollier[1] *and P. Zimak*[2]

[1]Institute of Theoretical Physics, University of Tübingen,
D-7400 Tübingen, Fed. Rep. of Germany, and
Institute of Theoretical Physics and Astrophysics,
University of Cape Town, South Africa
[2]Institute of Theoretical Physics, University of Basel, Switzerland
*Lecture presented by R.D. Viollier

It is shown that the ratio of the isoscalar and isotensor amplitudes, as determined from the $K \rightarrow \pi\pi$ decay data, can be understood in terms of a resonant decay mechanism that proceeds via the weak formation of a strongly interacting resonant two quark - two antiquark system which subsequently breaks up into two pions through rearrangement.

The $\Delta I = \frac{1}{2}$ selection rule of the hadronic weak interactions [1] has been introduced to account for the rather surprising observation that the $K_S^0 \rightarrow \pi^+\pi^-$ decay rate is orders of magnitude larger than the $K^+ \rightarrow \pi^+\pi^0$ decay rate [2]. Although this remarkably successful and universal selection rule was postulated more than three decades ago, it has never been understood in a firm theoretical framework. Today, it is therefore still considered as one of the most challenging and longstanding mysteries of low-energy particle physics ([3],[4]).

Neglecting the strong interaction, the driving terms of the $K \rightarrow \pi\pi$ decays are the spectator diagrams, in which the $\bar{s}$-quark of the K- meson decays into $\bar{u}, u$ and $\bar{d}$ quarks through the weak interaction, while the u-quark or d-quark in the K-meson remain a spectator. Due to the short-range nature of the weak interaction, the decay particles $\bar{u}, u$ and $\bar{d}$ are produced at or nearly at the position of the $\bar{s}$-quark, well inside the hadronic bubble of the K-meson. Moreover, this decay process is very soft, since the energy released is of the order of the hadronic scale, i.e. 220 MeV.

It is therefore conceivable that the *doorway* state, shortly after the nearly instantaneous weak interaction, is a hadronic state consisting of two pairs of strongly interacting massless quarks and antiquarks, trapped in the hadronic bubble that is left by the decaying K-meson. The lifetime of this dimeson state should be at least $3 \cdot 10^{-24}$ sec, corresponding to the time a relativistic quark takes to propagate a distance of the size of the hadron, implying a resonance width of about 300 MeV or less.

Although this time scale is relatively short, it is long enough to allow the quarks and antiquarks in the hadronic bubble to exchange or to annihilate into gluons. This conjecture is supported *a posteriori* by the fact that the strong interaction shifts of the various dimeson levels are indeed much larger than their widths. Thus, depending on quantum numbers and masses, some resonances will play a prominent role as intermediate states in the $K \to \pi\pi$ decays.

The purpose of this note is to show that the $\Delta I = \frac{1}{2}$ selection rule in $K \to \pi\pi$ decays can indeed be understood in terms of this resonant dimeson decay mechanism. A similar idea, though in less quantitative detail, has been put forward recently by Stech [5], who argued that the $\Delta I = \frac{1}{2}$ enhancement is due to virtual diquark-antidiquark states which subsequently decay into two pions.

The part of the amplitude relevant for the $K \to \pi\pi$ decay via the virtual excitation and subsequent break-up of a dimeson state N is given by

$$M(K \to ab) = \sum_N \frac{\langle \chi_{ab}\Psi_a\Psi_b \mid \mathcal{B} \mid \Psi_N\rangle\langle\Psi_N \mid \mathcal{W} \mid \Psi_K\rangle}{m_K - m_N + i\Gamma_N/2}. \tag{1}$$

Here, $\mid \Psi_n\rangle$ are the internal wave functions and m_n the masses of the various composite particles, $n = a, b, N$ and K. The plane wave χ_{ab} stands for the relative motion of the distinguishable decay products, a and b. Γ_N is the total width of the intermediate state N. The isoscalar operator $\mathcal{B}$ describes the isospin and parity conserving strong interaction, while the isospin and parity breaking weak interaction $\mathcal{W}$ is a sum of two tensor operators in isospin space.

Using the Wigner-Eckart theorem, we can easily factor out the trivial dependence on the isospin projections of the matrix elements in eq. (1) yielding

$$\langle \chi_{ab}\Psi_a\Psi_b \mid \mathcal{B} \mid \Psi_N\rangle = \langle I_a i_a I_b i_b \mid I_N i_N\rangle B^N_{ab}. \tag{2}$$

and

$$\langle \Psi_N \mid \mathcal{W} \mid \Psi_K\rangle = \sum_I \langle I i I_K i_K \mid I_N i_N\rangle W^K_N(I,i). \tag{3}$$

Here I_n and i_n denote the isospin and isospin projection of the spinless particles $n = a, b, N$ and K, while the isospin I and isospin projection i stand for the difference of the quantum numbers of the final and initial states. We can expand the amplitude (1) in terms of isoscalar and isotensor amplitudes defined as

$$A_\Lambda = \sum_N \frac{B^N_{ab}\ W^K_N(\frac{1}{2}(\Lambda+1),\frac{1}{2})}{m_K - m_N + i\Gamma_N/2}\ \delta_{I_N,\Lambda}, \quad \Lambda = 0, 2. \tag{4}$$

In order to determine the complex amplitudes A_0 and A_2, we need a reliable estimate of the spectrum and eigenstates of a dimeson system consisting of two pairs of strongly interacting massless quarks and antiquarks. In particular, we are interested in the colour singlet states with the quantum numbers $I^G(J^{PC}) = 0^+(0^{++})$ and $I^G(J^{PC}) = 2^+(0^{++})$, since these are the states that can be excited in $K \to \pi\pi$ decays.

The properties of dimesons have been calculated [6] in the framework of the M.I.T. bag model [11-13], with the quarks and antiquarks occupying the $1s_{\frac{1}{2}}$-mode of a spherical cavity and their interactions given by the one-gluon exchange diagram. The one-gluon annihilation graph, however, which plays an important role in lifting the isospin degeneracy of the dimeson levels [7-9], has been neglected in the previous analysis [6]. Dimeson resonances are in general deformed states which should be described using a deformed basis [10]. However, since we are interested in the spinless resonances, we will work in a spherical basis, including all two-body interactions to order α_s, i.e. the one-gluon exchange *and* annihilation diagrams. This means that our interacting $q^2\bar{q}^2$ states will be coupled to $q^2\bar{q}^2g$ and $q\bar{q}g$ intermediate states.

Details of the calculation will be presented elsewhere [14]; here we merely quote the results. Using the parameter set of the original M.I.T. bag model, which fits the ordinary hadron sector reasonably well, we arrive at the spectrum shown in Fig.1, in which the lowest of the dimeson resonances carries the quantum numbers of the vacuum. While

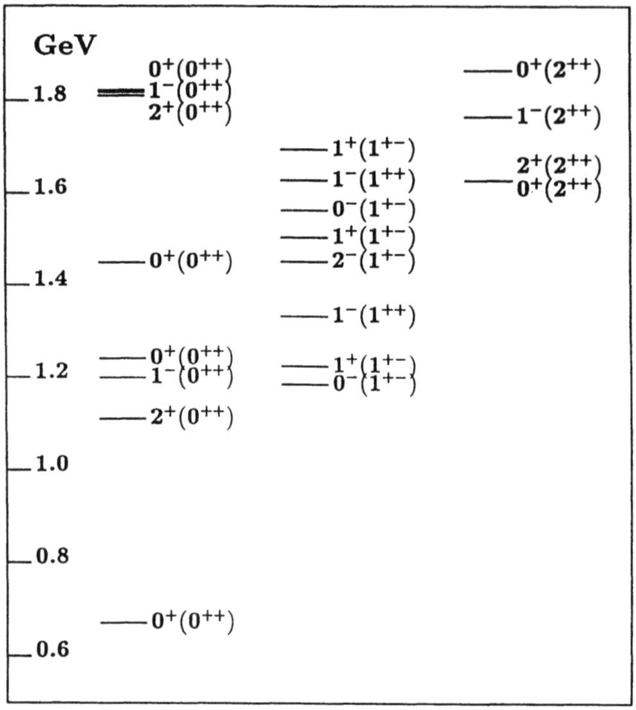

Figure 1: The mass spectrum of the $q^2\bar{q}^2$ system interacting through the one-gluon exchange and annihilation diagrams. The spectrum is calculated using the parameters of the M.I.T. bag model. The states are labelled with the quantum numbers $I^G(J^{PC})$.

this low-lying intermediate state, which we interpret as the unobservable σ-meson of the nucleon-nucleon force, contributes dominantly to the K_S^0 decay, it is inaccessible in the $K^\pm$ decays. The resonant dimeson decay mechanism offers thus a natural dynamical explanation for the longstanding puzzle of the well-established $\Delta I = \frac{1}{2}$ selection rule in $K \to \pi\pi$ decays.

The eigenfunctions of the dimeson system can be expanded in a two-meson basis

$$|\Psi_N\rangle = \sum_{ab} \langle K_a\kappa_a K_b\kappa_b | K_N\kappa_N\rangle \bar{D}^N_{ab} |\Phi_a\rangle |\Phi_b\rangle, \tag{5}$$

where K_n stands for the Casimir and multiplicity quantum numbers, while κ_n denotes the projection quantum numbers of the semi-simple group $SU(3)_{colour} \otimes SU(2)_{spin} \otimes$

Table I: Coefficients of fractional parentage for the interacting $q^2\bar{q}^2$ system with the quantum numbers $I^G(J^{PC}) = 0^+(0^{++})$ and $I^G(J^{PC}) = 2^+(0^{++})$. The masses are given in GeV, the bag radii in fm.

$I^G(J^{PC})$	$0^+(0^{++})$				$2^+(0^{++})$	
mass	0.668	1.237	1.445	1.816	1.107	1.805
radius	0.881	1.082	1.139	1.230	1.042	1.227
$\|\eta^1;\eta^1\rangle$	0.4514	0.4608	−0.1949	0.0655	0	0
$\|\pi^1;\pi^1\rangle$	0.5940	−0.4306	−0.0295	−0.0504	0.6442	0.0408
$\|\omega^1;\omega^1\rangle$	0.0010	0.1958	0.3094	0.6385	0	0
$\|\rho^1;\rho^1\rangle$	−0.0699	0.0809	0.5371	−0.3980	0.1770	0.7430
$\|\eta^8;\eta^8\rangle$	0.0496	−0.4573	−0.1781	0.5462	0	0
$\|\pi^8;\pi^8\rangle$	0.1862	0.0568	−0.5797	−0.2905	−0.4068	0.6464
$\|\omega^8;\omega^8\rangle$	−0.2332	0.4997	−0.3401	0.1965	0	0
$\|\rho^8;\rho^8\rangle$	−0.5891	−0.3087	−0.3058	−0.0759	0.6230	0.1688

$SU(2)_{isospin}$. The dynamics of the dimeson system is contained in the coefficients $\bar{D}^N_{ab}$, which represent the probability amplitudes for finding two distinguishable, preformed and spatially uncorrelated clusters, a and b, in the dimeson resonance N. The non-vanishing coefficients of fractional parentage $\bar{D}^N_{ab}$ are shown in Table I for the interacting dimeson system with the quantum numbers $I^G(J^{PC}) = 0^+(0^{++})$ and $I^G(J^{PC}) = 2^+(0^{++})$. Here, the index $c = 1,8$ of the quark-antiquark pairs η^c, π^c, ω^c and ρ^c denote the dimensionality of the colour representation.

The reduced matrix element of the strong interaction can be approximated by a product of the probability amplitude $\bar{D}^N_{ab}$, a universal strength parameter B_0 and a form factor which describes the momentum dependence of the vertex function. Assuming that the mesons are produced at the surface of the dimeson resonance with radius R_N, we thus find for a decay in a s- wave

$$B_{ab}^{N} = B_0 \; \bar{D}_{ab}^{N} \; \frac{\sin q_{ab}^{N} R_N}{q_{ab}^{N} \; R_N}, \tag{6}$$

where q_{ab}^{N} denotes the relative momentum of the particles in the decay channel $N \rightarrow ab$.

The partial width for the decay of the dimeson resonance N into two mesons, a and b, in a relative s-wave is given by

$$\Gamma(N \rightarrow ab) = \left[\bar{D}_{ab}^{N}\right]^2 \; B_0^2 \left[\frac{\sin q_{ab}^{N} R_N}{q_{ab}^{N} R_N}\right]^2 \; \frac{E_a \, E_b \, q_{ab}^{N}}{\pi \; m_N}, \tag{7}$$

where the $E_n (n = a, b)$ denote the energies of the decay particles in the centre-of-mass system. Summing up the partial widths of the open channels, we arrive at the full two-meson decay width

$$\Gamma_N \; = \sum_{ab} \Gamma(N \rightarrow ab) \Theta \left((q_{ab}^{N})^2\right). \tag{8}$$

Similarly, the reduced matrix element of the weak interaction can be expanded in a two-meson basis yielding

$$W_N^K(I, i) \; = \sum_{ab} W_{ab}^K(I, i) \; \bar{D}_{ab}^{N}, \tag{9}$$

where the elementary weak amplitudes contributing to this decay are

$$W_{\lambda\lambda}^K, (\tfrac{1}{2}, \tfrac{1}{2}) = \frac{5}{3\sqrt{3}} a_\lambda \;\; = c_0 a_\lambda, \tag{10}$$

$$W_{\lambda\lambda}^K (\tfrac{3}{2}, \tfrac{1}{2}) = \frac{4}{3}\sqrt{\frac{2}{3}} a_\lambda \;\; = c_2 a_\lambda, \;\; \lambda = \pi, \rho. \tag{11}$$

The straightforward calculation of the Fermi and Gamow-Teller amplitudes, $a_\pi = -7.623 \cdot 10^{-9}$ GeV and $a_\rho = -8.133 \cdot 10^{-9}$ GeV, respectively, which describe the external W-emission in the hadronic bubble of the K-meson in absence of any strong interaction, will be reported elsewhere [14].

Inserting eqs. (6) and (9) to (11) into eq. (4), we arrive at the isospin amplitudes

$$A_\Lambda = B_0 c_\Lambda \sum_N \frac{\bar{D}_{\pi\pi}^{N} \sin(q_{\pi\pi}^{N} R_N) \left(\bar{D}_{\pi\pi}^{N} a_\pi + \bar{D}_{\rho\rho}^{N} a_\rho\right)}{q_{\pi\pi}^{N} R(m_K - m_N + i\Gamma_N/2)} \; \delta_{I_N,\Lambda}, \;\; \Lambda = 0, 2. \tag{12}$$

Due to the form factor and the energy denominator, the isospin amplitudes are largely dominated by the contribution of the first $I^G(J^{PC}) = 0^+(0^{++})$ and $I^G(J^{PC}) = 2^+(0^{++})$

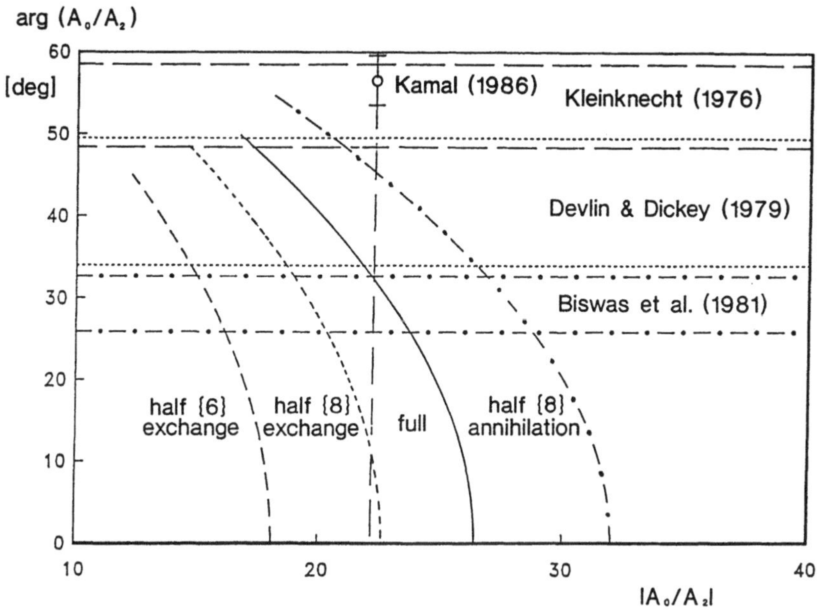

Figure 2: The phase as a function of the modulus of the ratio of the amplitudes A_0 and A_2.

resonances. Since the lowest isoscalar resonance is much lower than the lowest isotensor resonance, $|A_0|$ is much larger than $|A_2|$. Moreover, $|\bar{D}^N_{\pi\pi}|$ dominates over $|\bar{D}^N_{\rho\rho}|$ for these low-lying states, leading to a dominance of the Fermi over the Gamow-Teller contribution.

Using the parameter set of the original M.I.T. bag model [11-13], which fits the low-lying mesons and baryons reasonably well, we can by varying B_0 plot the phase of the ratio A_0/A_2 as a function of its modulus (Fig.2).

If the sum over the resonances in eq. (4) includes the dominant contributions from the intermediate states, the phases $\delta_\lambda = \arg A_\lambda$ should not differ too much from the $\pi\pi$ s-wave phase shifts. The solid line in Fig.2, representing the resonant dimeson decay mechanism, misses the point favoured by an analysis of the K-decays [15], a well as the error band of the $\pi\pi$ s-wave phase shifts suggested by Kleinknecht [16], at the value representing the experimental ratio $|A_0/A_2|$. However, the model is in general agreement with the $\pi\pi$ s-wave phase shifts, as determined by Devlin and Dickey [17] and Biswas et al. [18], at the ratio $|A_0/A_2| = 22.2$. Using for instance phases favoured by Devlin and Dickey, $\delta_0 - \delta_2 = (41.4 \pm 8.1)^0$, which fix the only free parameter of our model to $B_0 = (10.3 \pm 1.5)$ GeV$^{-\frac{1}{2}}$, we obtain

$$\left|\frac{A_0}{A_2}\right|_{theor} = 19.5 \pm 2.5, \tag{13}$$

which is surprisingly close to the experimental ratio [3] of

$$\left|\frac{A_0}{A_2}\right|_{exp} = 22.2 \pm 0.1. \tag{14}$$

Moreover, based on the quoted value for B_0, we arrive at an acceptable width of the σ-meson of $\Gamma(\sigma \to \pi\pi) = (310 \pm 85)$ MeV, which is one of the reasons why the σ-meson is so difficult to observe.

In this context it is interesting to note that the observed low-lying axial vector resonances $h_1(1170)$, $b_1(1235)$ and $a_1(1260)$ [19], can be interpreted as dimeson resonances (Fig.1), as well. For instance, for the $b_1(1235)$ resonance, which is observed with a mass of (1233 ± 10) MeV, our model yields 1221 MeV. The coefficient of fractional parentage for the decay mode $b_1 \to \omega\pi$, which is the only open two- meson decay channel, is $\overline{D}^{b_1}_{\omega\pi} = -.3173$. Fitting the observed width of $\Gamma(b_1 \to \omega\pi) = (150 \pm 10)$ MeV, the strength parameter is $B_0 = (9.0 \pm 0.3)$ GeV$^{-\frac{1}{2}}$ which corresponds to $|A_0/A_2| = 21.7 \pm 0.5$ and $\delta_0 - \delta_2 = (35.0 \pm 1.5)^0$. We thus conclude that the resonant dimeson decay mechanism is indeed able to describe the $\Delta I = \frac{1}{2}$ selection rule in $K \to \pi\pi$ decays.

The theoretical value for the modulus $|A_0/A_2|$ rests largely upon the form factors and the energy dominators, although the Gamow-Teller amplitude a_ρ contributes a factor of about 1.5, as well. The other factors in eq. (12) are approximately of unit value. The modulus and the phase are therefore very sensitive to our ability to describe the dimeson resonances properly, in particular the masses of the low-lying resonances.

It is difficult to estimate the systematic error in the calculation of the ratio of the isospin amplitudes, as we must be careful not to spoil the agreement between theory and experiment in the ordinary meson and baryon sector. While the form factors can be slightly changed without contradicting our experience, any modification the $\{\bar{3}\}$ quark-quark, the $\{1\}$ quark- antiquark interaction or the bag parameters will destroy this agreement at once. However, we can arbitrarily modify the $\{6\}$ quark-quark interaction and the $\{8\}$ quark-antiquark interaction, since these cannot contribute to the ordinary mesons and baryons.

In fact, if we reduce the one-gluon-exchange interaction of a quark pair in the colour $\{6\}$ state by 50%, we obtain the dashed curve in Fig.2. Reducing the colour $\{8\}$ one-gluon-exchange interaction of a quark-antiquark pair by 50%, we arrive at the dotted curve. Lastly, if we reduce the quark-antiquark interaction via annihilation which contributes only to the colour $\{8\}$ state, we get the dashed-dotted curve in Fig.2.

Turning the argument around, the remarkable agreement between the theoretical and experimental ratio of the isospin amplitudes can be interpreted as a sensitive test for the validity of the colour $\{6\}$ quark-quark and the colour $\{8\}$ quark-antiquark interaction,as given by the one-gluon exchange and the annihilation diagrams, which cannot be tested in ordinary hadrons. The agreement can also be seen as evidence for the presence of dimeson resonances in intermediate states of $K \to \pi\pi$ decays.

Acknowledgements

We have benefitted from many valuable comments and discussions with Amand Faessler, G. Backenstoss, R. Brandenburg, C.A. Dominguez, T.E.O.Ericson, H. Fritzsch, H. Müther, N. Paver and W. Sandhas. One of us (R.D.V) would like to thank Amand Faessler for the warm hospitality offered at the University of Tübingen, where part of this work was performed. Financial support from the Swiss National Science Foundation, FRD/CSIR and Deutsche Forschungsgemeinschaft are gratefully acknowledged.

References

[1] M. Gell-Mann, A. Pais, *International Conf. on High Energy Physics,* Glasgow (Pergamon Press, London 1955)

[2] R.W. Birge et al. *Nuovo Cimento 4*, 834 (1956); G. Alexander et al. *Nuovo Cimento 6*, 478 (1957)

[3] H.Y. Cheng, *Status of the $\Delta I = \frac{1}{2}$ Rule in Kaon Decay*, Indiana University, Bloomington, preprint IUHET-132, 1987 (to be published)

[4] L.L. Chau, *Phys. Reports 95*, 1 (1983)

[5] B. Stech, *Phys, Rev. D36*, 975 (1987)

[6] R.L. Jaffe, *Phys. Rev. D15*, 267 (1977); R.L. Jaffe, *Phys. Rev. D15*, 281 (1977)

[7] R.F. Buser, R.D. Viollier, P. Zimak, *Int. Journal of Theor. Phys. 27*, 925 (1988)

[8] P. Zimak, *Das $q^2\bar{q}^2$ - System im Rahmen des M.I.T. Bagmodells*, Masters Thesis, University of Basel 1983 (unpublished)

[9] A.J. Stoddart, R.D. Viollier, *Phys. Letters 208B*, 65 (1988)

[10] R.D. Viollier, S.A. Chin, A.K. Kerman, *Nucl. Phys. A407*, 269 (1983)

[11] A. Chodos, R.L. Jaffe, K. Johnson, C.B. Thorn, V.F. Weisskopf, *Phys. Rev. D9*, 3471 (1974)

[12] A. Chodos, R.L. Jaffe, K. Johnson, C.B. Thorn, *Phys. Rev. D10*, 2599 (1974)

[13] T. DeGrand, R.L. Jaffe, K. Johnson, J. Kiskis, *Phys. Rev. D12*, 2060 (1975)

[14] R.D. Viollier, P. Zimak, C.A. Dominguez (to be published)

[15] A.N. Kamal, *J. Phys. G12*, L43 (1986)

[16] K. Kleinknecht, *Ann. Rev. Nucl. Sci. 26*, 1 (1976)

[17] T.J. Devlin, J.O. Dickey, *Rev. Mod. Phys. 51*, 237 (1979)

[18] N.N. Biwas et al. , *Phys. Rev. Letters 47*, 1378 (1981)

[19] Particle Data Group, *Phys. Letters 204B*, 1 (1988)

All That RAZ – An Overview of Radiation Amplitude Zeros*

J.H. Reid[1] *and G. Tupper*

Institute of Theoretical Physics and Astrophysics,
University of Cape Town, South Africa
[1]**On leave from the Physics Department, University of Tulsa,**
600 S. College Ave., Tulsa, OK 74104, USA
***Lecture presented by J.H. Reid**

1 INTRODUCTION

Interference phenomena such as the celebrated double slit experiment in optics are familiar to every first year physics student. Equally familiar although much overlooked is the text book problem of the scattering of a system of nonrelativstic charged particles, where one finds that dipole radiation vanishes identically if the colliding particles all have equal charge to mass ratio.

A little less than a decade has past since it was first realized [1] that the above quoted result has an extension to relativistic quantum field systems: the tree amplitude for radiation of a single photon of 4-momentum k in a scattering process involving a total of n initial and final particles of spin ≤ 1 with charges (4-momenta) $Q_i(p_i)$ vanishes identically when $Q_i/k_i \cdot p_i$ is the same for $i = 1, \cdots, n$ provided all couplings are minimal, i.e. of the Yang- Mills type. Although the name amplitude zero and radiation zero has variously been applied in the literature we will follow the practice of refering to this as the Radiation Amplitude Zero (RAZ).

The importance of the RAZ to the phenomenologist is not to be discounted for a null experiment is after all the ideal. In practical terms the RAZ provides for (i) a test of the Standard Model (SM) by virtue of its existence (ii) a measure of the quark charges through the "null zone" defined by the condition $Q_i/k \cdot p_i = Q_j/k \cdot p_j \forall i,j$ and (iii) a comparatively clean way of bounding non-standard physics. Despite this and the many papers that have appeared on the topic the subject of the RAZ remains relatively obscure. Hence in this brief review we shall attempt to acquaint the the novice with what the RAZ is about and explore some of the applications of the phenomena to processes which test the SM physics.

In particular, in Section 2 we shall give the theoretical underpinnings of the RAZ and examine the conditions under which it exists or is destroyed. We shall then in Section 3 - 5 present results for applications which are pending experimental tests. Next in Section 6 we shall discuss the practical aspects of searching for RAZ's in realistic experimental situations. Finally in Section 7 we present some tentative conclusions on the future of the RAZ in phenomenology.

2 MUCH ADO ABOUT ZERO

Much of the workings of the RAZ can be understood by examining a comparatively simple subcase: a self interacting scalar field theory coupled to the electromagnetic field. Consider, for example, the scattering of m scalars into $n-m$ scalars at a single n-point vertex together with emission of a single photon. The scalars carry changes Q_i and have 4-momentu $p_i, i = 1, \cdots, n$, and the photon has 4-momentum k. The amplitude with radiation is proportional to the amplitude without radiation, the proportionality being

$$Z = \sum_{i=1}^{n} \eta_i \frac{Q_i}{k \cdot p_i} p_i \cdot \epsilon \tag{1}$$

where ϵ is the photon polarization and $\eta_i = -1(+1)$ for an initial (final) state particle. When $Q_i/k \cdot p_i$ is the same for all i then in turn

$$Z \propto \sum_{i=1}^{n} (\eta_i p_i) \cdot \epsilon = -k \cdot \epsilon = 0 \tag{2}$$

and the radiation amplitude vanishes.

The process $m \to (n-m) + \gamma$ at the tree level may be analysed in a similar way. A *source graph* is defined as a general graph for the process $m \to (n-m)$ the corresponding *radiation* graph is obtained by attaching a single real photon onto a specific line of the source graph and the *radiation amplitude* is the sum of all radiation graphs. The essential trick is to note that when a photon is attached to the $r-th$ internal line one uses the identity.

$$\begin{aligned} &\Delta_F(q'_r) Q_r (q'_r + q_r) \cdot \epsilon \Delta_F(q_r) \\ = \; &\Delta_F(q'_r) \left[\frac{Q_r}{k \cdot q_r} q_r \cdot \epsilon \right] + \left[-\frac{Q_r}{q'_r \cdot k} q'_r \cdot \epsilon \right] \Delta_F(q) \end{aligned} \tag{3}$$

when we have exploited $k \cdot q_r = k \cdot q'_r$ and $q_r \cdot \epsilon = q'_r \cdot \epsilon$. The radiation amplitude is therefore a sum over terms where each associates a Z-factor of the type of eqn.(1) to a single scalar vertex of the source graph. Now the radiation amplitude vanishes when all of the Zs are zero, which is to say

$$\frac{Q_i}{k \cdot p_i} = \frac{Q_r}{k \cdot q_r} \quad \forall_{i,r} \tag{4}$$

but in fact since we have a tree amplitude (no loops) all of the internal 4- momena and charges are fixed in terms of the external ones

$$\begin{aligned} q_r &= \sum_{j=1}^{n} C_{rj} p_j, \\ Q_r &= \sum_{j=1}^{n} C_{rj} Q_j \end{aligned} \tag{5}$$

so if all of the $Q_i/k \cdot p_i$ are equal

$$k \cdot q_r = \sum_{j=1}^{n} \left(\frac{k \cdot p_j}{Q_j} \right) C_{rj} Q_j = \frac{k \cdot p_i}{Q_i} \sum_{j=1}^{n} C_{rj} Q_j = \left(\frac{k \cdot p_i}{Q_i} \right) Q_r$$

and eqn.(4) holds automatically! The necessary conditions

$$\frac{Q_i}{k \cdot p_i} = \frac{Q_j}{k \cdot p_j} \quad \forall_{i,j} \tag{6}$$

define the *null zone*, i.e. the (not necessarily allowed) region of phase space where the radiation amplitude vanishes.

We may immediately make some observations on this result; first since $k \cdot p_i \geq 0$ all of the charges must be like-sign. Second, neutral particles must be massless and collinear with the photon such that $k \cdot p_i = 0$. Third, loop effects destroy the radiation zero because, per definition, for a loop graph egn.(5) cannot hold for all r. Finally, in the non-relativistic limit the zero conditions reduce to Q_i/M_i being equal for all particles which is just the classical result for dipole radiation; by this we recognise that the RAZ is the statement that when eqn.(6) are fulfilled the amplitude vanishes due to complete destructive interference of the radiation pattern.

Remarkably, Brodsky, Brown and Kowalski have proven that the RAZ occurs in the more general setting of an arbitrary field theory, spins ≤ 1, provided all couplings are of the Yang-Mills - i.e. minimal-form. The proof is quite involved and will not be detailed here, rather we refer the reader to their work [1]. Suffice it to say, however, that (i) the requirement of minimal coupling derives from the use of Ward identities in the form of relations of sort in eqn.(3) and (ii) the complications of spin are neatly handled by noting the connection of spin to Lorentz transformations. The Brodsky-Brown-Kowalski theorem states: *"The radiation amplitude generated by the tree source graphs with gauge theoretic vertices is zero when all ratios $Q_i/k \cdot p_i$ are equal".*

One might well be concerned that this theorem is physically empty since it is restricted to tree amplitudes whereas nature is not. Loop effects in simple models have been examined in [2]. In the context of perturbative calculations within the Standard Model such worries turn out to be mollified, however; by far the most interesting processes involve quarks so the most important loop corrections arise out of quantum chromodynamics (QCD) in the persona of gluon exchange, thus contributing in relative order α_s/π. While loop connections are important for the radiation amplitude in the null zone elsewhere they are insignificant and moreover the 1-loop amplitude only appears in interference with the tree amplitude which vanishes on the null zone. Thus, as first stressed by one of us [3], the important effect is not that of loops but rather that of real gluon emission for the gluon cannot be restricted to be collinear with photon instead being combined with the other parton in some *'jet dressing'* scheme. The physical situation then involves not a zero but still a very large dip in the distribution - a point we shall return to in Section 6.

It is amusing at least to inquire what transpires if the correct theory is not the Standard Model being instead it supersymmetric (SUSY) extension, the Supersymmetric Standard Model (SSSM) [4]. As an extension of the theorem above, Brown and Kowalski [1] have demonstrated the following: Consider a gauge theory with global (rigid) supersymmetry (all spins ≤ 1) where any derivative coupling present is minimal. Then all tree-approximated amplitudes for photino emission have the same radiation zero as those tor photon emission.

The basis for the supersymmetric theorem is easily understood: supersymmetry involves a charge which transforms particles to their supersymmetric partners, denoted by $\sim$, and vice versa, and which commutes with the scattering operator. It follows that the amplitude for photino, $\tilde{\gamma}$, emission is related to that for photon emission by

$$< a_1, a_2, \cdots |S| b_1, b_2, \cdots, \tilde{\gamma} >=$$
$$\sum_i A_i < a_1, \cdots, \tilde{a}_i, \cdots |S| b_1, \cdots, \gamma > - \sum_i B_j < a_1, \cdots |S| b_1, \cdots, \tilde{b}_j, \cdots \gamma > \quad (7)$$

The existence of the RAZ for the tree amplitudes on the right hand side of eqn.(7) has already been established whence the tree amplitude involving photino emission must vanish under the same conditions, namely eqn.(6), with now k the photino 4-momentum.

An example of this will be considered below.

3 ZEROS IN THE MAKING

Historically the RAZ was accidentally found by Mikaelian, Samuel and Sahdev [5] in the process

$$q_i(p_1) + \bar{q}_j(p_2) \to W^{\pm}(p) + \gamma(k) \quad (8)$$

spurring the theoretical efforts that subsequently led to the BBK theorem. The original interest in this process - and what makes it a current concern also - derives from the observation that one of the radiation graphs shown in Figure 1 involves the $W - \gamma - W$ interaction.

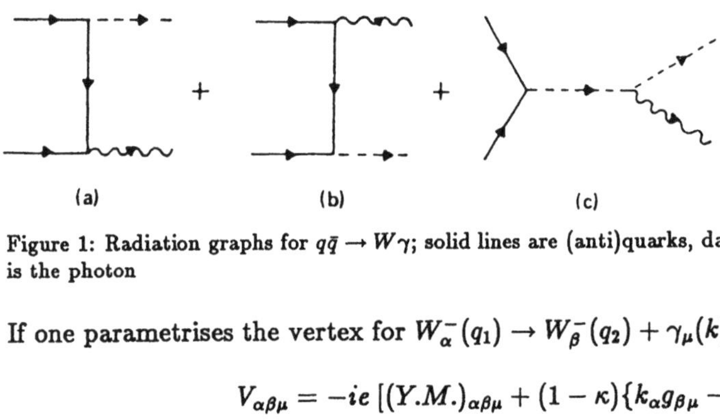

Figure 1: Radiation graphs for $q\bar{q} \to W\gamma$; solid lines are (anti)quarks, dashed line is the W and wavy line is the photon

If one parametrises the vertex for $W_\alpha^-(q_1) \to W_\beta^-(q_2) + \gamma_\mu(k)$ as

$$V_{\alpha\beta\mu} = -ie\left[(Y.M.)_{\alpha\beta\mu} + (1-\kappa)\{k_\alpha g_{\beta\mu} - k_\beta g_{\alpha\mu}\}\right] \quad (9)$$

$$(Y.M.) = (q_1 - q_2)_\mu g_{\alpha\beta} + (q_2 - k)_\alpha g_{\beta\mu} + (k - q_1)_\beta \; g_{\mu\alpha} \quad (10)$$

then the κ-parameter which measures the deviation from the (minimal) Yang Mills coupling is related to the Landé g-factor of the W by $g = 1 + \kappa$. Armed with hindsight provided by the BBK theorem one expects that a RAZ exists in the differential distribution for $d\bar{u} \to W^-\gamma$ and $u\bar{d} \to W^+\gamma$ iff $\kappa = 1$. Indeed one finds for the parton level differential cross-section

$$\frac{d\sigma}{dt} = \frac{\pi\alpha^2}{2s^2\sin^2\Theta_W} \; G_F \; K_{ij}^2 \left\{\left(Q_i + \frac{u}{t+u}\right)^2 \left(\frac{t^2+u^2+2sM_w^2}{tu}\right) + \right.$$
$$\left. +(\kappa-1)\left(Q_i + \frac{u}{t+u}\right)\left(\frac{t-u}{t+u}\right)\frac{+(\kappa-1)^2}{2(t+u)^2}\left(tu + \frac{s(t^2+u^2)}{4M_w^2}\right)\right\} \quad (11)$$

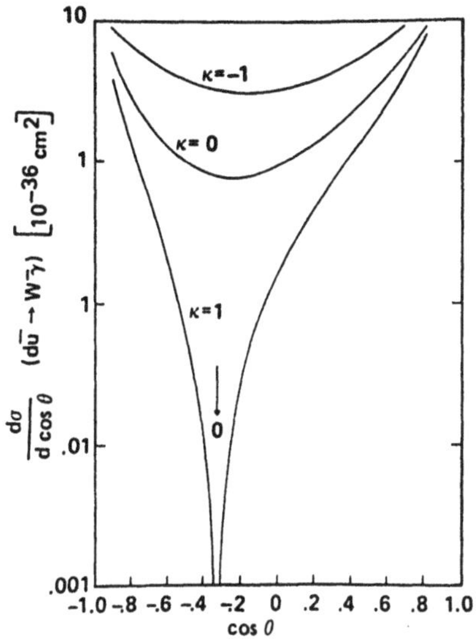

Figure 2: Differential cross-section for $d\bar{u} \to W^-\gamma$. θ is the angle between W^- and d, or between γ and $\bar{u}$, in the c.m. frame. $\sqrt{s} = 200$ GeV and $M_W = 85$ GeV/c^2

where $s = (p_1+p_2), t = (k-p_2)^2, u = (k-p_1)^2$, the quark charges $Q_{i,j}$ satisfy $Q_j = Q_i+1$, Θ_W is the Weinberg angle and K_{ij} is the Kobayashi-Maskawa mixing matrix element. Working in the $q\bar{q}$ center of momentum frame with Θ the angle between the photon and antiquark one sees that

$$Q_i + \frac{u}{t+u} = \frac{1}{2}\left[(1+2Q_i) + \cos\Theta\right] \tag{12}$$

which vanishes for $\cos\Theta = -(1+2Q_i)_j$ thus when $\kappa = 1$, $d\sigma/d\cos\Theta$ for $d\bar{u} \to W^-\gamma$ vanishes for $\cos\Theta = -\frac{1}{3}$. Note that the position of the RAZ in the angular distribution measures the quark charge!

In order to illustrate the point further, we show in Figure 2 the distribution for $d\bar{u} \to W^- - \gamma$. The essential observation is that large deviations from $\kappa = 1 (g = 2)$ lead to the distribution in the null zone being filled in (albeit the degree is somewhat distorted by the logarithmic scale). Of course by virtue of our earlier conclusions we may forecast that QCD correction produce effects not dissimilar to $\kappa \neq 1$. Only recently have the QCD corrections to $q\bar{q} \to W\gamma$ been fully evaluated [6], however these are inexticably linked to the imbedding of the partonic process into the hadronic one so that we postpone discussion of them to Section 7.

If one entertains the idea of supersymmetry then several other processes open up possibilities for RAZs in radiative production. Intrinsic to the SSSM is the existence of *two* Higgs doublets and consequently real charged Higgs scalars $H^{\pm}$. Equally well one could introduce multiple Higgs doublets without appeal to SUSY as in the model of Haber, Kane and Stirling (HKS) [7]. Both possibilities are neatly accounted for by writing the Yukawa couplings of the charged Higgs to quarks in the form

$$L_{YUK} = \frac{g}{2\sqrt{2}} K_{ij}\bar{u}_i(a_{ij} + b_{ij}\gamma_5)d_j H^+ + h.c. \tag{13}$$

where $a_{ij} = (am_i+bm_j)/M_W$, $b_{ij} = (am_i+bm_j)/M_W$ and in the Weinberg Supersymmetric Standard Model (WSSM) $b = 1/a$, $a = v_2/v_1$ while in the HKS model $a = b = v_2/v_1$; $v_{1,2} =< 0|\phi^0_{1,2}|0 >$ are the vacuum expectation values of the Higgs doublets.

The radiation amplitude for

$$q_i(p_1) + \bar{q}_j(p_2) \to H^{\pm}(p) + \gamma(k) \tag{14}$$

is again given by the graphs of Figure 1 although now the dashed line stands for the charged Higgs. Once more a RAZ is expected and the result for the parton differential cross-section, as obtained by He and Lew [8], is

$$\frac{d\sigma}{dt} = \frac{\pi \alpha^2 \, |K_{ij}|^2}{8 \sin^2\Theta_w} (a_{ij}^2 + b_{ij}^2)(Q_i + \frac{u}{t+u})^2 \left[\frac{1 + \frac{M_H^4}{S^2}}{tu} \right] \tag{15}$$

where all notations are as before. In eqn.(15) quark masses have been dropped, except of course in the a_{ij} and b_{ij}. In the case of $d\bar{u} \to H^- \gamma$ there is evident a RAZ at $\cos\Theta = -\frac{1}{3}$ but at the same time when embedded in the hadronic process such light valence quarks suppress the cross-section by $(m_{u,d}/M_W)^2 \simeq 10^{-10}$! Heavier sea quarks may somewhat relieve this situation at the price of washing out the RAZ; for a full discussion we refer the reader to He and Lew's paper [8].

Yet another SUSY process is that of wino-photino production

$$q_i(p_1) + \bar{q}_i(p_2) \to \tilde{W}(p) + \tilde{\gamma}(k) \tag{16}$$

Because the SUSY partner of the photon is involved, such supersymmetric radiation amplitude zeros are often referred to as SZEROs. The general graphical structure is once more as in Figure 1 but with the distinction that internal lines refer to the SUSY partners of the quark, antiquark and wino whose physical masses given by softly broken SUSY may be not at all negligible! The amplitude for this process has been given by Barger, Robinett, Keung and Phillips [9] (neglecting mixing):

$$A = \frac{e^2}{\sqrt{2}\sin\Theta_W} \left\{ \left[\frac{1}{S - M_W^2 + iM_W\Gamma_W} + \frac{Q_j}{t - \tilde{m}_j^2} \right]_1 (\bar{\tilde{W}}_R \gamma^\mu \tilde{\gamma}_R)(\bar{q}_{iL}\gamma_\mu q_{iL}) \right.$$
$$\left. + \left[\frac{1}{S - M_W^2 + iM_W\Gamma_W} - \frac{Q_i}{u - \tilde{m}_i^2} \right]_2 (\bar{\tilde{W}}_L \gamma^\mu \tilde{\gamma}_L)(\bar{q}_{jL}\gamma_\mu q_{iL}) \right\} \tag{17}$$

where Γ_W is the decay width of the W boson. In the limit of unbroken supersymmetry, where the Brown-Kowalski theorem applied $\tilde{m}_{i,j} = m_{i,j} = 0$, $M_{\tilde{W}} = M_W$ and $\Gamma_W = 0$ so that the factors in [ ] in eqn.(17) reduce as

$$\begin{aligned} [\ \ ]_1 &\longrightarrow \frac{1}{t}\left(Q_i + \frac{u}{t+u}\right) \\ [\ \ ]_2 &\longrightarrow -\frac{1}{u}\left(Q_i + \frac{u}{t+u}\right) \end{aligned} \tag{18}$$

and we see the by now familiar *"zero factor"*, $Q_i + \frac{u}{t+u}$ appearing.

Barger et al. have gone on to give the parton level and hadronic distributions for the (realistic) broken SUSY case. We will not reproduce their results but do wish to emphasize that for very large $\sqrt{s}$ all masses, non-SUSY and SUSY become negligable such that the distributions evidence a remnent of the SZERO in the form of a large dip. Rather we do want to stress a general problem with all SZERO processes: SUSY particles rapidly decay into the Lightest Supersymmetric Particle (LSP) and the LSP is very weakly interacting so it will escape detection. Thus SUSY events are characterized by large missing energy due to unseen particle - 'UFOs' - and, as we will discuss further in Section 6, when *even one ordinary* particle is unmeasured in an experiment the difficulty of interacting sufficient information for reconstruction of the null zone becomes only barely managable. Of course this limitation applies equally well to the SUSY processes $H^{\pm}, W_R^{\pm} \to \tilde{q}\bar{\tilde{q}}\gamma$ recently discussed in the literature [10,11].

4 A SCATTERING OF QUARKS AND LEPTONS

All of the processes examined in Section 3 share the feature that the source graph is a simple vertex. Even within the SM a variety of other channels exist with non-trivial source graphs, such as [12-14]

$$\begin{aligned} e^{\pm} + q &\rightarrow e^{\pm} + q + \gamma \\ e^{\pm} + \bar{q} &\rightarrow e^{\pm} + \bar{q} + \gamma \end{aligned} \tag{19}$$

and [12, 13, 15]

$$\begin{aligned} q_i + q_j &\rightarrow q_i + q_j + \gamma \\ q_i + \bar{q}_j &\rightarrow q_i + \bar{q}_j + \gamma \end{aligned} \tag{20}$$

For simplicity we only examine the former; the latter involves the possability of exchange terms for identical particles and, in any case, it portends of more serious problems for observing the RAZ.

All of the cases under consideration may be obtain from

$$f_i(p) + f_j(q) \rightarrow f_i(p') + f_j(q') + \gamma(k) \tag{21}$$

with appropriate choices for the charges, Q_i and Q_j. Defining the variables

$$\begin{aligned} s &= (p+q)^2 & s' &= (p'+q')^2 & & \\ t &= (p-p')^2 & t' &= (q-q')^2 & s+t+u+s'+t'+u' &= 0 \\ u &= (p-q')^2 & u' &= (p'-q)^2 & & \end{aligned} \tag{22}$$

one finds the differential cross-section is

$$d\sigma \;\propto\; \frac{\alpha}{8\pi^2} Q_i^2\, Q_j^2 \left(\frac{s^2 + s'^2 + u^2 + u'^2}{tt'}\right)\; Z, \tag{23}$$

were Z is the RAZ factor

$$Z = \left[\frac{tQ_i^2}{p\cdot k\, p'\cdot k} + \frac{t'Q_j^2}{q\cdot k q'\cdot k} + Q_iQ_j\left\{\frac{u}{p\cdot k\, q'\cdot k} + \frac{u'}{p'\cdot k\, q\cdot k} + \frac{s}{p\cdot k\, q\cdot k} + \frac{s'}{p'\cdot k\, q'\cdot k}\right\}\right] \tag{24}$$

Taking into account eqn.(22) one sees that Z vanishes when

$$\frac{Q_i}{p\cdot k} = \frac{Q_i}{p'\cdot k} = \frac{Q_j}{q\cdot k} = \frac{Q_j}{q'\cdot k} \tag{25}$$

In the center of momentum frame eqn.(25) leads to two useful conditions

$$\cos\Theta_\gamma = -\frac{(Q_i - Q_j)}{(Q_i + Q_j} \tag{26}$$

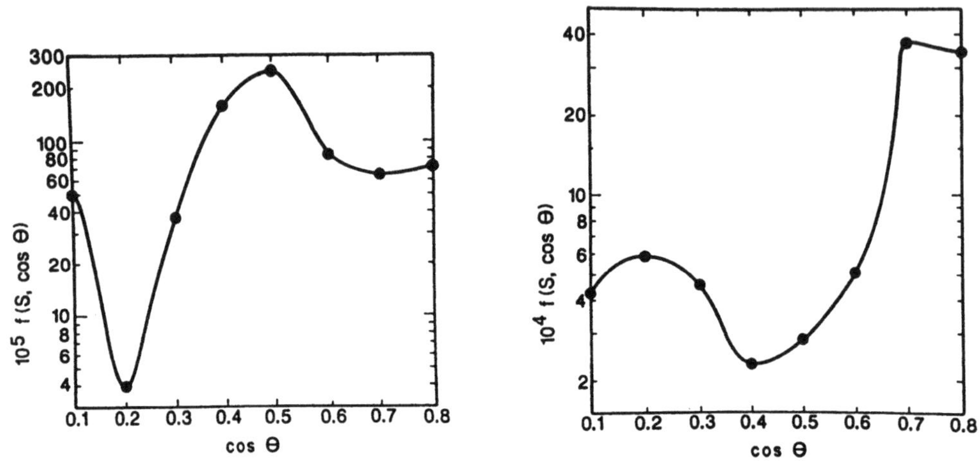

Figure 3: Representative angular distributions for e^+P (left) and e^-P (right) radiative scattering in $\cos\Theta = \cos\Theta_\gamma$

where Θ_γ is the angle between the photon and incident (anti)- fermion j 3-momenta, and

$$\frac{(1-x_j)}{(1-x_i)} = \frac{Q_i}{Q_j} \tag{27}$$

with $x_i(x_j)$ the energies of the final lepton and (anti)-quark scaled with respect of $\sqrt{s}$.

Reid, Li and Samuel [14] as well as Couture [16] have studied the corresponding hadronic process $e^\pm P \to e^\pm P\gamma$ as may be the subject of future experiments at DESY's HERA collider (30 GeV electrons on 820 GeV protons). In the case of e^+P scattering the (valence) process e^+u has a RAZ at $\cos\Theta_\gamma = 0.2$ while e^+d has no RAZ due to unlike sign charges. The contributions from both, however, must be added incoherently since the (light) quark will materalize as a hadron jet thereby washing out information on charge. The basic idea of Reid et al. and Couture is to integrate over a small bin around the condition in eqn.(27) leaving the hadronic cross-section differential in $\cos\Theta_\gamma$ only. Therein the distribution for e^+P was found to show a significant dip at $\cos\Theta_\gamma = 0.2$ whereas e^-P was discovered to display *no* corresponding RAZ associated dip at $\cos\Theta_\gamma = 0.5$ as illustrated in Figure 3. This finding may be easily understood qualitatively: the proyon after all has valence quark content uud so for $e^+(e^-)$ scattering the unlike sign process is (sub-)dominant.

Finally, it is useful to note that the competition between RAZ and non-RAZ and processes may be overcome if the quark or antiquark concerned is one of the heavy flavours, e.g. charm, pulled from the sea. In the case of e^+c scattering, one could look for jets containing a fast charmed particle. At the same time this gain must be weighed against the loss in event rate due to the relatively small sea quark distributions in the proton.

5 ZEROS FOR THE MASSES

Besides production and scattering, decay processes may also exhibit RAZs. Soon after the original work of Mikaelian, Samuel and Sahdev [5], Grose and Mikaelian [17] presented results for the crossed channel process $W \to q\bar{q}\gamma$. The doubly differential decay width for massless quarks is given by

$$\frac{1}{\Gamma_0}\frac{\partial^2\Gamma}{\partial y_1, \partial y_2} = \frac{\alpha Z^2}{2\pi}\left(\frac{(1-y_1)^2+(1-y_2)^2}{y_1 y_2}\right) \tag{28}$$

where $y_{1,2} = 1-2 \quad E_{1,2}/M_W$, $E_{1,2}$ are the $q, \bar{q}$ energies in the W rest frame with charges $Q_{i,j}$,

$$Z = \frac{Q_i y_1 - Q_j y_2}{y_1 + y_2} \tag{29}$$

and

$$\Gamma_0 = \frac{\alpha M_W}{4\sin^2\Theta_W} \tag{30}$$

is the lowest order width for $W \to q\bar{q}$. Phase space considerations constrain $y_{1,2}$ to lie in the triangle defined by $0 \le y_1, y_2, y_1 + y_2 \le 1$. The null zone assumes the form of a line $y_1 = Q_j y_2/Q_i$ cutting this triangle - specifically for $W^- \to (d\bar{u}, s\bar{c}, b\bar{t})\gamma$ one has a line of RAZs along $y_1 = 2y_2$.

In the history of the RAZ the Grose-Mikaelian [17] paper is of some importance since it was among the first to exhibit the factorization property of the radiation amplitude [18] and to demonstrate the spin independence of the zero condition. On the other hand for the heavy quark (bt) mode the masses are not negligible; indeed current indications are that the top quark may be so heavy as to close this channel! Equally, if one want to examine the crossed process to $H\gamma$ production, namely $H \to q\bar{q}\gamma$ then as has been stated in Section 3 if the coupling is to be non-negligible then so are the quark masses.

Recently Tupper, Reid, Li and Samuel have obtained expressions for radiative W [19] and H [20] decay with arbitrary quark masses. The essential idea relies on a generalization of variables that had been introduced previously for W decay into massless quarks [21]. Consider the process

$$A(p) \to f_i(p_1) + \bar{f}_j(p_2) + \gamma(k) \tag{31}$$

where A may stand for a W or H, and define

$$\begin{aligned} x &= 2(p_1+p_2)\cdot k/_{M_A^2} \\ y &= \frac{(p_1-p_2)\cdot k}{(p_1+p_2)\cdot k}; \end{aligned} \tag{32}$$

in the frame of A x is just the scaled photon energy. Let

$$\begin{aligned} \mu_{i,j} &= \frac{m_{i,j}}{M_A} \\ \varepsilon &= \mu_i^2 + \mu_j^2 \\ \Delta &= \mu_i^2 - \mu_j^2 \\ r &= (\mu_i + \mu_j)^2 \\ \delta &= (\mu_i - \mu_j)^2 \\ \bar{Q} &= \frac{Q_i - Q_j}{Q_i + Q_j} \end{aligned} \tag{33}$$

$$\Omega = 1 - x - 2\left[\frac{\mu_i^2}{1+y} + \frac{\mu_j^2}{1-y}\right]$$
$$\lambda = [(1-x-r)(1-x-s)]^{\frac{1}{2}} \tag{34}$$

in terms of which the physical region is described by the inequality

$$0 \leq \Omega \tag{35}$$

$$0 \leq x \leq 1-r,$$
$$\frac{\Delta - \lambda(x)}{1-x} \leq y \leq \frac{\Delta + \lambda(x)}{1-x}. \tag{36}$$

Now, in terms of these variables the doubly differential width, are for W- decay

$$\frac{1}{M}\frac{\partial^2\Gamma}{\partial x \partial y} = \frac{\alpha^2 C Q^2 (K_{ij})^2}{96\pi \sin^2\Theta_W}\frac{(\bar{Q}-y)^2}{(1-Y^2)}\left[\frac{2(2-\varepsilon-\Delta^2)\Omega}{x} + x(1+y^2+\varepsilon)\right] \tag{37}$$

and for H-decay

$$\frac{1}{M}\frac{\partial^2\Gamma}{\partial x \partial y} = \frac{C\alpha^2 Q^2 K_{ij}^2}{64\pi \sin^2\Theta_W}\frac{(y-\bar{Q})^2}{1-y^2}\left\{\frac{2}{x}\{a_{ij}^2(1-r) + b_{ij}^2(1-\delta)\}\Omega + x(a_{ij}^2 + b_{ij}^2)\right\}. \tag{38}$$

The salient feature here is that in each case the null zone is a line $y = \bar{Q}$ independent of x and the quark masses; this is to be contrasted to the use of ordinary scaling variables where different quark masses lead to shifted lines of zeros [22]. One may further observe that as the quark masses increase the phase space shrink such that the null zone exits the physical region *before* the channel closes. Indeed the condition for the zero line to intersect the physical region is

$$0 \leq x \leq 1 - (Q_i + Q_j)\left[\frac{\mu_i^2}{Q_i} + \frac{\mu_j^2}{Q_j}\right] \tag{39}$$

As an example, for the $b\bar{t}$ mode the RAZ lines outside of the physical region when $m_t \geq 67$ GeV. Since such a top quark mass is indicated, it appears that mass effects are only relevant for SUSY inspired decays.

6 THE QUEST FOR ZERO DEFECTS

If as we have seen RAZs appear in a variety of processes then it must also be said that only in the case of radiative W production/decay has the phenomenon found a potentially serious application: the measurement of/or more precisely the placing of bounds on - the W magnetic moment. In this regard there are two theatres of operation, hadronic W production and hadronic W decay.

The production of single Ws in the process $P\bar{P} \to W + X$ is well known so one might hope to examine $P\bar{P} \to W\gamma + X$, yet there is the problem that the W is not seen directly; rather it decays into hadrons which cannot be dug out of the enormous QCD background, or else it decays leptonically in which case the (anti)neutrino is unobserved. What is available is the signal $P\bar{P} \to e^-\gamma + X + \not{p}_T$ with large missing transverse momentum resulting from either

$$
\begin{gathered}
P\bar{P} \to W^-\gamma + X, \\
W^- \to e^-\bar{\nu}_e \\
\textbf{PRODUCTION}
\end{gathered}
$$

or

$$
\begin{gathered}
P\bar{P} \to W^- + X, \\
W^- \to e^-\bar{\nu}_e\gamma \\
\textbf{DECAY}
\end{gathered}
$$

If finite width of the W is taken into account the production and decay amplitudes interfere and cannot be so neatly separated, however the effect is small.

The issues of separating the production and decay and subsequently interacting the necessary information to reconstruct events have been examined by several authors [23 - 25]. In essence the idea as to first form the cluster transverse mass

$$
\begin{aligned}
m_T^2(e\gamma;\not{P}_T) &= \left[(m_\gamma^2 + |\vec{P}_{T\gamma} + \vec{P}_{Te}|^2)^{\frac{1}{2}} + |\not{P}_T|\right]^2 \\
&- |\vec{P}_{T\gamma} + \vec{P}_{Te} + \vec{\not{P}}_T|^2 \qquad (40)
\end{aligned}
$$

where $m_{e\gamma}$ denotes the invariant mass of the $e\gamma$ pair. In the case of W radiative decay $m_T(e\gamma;\not{P}_T)$ peaks at M_W and then drops rapidly such that a cut of $m_T(e\gamma,\not{P}_T) \lessgtr 90$ GeV is sufficient to distinguish the decay and production processes. Next, by requiring that for the production (decay) process the invariant mass of the $e\nu(e\nu\gamma)$ system matches the W mass the neutrino longitudinal 3-momentum can be obtained, albeit with a two-fold ambiguity. As a result, the appropriate angular distributions show a remanent of the RAZ as a dip, not unlike what was discussed in Section 4 for eP radiative scattering.

One might worry that QCD corrections undo this analysis, however explicit calculations [6] whow these to be small, as expected, and well approximated by an orthodox K-factor. A far more difficult question concerns the background from $P\bar{P} \to W\mathrm{Jet} + X$ with the jet misidentified as a photon. Baur and Berger [24] concludes that when all is said and done the $W - \gamma - W$ vertex can be probed to 25-40% accuracy through hadronic production. This is of no small importance since the CDF groups at FERMILABs TEVATRON collider has to date collected about 5000 [26] Ws together with a few $W\gamma$ events and a dozen or so radiative W decays [27]. Future runs will yield as many as $10^5 W$s with a corresponding increase in the radiative event sample.

A rather different approach has been examined by Reid, Tupper and van Zijl [28]. The intention of the CERN LEPII collider is to produce W pairs in e^+e^- annihilation, so one may look for

$$
\begin{aligned}
e^+e^- &\to W^+W^-; \\
W^+ &\to e^+\nu_e, \\
W^- &\to \mathrm{JET} + \mathrm{JET} + \gamma \qquad (41)
\end{aligned}
$$

with the total invariant mass of the γ-hadronic system matched to M_W; thus it suffices to study the decay

$$
\begin{aligned}
W^- &\to q\bar{q}\gamma; \\
q &\to \mathrm{JET}_1, \\
\bar{q} &\to \mathrm{JET}_2. \qquad (42)
\end{aligned}
$$

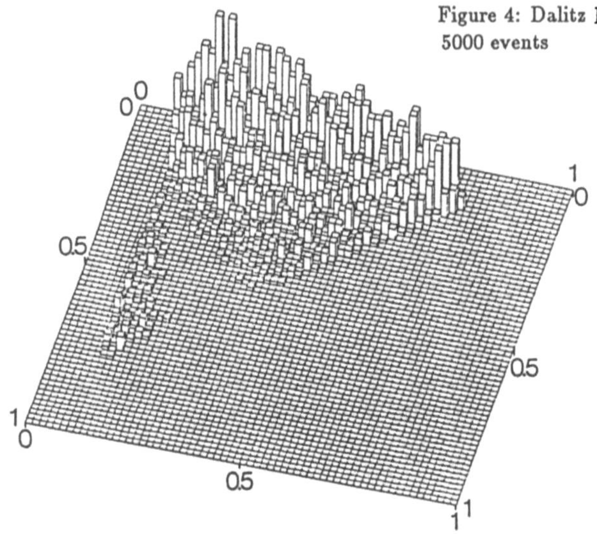

Figure 4: Dalitz Plot in y_1, y_2 for $W^- \to q\bar{q}\gamma$, $\kappa = 1$ with 5000 events

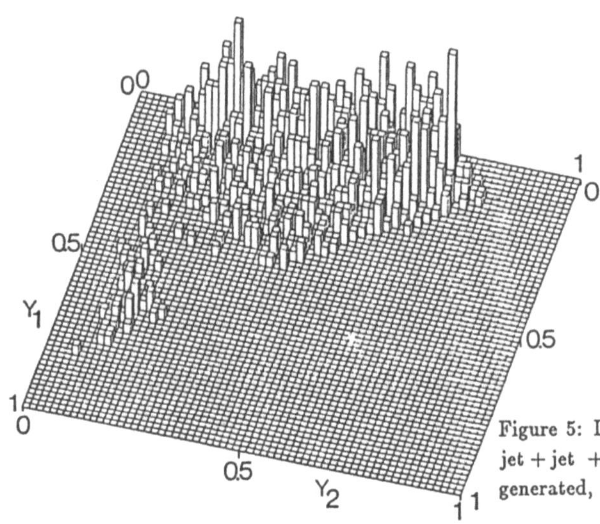

Figure 5: Dalitz Plot in y_1, y_2 for $W^- \to$ jet + jet $+\gamma$, $\kappa = 1$, 5000 parton events are generated, 1842 survive

Because the null zone does not bisect the Dalitz plot it is necessary to distinguish which jet is associated to the quark and which to the anti-quark. This, however may be elegantly achieved by focusing on the $s\bar{c}$ mode and tagging the anti-quark jet by the pressence of a fast anti-charmed particle. As in the case of hadronic W production, the QCD corrections to radiative W decay has been computed [3, 29, 30] and are known to be small.

The procedure then is to generate events according to the distribution of eqn.(28) and shower/fragment/hadronize using the LUND string model [31]. Resulting events with nentrinos are excluded and for the remaining sample a model detector is used to analysize the events and to reconstruct $y_{1,2}$ with (anti) charm tagging. The outcome is rather striking: Figure 4 shows the Dalitz plot for the parton level decay while the plot for real jet events, Figure 5, evidences a virtually indistinguishable distribution! By way of contrast if $\kappa = 0$ as in Figure 6 the null zone is well populated.

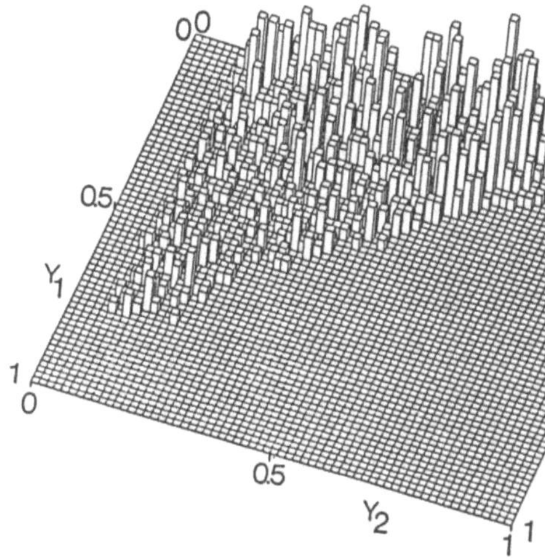

Figure 6: Dalitz Plot in y_1, y_2 for $W^- \rightarrow$ jet + jet + γ, $\kappa = 0$. 5000 events are generated, 1960 survive

What may be concluded is that radiative hadronic W decay is sensative to the W magnetic moment but *not* to QCD corrections or to the details of hadronization. Of course in a real experiment one will have not thousands of events but perhaps 250. In that case the variable x and y come into their own since one may bin in y only while accumulating event in x thereby increasing sensitivity [28].

7 ZERO PROSPECTS

Radiation amplitude zeros are by now an old topic with a large literature. Herein we have tried to briefly review the field the particular emphasis on the point that the RAZ, like group theory, should not be treated as an end in itself, but taken as a tool. From such a perspective one can see, that SZEROs are more a matter of theoretical than phenomenological interest. Similarly, radiative scattering of positrons and/or quarks may yield a direct measure of the quark charges, yet these can be inferred by other means.

The area where the RAZ really 'shines' is in the measurement/bounding of the W magnetic moment through radiative W production and decay. One expects that the next few years will see emerging the first measurement of κ through hadronic production at the TEVATRON. Moreover, in the not too distant future complementary measurements may be made through W hadronic decay at LEPII where flavour tagging will provide for a relatively clean list of this aspect of the standard model.

References

[1] S.J. Brodsky and R.W. Brown, *Phys. Rev. Lett.* 49 (1982) 966; R.W. Brown, K. Kowalski and S.J. Brodsky, *Phys. Rev.* D28 (1983) 624. For the extension to supersymmetric theories see R.W. Brown and K.L. Kowalski, *Phys. Lett.* B144 (1984) 235.

[2] M.L. Launsen, M.A. Samuel, A. Sen and G. Tupper, *Nuc. Phys.* B226 (1983) 429; M.L. Laursen, M.A. Samuel and A. Sen, *Phys. Rev.* D28 (1983) 650.

[3] G. Tupper, *Phys. Lett.* B156 (1985) 400.
[4] For review see, for example, H.P. Nilles, *Phys. Rep.* 110 (1984) 1; H. Habes and G. Kane, *Phys. Rep.* 117 (1985) 75.
[5] K. Mikaelian, M.A. Samuel and D. Sahdev, *Phys. Rev. Lett.* 43 (1979) 746.
[6] J. Smith, D. Thomas and W.L. van Neerven *Zeit. Phys.* C44 (1989) 267.
[7] H. Haber, S. Kane and T. Stirling, *Nucl. Phys.* B161 (1979) 493.
[8] Xiano-gang He and H. Lew, *Mod. Phys. Lett.* 3 (1988) 1199.
[9] V. Barger, R.W. Robinett, W.Y. Keung and R.J.N. Phillips, *Phys. Letts.* B131 (1983) 372.
[10] B. Mukhopadhyaya, M.A. Samuel, J. Reid and G. Tupper, *Oklahoma State University Research Note* 240 June 1990 (Submitted to Phys. Lett. B).
[11] G. Tupper, J.H. Reid and E.J.O. Gavin, *University of Cape Town Preprint* UCT-TP 137/90, April 1990 (Submitted to Phys. Lett. B).
[12] K. Hagiwara, F. Halzen and F. Huzog, *Phys. Lett.* B135 (1984) 324.
[13] E.A. Berends, R. Kleiss, P. De Consmaecken and R. Gastmans, *Phys. Lett.* B103 (1981) 124.
[14] J.H. Reid and M.A. Samuel, *Phys. Rev.* D35 (1987) 3305.
J.H. Reid, G. Li and M.A. Samuel, *Phys. Rev.* D41 (1990) 1675.
[15] J.H. Reid and M.A. Samuel, *Phys. Rev.* D39 (1989) 2046.
[16] G. Couture, *Phys. Rev.* D39 (1987) 2527.
[17] T.R. Grose and K.O. Mikaelian, *Phys. Rev* D23 (1981) 123.
[18] C.J. Goebel, F. Halzen and J.P. Leville, *Phys. Rev.* D23 (1982) 2682; Z. Dongpei, *Phys. Rev.* D22 (1980) 2266.
[19] G. Tupper, J. Reid, G. Li and M.A. Samuel, *Phys. Lett.* B237 (1990) 91.
[20] J. Reid, G. Tupper, G. Li and M. Samuel, *Phys. Lett.* B241 (1990) 105.
[21] M.A. Samuel and G. Tupper, *Prog. Theor. Phys. Lett.* 74 (1985) 1352.
[22] M.A. Samuel, *Phys. Rev.* D27 (1983) 2724.
[23] J. Cortes, K. Hagiwara and T. Herzog, *Nucl. Phys.* B278 (1986) 26.
[24] U. Baur and E.L. Berger, *Phys. Rev.* D41 (1990) 1476.
[25] U. Baur, *University of Wisconsin-Madison Preprint* MAD/PH/541 November 1989.
[26] Leon Lederman, private communication.
[27] Mark Timko and Darryl Dibetento, private communication.
[28] J. Reid, G. Tupper and M. van Zijl, *Phys. Letts.* B218 (1989) 473.
[29] G. Tupper and J. Reid, *Phys. Letts.* B166 (1986) 209.
[30] J. Smith and W.L. van Neerven, *Zeit. Phys.* C30 (1989) 267.
[31] T. Sjöstrand, *Comp. Phys. Comm.* 39 (1986) 347.
B. Andersson, G. Gustafson, G. Ingelman, T. Sjöstrand, *Phys. Rep.* 97 (1983) 31.

Quark and Gluon Structure of Nuclei*

J.P. Vary

Department of Physics, Iowa State University, Ames, IA 50011, USA

Using primarily the tool of deep inelastic lepton scattering, I outline some of the major features of our understanding of the quark and gluon structure of nuclei. The natural progression is from nucleons to deuterium to heavier nuclei. Among other topics, I address the quark-parton model, x-scaling, y-scaling, quark cluster model, and the nuclear equation of state at high density and low temperature.

1 Introduction and Motivation

Deep inelastic lepton-nucleus scattering (DIS) is a well-established and powerful tool for exploring hadronic and nuclear substructure. Results from these experiments are widely accepted as providing one of the major supports for quantum chromodynamics (QCD) as the theory of the strong interactions.[1] Furthermore, in recent years DIS has revealed that at high momentum transfers and at associated short-distance scales, the nucleus is *not* simply a collection of quasifree nucleons. The parton (i.e., quark and gluon) distributions are significantly altered from those of free nucleons. This conclusion from DIS results is forced upon us by the observation[2] (the "EMC Effect") of both enhanced and suppressed regions of scattering from nuclear targets relative to free nucleon targets.

There are several major issues raised by this nontrivial response of nuclei to DIS. For example, can these effects be understood within QCD and/or standard nuclear many-body theory? What are the implications of these results for nuclear theory, for QCD, for the interpretation of high energy particle-nucleus interactions, for the interpretation of high energy nucleus-nucleus interactions, for the nuclear equation of state at high density, etc.? I shall address some of these issues in these three lectures.

2 DIS—Kinematics and Definition of Variables

Cross sections in DIS are so small that most experiments are of the "single-arm" or "inclusive" type where only the scattered lepton is detected. Thus, the experiment sums over ("includes") all final states of the target consistent with a particular final state of the lepton. We are concerned with an experiment where the beam energies E range typically from 10 GeV to 300 GeV and the target absorbs energy ν in the lab frame ranging from a few hundred MeV to tens of GeV. Working in the single (virtual) photon exchange approximation in the lab frame with k, k' representing the initial and final lepton 4-vectors, with q signifying the virtual photon's 4-vector, and with P, P' representing the initial and final target 4-vectors, we have the relations:

---

* Based upon lectures delivered at the South African Summer School in Physics, Cape Town, South Africa, January 1990.

$$k = (\vec{k}, E) \simeq E(\hat{k}, 1) \quad (1a)$$

$$k' = (\vec{k}', E') \simeq E'(\hat{k}', 1) \quad (1b)$$

$$q = k - k' = (\vec{q}, E - E') = (\vec{q}, \nu) \quad (1c)$$

$$P = (0, M) \quad (1d)$$

$$P' = P + q = (\vec{q}, M + \nu) \quad (1e)$$

Now, let Q^2 be the (positive) invariant 4-momentum transfer squared, let θ be the lab scattering angle of the lepton, and let W^2 be the excess invariant final state mass squared:

$$Q^2 \equiv -q^2 \simeq 2kk' = 4EE' \sin^2 \theta/2 \quad (2a)$$

$$W^2 \equiv P'^2 - P^2 = 2P \cdot q + q^2 = 2M\nu - Q^2 \quad (2b)$$

The adjective "deep" of DIS is usually reserved for the kinematic region $Q^2 \geq 1\text{GeV}^2/c^2$ and $W^2 \geq 2\text{GeV}^2$, but these boundaries are not rigid. Below these values particle resonances contribute substantially to the inclusive inelastic cross sections. In more recent times there is a preference to move these boundaries to higher values. The kinematics of DIS are depicted in Fig. 1.

Since $W^2 \geq 0$ we have $2M\nu \geq Q^2$ from Eq (2b) with equality occurring at the elastic-scattering limit where the target recoils in its ground state. Thus, when $Q^2 = 2P{\cdot}q = 2M\nu$ we are measuring the elastic electromagnetic form factor of the target. Using m for the mass of a nucleon we define Bjorken x as

$$x \equiv \frac{Q^2}{2m\nu} \quad (3)$$

which has the property that, for lepton scattering from a free nucleon,

$$0 \leq x \leq 1 \quad (4)$$

with $x = 1$ corresponding to elastic lepton-nucleon scattering. The primary virtue of choosing x as a kinematic variable is that in QCD (or in any non-abelian gauge theory) as $Q^2 \to \infty$ and $\nu \to \infty$ the DIS cross section becomes a function *only of* x in the single

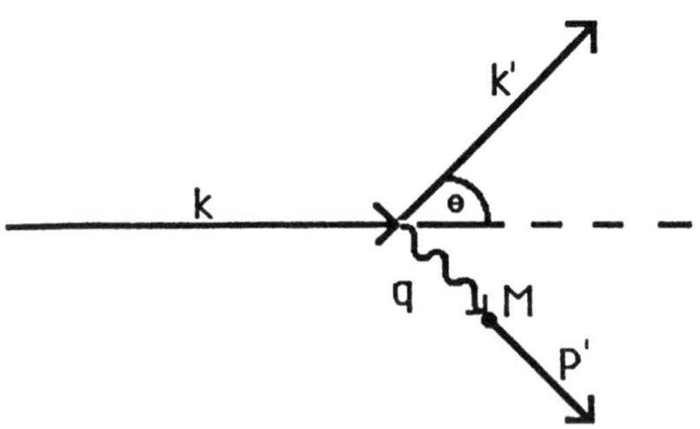

Figure 1.

photon exchange approximation. This is referred to as "Bjorken-scaling" or "x-scaling." It was the appearance of this simple scaling in the SLAC nucleon data in the late 1960s that prompted Feynman to introduce his parton model description of the hadrons. In this model (and in QCD) x represents the light-cone momentum fraction of the hadron's 4-momentum carried by the parton (quark) absorbing the photon. The feature that the cross section at high Q^2 depends only on the initial state of the system and that final state interactions are of no consequence is referred to as "asymptotic freedom."

To introduce QCD and quantize it on the light cone and then to derive the inclusive cross section would require an extensive presentation. I will resort to a more heuristic presentation. Consider the reaction when Q^2 gets very large, so large that we can neglect the motions of quarks transverse to the virtual photon direction. Now boost to a frame where the quark that absorbs the photon has its z-component reversed. In this frame the photon has no 4th component since the struck parton's energy is unchanged. As $Q^2 \to \infty$ this frame is equivalent to the "infinite momentum frame."[3] This gives us the parton picture in the "brickwall" or "Breit" frame as depicted in Fig. 2, where we use the notation that a 4-vector p is specified by $(\vec{p}_\perp, p_3, p_0)$ and where we have oriented the z-axis with the photon. Note that in Fig. 2a we specify that Q^2 is so high that all the partons are moving toward the photon. Although I have only depicted 3 partons in the target, the number of partons is actually arbitrary. Also, the "spectator" partons can be either quarks or gluons. From these considerations we conclude (working with Fig. 2):

$$Q = -2p_3 \tag{5a}$$

$$x = \frac{Q^2}{2P \cdot q} = \frac{Q^2}{-2P_3 Q} = \frac{p_3}{P_3} \tag{5b}$$

Thus we identify x as the longitudinal momentum fraction carried by the struck quark. Consequently, one sees that the inelastic cross section as a function of x maps out the x-distribution of partons in the target. Furthermore, at $x = 1$ all the spectator partons are essentially at rest and the struck parton carries all the momentum of the target. Reversing the struck parton's momentum at $x = 1$ puts the entire target in its time-reversed state in this frame, which is a manifest elastic-scattering situation. Viewing the reaction in the Breit frame gives a physical picture to the meaning of x, but the more precise definition in terms of the light-cone momentum fraction will be given below. As $Q^2 \to \infty$ the Breit

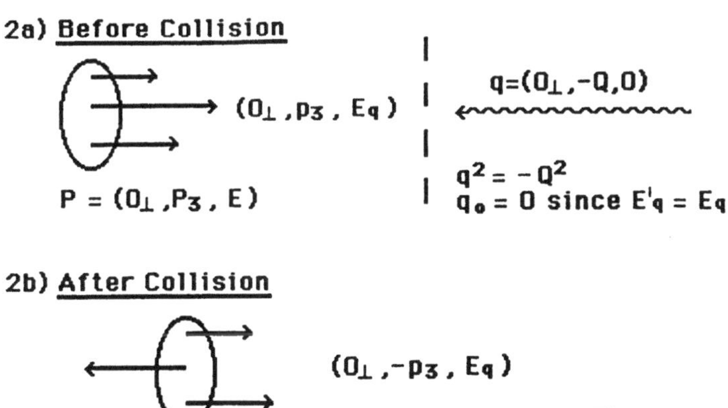

Figure 2.

frame becomes the infinite momentum frame in which the kinematics are equivalent to the light-cone kinematics.

To generalize the above discussion for arbitrary nuclear targets it is customary to retain the definition of x in Eq (3) but note that Q^2 extends up to $2P \cdot q = 2M\nu$ so that instead of Eq (4) we have

$$0 \le x \le \frac{2P \cdot q}{2m\nu} = \frac{M}{m} \tag{6}$$

If we write the $M = A(m - \epsilon)$ where ϵ is the average binding energy per nucleon in the nucleus with A nucleons then

$$0 \le x \le A(1 - \frac{\epsilon}{m}) \tag{7}$$

For experiments at fixed E there are two independent scattering variables. Frequent choices of these two variables are (Q^2, ν) and (x, Q^2). Other choices occur such as (y, Q^2), which is discussed at some length below.

3 Light-Cone Variables and Cross Section Expressions

It has been shown that for the purpose of applying perturbation theory there are some advantages to quantizing a field theory on the light cone.[4,5] For a review of the rules for evaluating perturbative QCD graphs on the light cone, see Ref. 6, and for a comparison see the lectures by Thews in this summer school.

Working in light-cone coordinates, the DIS cross section is dominated by the "handbag" diagram for the imaginary part of the forward Compton scattering amplitude as depicted in Fig. 3. These coordinates are defined by leaving the perpendicular components of vectors unchanged while making the following linear combinations of the third and fourth components.

$$k^+ = k^0 + k^3 \tag{8a}$$

$$k^- = k^0 - k^3 \tag{8b}$$

The quantity $k^+(k^-)$ is referred to as the light-cone momentum (energy). In quantizing a field theory on the light cone one quantizes on a surface of equal light-cone time $x_- = x_0 - x_3$. Note that

$$k^0 = \sqrt{m^2 + (k^3)^2 + (k^\perp)^2} \ge k^3 \tag{9a}$$

which implies

$$k^+ \ge 0 \tag{9b}$$

A hadron on the light cone is composed of partons ($i = 1, 2, \ldots$) and the total light-cone momentum K^+ of the particle is

$$K^+ = \sum_i k_i^+ \tag{10}$$

Therefore, if we define the light-cone momentum fraction of the ith parton

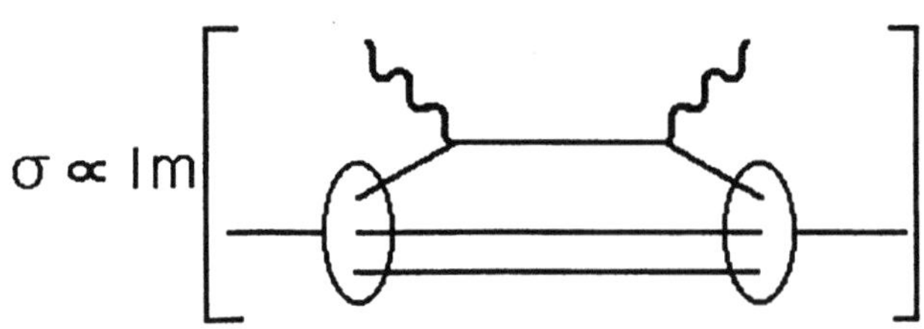

Figure 3.

$$x_i = \frac{k_i^+}{K^+} \tag{11}$$

then, clearly, this is a Lorentz-invariant quantity that satisfies

$$0 \leq x_i \leq 1 \tag{12}$$

and

$$\sum_i x_i = 1 \tag{13}$$

If we also introduce $f_i(x)$, the Lorentz-invariant distribution of light-cone momentum fraction for the ith constituent we can require a normalization

$$\int_0^1 f_i(x)dx = 1 \tag{14}$$

We can denote the mean light-cone momentum fraction as

$$< x_i >= \int_0^1 f_i(x)x dx \tag{15}$$

so that, in view of Eqs (13) and (14) the conservation of total momentum fractions is

$$\sum_i < x_i >= 1 \tag{16}$$

The connection between the light-cone momentum fraction and Bjorken x should be clear from the heuristic arguments developed in the Breit frame. That is, they are the same quantity. Application of the rules for perturbation theory reveals that Fig. 3 results in a cross section for small-angle scattering which behaves as:

$$\sigma(x, Q^2) \propto \sum_i e_i^2 x f_i(x, Q^2) \tag{17}$$

where e_i is the charge on the ith quark and $f_i(x, Q^2)$ is the momentum distribution of the ith quark as seen at a momentum transfer squared of Q^2. The higher Q^2 is the more the probe resolves the momentum distribution of the quark. Imagine the quark as radiating and reabsorbing gluons and $q\bar{q}$ pairs. As the Q^2 increases we resolve the shorter time components of these processes. Asymptotic freedom implies this Q^2 dependence becomes extremely weak and can be essentially neglected over a wide range of Q^2 values. This "x-scaling" or "Bjorken-scaling" was observed in the SLAC data in the late 1960s. The distributions $f_i(x)$ which emerge from these data are then interpreted as simply the light-cone momentum distributions of the quarks. The corresponding $f_i(x)$ for the gluons are not as readily accessible but are constrained, for example, by Eqs (14) and (16). These distributions are referred to variously as $f_q(x), f_g(x)$ for quarks and gluons, respectively. When flavor-specific information is retained we have $f_u(x), f_d(x), f_s(x), f_g(x)$, etc. We also speak of valence, $f_v(x)$, and ocean (or sea) distributions, $f_o(x)$, to distinguish the dominant high x from the dominant low x contributions, respectively.

The expression for the double differential cross section measured in the single-arm experiments when no spin information is retained is written:

$$\frac{d^2\sigma}{d\Omega dE'} = \sigma_M[W_2(x, Q^2) - 2W_1(x, Q^2)\tan^2\theta/2] \tag{18}$$

and it is valid for lepton scattering at any angle. Here W_1 and W_2 are the inelastic structure functions and σ_M is the Mott cross section

$$\sigma_M = \frac{4\alpha^2(E-\nu)^2 \cos^2\theta/2}{Q^4} \tag{19}$$

with α representing the fine structure constant. For small- angle scattering the second term in Eq (18) can be neglected. Then by using appropriate factors we can isolate a dimensionless function of (x, Q^2) which can be related to the quark distribution functions through evaluating the diagram in Fig. 3. This results in

$$\frac{\nu}{\sigma_M}\frac{d^2\sigma}{d\Omega dE'} = \nu W_2(x, Q^2) = \sum_i e_i^2 x f_i(x, Q^2) \tag{20}$$

Due to the charge weighting one can imagine that a sufficiently complete set of data on proton and deuterium targets could be used to extract the $f_i(x, Q^2)$ from the DIS data. This led to a large amount of experimental activity in elementary particle physics and to a number of sets of phenomenological parameterizations of these $f_i(x, Q^2)$. A key issue to which we will return in the second lecture is whether the corrections for strong internucleon interactions in the deuterium system are well enough known that we can have confidence in correcting the deuterium data to arrive at data from a "neutron target." It is important to note that since neutrinos couple to quarks through both charge and neutral currents, the neutrino cross sections are helpful in extracting the $f_i(x, Q^2)$ in spite of being statistically less significant.

Before looking at the results for the quark distribution functions from nucleon targets, I ask the question: How is the above discussion modified for a nuclear target consisting of A nucleons? For simplicity I will neglect binding and take $\epsilon \to 0$ in Eq (7) and Eqs (12–13) become

$$0 \le x_i \le A \tag{21}$$

and

$$\sum_i x_i = A \tag{22}$$

respectively. This implies that

$$x_{el} = \frac{Q^2}{2m\nu} = A \tag{23}$$

corresponds to the elastic-scattering limit and the region where x approaches A is referred to as the "threshold" region.

To more easily compare results from $A > 1$ targets with those from nucleon targets, it is customary to define the nuclear structure function per nucleon (again for small angle scattering by)

$$\frac{\nu W_2(x, Q^2)}{A} \equiv \frac{1}{A}\frac{\nu}{\sigma_M}\frac{d^2\sigma^A}{d\Omega dE'} \tag{24}$$

$$\equiv \frac{1}{A}\sum_i e_i^2 x f_i^A(x, Q^2) \tag{25}$$

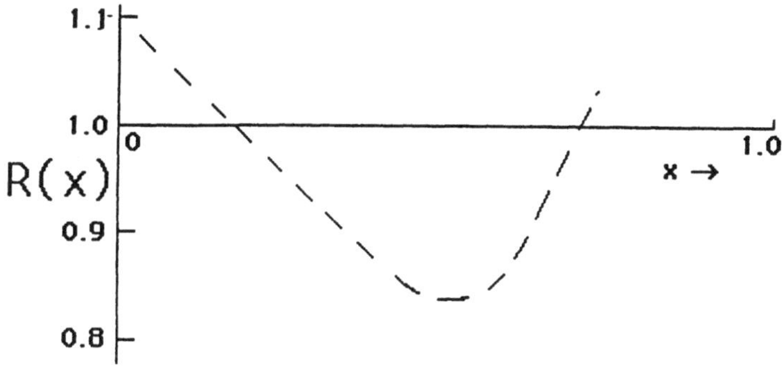

Figure 4.

Note that x/A is the light-cone momentum fraction of the *total* nuclear momentum carried by the struck quark. A major focus of recent research has been the issue of differences between $f_i(x,Q^2)$ from Eq (20) and $f_i^A(x,Q^2)$ of Eq (25). Typically, this difference is studied only after averaging over the flavor dependence since the data is not abundant enough to extract such detail. Thus, one extracts a flavor independent $f(x,Q^2)$ via Eq (20) (with averaging over the appropriate number of protons and neutrons) and Eq (25) from the data by dropping the subscript on $f(x,Q^2)$. The most celebrated example of this comparison is that of the EMC group[2] who measured the ratio of these two functions in iron and deuterium at very high Q^2

$$R(x) \equiv \frac{f^{^{56}Fe}(x,Q^2)}{f^D(x,Q^2)} \tag{26}$$

The EMC results, which have been confirmed in a range of other experiments, are illustrated schematically in Fig. 4.

The main thing to note is that there are enhanced and suppressed regions of scattering in iron relative to deuterium. The low x behavior has been shifted downward in more recent publications of the data but that region is more complicated, theoretically, and we shall not discuss it at any length here. The suppressed region of scattering at intermediate x values of $R(x)$ indicates the intermediate momentum valence quarks are depleted in iron relative to deuterium. The enhancement at large x is thought to be due at least, in part, to stronger Fermi motion in iron relative to deuterium.

It was the intermediate x-behavior that most surprised the particle physicists and spurred a great deal of theoretical effort. For one example, consult Ref. 7, which interprets the EMC effect in terms of the Quark Cluster Model (QCM).[8] I shall return to the QCM in more detail below where the largely unexplored $x > 1$ region will be seen to be of great interest.

The entire kinematic domain for high energy lepton-nucleus scattering is depicted in Fig. 5. Note that there is a domain between slope $x = 1$ and slope $x = A$ which is only accessible with nuclear targets. The line $x = 1$ corresponds to the elastic form factor measurements from nucleon targets and to the quasi-elastic peak (nucleon knockout) in inelastic lepton- nucleus scattering at low to moderate Q^2. This peak disappears with increasing Q^2 at the rate of decrease of the elastic nucleon form factor. I shall show this in some detail with the deuteron target in the second lecture.

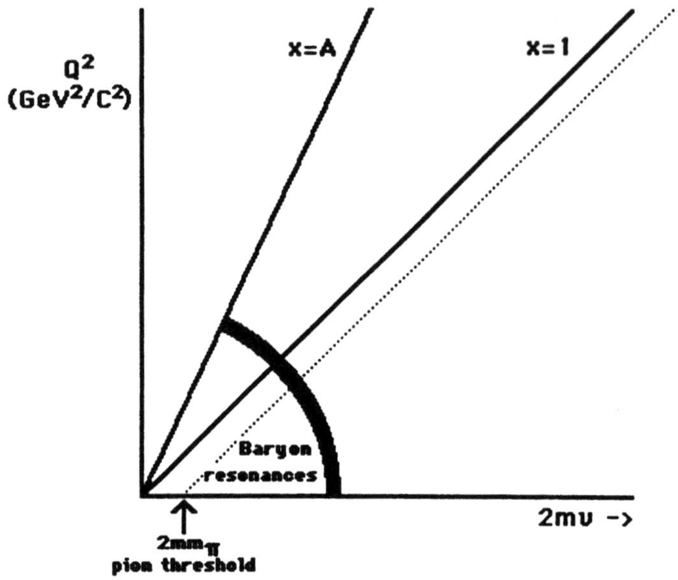

Figure 5.

4 Quark Distribution Functions for Free Nucleons

Some of the qualitative features of these quark distribution functions are easy to understand. Imagine an amplitude $\phi_i(k^\perp, x, k^-)$ for finding a quark i with the specified kinematic variables. This amplitude may be, for example, the solution of a covariant wave-mechanical treatment of QCD in light-cone quantization. Suppose the normalization is chosen appropriately so that on some scale of resolution $Q^2 = Q_0^2$ we have

$$f_i(x) = \int dk^- \int d^2k^\perp \left|\phi_i(\vec{k}^\perp, x, k^-)\right|^2 \tag{27}$$

with

$$\int_0^1 f_i(x)dx = 1 \tag{28}$$

This would provide a dynamical framework for theoretical predictions of $f_i(x)$. However, this requires a full nonperturbative solution of QCD which is not yet available in spite of more than 10 years of effort in lattice QCD.

Nevertheless, some of the qualitative features of the theoretical $f_i(x)$ can be obtained.[9] For example, the relativistic many-body phase space for N partons controls the behavior as $x \to 1$ (counting rules). The normalization of Eq (28) and endowing the structured particle with a Regge behavior for the mass spectra give further constraints.[9] These conditions lead to a flavor-independent form for the valence quarks:

$$f_v(x) \sim \text{norm} \cdot \frac{1}{\sqrt{x}}(1-x)^{2(N-1)-1+S} \tag{29}$$

where $N = 3$ for baryons and 2 for mesons, and $S = 1(0)$ for integer (half-odd integer) spin. This form is a reasonable first approximation to the measured $f_v(x)$ at $Q^2 \sim 10\ \text{GeV}^2/c^2$.

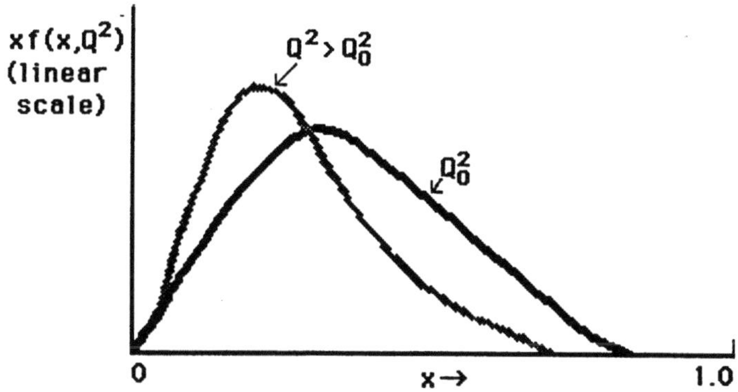

Figure 6.

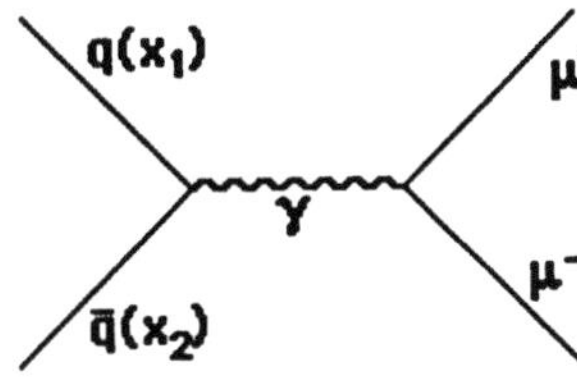

Figure 7.

Starting with the above considerations and adding the feature of QCD scale evolution (evolution in Q^2) which was studied thoroughly by Alterelli and Parisi,[10] Buras and Gaemers[11] developed one of the first flavor-dependent parameterizations which successfully fit a wide range of data. Their flavor-independent valence-quark parameterization is summarized:

$$f_v(x, Q^2) = (B[0.5,\ 2N + 1 + \alpha\overline{S}])^{-1} x^{-1/2}(1 - x)^{2(N-1)+\alpha\overline{S}} \tag{30}$$

where B is Euler's beta function, $\overline{S} = \ln[\alpha_s(Q_0^2)/\alpha_s(Q^2)]$, $Q_0^2 = 1.8\ \mathrm{GeV}^2/\mathrm{c}^2$, $\alpha = 2.4$ and α_s is the running coupling constant of QCD. The general character of the Q^2 dependence (scale violations) is sketched in Fig. 6.

Buras and Gaemers[11] also present sea quark distribution functions. The sea quarks are quark-antiquark ($q\overline{q}$) pairs which dominate the soft components (low x) of the nucleon. The $q\overline{q}$ content is accessible via the Drell-Yan process[12] wherein two hadrons collide at high energy and produce a massive $\mu^+\mu^-$ pair.

The elementary process involves q from one hadron fusing with $\overline{q}$ from the other and forming a virtual photon which then decays to the μ-pair as illustrated in Fig. 7. It is the involvement of the $\overline{q}$ which leads to information on the sea quarks in the nucleon. In lowest order, the cross section is written

$$\frac{d\sigma}{dM^2} = \frac{4\pi\alpha^2}{9M^2} \sum_i e_i^2 \int dx_1 dx_2 F(x_1, x_2)\delta(M^2 - x_1 x_2 s) \tag{31}$$

where M represents the μ-pair mass and S is the total 4 momentum squared of the colliding hadrons. In this expression

$$F(x_1, x_2) = q_i^A(x_1)\overline{q}_i^B(x_2) + \overline{q}_i^A(x_i)q_i^B(x_2) \tag{32}$$

and $q_i(x)$ represents the sum of valence and sea quark $f(x)$'s while $\overline{q}_i(x)$ represents the $\overline{q}$ distribution functions. Experimentally it is known that

$$\overline{q}_i(x) \sim norm \cdot (1 - x)^7 \tag{33}$$

where the exponent is rather uncertain and may be as high as 9 or 10.

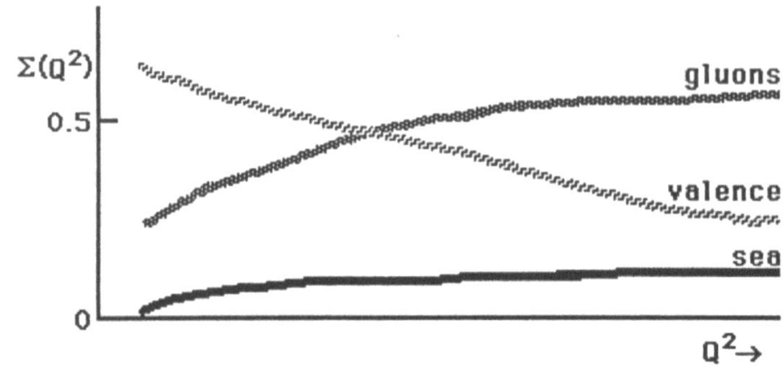

Figure 8.

The issue of how the sea quark distributions are altered in nuclei was raised in the EMC effect and has received considerable attention since then.[13] A Drell-Yan experiment has been performed with the explicit goal of studying this issue[14] and has found that the sea distributions in nuclei are largely A- independent and comparable with the sea distributions in nucleons. This would appear to raise serious difficulties for models of the EMC effect based on enhanced pion currents in nuclei. On the other hand this A-independence at low x is consistent with the QCM.[15]

The gluon distributions in nucleons are less well-known than the valence and sea distributions. However, gluon distribution functions are determined[11,16] through the use of constraints provided, for example, by sum rules and by requiring that QCD scale–evolution of quark distributions (where gluon distributions are inputs) fit available data.

From these distributions one can determine the fraction of the nucleon's light-cone momentum carried by the valence, gluon, and sea distributions according to Eq (15). The experimental results imply that at $Q^2 \simeq 30\ \mathrm{GeV}^2/\mathrm{c}^2$

$$\sum_{\text{valence } i} < x_i > \sim 0.5$$
$$\sum_{\text{sea } i} < x_i > \sim 0.1 \tag{34}$$
$$\sum_{\text{gluons } i} < x_i > \sim 0.4$$

Then, as a function of Q^2 these sums, referred to as $\sum(Q^2)$, evolve in a manner depicted in Fig. 8. For the purposes of the detailed examination of the DIS data on deuterium, I will present, here, a sample of some DIS data on a hydrogen target at rather high x-values.

In Fig. 9, I depict νW_2 versus x at various values of Q^2 from SLAC experiment E133.[17] The solid curve represents the SLAC parameterization[17] of their data and includes terms to accommodate the strong Q^2-dependent baryon-resonance contributions which are clearly seen as bumps in these data. For comparison, I also show the smooth Buras-Gaemers parameterization which is from a global fit to all DIS data and is designed to omit resonance contributions.

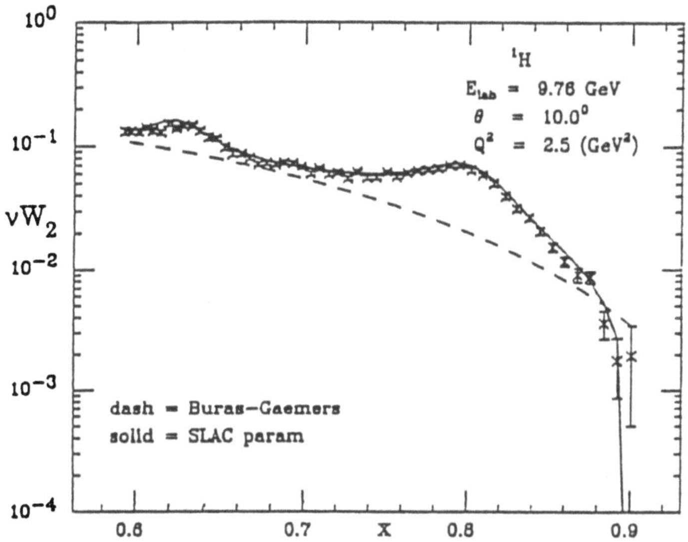

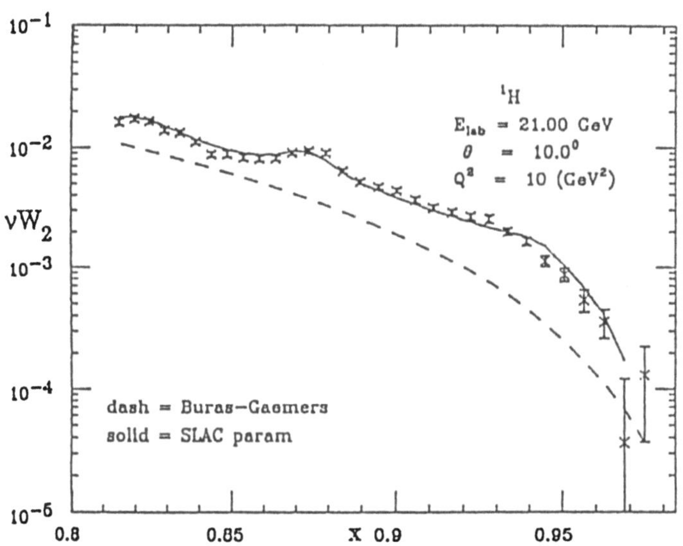

Figure 9. Electron-proton inclusive data from SLAC experiment E133[17] expressed as νW_2 versus Bjorken x. The solid curve represents the SLAC parameterization discussed in the text while the dash curve gives the results from the parameterization of Ref. 11.

5 Introduction to DIS from Deuterium Targets[18]

One might ask why we should study deuterium. Deuterium is the "elementary" target of choice for "neutron" data. Consequently, there is abundant deuterium data and there has been a large amount of theoretical effort to develop consistent procedures to "subtract" the proton contributions from the deuterium data. High quality deuterium wavefunctions are available which fit all the low energy properties of deuterium and arise from nucleon-nucleon (NN) interactions which fit the NN phase shifts. Since deuterium has the lowest average density of any nucleus one hopes that final state interactions (FSI's) are of minimal importance and can be neglected.

The thrust of this lecture is to build a case for novel effects in the nuclear response that can be attributed to the formation of a 6-quark cluster in deuterium with a certain probability attached. In the QCM this probability increases with density so one expects this probability is small in deuterium. Indeed, I will show this probability (called $\tilde{p}_6$) is 4.7% for the Reid Soft Core deuterium wavefunction[19] and 5.4% for the Bonn wavefunction.[20] Thus, to establish this effect I will concentrate on several issues revolving around these data. The first concerns the phenomenon of y-scaling, or, equivalently, the claim that the data I address is dominated by quasi-elastic nucleon knockout processes. If this is true, then inelastic (quark) processes are not relevant and Bjorken x-scaling is not observed. I will show that, in fact, both processes are contributing in the kinematic region where the data is most sensitive to clustering effects.

6 How Useful Is the Concept of y-Scaling?[21]

Consider elastic lepton-nucleon scattering from an off-shell nucleon moving in the lab frame with $\vec{k}$ and energy $E_1(\vec{k})$. In the impulse approximation this nucleon absorbs the virtual photon $(\vec{q}, \nu)$ leaving an on-shell spectator nucleon from the deuterium target. The kinematics are depicted in Fig. 10 and it clearly indicates the neglect of FSI. For this quasi-elastic nucleon knockout

$$\frac{d^2\sigma}{d\Omega dE'} = \sigma_M \frac{m}{E(|\vec{k}+\vec{q}|)} \frac{m}{E_1(|\vec{k}|)} \delta(\nu + E_1(|\vec{k}|) - E(|\vec{k}+\vec{q}|)) \cdot [W_2^{el} - 2W_1^{el} \tan^2\theta/2] \quad (35)$$

where

$$W_2^{el} = (F_1^2 + \frac{\kappa Q^2}{4m^2} F_2^2) \quad (36)$$

$$W_1^{el} = -\frac{Q^2}{4m^2}(F_1 + \kappa F_2)^2 \quad (37)$$

Here F_1 and F_2 are the Dirac and Pauli form factors and κ is the anomalous magnetic moment and the conventions are those of Bjorken and Drell.[22] In the above there is an average over the nucleon spin. To finish embedding this in the deuteron we treat the nucleon as moving with a laboratory momentum distribution $n(|\vec{k}|) = n(k)$ in the initial state given by any chosen dynamical model. Then

$$\frac{d^2\sigma}{d\Omega dE'} = (\sigma_n + \sigma_p) \int d^3k\, n(k) \frac{m}{E(|\vec{k}+\vec{q}|)} \delta(\nu + E_1(|\vec{k}|) - E(|\vec{k}+\vec{q}|)) \quad (38)$$

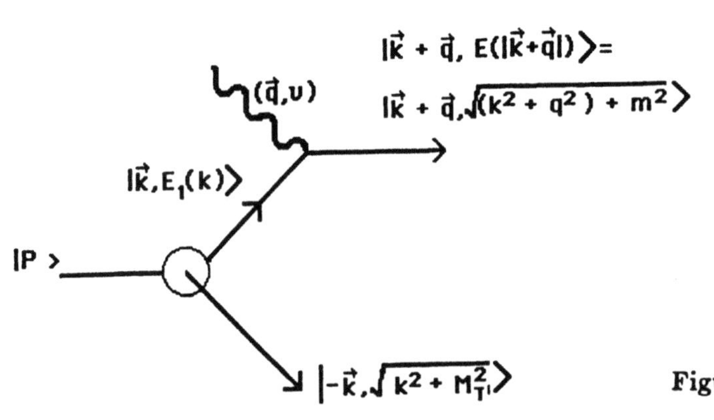

Figure 10.

with

$$\sigma_j = \sigma_M[W_2^j - 2W_1^j \tan^2\theta/2] \tag{39}$$

where j stands for n or p, and the normalization of $n(k)$ is given by

$$\int d^3k\, n(k) = 1 \tag{40}$$

It is important to note that a factor $\frac{m}{E_1(|\vec{k}|)}$ has been absorbed into $n(k)$ by virtue of the normalization chosen. Now, utilize the symmetry of the delta function to reduce the cross section to a single dimensional integral. This results in

$$\frac{d^2\sigma}{d\Omega dE'} = (\sigma_n + \sigma_p)\frac{2\pi m}{|\vec{q}|}\int_y^{y'} n(k)kdk \tag{41}$$

with

$$y = a - b \tag{42}$$

$$y' = a + b \tag{43}$$

$$b = \frac{|\vec{q}|}{2} \tag{44}$$

$$a = \frac{\nu + M_d}{2}\left[1 - \frac{4m^2}{(\nu + M_d)^2 - |\vec{q}|^2}\right]^{1/2} \tag{45}$$

In the above, M_d represents the deuterium mass. This discussion can easily be extended to an arbitrary nucleus and the resulting expressions for a and b are more complicated.[21] The nucleon cross sections σ_j are evaluated here and throughout this work with the parameterizations of Gari and Krümpelmann.[23]

In Eq (41) we can define

$$F(y, Q^2) = \int_y^{y'} n(k)kdk \tag{46}$$

and for $Q^2 \geq 0.25$ GeV^2/c^2 the upper limit, y', can be extended to infinity with little error for typical nuclear $n(k)$ distributions. In this limit F becomes dependent only on y ("y-scaling"). The variable y can be viewed in a manner analogous to Bjorken x. For negative values of y, it represents the momentum carried by the nucleon toward the photon it is absorbing.

The key issue of y-scaling is whether the data indicate that $F(y, Q^2)$ is independent of Q^2. In the impulse approximation the scaling function $F(y)$ is directly related to the nucleon momentum distribution function. Thus, if the experimental data exhibit y-scaling we have a model-independent determination of the nucleon momentum distribution in the nucleus. Of course, there is still a possible conspiracy that neglected effects cancel but more exclusive experiments could eventually test this. In fact, a recent report[24] on (e,e′p) measurements with ^{3}He targets indicates the invalidity of the y-scaling interpretation of the inclusive data. The main point here is to exhibit the lack of y-scaling in the inclusive deuterium data.

Consider results from two experiments[17,25]: one with scattering angle $\theta = 8°$, 6.519 GeV $\leq$ incident energy $E_{lab} \leq$ 18.257 GeV and the second with $\theta = 10°$, 9.761 GeV $\leq E_{lab} \leq$ 21.001 GeV. The data group into 11 sets characterized by $Q^2 \simeq$ 0.8, 1.0, 1.5, 1.75, 2.0, 2.5, 3, 4, 6, 8, and 10 (GeV/c)^2. In order to have a smooth representation of the $y \leq 0$ data we fit to a functional form of a gaussian centered at $y = 0$ plus two exponentials. The form $F(y, Q^2) = a\exp(-by^2) + c\exp(dy) + f\exp(gy)$ employs Q^2 dependent coefficients. Without Q^2 dependence the best fit gives χ^2/d.o.f = 15.1. With the restriction that the Q^2 dependence of the coefficients be of polynomial form with terms Q^{-4}, Q^{-2}, Q^2, and Q^4 and the restriction of only 10 parameters for the Q^2 dependence, the best fit (χ^2/d.o.f. = 1.26) was obtained with

$$\begin{aligned} a &= 0.01335(1.0 - 0.778\,Q^{-2} - 0.2\,Q^2 + 0.0151\,Q^4), \\ b &= 369.7, \\ c &= 0.0028(1.0 + 0.5209\,Q^{-2} + 0.406\,Q^2), \\ d &= 21.05(1.0 - 0.124\,Q^{-2}), \\ f &= 0.000365(1.0 - 0.390\,Q^2 + 0.0573\,Q^4), \\ g &= 1.979(1.0 + 1.053\,Q^{-2} + 0.39\,Q^2) \end{aligned} \tag{47}$$

which I refer to as fit A.

With the same functional dependence a fit (fit B) to the 8 data sets omitting the $Q^2 = 0.8$, 1.0, and 10.0 GeV$^2/c^2$ sets yielded χ^2/d.o.f. = 1.17 with different parameter values from those quoted above. The differences between fit A and fit B in the range $1.5 \le Q^2 \le 8$ GeV$^2/c^2$ can be interpreted as an indication of the uncertainty in the Q^2 dependence of $F(y, Q^2)$ dictated by the deuteron data.

In Fig. 11 I give a typical comparison of a data set with the fits. Here $Q^2 = 4.0$ (GeV/c)^2. Since inelastic contributions play a major role in the high energy side of the quasi-elastic peak (positive y), I consider only the negative y region. The results of fit A for $F(y, Q^2)$ are plotted versus y in Fig. 12 for certain Q^2 values in the range $1.75 \le Q^2 \le 8$ GeV$^2/c^2$. One finds systematic scaling violations for the *entire* range of y. $F(y, Q^2)$ from both fits is plotted versus Q^2 for three different values of y in Fig. 13.

For each value of y a different pattern emerges in Fig. 13. One observes that at any value of Q^2 the data do not indicate approach to scaling in the variable y. For $y = -0.1$ GeV/c, $F(y, Q^2)$ rises continuously with Q^2, while for $y = -0.4$ GeV/c it exhibits a minimum at $Q^2 \simeq 4$(GeV/c)^2 and for $y = -0.8$ GeV/c it falls uniformly over this range of Q^2. Thus, one concludes these data do not exhibit y-scaling and hence are not

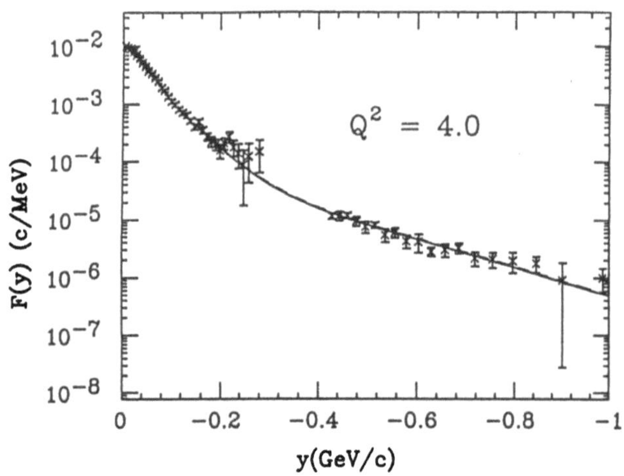

Figure 11. Comparision of fit A (solid line) and fit B (dashed line) to $F(y, Q^2)$ for $0 \le y \le -1$ at $Q^2 = 4.0$ (GeV/c)^2 with the data. The fits coincide so are not distinguishable in the figure.

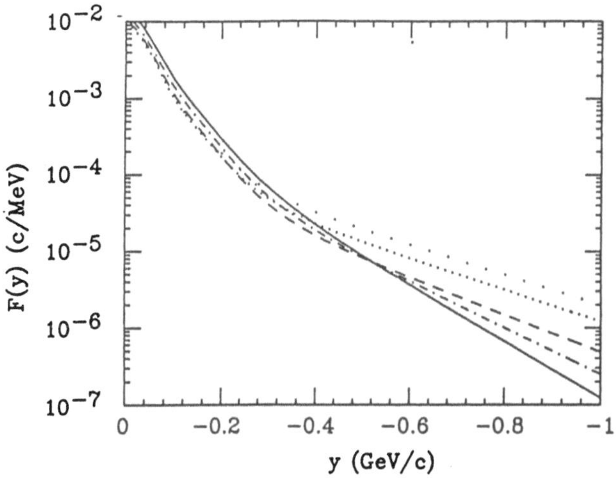

Figure 12. Fit A results for $F(y, Q^2)$ plotted versus y for the range $0 \leq y \leq -1$. $Q^2(\mathrm{GeV}/c)^2 = 1.75$ (rare dots), 2.5 (dense dots), 4.0 (dashes), 6.0 (dot-dashes), and 8.0 (solid).

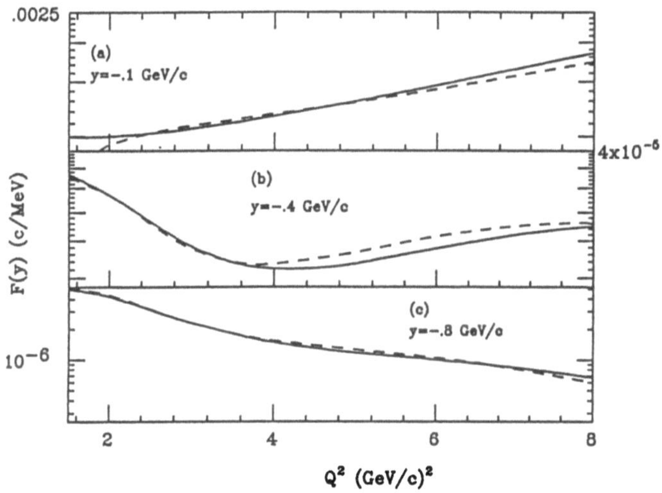

Figure 13. $F(y, Q^2)$ versus Q^2 for three different values of y. Fit A is the solid line and fit B is the dashed line.

explained by a simple quasi-elastic nucleon knockout picture. Similar conclusions about the inadequacy of the y-scaling picture have been made in connection with data from other nuclear targets.[26] The deviations from y-scaling might be assigned to effects of FSI, meson exchange currents or quark clustering. I would argue that the Q^2 dependence of each of these is expected to be quite different. In addition, different choices of variables could elucidate the origins of these deviations from y-scaling. For example, if the differences between the measured $F(y, Q^2)$ and a single $F(y)$ were shown to obey Bjorken x-scaling we would have conclusive evidence that incoherent inelastic scattering from quarks was the origin of the deviations from y-scaling. As I will show with detailed calculations this is indeed the case for the deuterium data.

Since the Q^2 dependence of the effects which violate y-scaling might be expected to be different, it is possible that additional theoretical and experimental effort could map out the regions where each effect produces its own characteristic signature.

7 Quasi-Elastic and Nucleon Inelastic Processes in Deuterium[21]

Based upon Eq. (41) the quasi-elastic process can be evaluated with a model for the nucleon momentum distribution, $n(k)$.

In Fig. 14 I present the $n(k)$ emerging from the ground state wavefunctions of the RSC[19] and of the Bonn[20] NN potentials. Fig. 14(a) presents the total $n(k)$ which is the sum of the S-state (u) and D-state (w) contributions. That is,

$$\begin{aligned} n(k) &= \tilde{u}^2(k) + \tilde{w}^2(k) \\ &= \frac{1}{2\pi^2}\left\{\left|\int j_0(kr)u(r)rdr\right|^2 + \left|\int j_2(kr)w(r)rdr\right|^2\right\} \end{aligned} \tag{48}$$

For convenience I use the simplified parametrizations of the deuterium ground state wavefunctions as in Refs. 19 and 20. Noticeable differences between the RSC and Bonn distributions are found for $k \geq 0.25$ GeV. Figs. 14(b) and 14(c) are intended to elucidate the origins of these differences by displaying separately the S-state and D-state contributions to the total distribution in the RSC and Bonn, respectively. Although the D-state constitutes only 6.48% of the RSC ground state and only 4.25% of Bonn ground state, the S-state and D-state contributions to the total momentum distributions are comparable around $k = 0.25$ GeV for both potentials. For $0.25 \leq k \leq 0.60$ GeV the D-state contribution dominates in both cases. For $0.60 \leq k \leq 1.0$ GeV the situation is more complex. In the case of Bonn the S- and D-state contributions are comparable. On the other hand,

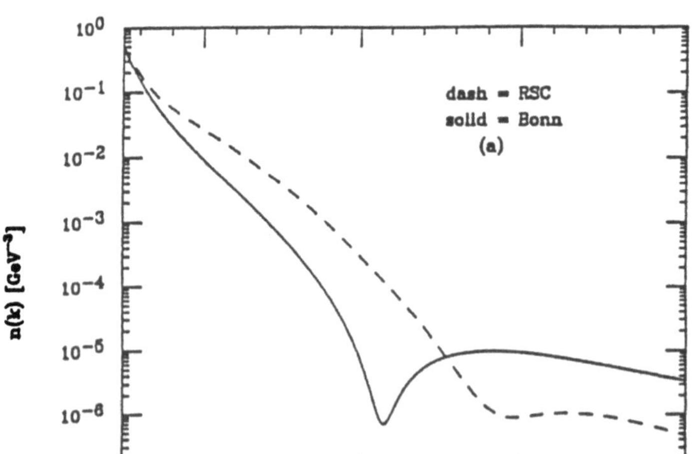

Figure 14. (a.) The total momentum distribution of a single nucleon in a deuterium nucleus for the RSC potential (dashes) and for the Bonn potential (solid).

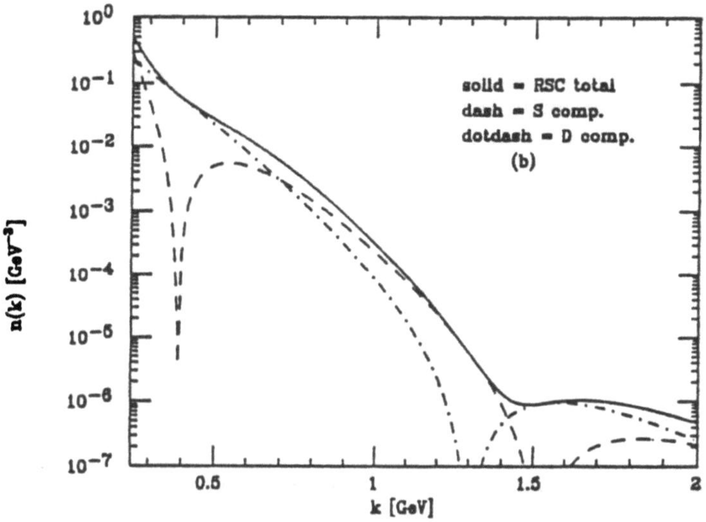

Figure 14. (b.) The RSC momentum distribution of a single nucleon in a deuterium nucleus (solid) and its S-state (dashes) and D-state (dotdashes) components.

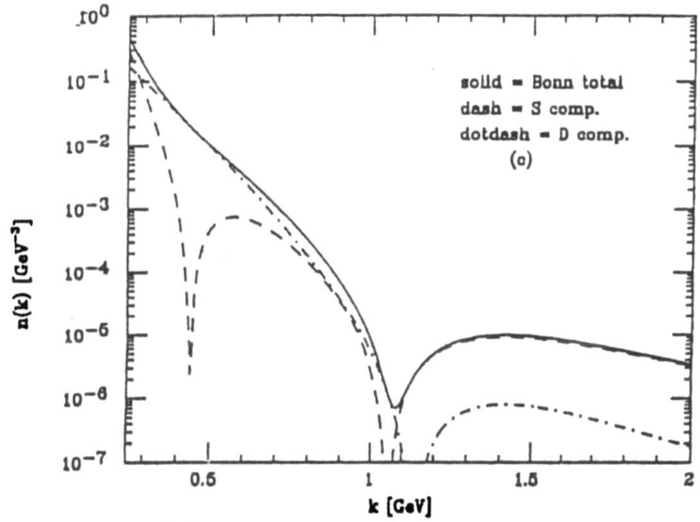

Figure 14. (c.) The Bonn momentum distribution of a single nucleon in a deuterium nucleus (solid) and its S-state (dashes) and D-state (dotdashes) components.

for RSC, the S-state contribution is rising relative to the D-state contribution. The net result is that the RSC S-state contribution at $k = 1$ GeV is nearly 100 times the S-state contribution from Bonn. On the other hand, the RSC D-state contribution at $k = 1$ GeV is only about 20 times the D-state contribution from Bonn.

The data I address are primarily sensitive to these details for k values up to about 0.75 GeV. One main point I wish to stress is that due to the importance of the D-state for momenta up to at least 1 GeV one must go beyond this momentum to isolate the role of short-range correlations. This is consistent with a naive picture based on the de Broglie relation where a momentum less than 1 GeV implies distance sensitivity greater than $1fm$. Thus, a major focus of this investigation is upon the intermediate momenta components of deuterium which are dominated by tensor correlations.

The other favorable process to occur in the data I address is inelastic scattering involving meson production off the bound nucleon. This proceeds by taking a representation of the inelastic lepton-nucleon data and "smearing" it over the nucleon's Fermi motion in the deuteron. Although there is a controversy[27–29] about how the smearing procedure is implemented there is no doubt that in a nucleon based picture of the nucleus there is a need for treating the Fermi motion.[29] Since the deuteron is not only the simplest but also a diffuse and barely bound nuclear system, one does not intuitively expect the total DIS cross section of the deuteron (σ_d) to be significantly different from the sum of the DIS cross sections of the free nucleons ($\sigma_p + \sigma_n$) in the region where the energy scale is many GeV. However, West[27,28] pointed out that the effect of smearing due to Fermi motion is not negligible.

The assumptions listed below are commonly made for the spin averaged inelastic cross section to be evaluated here. These assumptions are parallel to those made in evaluating the quasi-elastic process in the preceding section.

1. The deuteron is considered as a (p, n) bound state in which the isobar admixture, the six-quark degree of freedom, and the meson exchange current contribution are neglected. Later, I introduce the quark cluster model for the six-quark effects.
2. The "off-shell kinematics"–"on-shell dynamics" formalism is adopted. In other words, on-shell amplitudes are used with off-shell kinematics.
3. The impulse approximation is invoked and therefore shadowing corrections and FSI are ignored.

Before introducing the smearing procedure, I briefly explain how the nucleon inelastic structure functions were selected. In order to be realistic in addressing deuterium data at $Q^2 \leq 10$ GeV^2/c^2 we adopted[21] the SLAC parameterization[17] for the proton structure function which includes nucleon resonance effects. In order to incorporate the same resonance effects into the neutron structure function we multiplied the SLAC proton structure function by the ratio of the neutron to the proton structure functions in the Buras and Gaemers[11] (BG) parameterization, $\nu W_2^{SLAC(n)} = \nu W_2^{SLAC(p)} \times (\nu W_2^{BG(n)}/\nu W_2^{BG(p)})$. Throughout the remainder of these lectures I deal with theory and data where the W_1 contribution to the cross section is negligible.

With these assumptions and neglecting the role of the spin, the forward virtual Compton amplitude is given as

$$
\begin{aligned}
|\mathcal{M}_d(s)|^2 &= \sum_{i=p,n} \int \frac{d^4k}{(2\pi)^4} \frac{2\pi\phi^2(p^2)\delta(k^2-m^2)}{(p^2-m^2+i\epsilon)^2} |\mathcal{M}_i(s')|^2\, \theta(s'-m^2) \\
&= \sum_{i=p,n} \int \frac{d^3k}{k_0/m} 2|f(\vec{k})|^2 |\mathcal{M}_i(s')|^2 \theta(s'-m^2) \qquad (49)
\end{aligned}
$$

where $s = (\mathcal{P}+q)^2$, $s' = (p+q)^2$, and

$$
\left|f(\vec{k})\right|^2 = \frac{1}{4m(2\pi)^3} \frac{\phi^2(p^2)}{(p^2-m^2+i\epsilon)^2} \qquad (50)
$$

Now the question is how to recognize the nonrelativistic wavefunction in Eq. 49. Atwood and West[27,28,30] chose the following normalization:

$$\int \frac{d^3k}{k_0/m} \left| f(\vec{k}) \right|^2 = 1, \tag{51}$$

by identifying $n(k)$, the absolute square of the non-relativistic deuteron wavefunction, with $\left| f(\vec{k}) \right|^2 /(k_0/m)$. However, we followed[21] Kusno and Moravcsik,[30] as well as Frankfurt and Strikman,[29] who adopted a different normalization using arguments summarized below:

$$\int \frac{d^3k}{k_0/m} |f(\vec{k})|^2 \frac{2(E_1(|\vec{k}|) + k_3)}{M_d} \theta(E_1(|\vec{k}|) + k_3) = 1, \tag{52}$$

so that we identified

$$n(k) = \frac{m}{k_0} |f(\vec{k})|^2 \frac{2(E_1(|\vec{k}|) + k_3)}{M_d}. \tag{53}$$

Frankfurt and Strikman[29,30] criticized the normalization of Eq. 51 since it is inconsistent with the usual normalization dictated by requiring that the elastic electromagnetic form factor at $Q^2 = 0$ be equal to the total charge. In addition Kusno and Moravcsik derived the normalization of Eq. 52 by using light-front kinematics in Ref. 30 and showed it satisfied the Frankfurt and Strikman criterion. By comparing the normalization of the nonrelativistic wavefunction of Eq. 48 with Eqs. 52 and 53 one observes that they agree provided one ignores the θ function in Eq. 52. Putting in the θ function introduces an error of one part in 10,000. Thus the normalization here has this error since I retain the θ function throughout this work. Another parameterization of the nucleon momentum density in terms of the light cone variables is $G(x, \vec{k}_\perp)$ and is normalized

$$\int d^2\vec{k}_\perp \int_0^1 dx G(x, \vec{k}_\perp) = 1 \tag{54}$$

where x is the light cone fraction on one of the nucleons and $\vec{k}_\perp$ is the transverse momentum. The four-momenta are defined by:

$$\begin{aligned}
\mathcal{P} &= \left(P + \frac{M_d^2}{4P}, \vec{0}_\perp, P - \frac{M_d^2}{4P} \right) \\
q &= \left(\frac{P \cdot q}{2P}, \vec{q}_\perp, -\frac{P \cdot q}{2P} \right), \\
k &= \left((1-x)P + \frac{k^2 + \vec{k}_\perp^2}{4(1-x)P}, -\vec{k}_\perp, (1-x)P - \frac{k^2 + \vec{k}^2}{4(1-x)P} \right) \\
p &= \left(xP + \frac{(1-x)M_d^2 - k^2 - \vec{k}_\perp^2}{4(1-x)P}, \vec{k}_-, xP - \frac{(1-x)M_d^2 - k^2 - \vec{k}_\perp^2}{4(1-x)P} \right)
\end{aligned} \tag{55}$$

Here $P = \frac{1}{2}(\mathcal{P}_0 + \mathcal{P}_3)$ is an arbitrary parameter, and $x = 1 - (k_0 + k_3)/2P$. In the deuteron rest frame $P = \frac{1}{2} M_d$, and $x = 1 - (k_0 + k_3)/M_d$. Meanwhile the function $G(x, \vec{k}_\perp)$ is the new function for the integrand of Eq. 52 with the new variables.

Furthermore one identifies the function $G(x, \vec{k}_\perp)$ as:

$$G(x, \vec{k}_\perp) = \left[M_d + \frac{E_\perp^2}{(1-x)^2 M_d}\right] n(\vec{k}(x, \vec{k}_\perp))\theta(M_d - E_\perp) \tag{56}$$

where $E_\perp = \sqrt{m^2 + \vec{k}_\perp^2}$. Therefore, the deuteron inelastic structure obtained through this smearing[30] of the nucleon's (3-q) inelastic structure function over the motion given by $n(\vec{k})$ is

$$\nu W_2^{3-q}(Q^2, \nu) = \sum_{i=p,n} \int d^2\vec{k}_\perp \int_0^1 \frac{dx}{x} G(x, \vec{k}_\perp) B \frac{\nu}{\nu'} [\nu' W_2^{(i)}(Q^2, \nu')] \tag{57}$$

where $\nu' = (p \cdot q)/2m$ is different from $\nu = (\mathcal{P} \cdot Q)/2m$, and

$$\begin{aligned} B = & \frac{1}{2(q^2 M_d^2 - m^2\nu^2)^2} \{2x^2 m^4 \nu^4 - 4x m^3 \nu^3 (\vec{q}_\perp \cdot \vec{k}_\perp) \\ & + m^2\nu^2[q^2 C_3 + 2(\vec{q}_\perp \cdot \vec{k}_\perp)^2] + m\nu q^2 (\vec{q}_\perp \cdot \vec{k}_\perp) C_4 \\ & + q^4 C_5 + q^2 M_d^2 (\vec{q}_\perp \cdot \vec{k}_\perp)^2\} \end{aligned} \tag{58}$$

and where

$$\begin{aligned} C_3 = & \frac{1}{1-x}\left[2xm^2 - 2x(1-x^2)M_d^2 + (3x-1)\vec{k}_\perp^2\right] \\ C_4 = & \frac{1}{1-x}\{-3E_\perp^2 + [4-(1+x)^2]M_d^2\} \\ C_5 = & \frac{3\vec{k}_\perp^4}{4(1-x)^2} + \frac{\vec{k}_\perp^2}{2(1-x)^2}[3m^2 + (3x^2 - 2x - 1)M_d^2] \\ & + \frac{1}{4(1-x)^2}[3m^4 + 2m^2 M_d^2 (x^2 + 2x - 3) \\ & + M_d^4 (3x^4 - 4x^3 + 2x^2 - 4x + 3)] \end{aligned} \tag{59}$$

I will refer to the sum of the quasi-elastic and inelastic contributions evaluated in this section as the "conventional" approach to DIS data on deuterium.

8 Results from the "Conventional" Approach

For the deuteron data I will address here, the quasi- elastic single nucleon knockout mechanism contributes significantly. The data covers a range in Q^2 between 2.5 and 10 $\mathrm{GeV}^2/\mathrm{c}^2$. In order to describe the data, therefore, the deuteron inelastic structure function at least should contain the incoherent sum of the quasi-elastic νW_2^{q-el} and the smeared inelastic structure functions. That is, in the conventional approach

$$\nu W_2^{(D)} = \nu W_2^{q-el} + \nu W_2^{3-q} \tag{60}$$

This model is considered conventional since it relies in a major way on measured free-space elastic and inelastic properties of nucleons and on traditional models of nuclear structure.

Utilizing this approach we have presented in Ref. 21 comparisons with data.[17,25,31] I extract a selection of these comparisons in Figs. 15, 16, 17, and 18 where Q^2 falls in

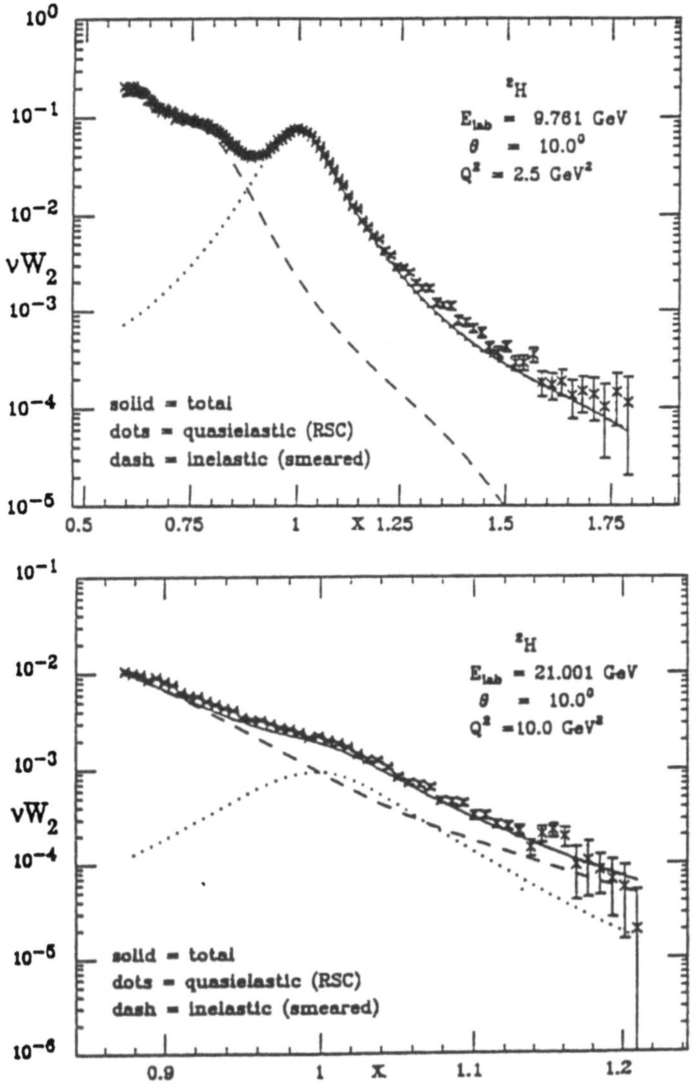

Figure 15. Electron-deuteron inclusive data from SLAC experiment E133[17] expressed as νW_2 versus Bjorken x compared with the conventional approach based on the RSC wavefunction. The quasi-elastic component is represented by the dotted curve and the 3-q smeared inelastic is represented by the dashed curve. The solid curve represents the incoherent sum of these contributions.

the range $2.5 \leq Q^2 \leq 10.0$ GeV^2/c^2. The results are plotted as a function of Bjorken $x = Q^2/2m\nu$ for $\nu W_2^{(D)}$, which, for the data, is *defined* as the double differential cross section multiplied by ν and divided by the Mott cross section σ_M. The results of Figs. 15 and 17 correspond to the RSC[19] potential while those of Figs. 16 and 18 correspond to the Bonn[20] potential.

The first major impression gained from Figs. 15–18 is that for the values of $x > 1$ there is a substantial contribution from quasi-elastic knockout evaluated with conventional nuclear models in lowest order (i.e., with the neglect of final state interactions.) This

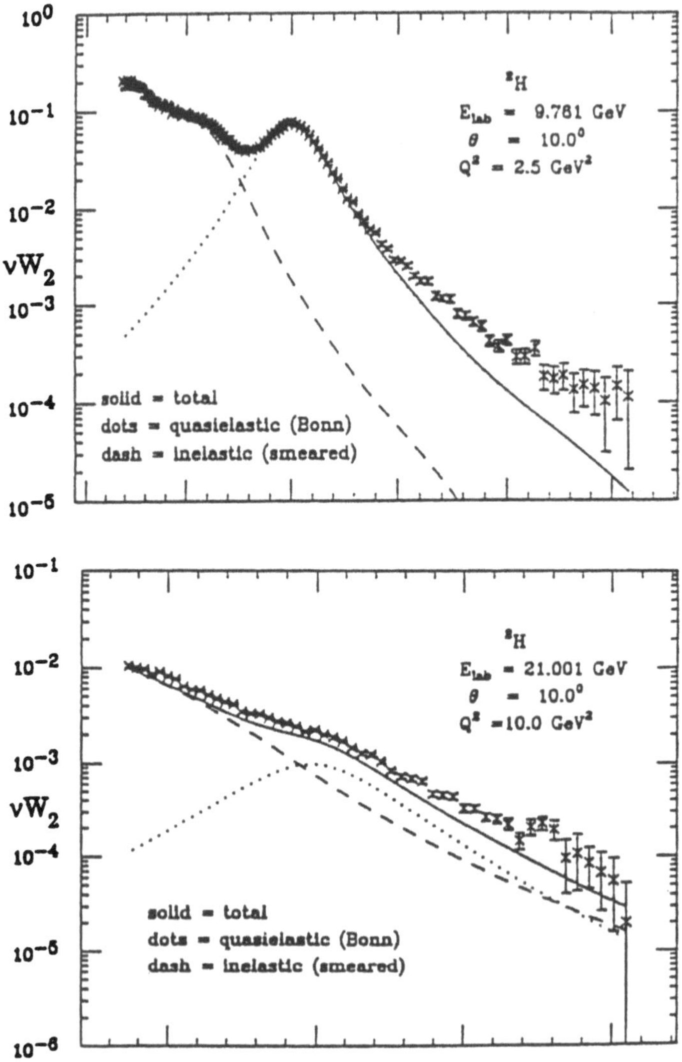

Figure 16. Electron-deuteron inclusive data from SLAC experiment E133[17] expressed as νW_2 versus Bjorken x compared with the conventional approach based on the Bonn wavefunction. The quasi-elastic component is represented by the dotted curve and the 3-q smeared inelastic is represented by the dashed curve. The solid curve represents the incoherent sum of these contributions.

conclusion is true for calculations based on RSC and on Bonn wavefunctions. The second major conclusion concerns the region $x < 1$ where one finds large contributions from the nucleon inelastic structure function. These inelastic contributions show approximate Bjorken scaling (independence of Q^2 at fixed x) and hence become more important with increasing Q^2 relative to the quasi-elastic nucleon knockout process. This is clearly seen in the data for νW_2 where there is approximate independence of Q^2 at $x = 0.75$ whereas the data at the quasi-elastic peak ($x = 1$) decrease by about a factor of 44 as Q^2 increases from 2.5 to 10.0 $\mathrm{GeV}^2/\mathrm{c}^2$.

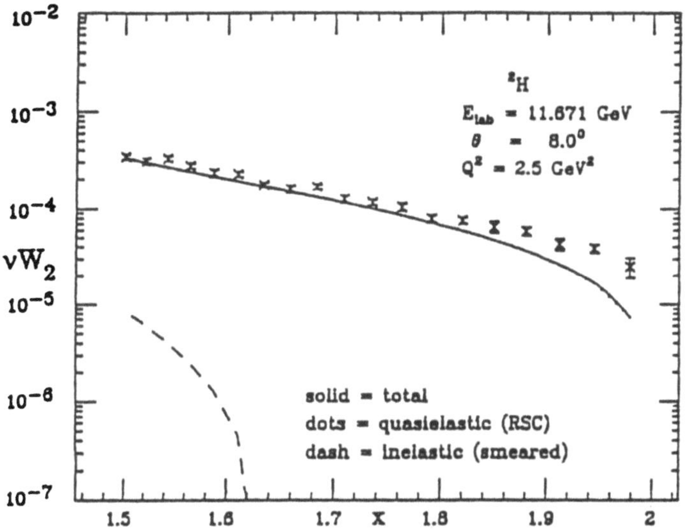

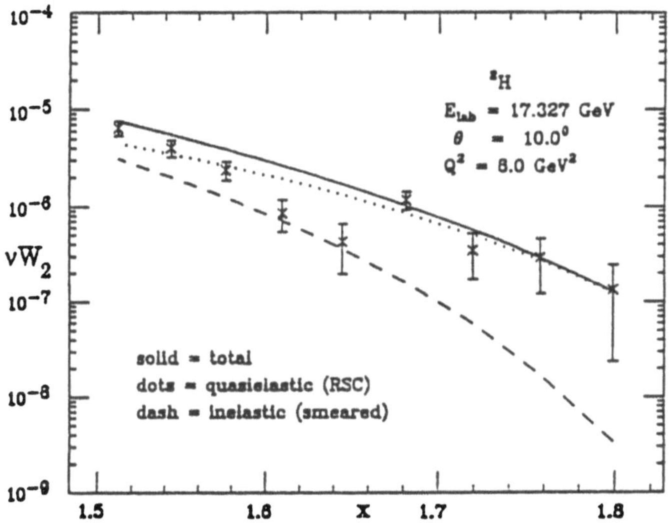

Figure 17. Electron-deuteron inclusive data from SLAC experiment E101[25] expressed as νW_2 versus Bjorken x compared with the conventional approach based on the RSC wavefunction. The quasi-elastic component is represented by the dotted curve and the 3-q smeared inelastic is represented by the dashed curve. The solid curve represents the incoherent sum of these contributions.

The calculated curves are instructive since the total result follows the trends in the data and since the quasi-elastic component reflects pure y-scaling while the inelastic component nearly exhibits Bjorken x-scaling. The trend in the calculated curves from $Q^2 = 2.5$ to 10.0 GeV^2/c^2 indicates the data are approaching a region where Bjorken scaling should set in and, hence, y-scaling should become invalid. The onset of x-scaling for $x > 1$ is about the same for the results obtained with both the RSC and the Bonn wavefunctions.

The sensitivity to the deuteron wavefunction for $x > 1.25$ is clearly seen by comparing the results in Figs. 15–18. The Bonn wavefunction significantly underpredicts the data

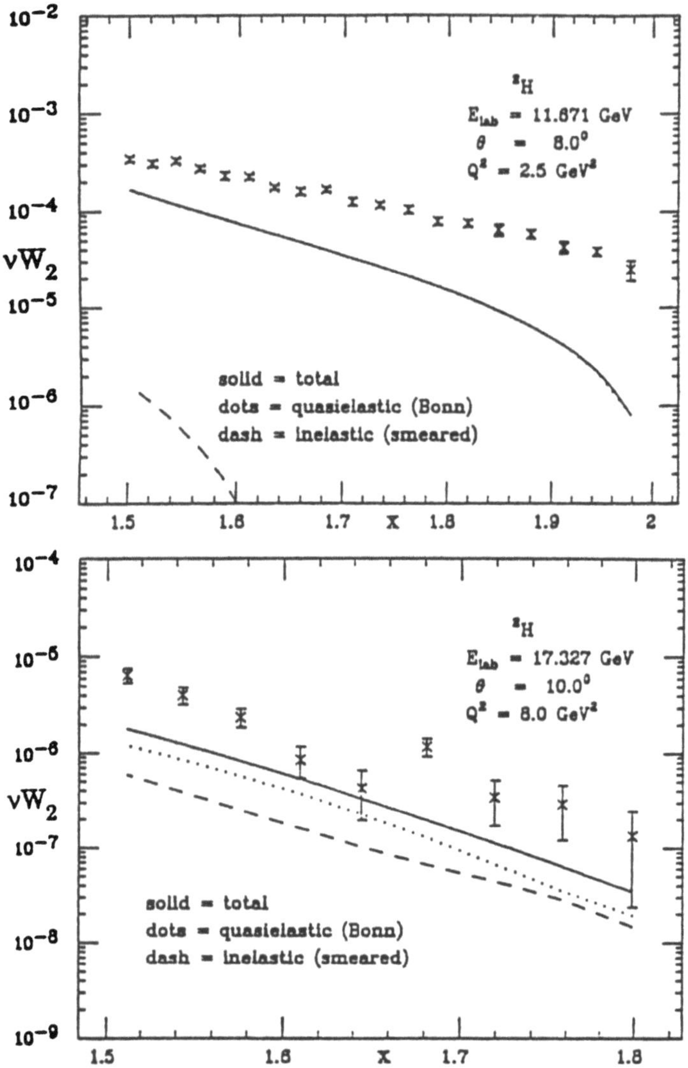

Figure 18. Electron-deuteron inclusive data from SLAC experiment E101[25] expressed as νW_2 versus Bjorken x compared with the conventional approach based on the Bonn wavefunction. The quasi-elastic component is represented by the dotted curve and the 3-q smeared inelastic is represented by the dashed curve. The solid curve represents the incoherent sum of these contributions.

while the RSC does reasonably well in this kinematic region. We have examined[21] the individual S-state and D-state contributions to νW_2^{q-el} and have found that in all cases shown it is the D-state contribution which dominates for $1.25 \leq x \leq 1.75$. For $x > 1.75$ the S-state contributions again become significant. The better agreement obtained with RSC is due to its enhanced momentum components in the range $0.25 \leq k \leq 1.0$ GeV, which were described above.

Some recent efforts[32] have obtained results for low-energy properties of nuclei which may favor the Bonn potential over the RSC potential. Up to this stage in my presentation our applications would tend to favor the RSC results over the Bonn results. However, we feel our results should not be viewed as providing a definitive test of these wavefunctions. Rather, additional effects such as the six-quark cluster effects evaluated below, should be included with a representative sample of realistic wavefunctions in order to obtain a true perspective of their potential significance.

A significant uncertainty in the overall magnitude of these results for νW_2^{q-el} in Figs. 15–18 arises from uncertainties in the nucleon elastic form factors. The detailed studies in Gari and Krümpelmann[23] indicate that these uncertainties increase substantially for $Q^2 > 4$ GeV^2/c^2. It is satisfying that the quasi-elastic peak in conjunction with νW_2^{3-q} reasonably tracks with Q^2 the measured νW_2 at $x = 1$. Some improvement of the agreement between the conventional approach and the data may be achieved by modifying the nucleon form factors but I have not elected to explore that challenging problem at this time.

9 The Quark Cluster Model

As shown in the figures in the previous section, in the high x region, especially with higher Q^2 there is still room for improvement in the model to obtain better agreement with the data. Therefore, I introduce the six-quark (6-q) cluster component of the Quark Cluster Model[8] (QCM) into the calculation in this section. The QCM was first introduced to fit certain $x > 1$ features of the ^{3}He DIS data.[33] A careful analysis based on the conventional nuclear physics[34] (only nucleonic degree of freedom considered) had been unsuccessful. Later, it was shown[7] that the QCM is also capable of explaining the EMC effect.[2] There are three main assumptions of the QCM. The first two are traditional assumptions for any parton model interpretation of deep inelastic lepton-hadron data while the third is particular to the nuclear environment. The assumptions of the QCM are:

1. A photon absorbed by a nucleus at high Q^2 is absorbed by a quark through a quasi-free process when viewed in the infinite momentum frame.
2. The quark is a constituent of a quasi-free color singlet cluster in the nucleus.
3. A 3-quark (3-q) cluster (nucleon) is assigned a critical radius R_C. Clusters with $i = 6, 9, 12, \ldots, 3A$ quarks are defined by the number of 3-q clusters joined by links of length $\leq 2R_C$.

The critical radius R_C is taken as a free parameter and adjusted to fit data. If $R_C = 0$ gives the best overall fit to the data, then no quark clusters other than nucleons (3-q) are formed, and the conventional approach presented in the previous sections survives. On the other hand, if R_C is large, say 1.5 fm, the nucleus has a high probability of being found in the $3A - q$ cluster configuration. R_C was found to be $0.50 \pm 0.05 fm$ from fitting the ^{3}He data. In what follows I summarize our results[21] that show that $R_C = 0.50 fm$ yields a 4.7% (5.4%) $6 - q$ cluster configuration in deuterium with the RSC (Bonn) wavefunction and I examine the consequences for the description of the DIS deuterium data. Applications of the QCM to heavier nuclei are also discussed further.

The model is easily visualized in the Breit frame as described earlier in these lectures. The inelastic nuclear structure function can be written, according to the parton model, as

$$\nu W_2(\nu, Q^2) = \sum_{\text{quarks } j} e_j^2 x f_j^A(x), \qquad (61)$$

where the sum is over all quarks in the nucleus, e_j is the charge on the quark j, and $f_j^A(x)$ is the probability of finding quark j carrying fraction x/A of the total nuclear momentum P. If one approximates with a properly weighted average of up and down quark distributions, $f^A(x)$, one can write

$$\nu W_2(\nu, Q^2) \simeq \sum_{\text{quarks } j} e_j^2 x f^A(x) \tag{62}$$

According to the model assumptions[8] the quarks are found in an i-quark cluster ($i = 3, 6, \ldots, 3A$) within the nucleus with probability $\tilde{p}_i$ so that

$$f^A(x) = \sum_{\text{clusters } i} \tilde{p}_i \overline{P}_i(x), \tag{63}$$

where $\overline{P}_i(x)$ is the x distribution of quarks from an $i-q$ cluster in the nucleus. Therefore, by writing

$$\nu W_2^{3-q} = \sum_{j=1}^{3} e_j^2 x \overline{P}_3(x), \tag{64}$$

$$\nu W_2^{6-q} = \sum_{j=1}^{6} e_j^2 x \overline{P}_6(x), \tag{65}$$

then the deuteron inelastic structure function is written as

$$\nu W_2^{(D)} = \tilde{p}_3(\nu W_2^{3-q} + \nu W_2^{q-el}) + \tilde{p}_6(\nu W_2^{6-q}) \tag{66}$$

where

$$\tilde{p}_3 = \int_{2R_C}^{\infty} dr \left[u^2(r) + w^2(r)\right] \tag{67}$$

$$\tilde{p}_6 = \int_{0}^{2R_C} dr \left[u^2(r) + w^2(r)\right] \tag{68}$$

with $\tilde{p}_3 + \tilde{p}_6 = 1$. With $R_C = 0.50 fm$ one obtains $\tilde{p}_3 = 0.953$ and $\tilde{p}_6 = 0.047$ for RSC and $\tilde{p}_3 = 0.946$ and $\tilde{p}_6 = 0.054$ for Bonn. The larger 6-q cluster probability for Bonn reflects somewhat weaker short-range correlations.

The above description has been qualitative in that it neglects the role of transverse motion. The more complete description of Sec. 7 will be adopted for νW_2^{3-q}. Since the 6-q cluster is at rest in the nuclear cm frame for deuterium there is no need to consider its transverse motion. In nuclei with $A > 2$ our practice is to introduce Fermi motion smearing for the heavier clusters.[8] Here, the "6-q structure function," νW_2^{6-q}, with the modification of replacing x by ξ, the Nachtmann quark variable,[35] is argued[7,8] to be

$$\nu W_2^{6-q} = \frac{\xi}{2}\left(\sum_{i=1}^{6} e_i^2\right)\overline{P}_6(\xi), \tag{69}$$

where, using the counting rules and other constraints discussed in Sec. 4

$$\overline{P}_6(\xi) = \frac{1.850069}{\sqrt{\xi/2}}(1-\xi/2)^{10} \tag{70}$$

with

$$\xi = \frac{2}{1+\sqrt{1+Q^2/\nu^2}} \tag{71}$$

Much of the $x \sim 2$ data were acquired with the final state invariant mass only a few tens of MeV above breakup threshold. The 6-q inelastic structure function should be modified to respect at least the phase space limitations just above this threshold.[36] Therefore, a threshold factor $\mathcal{R}_{ps}$ is introduced to modify the 6-q structure function (Eq. 69):

$$\nu W_2^{6-q} = \mathcal{R}_{ps}\frac{\xi}{2}\left(\sum_{i=1}^{6} e_i^2\right)\overline{P}_6(\xi), \tag{72}$$

where

$$\mathcal{R}_{ps} = \begin{cases} \sqrt{1-(2m/W)}, & \text{if } W > 2m \\ 0, & \text{otherwise} \end{cases} \tag{73}$$

Here $W^2 = s = (\mathcal{P}+q)^2$

10 Results with the Full Quark Cluster Model

Now, let us proceed to compare the same data sets as above with the results of the full QCM. In Fig. 19 (21) we see the QCM results for the RSC wavefunction in comparison with the SLAC E133 (E101) data. The 3-q and the 6-q cluster contributions at fixed $x > 1$ clearly rise together with Q^2 relative to the quasi-elastic contribution. However, excluding the E101 data in Fig. 21 for $Q^2 = 8.0$ GeV^2/c^2, these inclusive data do not go to sufficiently high x and sufficiently high Q^2 to be sensitive to the 6-q cluster contribution when the model is based on the RSC wavefunction. For the E101 data at $Q^2 = 8.0$ GeV^2/c^2 the model calculations show large contributions from the 6-q cluster but the present uncertainties in the data prevent us from making definitive conclusions.

As a consequence we can say the conventional approach (Figs. 15 and 17) and the QCM (Figs. 19 and 21) when based on the RSC wavefunction provide an equally acceptable description of these data.

In Figs. 20 and 22 we see the QCM results for the Bonn wavefunction in comparison with the same data sets. Here, we observe rather dramatic consequences due to the 6-q cluster contributions. The overall agreement between theory and experiment is considerably improved by the QCM in the $x > 1$ region over the conventional results of Figs. 16 and 18.

11 Conclusions from the Deuterium Data

The contribution of quasi-free nucleon knockout and of inelastic lepton-nucleon scattering in inclusive electron-deuteron reactions at large momentum transfer have been evaluated.[21] The degree of quantitative agreement with deuteron wavefunctions from the RSC and Bonn realistic nucleon-nucleon interactions have been examined. For the range of data available we saw there is strong sensitivity to the tensor correlations which are distinctively different in these two deuteron models. At this stage the RSC wavefunction provided a reasonable description of the data while the Bonn wavefunction did not. After introducing a 6-q cluster component whose relative contributions is based on an overlap criteria, we saw a good description of all the data with both interactions. The critical separation at which overlap occurs (formation of 6-q clusters) was taken to be $1.0fm$ and the 6-q cluster probability was 4.7% for RSC and 5.4% for Bonn.

The description of the DIS deuterium data with the QCM is as good as the conventional description when both are based on the RSC wavefunction. The description of the same data with the QCM is definitely superior to the conventional description when

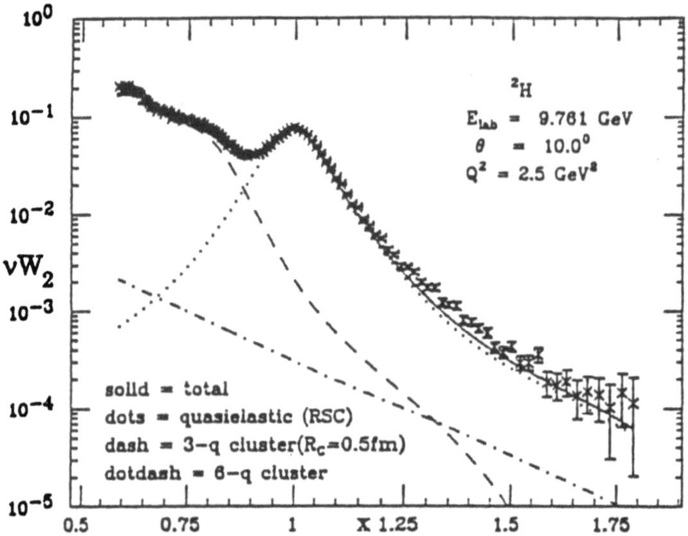

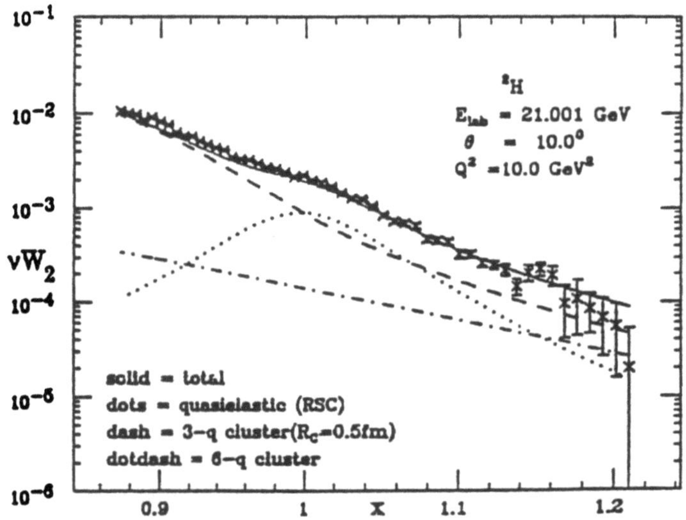

Figure 19. Electron-deuteron inclusive data from SLAC experiment E133[17] expressed as νW_2 versus Bjorken x compared with the results from the Quark Cluster Model based on the RSC wavefunction. The quasi-elastic component is represented by the dotted curve and the 3-q smeared inelastic is represented by the dashed curve. The dotdashed curve stands for the 6-q inelastic contribution. The solid curve represents the incoherent sum of these contributions.

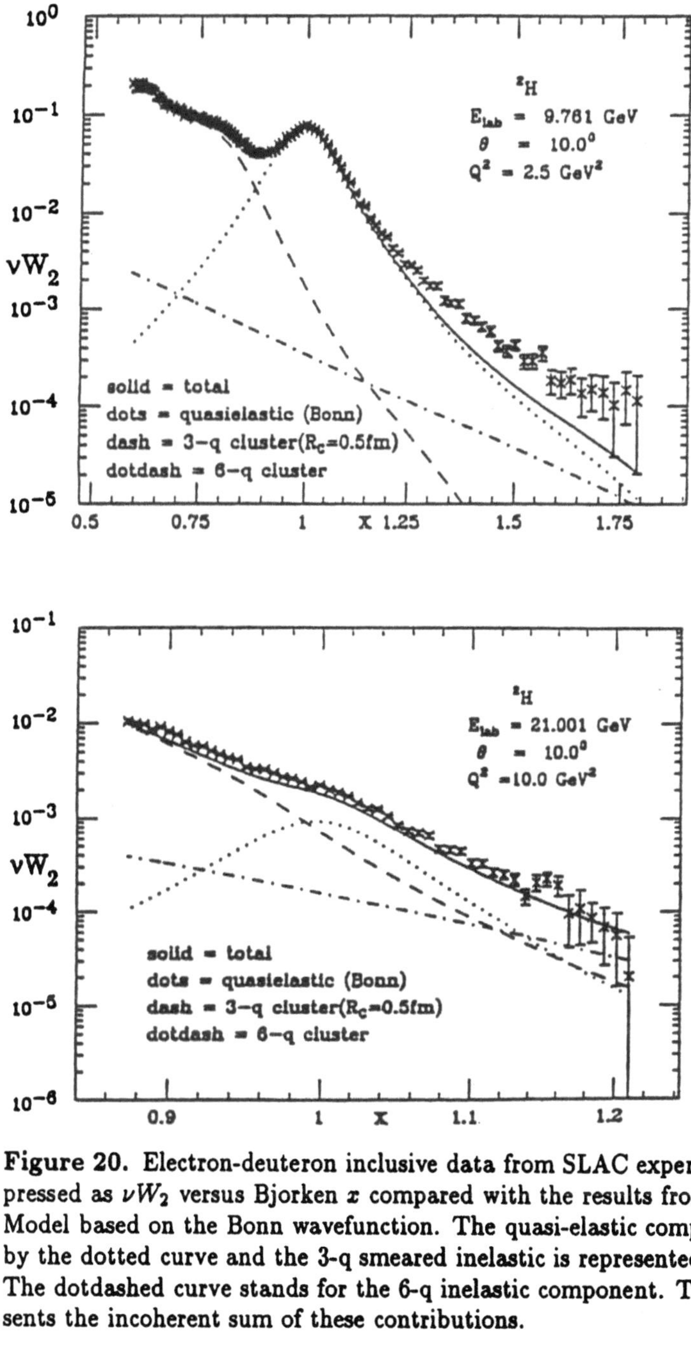

Figure 20. Electron-deuteron inclusive data from SLAC experiment E133[17] expressed as νW_2 versus Bjorken x compared with the results from the Quark Cluster Model based on the Bonn wavefunction. The quasi-elastic component is represented by the dotted curve and the 3-q smeared inelastic is represented by the dashed curve. The dotdashed curve stands for the 6-q inelastic component. The solid curve represents the incoherent sum of these contributions.

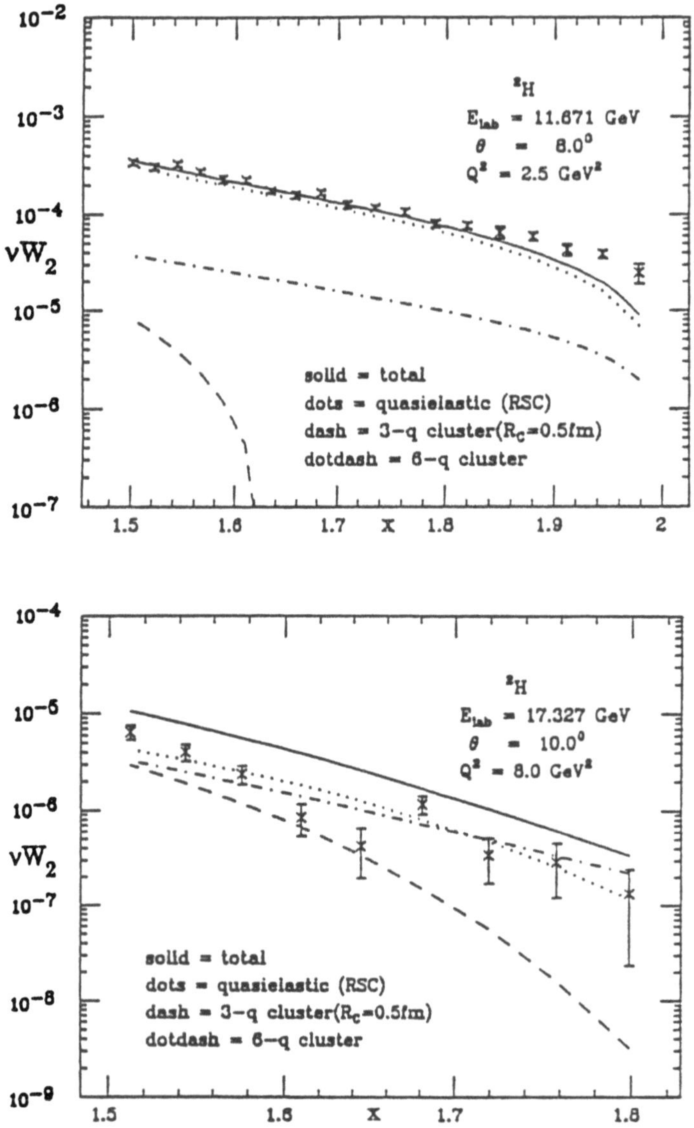

Figure 21. Electron-deuteron inclusive data from SLAC experiment E101[25] expressed as νW_2 versus Bjorken x compared with the results from the Quark Cluster Model based on the RSC wavefunction. The quasi-elastic component is represented by the dotted curve and the 3-q smeared inelastic is represented by the dashed curve. The dotdashed curve stands for the 6-q inelastic component. The solid curve represents the incoherent sum of these contributions.

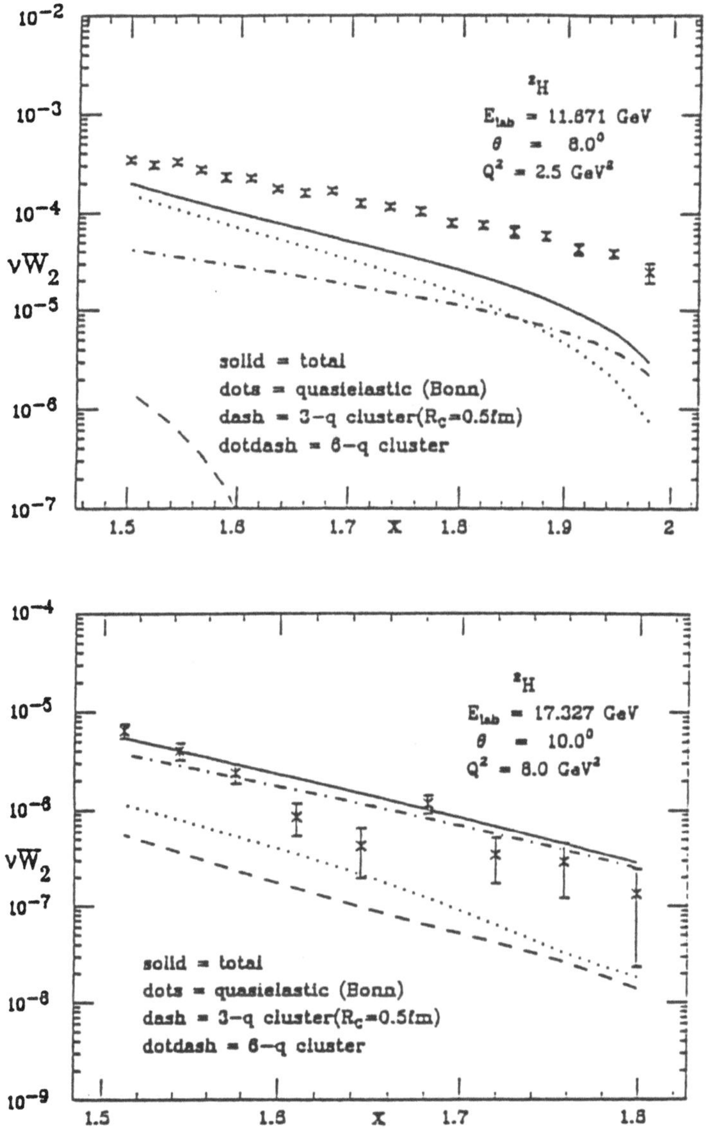

Figure 22. Electron-deuteron inclusive data from SLAC experiment E101[25] expressed as νW_2 versus Bjorken x compared with the results from the Quark Cluster Model based on the Bonn wavefunction. The quasi-elastic component is represented by the dotted curve and the 3-q smeared inelastic is represented by the dashed curve. The dotdashed curve stands for the 6-q inelastic component. The solid curve represents the incoherent sum of these contributions.

both are based on the Bonn wavefunction. The QCM results with the Bonn wavefunction are similar to the QCM results with RSC even though their respective results in the conventional model were very different. This implies that the QCM is more "robust" (i.e., has greater independence of the adopted deuterium wavefunction) than the conventional approach. It is easy to see why this is true. For fixed R_C, as one diminishes the short and intermediate range correlations the large x contributions from the quasi- elastic and 3-q inelastic processes decrease. At the same time $\tilde{p}_6$ will increase and therefore larger 6-q cluster contributions will emerge with the trend to offset the decreases in the other processes.

It is worthwhile to recall that there are a number of approximations made in developing the models presented herein. Two approximations which stand out as limiting the strength of these conclusions are the use of deuteron wavefunctions from the non-relativistic formalism and the neglect of final state interactions. It is hoped that these limitations can be removed in future efforts.

In view of these results it is worth speculating on experiments which could resolve these uncertainties. Inclusive data at $x > 1$ and at higher Q^2 would be one avenue since RSC and Bonn still differ somewhat in that region. More exclusive data, such as (e, e′ π) for $x > 1$ even at somewhat lower Q^2 could resolve between these models and wavefunctions since the integrated yield could be compared with the sum of the νW_2^{3-q} and νW_2^{6-q} presented in the various models here. It is hoped that these theoretical results will strongly motivate new experiments in these directions.

12 The QCM for $A = 3$

It was the puzzling ^{3}He data which first appeared in 1979[33] that prompted the introduction of the QCM[8]. Since then numerous improvements in the model ingredients have been developed. For example, realistic nuclear wavefunctions have been employed to calculate the quark cluster probabilities $\tilde{p}_i$ as a function of R_C.[37] For $A = 3$, a treatment of the quasi-elastic process at the same level of sophistication as for deuterium has been presented.[34] In this treatment, the ground state wavefunction is obtained exactly for realistic NN potentials and the quasi-elastic process is then evaluated in the impulse approximation. The 3-q inelastic structure function is smeared over $n(k)$ obtained from this same wavefunction (RSC). The 6-q and 9-q inelastic structure functions are chosen according to the forms given in Sec. 4. In addition, the center of mass motion of the 6-q cluster if Fermi smeared.[8] The resulting description of the ^{3}He data is presented in Fig. 23.

Note that the 6-q and 9-q contributions are appreciable at high x values where there are data to compare with the theory. The curves in Fig. 23 are obtained for $R_C = 0.50 fm$ as in the original efforts.[8] The quark cluster probabilities are $(\tilde{p}_3, \tilde{p}_6, \tilde{p}_9) = (0.88, 0.11, 0.01)$, respectively.

As in the case with deuterium it would be highly desirable to have additional $x > 1$ data at higher Q^2 to verify the QCM predictions in a domain without the strong contributions from the quasi-elastic peak. In other words, a crucial test of the QCM would be to observe x-scaling behavior in the $x > 1$ region.

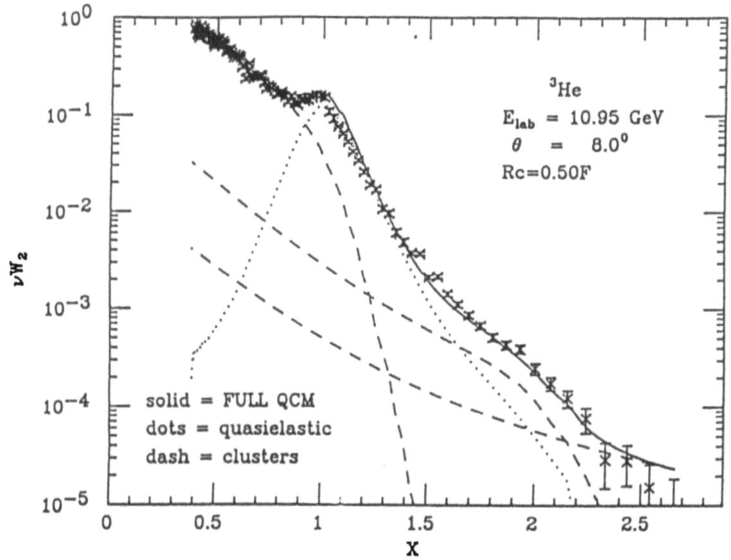

Figure 23. QCM contributions to νW_2 versus Bjorken x for ^{3}He compared with the data from Ref. 33. The 3- q, 6-q, and 9-q cluster contributions are signified by dashed lines with the 3-q being the strongest at low x while the 9-q is the weakest at low x. At high x the situation is reversed. The dotted line gives the quasi-elastic contribution and the solid line is the incoherent sum of all contributions to the QCM.

13. Further Tests of QCM with $x > 1$ Data

This brings us to the question of how may we best determine the correct model among the many proposed for the quark structure of nuclei. Experiments in the $x > 1$ region should be decisive. Fig. 24 presents a few predictions for an extended range of x for the characteristic behavior of the ratio of structure functions of a heavy nucleus A to a light nucleus B with $B > 4$. The QCM predicts a sequence of steps in a staircase where the height of a step in the region $n - 1 < x < n$ with $n > 1$ is the ratio of $3n$-q cluster probabilities of the heavier to the lighter nucleus. By contrast rather smooth behavior is predicted so far by other models. The dashed curve is the type of behavior expected from the relativistic wavefunction model of Frankfurt and Strikman[39] where short range correlations give rise to a shoulder in the $1 < x < 2$ region. Another relativistic wavefunction treatment by Garsevanishvili and Menteshashvili[40] and the color dielectric model of Pirner and collaborators[41] predict behavior indicated by the dash-dot curve. In the color dielectric model the quarks at very high Q^2 are free to move essentially throughout the volume of the entire nucleus. This naturally leads to softer momentum distributions of quarks in larger nuclei. Exactly what values of Q^2 for which the color dielectric model is expected to be valid must yet be specified. If the color dielectric picture is valid at high Q^2 the QCM will still be valid if we introduce a Q^2 dependence for R_C. Then, as Q^2 increases R_C will increase in a manner predicted by the color dielectric model. In this case the steps in Fig. 24 for $x > 1$ will drop with increasing Q^2 and will eventually fall below unity until the curve reaches the smooth prediction shown for the color dielectric model. This union of the QCM with the color dielectric model produces a result which contrasts the work of Refs. 41 and 42 where the 6-q cluster probability always rises with

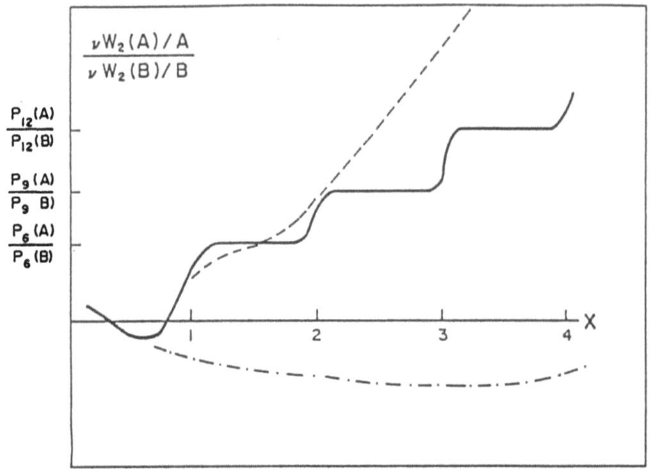

Figure 24. Characteristic behavior of the ratio of nuclear structure functions per nucleon for different models over a wide kinematic range of x. The QCM gives the solid curve. The dashed curve is due to the model of Ref. 38. The dashed-dot curve approximates the predictions of Refs. 39 and 40.

Q^2 and the probability in a heavy nucleus never equals that of a lighter nucleus. Thus the QCD evolution of the model in Refs. 41 and 42 will produce a curve for Fig. 24 which always remains above unity for $x > 1$.

Clearly, the wide range of behavior predicted and, in particular, the dramatic signature of the QCM motivates experiments in the $x > 1$ region.

14. Implications of the QCM for the Equation of State of Dense Baryonic Matter

I would like to conclude these lectures with some speculative remarks on the equation of state (EOS) at high density and low temperature. In particular, I refer to the point on the $T = 0$ axis of the phase diagram where we expect a transition to the quark-gluon plasma to occur. A sketch of a speculated version of this phase diagram is given in Fig. 25.

The basis for the speculations arises from an interpretation of the QCM. I argue that cluster formation in the nuclear ground state is a fluctuation phenomenon which is discernible through DIS. These fluctuations are signatures of the transition to the quark-gluon phase at high density. Let us make a simple estimate based on the critical length parameter $2R_C = 1.0fm$ for the critical density. Since the critical length parameter is "measured" at $T = 0$ we will only use this estimate of the critical density at $T = 0$. We assign each nucleon in nuclear matter to a Wigner-Seitz cell of radius R_0. Then, normal or saturation density gives

$$\rho_0 = \frac{0.17\text{baryons}}{fm^3} = \frac{1}{(\frac{4\pi}{3}R_0)^3} \tag{74}$$

which implies $R_0 = 1.12fm$. Then the ratio of the critical density ρ_C to normal density is given by

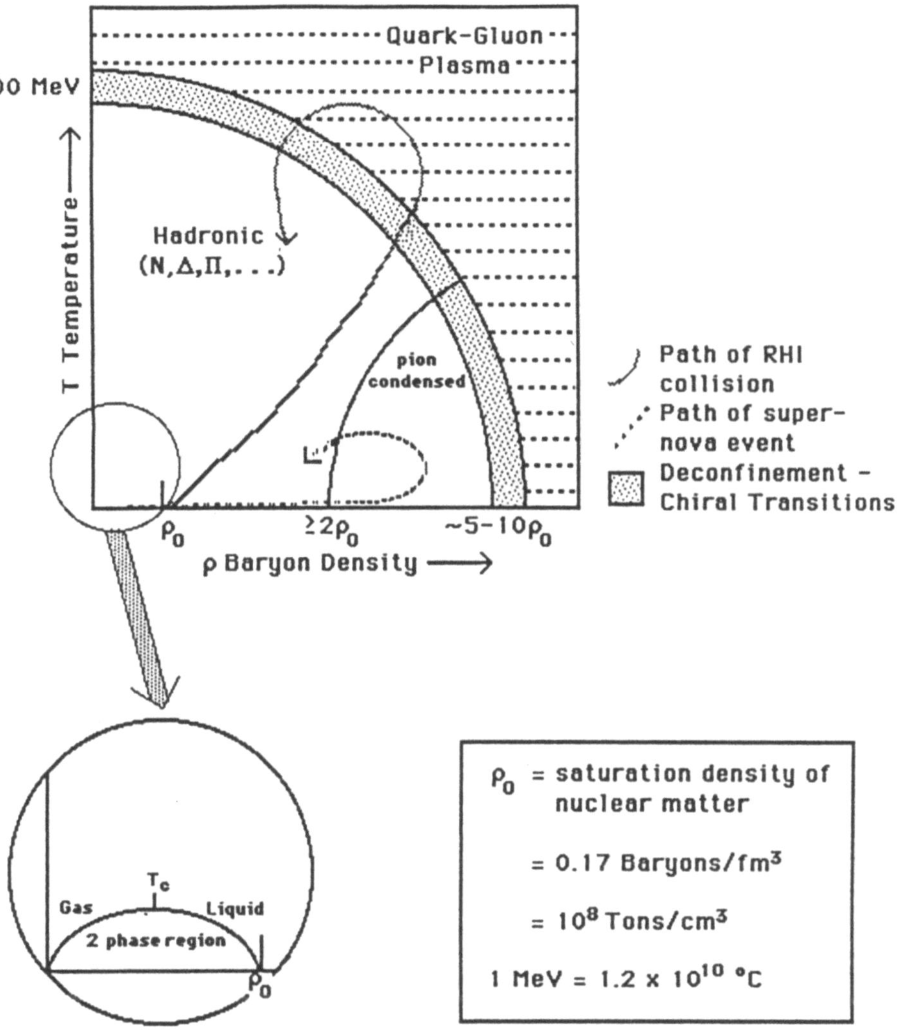

Figure 25.

$$\frac{\rho_C}{\rho_0} = \left(\frac{R_0}{R_C}\right)^3 = \left(\frac{1.12}{0.5}\right)^3 = 11.2 \tag{75}$$

This estimate is consistent with many estimates in the literature and should provide guidance to the experimentalists embarked upon ambitious quark-gluon plasma searches.

Acknowledgments

I would like to thank my collaborators A. Harindranath, H. J. Pirner, and G. Yen for their extensive efforts which led to much of the material I have had the pleasure to present at this summer school. This work is supported in part by the U.S. Department of Energy under contract No. DE-FG02-87ER40371, Division of High Energy and Nuclear Physics.

References

1. F. E. Close, *An Introduction to Quarks and Partons,* Academic Press (London, 1979).
2. European Moon Collaboration, J. J. Aubert et al. *Phys. Lett.* **123B**, 275 (1983).
3. S. Weinberg, *Phys. Rev.* **150**, 1313 (1966).
4. S.-J. Chang and S.-K. Ma, *Phys. Rev.* **180**, 1506 (1969); J. B. Kogut and D. E. Soper, *Phys. Rev. D* **1**, 2901 (1970); J. D. Bjorken, J. B. Kogut and D. E. Soper, *Phys. Rev. D* **3**, 1382 (1971); S.-J. Chang, R. G. Root, and T.-M. Yan, *Phys. Rev. D* **7**, 1133 (1973).
5. For a good review, see J. B. Kogut and L. Susskind, *Phys. Rep.* **8**, 75 (1973).
6. S. J. Brodsky, in *Short Distance Phenomena in Nuclear Physics,* D. H. Boal and R. M. Woloshyn, eds., (Plenum, 1983) p. 141 and references therein.
7. C. E. Carlson and T. J. Havens, *Phys. Rev. Lett.* **51**, 261 (1983).
8. H. J. Pirner and J. P. Vary, *Phys. Rev. Lett.* **46**, 1376 (1981); *Nucl. Phys.* **A358**, 413c (1981); J. P. Vary, *Nucl. Phys.* **A418**, 195c (1984); J. P. Vary, in *AIP Conf. Proc.* No. **110**, W-Y. P. Hwang and M. H. Macfarlane, eds. (AIP, New York, 1984); H. J. Pirner, *International Reviews of Nuclear Physics,* Vol. 1 (World Scientific, Sinapore, 1984), p. 409.
9. D. Sivers, S. J. Brodsky, and R. Blankenbecler, *Phys. Rep.* **23C**, 1 (1976), and references therein.
10. G. Altarelli and G. Parisi, *Nucl. Phys.* **B126**, 298 (1977).
11. A. J. Buras and K. J. F. Gaemers, *Nucl. Phys.* **B132**, 249 (1978).
12. S. D. Drell and T.-M. Yan, *Phys. Rev. Lett.* **25**, 316 (1970).
13. See for example, A. Harindranath and J. P. Vary, *Phys. Rev. D* **34**, 3378 (1986), and references therein.
14. D. M. Alde, et al., *Phys. Rev. Lett.* **64**, 2479 (1990).
15. K. E. Lassila and V. P. Sukhatme, *Phys. Lett.* **209B**, 343 (1988); in *Nuclear and Particle Physics on the Light Cone,* Proceedings of the LAMPF Workshop, Los Alamos, New Mexico, 1988, edited by M. B. Johnson and L. S. Kisslinger (World Scientific, Singapore, 1989), p. 309.
16. D. W. Duke and J. F. Owens, *Phys. Rev. D* **30**, 49 (1984).
17. S. Rock, R. G. Arnold, P. Bosted, B. T. Chertok, B. A. Mecking, I. Schmidt, Z. M. Szalata, R. C. York, and R. Zdarko, *Phys. Rev. Lett.* **49**, 1139 (1982); S. Rock, private communication.
18. The basis for this lecture stems from: Granddon Yen, Ph.D. thesis, Iowa State University, 1988; G. Yen, J. P. Vary, A. Harindranath, and H. J. Pirner, submitted for publication.
19. R. V. Reid, *Ann. Phys.* (N.Y.) **50**, 411 (1968); J. M. Greban, Department of Theoretical Physics, Council for Scientific and Industrial Research, South Africa, private communication.
20. R. Machleidt, K. Holinde, and Ch. Elster, *Phys. Rep.* **149**, 1 (1987) and references therein.
21. The principal results of this section are taken from G. Yen, A. Harindranath, J. P. Vary, and H. J. Pirner, *Phys. Letts.* **218B**, 408 (1989).
22. J. D. Bjorken and S. D. Drell, *Relativistic Quantum Mechanics,* (McGraw-Hill, New York, 1964).
23. M. Gari and W. Krümpelmann, *Z. Phys.* **A322**, 689 (1985); *Phys. Lett.* **B173**, 10 (1986)
24. C. Marchand et al., *Phys. Rev. Lett.* **60**, 1703 (1988).

25. W. P. Schütz, R. G. Arnold, B. T. Chertok, E. B. Dally, A. Grigorian, C. L. Jordan, and R. Zdarko, *Phys. Rev. Lett.* **38**, 259 (1977).
26. M. N. Butler and R. D. McKeown, *Phys. Lett. B* **208**, 171 (1988).
27. G. B. West, *Phys. Lett.* **37B**, 509 (1971); Ann. Phys. (N.Y.) **74**, 464 (1972).
28. W. B. Atwood and G. B. West, *Phys. Rev.* **D7**, 773 (1973).
29. L. L. Frankfurt and M. I. Strikman, *Phys. Lett.* **64B**, 433 (1976); **65B**, 51 (1976); **76B**, 333 (1978).
30. D. Kusno and M. J. Moravcsik, *Phys. Rev. D* **20**, 2734 (1979); *Phys. Rev. C* **27**, 2173 (1983).
31. P. Bosted, R. G. Arnold, S. Rock, and Z. Szalata, *Phys. Rev. Lett.* **49**, 1380 (1982).
32. R. A. Brandenburg, G. S. Chulick, Y. E. Kim, D. J. Klepacki, R. Machleidt, A. Picklesimer, and R. M. Thaler, *Phys. Rev. C* **37**, 781 (1988).
33. D. Day, J. S. McCarthy, I. Sick, R. G. Arnold, B. T. Chertok, S. Rock, Z. M. Szalata, F. Martin, B. A. Mecking, and G. Tamas, *Phys. Rev. Lett.* **43**, 1143 (1979).
34. H. Meier-Hajduk, Ch. Hajduk, P. U. Sauer, and W. Theis, *Nucl. Phys.* **A395**, 332 (1983).
35. O. Nachtmann, *Nucl. Phys.* **B63**, 237 (1973).
36. E. Byckling and K. Kajantie, *Particle Kinematics,* (John Wiley & Sons, London, 1973).
37. M. Sato, S. A. Coon, H. J. Pirner, and J. P. Vary, *Phys. Rev. C* **33**, 1062 (1986).
38. L. L. Frankfurt and M. I. Strikman, *Phys. Rep.* **76**, 215 (1981).
39. V. Garsevanishvili and Z. Menteshashvili, JINR, E2–84–314, Dubna (1984).
40. G. Chanfray, O. Nachtmann, and H. J. Pirner, *Phys. Lett.* **147B**, 249 (1984).
41. F. E. Close, R. G. Roberts, and G. G. Ross, *Phys. Lett.* **129B**, 346 (1983).
42. R. Jaffe, F. E. Close, R. G. Roberts, and G. G. Ross, *Phys. Lett.* **134B**, 449 (1984).

Index of Contributors

O. Nachtmann, University of Heidelberg

Elementary Particle Physics

Concepts and Phenomena

Translated from the German by A. Lahee, W. Wetzel

1990. XIX, 559 pp. 171 figs. (Texts and Monographs in Physics) Hardcover DM 136,– ISBN 3-540-50496-6 Softcover DM 98,– ISBN 3-540-51647-6

This thoroughly written textbook emphazises the fundamental concepts and their phenomenological consequences in the physics of elementary particles.
After an introduction to the theory of quantized fields the author deals in the second part with quantum electrodynamics and in the third part with quantum chromodynamics. In the fourth part the unifying principle of working with gauge groups is applied to explain the electroweak interaction. With this book the student can learn theoretical particle physics from its very roots, study hadrons and their interactions and become familiar with the Higgs mechanism. The author's main goals are to present the standard model and to make a detailed comparison between theoretical and experimental results in particle physics. The book is meant for graduates and postgraduates in physics.

P. Schmüser, Universität Hamburg

Feynman-Graphen und Eichtheorien für Experimentalphysiker

1988. VI, 217 S. (Lecture Notes in Physics, Vol. 295) Geb. DM 39,– ISBN 3-540-18797-9

Inhaltsübersicht: Relativistische Wellengleichungen.– Relativistische Invarianz der Dirac-Gleichung.– Interpretation der Lösungen negativer Energie.– Feynman-Graphen.– Anwendung der Feynman-Graphen.– Schwache Wechselwirkungen.– Lepton-Quark-Wechselwirkungen, Parton-Modell.– Schwierigkeiten der Theorie der schwachen Wechselwirkungen.– Eichinvarianz als dynamisches Prinzip.– Eichinvarianz bei massiven Vektor-Feldern.– Das Standard-Modell der elektroschwachen Wechselwirkung.– Quanten-Chromodynamik.

Springer-Verlag
Berlin
Heidelberg
New York
London
Paris
Tokyo
Hong Kong
Barcelona

MIX
Papier aus verantwortungsvollen Quellen
Paper from responsible sources
FSC® C105338

If you have any concerns about our products,
you can contact us on
ProductSafety@springernature.com

In case Publisher is established outside the EU,
the EU authorized representative is:
Springer Nature Customer Service Center GmbH
Europaplatz 3, 69115 Heidelberg, Germany

Printed by Libri Plureos GmbH
in Hamburg, Germany